Animal Personalities

ANIMAL PERSONALITIES

Behavior, Physiology, and Evolution

EDITED BY

CLAUDIO CARERE AND DARIO MAESTRIPIERI

THE UNIVERSITY OF CHICAGO PRESS

Chicago and London

Claudio Carere is adjunct professor in the Department of Ecological and Biological Sciences, University of Tuscia, Italy. Dario Maestripieri is professor of comparative human development, evolutionary biology, and neurobiology at the University of Chicago.

The University of Chicago Press, Chicago 60637
The University of Chicago Press, Ltd., London
© 2013 by The University of Chicago
All rights reserved. Published 2013.
Printed in the United States of America

22 21 20 19 18 17 16 15 14 13   1 2 3 4 5

ISBN-13: 978-0-226-92205-8 (cloth)
ISBN-13: 978-0-226-92197-6 (paper)
ISBN-13: 978-0-226-92206-5 (e-book)
ISBN-10: 0-226-92205-7 (cloth)
ISBN-10: 0-226-92197-2 (paper)
ISBN-10: 0-226-92206-5 (e-book)

Library of Congress Cataloging-in-Publication Data

Animal personalities : behavior, physiology, and evolution / edited by Claudio Carere and Dario Maestripieri.
    pages. cm.
    Includes bibliographical references and index.
    ISBN-13: 978-0-226-92205-8 (cloth : alk. paper)
    ISBN-10: 0-226-92205-7 (cloth : alk. paper)
    ISBN-13: 978-0-226-92197-6 (paper)
    ISBN-10: 0-226-92197-2 (paper)
    ISBN-13: 978-0-226-92206-5 (e-book)
    ISBN-10: 0-226-92206-5 (e-book)
    1. Animal behavior. I. Carere, Claudio. II. Maestripieri, Dario.
QL751.A667 2013
591.5—dc23

2012022902

⊗ This paper meets the requirements of ANSI/NISO Z39.48–1992 (Permanence of Paper).

Contents

Contributors

Mark J. Adams, Department of Animal and Plant Sciences, University of Sheffield, Sheffield SIO 2TN, UK, mark.adams@sheffield.ac.uk

Alison M. Bell, School of Integrative Biology, University of Illinois Urbana-Champaign, Urbana, IL 61801, USA, alisonmb@life.uiuc.edu

Daniel T. Blumstein, Department of Ecology and Evolutionary Biology, University of California Los Angeles, Los Angeles, CA 90095, USA, marmots@ucla.edu

Igor Branchi, Section of Behavioural Neurosciences, Dipartimento di Biologia Cellulare e Neuroscienze, Istituto Superiore di Sanità, 00161 Roma, Italy, branchi@iss.it

Doretta Caramaschi, Research Unit on Children's Psychosocial Maladjustment, Research Centre, Sainte-Justine Hospital, Montreal H3T 1C5, Quebec, Canada, dcaramaschi@gmail.com

Claudio Carere, Ichthyogenic Experimental Marine Center (CISMAR), Department of Ecological and Biological Sciences, Università degli Studi della Tuscia, Borgo Le Saline, 01016 Tarquinia, Italy, claudiocarere@unitus.it

Sonia A. Cavigelli, Department of Biobehavioral Health, Penn State University, University Park, PA 16802, USA, s-cavigelli@psu.edu

James P. Curley, Department of Psychology, Columbia University, New York, NY 10027, USA, jc3181@columbia.edu

Niels J. Dingemanse, Department of Behavioral Ecology and Evolutionary Genetics, Max Planck Institute for Ornithology, Seewiesen, Germany, ndingemanse@orn.mpg.de

G. Sander van Doorn, Department of Behavioral Ecology, Institute of Ecology and Evolution, University of Bern, Hinterkappelen, CH-3032, Switzerland, sander.vandoorn@iee.unibe.ch

Susan A. Foster, Department of Biology, Clark University, Worcester, MA 01610, USA, sfoster@clarku.edu

Samuel D. Gosling, Department of Psychology, The University of Texas at Austin, Austin, TX 78712, USA, samg@mail.utexas.edu

Ton G. G. Groothuis, Department of Behavioural Biology, University of Groningen, 9750 AA Haren, The Netherlands, a.g.g.groothuis@rug.nl

Felicity Huntingford, Ecology and Evolutionary Biology Group, Faculty of Biomedical and Life Sciences, University of Glasgow, Glasgow G12 8QQ, UK, F.Huntingford@bio.gla.ac.uk

Sunil Kadri, Ecology and Evolutionary Biology Group, Faculty of Biomedical and Life Sciences, University of Glasgow, Glasgow G12 8QQ, UK, Sunil.Kadri@glasgow.ac.uk

Jaap M. Koolhaas, Department of Behavioral Physiology, University of Groningen, 9750 AA Haren, The Netherlands, j.m.koolhaas@biol.rug.nl

Olof Leimar, Department of Zoology, Stockholm University, Stockholm, SE-106 91, Sweden, olof.leimar@zoologi.su.se

David Logue, Department of Biology, University of Puerto Rico at Mayagüez, Mayagüez, PR 00681, USA, david.logue@upr.edu

Dario Maestripieri, Department of Comparative Human Development, The University of Chicago, Chicago, IL 60637, USA, dario@uchicago.edu

Jennifer A. Mather, Department of Biological Sciences, University of Lethbridge, Lethbridge, Alberta T1K 3M4, Canada, mather@uleth.ca

Pranjal H. Mehta, Department of Psychology, University of Oregon, Eugene, OR 97403, USA, mehta@uoregon.edu

Flavia Mesquita, Ecology and Evolutionary Biology Group, Faculty of Biomedical and Life Sciences, University of Glasgow, Glasgow G12 8QQ, UK, f.mesquita.1@research.gla.ac.uk

Kerry C. Michael, Department of Psychiatry, University of Maryland School of Medicine, Baltimore, MD 21201, USA, kerry.c.michael@gmail.com

Marc Naguib, Netherlands Institute of Ecology, NL-6666 ZG Heteren, The Netherlands, M.Naguib@nioo.knaw.nl

Kees van Oers, Netherlands Institute of Ecology, NL-6666 ZG Heteren, The Netherlands, K.vanOers@nioo.knaw.nl

Christina M. Ragan, Department of Psychology, Michigan State University, East Lansing, MI 48824, USA, raganch1@msu.edu

Denis Réale, Département des Sciences Biologiques, Université du Québec à Montréal, Montréal, Québec H3C 3P8, Canada, reale.denis@uqam.ca

Andrea Sgoifo, Department of Evolutionary and Functional Biology, University of Parma, Parma, 43100, Italy, andrea.sgoifo@unipr.it

Andrew Sih, Department of Environmental Science and Policy, University of California at Davis, Davis, CA 95618, USA, asih@ucdavis.edu

David L. Sinn, School of Zoology, University of Australia, Hobart, Tasmania, Australia, david.sinn@utas.edu.au

Brian R. Smith, Department of Ecology and Evolutionary Biology, University of California Los Angeles, Los Angeles, CA 90095, USA, smithbr33@gmail.com

Alexander Weiss, Department of Psychology, School of Philosophy, Psychology and Language Sciences, The University of Edinburgh, Edinburgh EH8 9JZ, UK, alex.weiss@ed.ac.uk

Franz J. Weissing, Theoretical Biology Group, Centre for Ecological and Evolutionary Studies, University of Groningen, 9751 NN Haren, The Netherlands, F.J.Weissing@rug.nl

Max Wolf, Department of Biology and Ecology of Fishes, Leibniz-Institute of Freshwater Ecology and Inland Fisheries, 12587 Berlin, Germany, m.wolf@igb-berlin.de

Matthew Wund, Department of Biology, The College of New Jersey, Ewing, NJ 08628, USA, wundm@tcnj.edu

Introduction

Animal Personalities: Who Cares and Why?

CLAUDIO CARERE, DARIO MAESTRIPIERI

Everyone who has experienced close, long-term relationships with animals, such as pet owners or farmers, probably believes that domesticated animals show personality variation, the way people do. In fact, there are studies suggesting that pets' personalities appear to match those of their owners (Gosling 2001). It is possible that pet owners project their own personalities onto their pets, or expect that their pets will complement some aspects of their own personalities (Gosling et al. 2003). In this view, animal personalities would be a product of pet owners' imagination and social-emotional needs. It turns out, however, that personalities are not observed or suspected only in domesticated animals. They have been scientifically demonstrated in nondomesticated animals as well—from invertebrates to monkeys and apes—and in a variety of different environments. In fact, the study of animal personality is one of the fastest growing areas of research in behavioral biology and behavioral ecology.

In animal research, the concept of personality has been used to refer to the existence of behavioral and physiological differences among individuals of the same species, which are stable over time and across different contexts or situations. The behavioral and physiological traits that represent personality are often intercorrelated with one another (i.e., individuals have different clusters of these traits). In animal research, personalities have also been referred to as temperament, behavioral syndromes, coping styles, or simply predispositions (e.g., Wilson et al. 1994; Gosling and John 1999; Gosling 2001; Sih et al. 2004; Groothuis and Carere 2005; Nettle and Penke 2010; Stamps and Groothuis 2010).

A large number of animal studies, particularly in the last decade, have shown that individuals of the same species, often independent from sex or age, differ from each other in their behavior and underlying physiology, even under standardized laboratory conditions. Most of this variation is nonrandom and is consistent across contexts and over time. These individual differences often become conspicuous and easily measurable when

1

individuals have to cope with challenges in their environment, both social and nonsocial (Broom 2001). In the past, such differences were rarely viewed as an expression of biologically meaningful variation; instead, they were often interpreted either as the consequence of inaccurate measurements or as nonadaptive variation around an adaptive mean (Wilson 1998). In contrast, in humans such variation is interpreted as reflecting consistent individual variation in personality or temperament, the science of human personality being already more than one century old (Galton 1883). Personality characteristics in humans have a significant heritable component and a proximate basis in genetic polymorphisms and associated neurobiology; they have also been shown to predict important life outcomes, such as physical and mental health, as well as social and reproductive functioning (Nettle 2005; Nettle and Penke 2010).

Behavioral ecologists have long recognized the existence, within the same species, of individuals with different reproductive and foraging strategies, which have been referred to as "residents and satellites" or "producers and scroungers" (Barnard and Sibly 1981; Clark and Ehlinger 1987). In the past decade, a growing number of studies have shown that consistent individual differences in one trait covary with other behavioral and physiological traits, and that suites of traits may evolve and be maintained by natural selection (e.g., Dingemanse et al. 2004; Carere and Eens 2005; Wolf et al. 2007; Dingemanse and Wolfe 2010; Wolf and Weissing 2010) or even sexual selection (Schuett et al. 2010). Thus, the study of animal personality has become firmly grounded in evolutionary biology and life-history theory.

A major challenge for animal researchers has been to understand the origin of interindividual variation in personalities and the mechanisms responsible for the maintenance of this variation and its transmission across generations. The study of animal personalities is now at the forefront of contemporary behavioral biology particularly because of its integrative and holistic approach (Sih et al. 2004) and also, admittedly, because of its strong appeal to public opinion and mass media (e.g., Dall et al. 2004; Pennisi 2005). The recent surge of interest in phenotypic plasticity and reaction norms among evolutionary biologists can be viewed as arising from the same need for integrative approaches to the study of phenotypic variation (Schlichting and Pigliucci 1998; Bateson 2003; Stamps 2003; Dingemanse et al. 2010).

The study of animal personalities is important for several reasons: (i) it is conducted with an interdisciplinary approach that integrates proximate mechanisms with ontogenetic, functional, and phylogenetic analyses; (ii) it has important implications for evolutionary theory because different but correlated behaviors do not evolve in isolation, but often as a package and this can generate tradeoffs and canalizing effects, which set boundaries to unlimited plasticity (e.g., Sih et al. 2004; Wolf et al. 2007); (iii) personality has to be taken into account in both field and laboratory animal behavioral studies because different personality types may react differently to similar environments or different experimental treatments (e.g., Coppens et al. 2010; Luttbeg and Sih 2010); (iv) individuals may show differential vulnerability to stress and artificial housing conditions, with important implications for animal welfare (e.g., Huntingford and Adams 2005); (v) a better knowledge of the mechanisms underlying animal personalities and of the evolutionary causes and consequences of personalities may be extrapolated to humans and help to provide a better understanding of the nature and evolution of human personalities (e.g., Gosling 2001; Nettle and Penke 2010); (vi) the realization by the general public and policymakers that animals have personalities may help create greater empathy for wild animals and enhance conservation efforts.

For instance, studies in birds have shown not only that individual differences in behavior are heritable, but also that they are systematically related to fitness, with different optima occurring under different environmental conditions (Dingemanse and Réale 2005; Réale et al. 2007). For humans, such a demonstration is necessarily more indirect, both for methodological reasons and because of the profound differences between the contemporary environment and the presumed environment in which we evolved (Nettle 2005).

Research on animal personalities poses several theoretical and empirical challenges. Theoretically, it requires an understanding of the evolutionary mechanisms that may account for the origin and maintenance of clusters of interrelated phenotypic traits. Empirically, this research sometimes requires descriptive longitudinal studies, including studies of relationships between different behaviors, and their consistency across situations; studies on genetic and physiological mechanisms underlying the clustering of behavioral traits, such as pleiotropy, gene linkage, or common neuroendocrine substrates; ontogenetic studies on plasticity and environmental malleability; and field studies on survival and reproduction to understand how different personality profiles are maintained and under what circumstances they can be selected for.

In the past decade, growing interest in the study of animal personalities has resulted in the publication of many theoretical and empirical articles, special issues of journals entirely devoted to this topic (e.g., Carere and Eens 2005; Réale et al. 2010), as well as the publication of literature reviews in major scientific journals such as *Nature, Science, Proceedings of the National*

Academy of Sciences U S A, Current Biology, Biological Reviews, Quarterly Review of Biology, Philosophical Transactions of the Royal Society of London, and *Trends in Ecology and Evolution* (e.g., Dall et al. 2004; Sih et al. 2004; Pennisi 2005; Bell 2007; Blas et al. 2007; Wolf et al. 2007; Bergmüller and Taborsky 2010; Dingemanse et al. 2010; Réale et al. 2010; Schuett et al. 2010). Finally, the most recent editions of all major animal behavior textbooks dedicate entire chapters to the topic of animal personalities (e.g., Alcock 2009; Dugatkin 2009).

The goal of this edited volume is to synthesize and integrate recent research on animal personalities. We aim to provide a comprehensive overview of research on animal personalities in a wide range of taxa, including humans. We also illustrate the integrative and multidisciplinary nature of animal personality research. Evolutionary biology is clearly the discipline that provides a general framework for the study of animal personalities. Important contributions to personality research, however, are also made by ethology, ecology, genetics, endocrinology, neuroscience, and psychology. In assembling the chapters for this edited volume, we made an effort to address both the *how* and *why* questions about personality, and also to include descriptive and experimental studies from different animal taxa. The chapters in this volume illustrate how personalities vary along multiple dimensions; how they are influenced during ontogeny and in adulthood by genetic, physiological, and environmental factors; what is their functional significance, in terms of how they contribute to reproduction and survival; and what is their relevance for animal conservation in the wild and welfare in captivity.

Structure and content of the volume

The book is organized into four sections. The first section (Personalities across Animal Taxa) includes 5 chapters that illustrate the occurrence and the behavioral expression of personalities in different taxa, from invertebrates to humans. In the first chapter of this section, Mather and Logue review a number of descriptive, physiological, genetic-linkage, ontogenetic, and ecological studies to determine the degree to which invertebrates, often viewed as animals of limited behavioral repertoires, exhibit personalities. Bell, Foster, and Wund examine personalities in stickleback fishes. By comparing groups of sticklebacks that either show or do not show consistent individual differences in behavior across contexts, stickleback researchers have an opportunity to understand and experimentally investigate the selective factors that can favor the evolution of personality in fishes. Van Oers and Naguib provide an overview of research on personality in birds, including studies that have addressed behavioral variation, its underlying genetic bases, and its fitness consequences in natural avian populations. Weiss and Adams examine personality studies in nonhuman primates and integrate the findings obtained with different approaches, such as ecology-based and life-history–based approaches, human personality assessment procedures, and multivariate and behavior genetic approaches. The chapter by Gosling and Mehta concludes this section by discussing the value of animal personality research for understanding human personality. In doing so, these authors provide a broad review of animal personality studies across different taxa and discuss the parallels between their findings and those of human personality studies.

The second section (Genetics, Ecology, and Evolution of Animal Personalities) includes both theoretical and empirical chapters on animal personalities, which address the relation between genetic variation, phenotypic plasticity, ecological factors, and the selective mechanisms favoring the evolution and maintenance of personalities in natural animal populations. In the first chapter of this section, Van Oers and Sinn address the quantitative and molecular genetics of animal personality, discussing the role of direct genetic effects, maternal effects, and gene-environment interactions in the evolution and expression of animal personality differences and their transmission across generations. In the following chapter, Dingemanse and Réale address the question of whether animal personality differences represent different social and reproductive strategies, and the role of different selective processes in their evolution. In addition, they examine the reaction norm and character state view of animal personality and discuss how selection acting on personality could be studied from each perspective. In his chapter, Sih focuses on evolutionary studies of animal personality in a socio-ecological context. Specifically, he examines how variation in animal personalities relates to predation, mating, and cooperation as well as how variation in social conditions (e.g., availability of different social partners) affects plasticity in behavioral aspects of personality. The last chapter in this section is by Wolf, Van Doorn, Leimar, and Weissing, who provide a broad overview of the selective pressures favoring the evolution of animal personalities and a discussion of how these pressures affect the structure (e.g., the type of phenotypic traits that cluster together) and the developmental stability of individual differences in personalities.

The third section (Development of Personalities and Their Underlying Mechanisms) addresses the ontogenetic trajectories of different personality types, how they arise as a result of early parental influences, and how they are controlled and regulated by different neuroendocrine mechanisms. The chapters in this section integrate empirical research on behavioral and physiological aspects of animal personality conducted both in the field and in the laboratory. Curley and Branchi review studies of laboratory rodents illustrating the mechanisms through which stable individual differences in neurobiology and behavior emerge during development. In particular, they address the role of gene-environment interactions and epigenetic mechanisms in personality development of laboratory rodents. Maestripieri and Groothuis explore maternal environmental effects on offspring personality development and their underlying mechanisms in both oviparous vertebrates (fish, reptiles, and birds) and placental mammals (rodents and primates). Specifically, they discuss how maternal behavior, maternal stress, and prenatal exposure to varying amounts of maternal steroid hormones, both androgens and glucocorticoids, can result in stable individual differences in offspring physiology and behavior later in life. Finally, Caramaschi, Carere, Sgoifo, and Koolhaas review research on the relation between physiological and behavioral traits commonly considered in animal personality assessments, with particular regard to behavioral reactivity to stress and the activity of the hypothalamic-pituitary-adrenal axis, the hypothalamic-pituitary-gonadal axis, and the autonomic nervous system. They also discuss evidence linking the neurotransmitters serotonin and dopamine, as well as cortical brain structures such as the hippocampus, to variation in animal personality.

The fourth section (Implications of Personality Research for Conservation Biology, Animal Welfare, and Human Health) examines applied aspects of animal personality research. In their chapter, Smith and Blumstein emphasize that behavioral diversity, including personality variation, is an important component of biological diversity and therefore plays a significant role in the long-term persistence of animal populations. They then address how the study and management of animal personalities may play a key role in conservation biology, by arguing that anthropogenic activities can reduce behavioral diversity, and that personality traits can be useful in identifying potentially invasive species. In their chapter, Huntingford, Mesquita, and Kadri emphasize that although most research on fish personality is conducted on species such as sticklebacks and zebrafish, personality profiles have been identified also in many species of cultured fish, particularly salmonids. Since different personalities persist within populations because they allow individuals to grow and survive well in different environmental conditions, the study of behavioral and physiological personality traits in farmed fish can contribute to the scientific understanding of animal personality as well as to the maximization of production and welfare. In the last chapter, Cavigelli, Michael, and Ragan address the importance of research with rodent models of human personality for understanding the relationship between genes and environment, and behavior and physiology, in both health and disease processes. The authors review studies conducted with different strains of laboratory rodents to determine whether some of these strains have behavioral and physiological traits that would permit certain personality types to be resilient or susceptible to specific disease processes. They also compare differential behavioral profiles associated with health trajectories in laboratory rodents to potentially analogous personality traits and associated health and disease trajectories in humans.

It should be obvious from the breadth of topics included in this volume that knowledge of the findings of animal personality research is potentially interesting to many different people, such as scientists (e.g., evolutionary biologists, comparative psychologists, and physiologists), people involved in the conservation of wild animal populations, people involved in the production and welfare of farm animals, and people who own pets or who are simply fascinated by biodiversity, wildlife, and animal behavior.

References

Alcock, J. 2009. *Animal Behavior: An Evolutionary Approach*. 9th ed. Sunderland, MA: Sinauer.

Barnard, C. J., and Sibly, R. M. 1981. Producers and scroungers: a general model and its application to captive flocks of house sparrows. *Animal Behaviour*, 29, 543–550.

Bateson, P. 2003. The promise of behavioural biology. *Animal Behaviour*, 65, 11–17.

Bell, A. M. 2007. Future directions in behavioural syndromes research. *Proceedings of the Royal Society of London B*, 274, 755–761.

Bergmüller, R., and Taborsky, M. 2010. Animal personality due to social niche specialisation. *Trends in Ecology and Evolution*, 25, 504–511.

Blas, J., Bortolotti, G. R., Tella, J. L., Baos, R., and Marchant, T. A. 2007. Stress response during development predicts fitness in a wild, long-lived vertebrate. *Proceedings of the National Academy of Sciences U S A*, 104, 8880–8884.

Broom, D. M. 2001. *Coping with Challenge: Welfare in Animals Including Humans*. Berlin: Dahlem University Press.

Carere, C., and Eens, M. 2005. Unravelling animal personalities: how and why individuals consistently differ. *Behaviour*, 142, 1149–1157.

Clark, A. B., and Ehlinger, T. J. 1987. Pattern and adaptation in individual behavioral differences. In: *Perspectives in Ethology* (Bateson, P. P. G., and Klopfer, P. H., eds.), pp. 1–47. New York: Plenum Press.

Coppens, C. M., de Boer, S. F., and Koolhaas, J. M. 2010. Coping styles and behavioural flexibility: towards underlying mechanisms. *Philosophical Transactions of the Royal Society of London B*, 365, 4021–4028.

Dall, S. R. X., Houston, A. I., and McNamara, J. M. 2004. The behavioural ecology of personality: consistent individual differences from an adaptive perspective. *Ecology Letters*, 7, 734–739.

Dingemanse, N. J., Both, C., Drent, P. J., and Tinbergen, J. M. 2004. Fitness consequences of avian personalities. *Proceedings of the Royal Society of London B*, 271, 847–852.

Dingemanse, N. J., Kazem, A. J. N., Réale, D., and Wright, J. 2010. Behavioural reaction norms: animal personality meets individual plasticity. *Trends in Ecology and Evolution*, 25, 81–89.

Dingemanse, N. J., and Réale, D. 2005. Natural selection and animal personality. *Behaviour*, 142, 1159–1184.

Dingemanse, N. J., and Wolf, M. 2010. Recent models for adaptive personality differences: a review. *Philosophical Transactions of the Royal Society of London B*, 365, 3947–3958.

Dugatkin, L. A. 2009. *Principles of Animal Behavior*. 2nd ed. Boston: W. W. Norton.

Galton, F. 1883. *Inquiries into Human Faculty and Its Development*. London: Macmillan.

Gosling, S. D. 2001. From mice to men: what can we learn about personality from animal research? *Psychological Bulletin*, 127, 45–86.

Gosling, S. D., and John, O. P. 1999. Personality dimensions in nonhuman animals: a cross-species review. *Current Directions in Psychological Science*, 8, 69–75.

Gosling, S. D., Lilienfeld, S. O., and Marino, L. 2003. Personality. In: *Primate Psychology* (Maestripieri, D., ed.), pp. 254–288. Cambridge, MA: Harvard University Press.

Groothuis, T. G. G., and Carere, C. 2005. Avian personalities: characterization and epigenesis. *Neuroscience and Biobehavioral Reviews*, 29, 137–150.

Huntingford, F. A., and Adams, C. E. 2005. Behavioural syndromes in farmed fish: implications for production and welfare. *Behaviour*, 142, 1207–1221.

Luttbeg, B., and Sih, A. 2010. Risk, resources, and state-dependent adaptive behavioural syndromes. *Philosophical Transactions of the Royal Society of London B*, 365, 3977–3990.

Nettle, D. 2005. An evolutionary approach to the extraversion continuum. *Evolution and Human Behavior*, 26, 363–373.

Nettle, D., and Penke, L. 2010. Personality: bridging the literatures from human psychology and behavioural ecology. *Philosophical Transactions of the Royal Society of London B*, 365, 4043–4050.

Pennisi, E. 2005. Strong personalities can pose problems in the mating game. *Science*, 309, 694–695.

Réale, D., Dingemanse, N. J., Kazem, A. J. N., and Wright, J. 2010. Evolutionary and ecological approaches to the study of personality. *Philosophical Transactions of the Royal Society of London B*, 365, 3937–3946.

Réale, D., Reader, S. M., Sol, D., McDougall, P. T., and Dingemanse, N. J. 2007. Integrating temperament in ecology and evolutionary biology. *Biological Reviews*, 82, 291–318.

Schlichting, C., and Pigliucci, M. 1998. *Phenotypic Evolution: A Reaction Norm Perspective*. Sunderland, MA: Sinauer.

Schuett, W., Tregenza, T., and Dall, S. R. X. 2010. Sexual selection and animal personality. *Biological Reviews*, 85, 217–246.

Sih, A., Bell, A. M., Chadwick Johnson, J., and Ziemba, R. E. 2004. Behavioral syndromes: an integrative overview. *Quarterly Review of Biology*, 79, 241–277.

Stamps, J. A. 2003. Behavioural processes affecting development: Tinbergen's fourth question comes of age. *Animal Behaviour*, 66, 1–13.

Stamps, J. A., and Groothuis, T. G. G. 2010. Developmental perspectives on personality: implications for ecological and evolutionary studies of individual differences. *Philosophical Transactions of the Royal Society of London B*, 365, 4029–4041.

Wilson, D. S. 1998. Adaptive individual differences within single populations. *Philosophical Transactions of the Royal Society of London B*, 353, 199–205.

Wilson, D. S., Clark, A. B., Coleman, K., and Dearstyne, T. 1994. Shyness and boldness in humans and other animals. *Trends in Ecology and Evolution*, 9, 442–445.

Wolf, M., Van Doorn, S., Leimar, O., and Weissing, F. J. 2007. Life-history trade-offs favour the evolution of animal personalities. *Nature*, 447, 581–585.

Wolf, M., and Weissing, F. J. 2010. An explanatory framework for adaptive personality differences. *Philosophical Transactions of the Royal Society of London B*, 365, 3959–3968.

I

PERSONALITIES ACROSS ANIMAL TAXA

The Bold and the Spineless

Invertebrate Personalities

JENNIFER A. MATHER AND DAVID M. LOGUE

Introduction

Reviewing the study of personalities of invertebrates offers a series of challenges. First, there is a huge number of invertebrate species, sometimes estimated to represent 98% of the animal species on the planet (Pechenik 2000). Second, invertebrates exhibit a tremendous array of life history strategies, developmental trajectories, modes of reproduction, and physiological bases of behavior, many of which are poorly known. A further challenge arises from the diversity of perspectives and research backgrounds that characterizes invertebrate personality researchers.

One of the goals of this review is to determine the degree to which invertebrates, often viewed as animals of limited behavioral repertoires, exhibit personality as defined in the introduction and throughout this volume. We begin with a survey of reports that relate to personality in invertebrates. We categorize these as (1) descriptive reports, (2) physiological/genetic linkages (see Van Oers et al. 2005), (3) ontogenetic studies (*sensu* West-Eberhard 2003), and (4) ecological/selection studies (Groothuis and Carere 2005; Smith and Blumstein 2008). There will be some bias toward studies of poorly known taxa, even if their evidence is fragmentary. We then evaluate several particularly thorough and influential research programs in depth. In the final section, we provide recommendations for future research directions and attempt to summarize the current state of the field.

This review does not evaluate the division of labor (polyethism) in colonies of social insects, or the discrete morphologies (polyphenism) found in many invertebrates. Although we recognize that both of these phenomena may relate to personality (e.g., Bergmüller and Taborsky 2010; Bergmüller et al. 2010), we have chosen to focus on subtler forms of personality (i.e., those that form continuous rather than discrete distributions). We refer readers interested in polyethism and polyphenism to reviews by Beshers and Fewell (2001) and Emlen and Nijhout (2000), respectively.

General survey

A literature search of the papers that assess individual differences and/or correlated behavior in invertebrates revealed thirty-two papers. These range widely across taxa, one is on *Caenorhabdita* worms and one on snails and a small collection on cephalopod mollusks. Arthropods are well represented, with two on crustaceans, eight on spiders, and fourteen on insects, three on individuals in the larval stage, and one spanning developmental stages. The scope of these studies varies extremely widely, and only those on the fruit flies, grass spiders, fishing spider, field crickets, and cephalopods fall into a group of papers that merits further discussion. In Table 1.1 each paper is

Table 1.1. Studies of invertebrates that assess individual differences and/or correlated behaviors.

Authors	Species/ Group	Category	Study design	Results	Comments
de Bono and Bargmann 1998	*Caenorhabdita* worm	physiological/ genetic	1 measure of activity in 2 morphotypes	Activity is controlled by *mpr-1* gene	1 behavior
Turner et al. 2006	*Physa* snail	ecological/ selection	2 measures of antipredator behavior across experience levels	Mostly inherited, a little learned	1 situation
Mather and Anderson 1993	*Octopus* cephalopod	descriptive	19 behaviors in 3 contexts over 2 weeks	3 dimensions (activity reactivity avoidance)	Descriptive
Sinn et al. 2001	*Octopus* cephalopod	ontogenetic	15 behaviors in 3 contexts over 6 weeks	4 components, development program	Early lifespan
Sinn and Molschaniwskyj 2005	*Euprymna* squid	physiological/ genetic	12 behaviors in 3 contexts and morphological characteristics	Four axes of variation, all context-specific	Lab
Sinn et al. 2006	*Euprymna* squid	ecological/ selection	12 behaviors in 2 contexts, and heritability of behavior	Heritability greater for antipredator behavior than foraging	Lab

Table 1.1. (*continued*)

Authors	Species/ Group	Category	Study design	Results	Comments
Sinn et al. 2008	*Euprymna* squid	ontogenetic	10 behaviors relating to shy-bold dimension in 2 contexts over 13 weeks	Different correlations across situations/time	1 dimension
Sinn et al. 2010	*Euprymna* squid	ecological/ selection	10 behaviors relating to shy-bold dimension in 2 contexts in 2 populations over 2 years	Personality structure varied by population and year	1 dimension
Reaney and Backwell 2007	*Uca* fiddler crab	ecological/ selection	Activity, aggression, and courtship success, over variable time scales	Risk taking predicts courtship success	Males grouped shy or bold
Briffa et al. 2008	*Pagurus* hermit crab	descriptive	1 behavior relating to startle response in 5 contexts over 5 days	Plasticity and consistency	1 measure, females
Johnson and Sih 2005	*Dolomedes* spider	ecological/ selection	2 behaviors in 1 and 2 contexts, respectively, over variable time scales	Correlations between contexts and over time suggest behavioral spillover	Females
Johnson and Sih 2007	*Dolomedes* spider	descriptive	1 measure of boldness in 4 contexts and 2 developmental stages	Structure of syndrome changes over development, behaviors correlated between contexts in adults	1 female measure
Hedrick and Riechert 1989	*Agelenopsis spider*	ecological / selection	1 measure of fearfulness in 2 populations, in field and in lab	Inherited population differences	Females, one behavior
Riechert and Hall 2000	*Agelenopsis spider*	ecological / selection	1 measure of fearfulness in 2 field populations, reciprocal transplant, and lab-reared	Past selection has produced habitat-specific risk taking	Females, one behavior

(*continued*)

Table 1.1. (*continued*)

Authors	Species/ Group	Category	Study design	Results	Comments
Riechert and Hedrick 1990	*Agelenopsis* spider	ecological/ selection	1 behavior each in antipredator and feeding contexts in 2 field populations, and lab-reared	Population differences, between-context correlations in one population	Mostly at population level, females
Riechert and Hedrick 1993	*Agelenopsis* spider	descriptive	1 behavior each in antipredator and agonistic contexts in 2 field populations and lab reared	Contest winners exhibit shorter latencies	Females
Maupin and Riechert 2001	*Agelenopsis* spider	physiological/ genetic	1 behavior in feeding context across 4 populations in 2 habitats, planned crosses	Population differences, sex-linked inheritance	1 behavior, mostly at population level
Pruitt and Riechert 2012	*Anelosimus* spiders	ecological/ selection	11 behaviors in 4 contexts and index of sociality in 2 populations	Sociality covaried with behaviors across contexts	Females
Hedrick 2000	*Gryllus* crickets	ecological/ selection	1 behavior in calling context, 2 behaviors in antipredator context, field caught and lab-raised	Tradeoff between calling and antipredator behavior	Males
Kortet et al. 2007	*Gryllus* crickets	physiological/ genetic	3 antipredator behaviors, 2 physiological measures, 3 populations	Correlation pattern in different populations, mixed evidence for immune/ antipredator tradeoff	Males mostly population
Kortet and Hedrick 2007	*Gryllus* crickets	ecological/ selection	2 behaviors in antipredator context, success in agonistic context, all lab reared	Positive correlation between boldness and contest winning	Males mostly population
Despland and Simpson 2005	*Schistocerca* locust	ecological/ selection	1 behavior in feeding context, 2 phases (solitarious vs. gregarious)	Phases differ in food choice	1 behavior

Table 1.1. (*continued*)

Authors	Species/ Group	Category	Study design	Results	Comments
Rutherford et al. 2007	*Enallagma* damselflies larval	ecological/ selection	5 behaviors in antipredator, antiparasite, and mixed contexts	1 behavior repeatable across contexts, limited plasticity	Not across time
Weinstein and Maelzer 1997	*Perga* sawfly larva	descriptive	1 behavior in dispersal context, over 12–15 days	20% are "leaders"	1 behavior
Nemiroff and Despland 2007	*Malacosoma* tent caterpillars	descriptive	4 behaviors in activity context over 4 days	2 somewhat discrete levels of activity	Lab
Sokolowski 2001	*Drosophila* fly	physiological/ genetic	Genetics and individual behavior review	Review	Gene influence
Higgins et al. 2005	*Drosophila* fly	physiological/ genetic	5 behaviors summed into 1 measure of activity in 9 isogenic lines	Within-group (20%) and between-group differences in activity	Behavior observed at group level
Blankenhorn 1991	*Gerris* water striders	ecological/ selection	2 composite measures in foraging context and 2 composite measures in agonistic context over 17 weeks, plus fecundity	Foraging behavior correlated to dominance rank; foraging behavior and to a lesser degree dominance predicted fecundity	Females
Logue et al. 2009	*Gromphadorina* cockroach	ecological/ selection	25 behaviors in 5 contexts, morphology and reproductive success	2 behavioral syndromes, 1 of which predicts reproductive success	Males
Starks and Fefferman 2006	*Polistes* wasp	ecological/ selection	Game-theoretic model of nest founding	Predicts frequency of alternative tactics given starting conditions	Females, not stable
Retana and Cerda 1991	*Categlyphis* ants	descriptive	20 measures from 5 nests over 6 months	Activity was variable across group and over time	Observational

described as within one of four general categories: descriptive, ecological/ selection, physiological/genetic, and ontogenetic. Next, a brief comment is made about the study design, then major results of the study are summarized, and last, comments are made that allow the reader to understand the study better.

The genetic control of behavior in the fruit fly, *Drosophila*

Research on the behavior of *Drosophila* fruit flies has made a considerable contribution to our knowledge of individual differences in behavior, even though the researchers conducting this work did not frame their research in terms of personality. Rather, their work attempted to describe genetic influences on physiology and behavior. Sokolowski's (2001) review of the behavioral genetics of *Drosophila* is particularly informative to researchers interested in the genetic underpinnings of personality.

Sokolowski (2001) makes it clear that even in flies there are no genes "for" behavior. Rather, genes and the environment interact to determine anatomy and physiology, which, in turn, influence behavior. Variation in individual genes can influence more than one aspect of an animal's phenotype—a phenomenon known as pleiotropy (Cheverud 1996). Variation at pleiotropic loci can result in consistent individual differences across multiple contexts—in other words, personalities (Sih et al. 2004). Much of the investigation of the behavior of *Drosophila* has focused on linking genetic differences to differences in behavior. Researchers identify a variable locus in the fly genome that covaries with behavior and investigate the physiological pathway that links the genetic variation to differences in behavior. Of the many examples of research along these lines, the following two are particularly good demonstrations of the relationship between genetics and "personality" in fruit flies.

In fruit fly larvae, scientists have found a gene, *for*, which influences foraging behavior. The R variant of the gene is the dominant allele and the s variant is the recessive one. In the presence of food, rover (Rs or RR) larvae tend to leave a food patch, but sitter larvae (ss) tend to remain (Sokolowski 2001). Natural selection favors rovers in crowded situations and sitters in less crowded ones. In other words, selection for a behavioral strategy is density dependent. Environmental variation can also affect behavior, as food deprivation turns rovers into sitters and other environmental factors can transform sitters into rovers. To define the personality of an individual, its behavior should be consistent through all contexts; so, rovers should be always rovers. This example shows a relevant plasticity of behavioral response, which does not match with personality. The difference between the two behavioral types is mediated by a small difference in cGMP-dependent protein kinase activity—on average, rovers have 12% more activity than sitters. Rovers have better short-term but poorer long-term memory than sitters (Mery et al. 2007). The rover-sitter example elegantly demonstrates how one gene can influence individual differences in an important behavior pattern linked to activity.

Several mutations cause male fruit flies to terminate courtship prematurely. One of the most interesting of these is the *dunce* mutation, which alters learning (Sokolowski 2001). Normal males that have encountered a mated female suppress courtship for three hours. The mechanism of this courtship suppression is olfactory-based avoidance learning. *Dunce* males, however, do not exhibit olfactory-based avoidance associative learning and so do not suppress courtship. Learning suppression by this mutation (likely due to alteration in the mushroom bodies of the brain) could have pleiotropic effects. If, as seems likely, these effects influence behavior in more than one context and persist over time, pleiotropy would be causally linked to personality.

Studies of the *for* and *dunce* loci illustrate how relatively minor genetic variation can cause persistent, context-general differences in behavior, which might be measured as personality. Findings from such studies, however, should not be taken as evidence that genetic variation, rather than environmental variation or gene-by-environment interaction, is the primary determinant of personality in fruit flies. Higgins et al. (2005) studied "activity" in groups of flies derived from nine isogenic (i.e., genetically identical except for sex) lines. They found that less than 15% of variation in summed activity was due to lineage. A group-within-lineage effect accounted for over 20% of the variation in activity, suggesting an important contribution of social environment to activity. The interaction between genes and the environment explained another 11% of the variation in activity, meaning that various lineages responded differently to different environments. Interestingly, there were significant differences in the specific behaviors that contributed to activity (feeding, grooming, resting, and walking) among different lines, so a summed activity measure may not do justice to the complexity of the problem.

Foraging-predation tradeoffs in the
Western grass spider, *Agelenopsis aperta*

The field of behavioral ecology is defined by its hypothesis-driven, evolutionary approach to behavior. Whereas personality psychologists emphasize holism, behavioral ecologists often focus on specific behavioral traits and develop models that predict an individual's optimal behavior in a given set of circumstances. For example, a well-known optimal foraging model describes optimal prey choice (i.e., the prey type that will maximize a predator's energy intake) given the profitability, search time, and handling time of each potential prey type (Stephens and Krebs 1986). Behavioral ecologists' interest in optimal behavior stems from the fact that natural selection shapes behavior in ways that tend to maximize an individual's fitness over evolutionary time.

Using optimal foraging theory as an example, suppose that foraging preference (for prey A versus prey B) differs among individuals and covaries with boldness in the presence of predators, such that bolder individuals tend to prefer prey A. Optimal foraging theory does not consider individual differences because it assumes that all individuals behave optimally; if A is the optimal prey, all individuals should choose A until it is depleted to the point that B becomes the optimal choice. Further, optimal foraging theory treats foraging as a context-specific behavior, whose fitness consequences are independent of behavior in other contexts. If prey choice is linked (e.g., by pleiotropy, by genetic linkage) to boldness, the fitness consequences of boldness will affect the evolution of prey choice, and vice versa.

The behavioral syndromes paradigm incorporates individual differences and between-context correlations in behavior into an evolutionary framework (Sih et al. 2004; Bell 2007). A series of studies on the Western grass spider, *Agelenopsis aperta*, is an early example of the behavioral syndromes approach. The Western grass spider is a funnel web spider that occurs in both dry grassland and moist riparian habitats. Grassland dwellers experience lower prey availability, higher competition for webs, and lower predation from birds (Riechert and Hedrick 1990). The authors asked how the very different selection regimes in these two habitats affected spiders' behavior.

Riechert and Hedrick's (1990, 1993) work contributes to the personality literature because it measures multiple behavioral traits and quantifies the relationships between them. The traits that they examined—boldness, latency to forage, success in agonistic contests—might be said to constitute the "shy-bold" axis of personality widely studied in vertebrates (Gosling 2001). The authors, however, call these behaviors manifestations of "aggression" and "fear," which they view as separate but correlated axes (Smith and Riechert 1984). Regardless of the terminology or the underlying mechanisms, these traits are evolutionarily important because they affect predator avoidance, foraging, and other factors that determine fitness.

Riechert and Hedrick (1990) reasoned that natural selection should favor more fearful spiders in riparian habitats, and more aggressive spiders in grasslands. Because riparian spiders experience high predation risk and high food availability, spending a lot of time hiding (an expression of fearfulness) should provide high benefits in the form of reduced risk of predation at a relatively low cost—if they miss a meal because they are hiding, another meal will probably be available soon. In contrast, grassland spiders are under low predation risk, but they must maintain large territories and be vigilant for potential food because prey is scarce. Grassland spiders would benefit from low fearfulness and high aggressiveness if these traits help them to defend large territories and capture more prey. Because predation is relatively uncommon in the grassland population, the costs of being exposed on the web should be lower than in the riparian population. The investigators measured fearfulness by simulating an approaching avian predator and recording whether the resident spider retreated into its funnel and, if so, how long it took to re-emerge. Their predator stimulus was simple but appropriate to their subject's *umwelt*; experimenters blew puffs of air onto the web sheet with a rubber bulb designed to clean camera lenses. As predicted, free-living spiders from the riparian habitat exhibited significantly longer latencies to re-emerge from the funnel than their grassland counterparts (Riechert and Hedrick 1990). The same pattern was found among second-generation laboratory-raised spiders, showing that genetic factors are responsible for at least part of the difference in latency to emerge between the two populations.

Emergence time was not the only trait that differed between the populations. The researchers deposited standardized prey items into spiders' webs and measured their latency to attack. They found that riparian spiders took longer than grassland spiders to attack prey that had been deposited in their webs (14.1 vs. 6.6 sec, on average; Hedrick and Riechert 1989). The hypothesis that latency to attack and latency to recover from disturbance are constrained to evolve together can be used to generate the prediction that the two traits will covary within a given population. Indeed, recovery from disturbance and latency to attack prey were related among individuals in the riparian population, suggesting such a constraint. Further, when spiders exhibiting short latencies to recover from disturbance were pitted against spiders with long latencies in agonistic contests, the short-latency spiders tended to win (Riechert and Hedrick 1993). Thus, behaviors associated with fighting, recovery from disturbance, and predation all covary in *A. aperta*. Genetic causes of behavioral syndromes include pleiotropy (i.e., multiple effects attributable to a given genetic variant) or linkage disequilibrium (i.e., when two or more genes tend to be inherited together; Riechert and Hedrick 1990; Sih et al. 2004).

The researchers then asked whether the behavioral differences between the two populations resulted from adaptation to the local environments. Reciprocal transplant experiments revealed that local spiders fared better than transplants, supporting the hypothesis that spiders are adapted to their local environment (Riechert and Hall 2000). A natural experiment on local adaptation occurred at an Arizona site where a riparian zone intersects an arid habitat and migration between habitats resulted in substantial gene flow between the two populations (Riechert 1993). Although selection presumably continued to favor the individuals best suited to each environment, immigrants from the other environment continuously added maladaptive genes to the population. When she eliminated gene flow with experimental enclosures, Riechert (1993) found that the populations rapidly evolved to become more adapted to their environments.

As with most of the behavioral syndromes research that followed it, the Western grass spider project emphasized environmental influences on the evolution of correlated patterns of behavior. This approach offers insights into the fundamental significance of personality in animals: Why do personalities exist? Why is group A different from group B? How does the existence of personality affect the evolution of animal populations? One criticism of the behavioral syndromes approach is that it tends to focus on a small set of specified traits. For example, the Western grass spider project measured a small number of behaviors, and identified only one or two axes of covariation. Measuring more behaviors in more contexts would have allowed the researchers to more fully define these axes and perhaps to discover additional axes relevant to natural selection (see Pruitt et al. 2008; Logue et al. 2009). Such criticisms notwithstanding, the Western grass spider studies represent an important early step toward understanding the evolutionary basis of individual differences in behavior.

Behavioral spillover in the fishing spider, *Dolomedes triton*

Suppose that in a particular species, males with aggressive personalities exhibit high levels of aggression toward both male competitors and female prospective sex partners, whereas passive males are not aggressive toward either sex. Relative to passive males, aggressive males would enjoy an advantage in the context of male-male competition, but they would suffer a disadvantage in the context of courtship. Behavioral ecologists have adopted the term "behavioral spillover" to describe cases in which positive selection on a trait in one context results in maladaptive behavior in another context due to covariation between traits in the two contexts (Sih et al. 2004). Behavioral spillover generates an evolutionary tradeoff, such that the optimal behavioral type (i.e., the type favored by natural selection) is a function of the costs and benefits of behavioral tendencies in all of the affected contexts and of the frequency with which individuals experience each context. In the example given above, we would expect males to exhibit intermediate levels of aggression that would not be optimal in either context.

Johnson and Sih (2005) examined a potential case of behavioral spillover in the fishing spider, *Dolomedes triton*. Fishing spiders do not catch their prey in webs; rather, they hunt by ambushing prey on the surface of ponds or lakes. Female *D. triton* often eat courting males prior to insemination, a behavior termed "precopulatory cannibalism." In the laboratory, some females kill all of the males that approach them. From a Darwinian perspective, universal precopulatory cannibalism is an enigma. Although females may gain valuable nutrients from eating males (Arnqvist and Henriksson 1997), they miss a mating opportunity when they cannibalize males prior to copulation (as opposed to cannibalizing after copulation), resulting in a probable decrease in their lifetime fecundity. Applying the logic of behavioral spillover to the case of precopulatory cannibalism, Arnqvist and Henriksson (1997) hypothesized that natural selection favors females who exhibit high levels of "voracity," or tendency to attack, as juveniles, but evolution has not come up with a sufficiently low-cost mechanism of dramatically reducing voracity in adulthood. Thus, voracious females are stuck paying the costs of excessive voracity in the courtship domain.

Johnson and Sih (2005) tested for behavioral spillover by measuring captive females' latency to feed across life stages, latency to attack courting males, latency to recover from a disturbance, and reproductive success. Fishing spiders dive under water after a disturbance, so the authors' measure of latency to recover was the time that females spent underwater after being subjected to a standardized disturbance. Johnson and Sih (2005) found the positive correlations predicted by the behavioral spillover hypothesis. High feeding voracity was correlated with females' fecundity ($r = 0.50$), and juvenile voracity was correlated with adult voracity ($r = 0.38$). Moreover, feeding voracity at both life stages correlated positively with precopulatory attacks and cannibalism ($r = 0.35$ and $r = 0.39$, respectively).

This study also suggested that feeding voracity may contribute to a broad shy-bold continuum in fishing spiders, since a spider's latency to recover from disturbance (a common measure of boldness) correlated positively with voracity toward males ($r = 0.24$) and voracity toward heterospecific prey ($r = 0.32$). The findings that an individual's behavior in each of these contexts is linked to its behavior in the other two, and that voracity is stable over development, demonstrate the broad linkages and individual differences that define personality.

The authors of this study recognized that constraints on an animal's ability to change behavior at maturation could have profound evolutionary consequences because optimal behavior in juveniles is often very different from optimal behavior in adults. For example, juveniles should behave in ways that maximize growth, survival, and (in some cases) dispersal, whereas adults should maximize reproduction. Johnson and Sih (2007) explored the idea of behavioral continuity across the lifespan with a second longitudinal study of behavioral syndromes in *D. triton*. The researchers measured females' boldness as juveniles, as juveniles with prey in the environment, as adults, as adults with prey, as adults with a male in the environment, and as adults guarding an egg sac. An individual female's behavior was generally highly repeatable within contexts but was not fixed within individuals. For instance, spiders were bolder in foraging than control trials. Individuals were not infinitely plastic, however, as evidenced by the strong positive correlations ($r = 0.50$) between boldness behaviors in several contexts.

Johnson and Sih (2007) found that juvenile behavior predicted adult behavior in the "prey present" context, suggesting that individual differences persist across developmental stages in spiders. This pattern may be a maladaptive product of constrained plasticity. Alternatively, it may benefit certain individuals to be bolder (or less bold) than the population average throughout their lives (see Reaney and Backwell 2007; Wolf et al. 2007). Consider a scenario in which some individuals in a population are malnourished in infancy and others are well fed. In the malnourished group, no shy individuals obtain sufficient resources to reproduce but some bold individuals do. Individuals in the well-fed group can obtain sufficient resources to reproduce without being very bold, such that the costs of boldness (e.g., increased predation) exceed its benefits throughout the life cycle. In this hypothetical scenario, bold individuals are "risk prone" and shy individuals are "risk averse," but both strategies are "optimal" given the starting conditions of the individuals.

Johnson and Sih (2007) also found between-context correlations in adults, but not in juveniles. They suggest that the behavioral syndromes arise in adulthood, and propose that the adult molt serves as a "switch point" allowing for "a significant reorganization of behavior" (p. 7). This would be a contrast to personality continuity across the lifespan in vertebrates (see

Caspi 1998 for humans; Sinn et al. 2008 for a similar correlation in cephalopods). Perhaps in groups with major life transitions, continuity is difficult to maintain.

Behavioral tradeoffs in the field cricket, *Gryllus sp.*

In theory, behavioral correlations may be costly or beneficial at the level of the individual. For example, risk-sensitive life history theory predicts that individuals that experience poor rearing conditions will tend to be risk prone, whereas those with good rearing conditions will tend to be risk averse (see Stearns 1992; Hill et al. 1997). Risk-prone individuals are expected to take risks across contexts, and risk-averse individuals are expected to avoid risk across contexts, generating population-wide correlations between similar behaviors in different contexts (Wolf et al. 2007). Another kind of beneficial behavioral combination occurs when one type of behavior compensates for inherent costs of another type of behavior. In this case, individuals that engage in high levels of the costly behavior also exhibit higher levels of the compensating one, resulting in a linkage between the two types of behavior.

Hedrick (2000) identified an example of behavioral compensation in the field cricket, *Gryllus integer*. Male field crickets produce loud calls to attract females. Calling has the unintended consequence of attracting predators (Zuk and Kolluru 1998) and parasites (Cade 1975). Hedrick (2000) predicted that males that produce long calling bouts compensate for the heightened risk of calling with heightened shyness, and vice versa. She examined the relationship between calling-bout duration and predator-avoidance behavior in both lab-raised and wild-caught *G. integer* males. She found that mean calling bout duration correlated positively with latency to emerge from a shelter ($r = 0.38$) and to resume calling after a disturbance ($r = 0.38$). These findings support the hypothesis that males trade off hiding time with calling time to mitigate the risk of predation incurred by calling. According to this hypothesis, all individuals are compromising: in the absence of a cost to calling, each male would be best off calling a lot and hiding very little, but the "constraint" of acoustically orienting predators and parasites appears to have driven the evolution of a spectrum of compromise strategies.

Kortet and Hedrick (2007) looked deeper into boldness and aggression in *G. integer* males. Aggression and boldness covary in other species, including stickleback fish (Huntingford 1982; Bell 2005) and fishing spiders (Johnson and Sih 2005). If such distantly related organisms as fish, spiders, and crickets exhibit similar personality structures, we might tentatively conclude that this structure is fundamental (Gosling 2001), in the sense that it is a product either of phylogeny or of convergent evolution, and so we can begin to ask general questions about its origin and consequences. Kortet and Hedrick (2007) subjected males to pair-wise agonistic contests and measured their latency to move and latency to emerge from a shelter. They found that contest winners moved and emerged from shelter more quickly than losers, and concluded that their data "suggest that there is an aggressiveness/ activity behavioral syndrome in *G. integer*" (p. 475). Notably, however, they measured proportion of fights won and assigned each male a "dominant" or "subordinate" status but did not measure aggressiveness *per se*.

Given that behavior is the result of anatomy and physiology interacting with the environment, it would come as little surprise if physiological or morphological properties covaried with behavioral properties to generate "behavioral-physiological" syndromes (see Kagan 1994 for humans). A good example of this type of syndrome is described by Kortet et al. (2007). Their study compared male crickets' latencies to become active ("freezing time") and emerge from a shelter ("hiding time") to two measures of immune response: the encapsulation response and the lytic activity of haemolymph. Males from three populations were included in the study: *Gryllus integer* from California, *G. integer* from Arizona, and "*G. 15*" (an unnamed morphospecies) from Arizona. Males from Arizona had higher levels of predation and parasitization than California males. The authors hypothesized that increasing antipredator defense incurs a cost in immune response, as there is a finite amount of energy to invest in all systems. They therefore predicted that Arizona males would have higher levels of antipredator behavior and lower levels of immune response. Arizona males hid for a longer time and exhibited a stronger encapsulation response. In support of the hypothesis that immune response trades off with predator avoidance, the authors found that lytic activity was negatively related to freezing time ($r = -0.29$). In Arizona *G. integer* males, however, encapsulation was positively associated with hiding times ($r = 0.42$). Studies such as this one provide an important link between individual differences in behavior and underlying physiology.

Personality in cephalopods

Although cephalopod mollusks are phylogenetically distant from vertebrates, they exhibit high levels of behavioral variability (Mather 1995). This variability has led the researchers studying individual differences in cephalopod behavior to assume the bottom-up approach more commonly seen in assessment of personality in "higher" vertebrates (Gosling 2001). Studies of cephalopod mollusks have rarely addressed the fitness ramifications of traits or syndromes. Rather, researchers have tested animals in common situations and used multivariate analyses to describe personality axes. *Octopus rubescens* (Mather and Anderson 1999) and *O. bimaculoides* (Sinn et al. 2001) were evaluated in alerting, threat, and feeding contexts. Initial work on the sepiolid squid, *Euprymna tasmanica*, examined behavior in the contexts of threat and feeding (Sinn and Moltschaniwskyj 2005; Sinn et al. 2006). Further work on *Euprymna* focused on the shy-bold axis of personality (Sinn et al. 2008).

The main finding of these studies was that cephalopods exhibit a set of personality axes similar to those found in vertebrates (see Gosling 2001). *Octopus rubescens* (Mather and Anderson 1993) had three axes of personality (activity, reactivity, and avoidance) accounting for 44.9% of the overall variance in behavior. Young *O. bimaculoides* exhibited four axes (active engagement, arousal/readiness, aggression, and avoidance/disinterest) accounting for 53% (Sinn et al. 2001). *E. tasmanica* had four axes (shy/bold, activity, reactivity, and bury persistence) accounting for 75.2% of variance in the threat and 78.4% in the feeding test (Sinn and Moltschaniwskyj 2005; see Gosling 2001 for comparable data from vertebrates).

Testing of *O. bimaculoides* across six weeks allowed researchers to examine the stability of personality types over time (note that the life span of this octopus species is less than one year; Sinn et al. 2001). Consistency of behavior over time was generally high, and comparisons between different components both within and across time were low. In accordance with the vertebrate pattern of continuity and change (Hogan and Bolhuis 1994), however, there were some differences in the pattern of temperament traits across time. Notably, significant change occurred in arousal/readiness and aggression from week 3 to week 6.

Assessment of the shy/bold dimension over the five-month lifespan (weeks 3 to 16) of *E. tasmanica* was more complex (Sinn et al. 2008). Within-test consistency paired across adjacent time periods increased through the juvenile periods, but significantly so only in weeks 6 and 9. There was a major reorganization between weeks 9 and 12, the period of sexual maturity, after which behavior stabilized in threat tests but not feeding tests (compare to Johnson and Sih 2007). In the threat context, shy squid remained shy across the lifespan but bold ones became bolder. In the feeding context, bold ones remained bold but shy ones became shyer. Again, personality structure was characterized by both continuity and change.

For *E. tasmanica*, the shy/bold dimension was not related between threat and feeding tests (Sinn and Moltschaniwskyj 2005) and thus dimensions tended to be context-specific rather than context-general. This lack of correlation of the shy/bold dimension persisted across the lifespan (Sinn et al. 2008). A test of personality across two years and in two different populations, however, revealed a significant correlation in one of the four groups, suggesting that the relationship may also be environmentally dependent (Sinn, Gosling, and Moltschaniwskyj, unpublished data).

For the young *O. bimaculoides*, heritability of personality development is suggested by the significant differences in developmental pattern of the arousal/readiness dimension between half-sibling broods (Sinn et al. 2001). There is clearer evidence for heritability in *E. tasmanica* (Sinn et al. 2006), with significant heritabilities in behaviors contributing to the shy/bold dimension in the threat context, but not in the feeding context. The fitness ramifications of personality were examined in *E. tasmanica* (Sinn et al. 2008). Squids that were intermediate on the shy/bold continuum were the heaviest (weight is the best measure of size in the flexible cephalopods). Female personality did not predict fecundity, but their boldness in feeding situations covaried with size and did explain a significant amount of variation in hatching success.

Sex is expected to affect personality, and there is a rich history of the study of this relationship in humans (e.g., Phares and Chaplin 1997). Many invertebrate studies concern only one sex, and so do not address the issue of sex differences (e.g., Johnson and Sih 2007; Kortet and Hedrick 2007). Cephalopods are interesting subjects for studies of sex differences because most sexual maturation occurs relatively late in life (Mather 2006). For both octopus species studied, sex was not an important variable in the sorting of personality (Mather and Anderson 1993; Sinn et al. 2001). In general this was true for the squid as well during the three in-depth studies of personality variables. When, however, Sinn, Gosling, and Moltschaniwskyj (unpublished data) evaluated two wild populations of squid over two generations, they found that location on the shy/bold dimension was affected by sex. Specifically, they found differences in shy/bold scores attributable to population, sex, and body condition, underlining the importance and adaptive value of this dimension.

Invertebrates in personality research

One important reason to study personalities in the invertebrates is that they comprise the vast majority of animal species (and indeed, phyla) on our planet (Pechenik 2000). Research on invertebrates is therefore critical to advancing understanding of the broad patterns of individual differences in behavior. Invertebrates also offer many inherent advantages to investigators interested in animal personality. We conclude this chapter by reviewing some of these advantages, with reference to the studies described earlier in the review where applicable. We hope to stimulate future research into individual differences and correlated traits in these understudied animals.

Many invertebrates are amenable to laboratory study, where much personality research takes place. Invertebrates often have relatively short life cycles, facilitating research on fitness-relevant phenomena, such as survival and reproductive success, over multiple generations. This advantage of invertebrate species has been largely overlooked, however, as nearly all studies linking personality to fitness have focused on vertebrates (reviewed in Smith and Blumstein 2008). Laboratory studies of invertebrates could include manipulations of environmental conditions such as temperature, diet, and predator pressure, allowing researchers to examine the effects of these environmental factors on personality. Environmental manipulation coupled with measurements of fitness proxies (e.g., reproductive success, survivorship) could be used more widely to test hypotheses about adaptive plasticity in personality structure. This might be modulated by cognitive abilities, but not all vertebrates are highly intelligent and some invertebrates (such as octopuses) are.

The small size and simple husbandry needs of many invertebrates allow researchers to obtain larger samples than would be possible with most vertebrate species. A high sample size is important in personality studies that examine many contexts because multiple comparison corrections reduce the power to detect correlations (Bell 2007). Researchers interested in the effects of individual behavioral differences on social behavior could maintain entire colonies of invertebrates in the laboratory (see Pruitt et al. 2008). This type of research might ask how individuals' personalities and the mix of personalities in a group affect such group-level phenomena as dominance hierarchies, mating systems, and reproductive success. Initial research along these lines has begun on insect larvae (Weinstein and Maelzer 1997; Dussutour et al. 2008) and adult water striders (Sih and Watters 2005).

From a practical standpoint, institutional animal care and use committees typically have different standards for vertebrate and invertebrate research. It is thus simpler to obtain ethical approval to perform invasive manipulations on invertebrates that might provide insights into the physiological correlates of personality (e.g., Kortet et al. 2007). Further, certain ecologically important behaviors that have shaped animal personalities can be considered for study in invertebrates but not in vertebrates. For example, it is generally considered unethical to stage male-male combat, but staging such behaviors may be justified in certain invertebrate taxa in which such conflicts do not result in injury (e.g., hissing cockroaches, *Gromphadorhina portentosa*, Logue et al. 2009). We emphasize, however, that animals from all taxa deserve ethical consideration (Mather and Anderson 2007), and a distant taxonomic relationship to humans is not sufficient justification to perform invasive or stressful procedures.

Many areas of personality research could benefit from invertebrate studies. Researchers interested in the physiological mechanisms underlying personality would do well to focus on the relatively simple nervous and endocrine systems of invertebrates such as arthropods (Scharrer 1987) and cephalopods (Boyle 1987). Those interested in the genetic control of behavior might look to model organisms such as the fruit fly *Drosophila melanogaster* (Sokolowski 2001) and the worm *Caenorhabditis elegans* (deBono and Bargmann 1998) or to clonal groups such as cnidarians (Ayre 1982). Researchers interested in the development of personalities should consider the holometabolous insects, which undergo massive physical reorganization at metamorphosis. For example, we might predict that the behavioral tendencies of caterpillars (Weinstein and Maelzer 1997) would be drastically reorganized during metamorphosis, in contrast to the relative constancy of behavior throughout development exhibited by squid (Sinn et al. 2008), fishing spiders (Johnson and Sih 2007), and many vertebrates (Taborsky 2006). Surprisingly, this is not true. Brodin (2009) found that the activity of larval damselflies correlated with their position on dimensions of boldness and activity as adults. Finally, researchers wishing to use personality research to solve applied problems might consider the role of individual differences in organizing the group-level behavior of pest species (Despland and Simpson 2005; Nemiroff and Despland 2007).

Invertebrates pose a distinct set of problems for personality researchers. In contrast to the relatively well-studied vertebrates, most invertebrates are poorly known, with perhaps half as many awaiting description as have been classified (Pechenik 2000). Not just species but higher taxa are in need of revision, and studying the behavior of an animal whose phylogenetic relationships are in question poses obvious problems. The invertebrates not only are very diverse but also have physiologies that are totally unlike those of their human researchers. While this offers opportunities for comparative physiological linkage, it challenges us not to assume underlying system functioning (such as visual dominance). In particular, the invertebrates' sensory world, the *umwelt* (von Uexküll [1934] 1957), may be so different that experimenters must take care to avoid presenting subjects with stimuli that are incomprehensible or trivial to these organisms. We urge researchers to follow the lead of Riechert and Hedrick (1990) by choosing stimuli (in their case, vibrations transmitted through the web) that are salient to the subjects. The invertebrates' small size is a benefit for laboratory research but a difficulty in field work, where whole populations may not even be visible and also are drastically affected by the visiting observer. Finally, many invertebrate phyla are totally marine, presenting unique challenges to study, particularly in the field.

Conclusions

Does the information presented in this chapter demonstrate that invertebrates show personality variation? We argue that it does. In spite of the great variety of theoretical approaches employed by the authors that we review, their research clearly shows differences between individuals that are stable over time and affect more than one behavioral context (Groothuis and Carere 2005). Work on fruit fly genes and behavior has explored the genetic and environmental underpinning of individual differences (Sokolowski 2001). Research spearheaded by Riechert and Sih has linked evolutionary processes to correlated behaviors and individual differences in spiders. Hedrick's team demonstrated that behavioral tradeoffs can promote the evolution of behavioral syndromes, and pointed the way for studies linking personality and physiology. Sinn and his associates have found three or four axes of personality in cephalopods, which are similar in character to axes found in vertebrates (see Gosling 2001). We suggest that the future of personality research will incorporate both the behavioral ecological and personality psychology approaches (see Pruitt et al. 2008). Researchers pursuing this type of synthesis will examine a wide range of ecologically salient behaviors to define broad patterns of personality and then test evolutionary hypotheses pertaining to these patterns. Continued research on invertebrates promises to contribute to an improved understanding of the causes and consequences of correlated traits and individual differences in animals.

References

Arnqvist, G., and Henriksson, S. 1997. Sexual cannibalism in the fishing spider and a model for the evolution of sexual cannibalism based on genetic constraints. *Evolutionary Ecology*, 11, 255–273.

Ayre, D. J. 1982. Intergenotype aggression in the solitary sea anemone *Actinia tenebrosa*. *Marine Biology*, 68, 199–205.

Bell, A. M. 2005. Behavioural differences between individuals and two populations of stickleback *(Gasterosteus aculeatus)*. *Journal of Evolutionary Biology*, 18, 464–473.

———. 2007. Future directions in behavioural syndromes research. *Proceedings of the Royal Society of London B*, 274, 755–761.

Bergmüller, R., Schürch, R., and Hamilton, I. M. 2010. Evolutionary causes and consequences of consistent individual variation in cooperative behaviour. *Philosophical Transactions of the Royal Society of London B*, 365, 2751–2764.

Bergmüller, R., and Taborsky, M. 2010. Animal personality due to social niche specialisation. *Trends in Ecology and Evolution*, 25, 504–511.

Beshers, S. N., and Fewell, J. H. 2001. Models of division of labor in social insects. *Annual Review of Entomology*, 46, 413–440.

Blankenhorn, W. V. 1991. Fitness consequences of foraging success in water striders *(Gerris remigis;* Heteroptera: Gerridae*). Behavioral Ecology*, 2, 46–55.

Boyle, P. R. 1987. *Comparative Reviews*. Vol. 2 of *Cephalopod Life Cycles*. London: Academic.

Briffa, M., Rundle, S. D., and Fryer, A. 2008. Comparing the strength of behavioural plasticity and consistency across situations: animal personalities in the hermit crab *Pagurus bernhardus*. *Proceedings of the Royal Society of London B*, 275, 1305–1311.

Brodin, T. 2009. Behavioral syndrome over the boundaries of life-carryovers from larvae to adult damselfly. *Behavioral Ecology*, 20, 30–37.

Cade, W. 1975. Acoustically orienting parasitoids—fly phonotaxis to cricket song. *Science*, 190, 1312–1313.

Caspi, A. 1998. Personality development across the life course. In: *Social, Emotional, and Personality Development*. Vol. 3 of *Handbook of Child Psychology* (5th ed., Eisenberg, N., ed.), pp. 311–388, New York: Wiley.

Cheverud, J. M. 1996. Developmental integration and the evolution of pleiotropy. *American Zoologist*, 36, 44–50.

de Bono, M., and Bargmann, C. I. 1998. Natural variation in a neuropeptide Y receptor homolog modifies social behavior and food response in *C. elegans. Cell*, 94, 679–689.

Despland, E., and Simpson, S. J. 2005. Food choices of solitary and gregarious locusts reflect cryptic and aposematic antipredator strategies. *Animal Behaviour*, 69, 471–479.

Dussutour, A., Nicolis, S. C., Despland, E., and Simpson, S. J. 2008. Individual differences influence collective behaviour in social caterpillars. *Animal Behaviour*, 76, 5–16.

Emlen, D. J., and Nijhout, H. F. 2000. The development and evolution of exaggerated morphologies in insects. *Annual Review of Entomology*, 45, 661–708.

Gosling, S. D. 2001. From mice to men: what can we learn about personality from animal research? *Psychological Bulletin*, 127, 45–86.

Groothuis, T. G. G., and Carere, C. 2005. Avian personalities: characterization and epigenesis. *Neuroscience and Biobehavioral Reviews*, 29, 137–150.

Hedrick, A. V. 2000. Crickets with extravagant mating songs compensate for predation risk with extra caution. *Proceedings of the Royal Society of London B*, 267, 671–675.

Hedrick, A. V., and Riechert, S. E. 1989. Genetically based variation between two spider populations in foraging behavior. *Oecologia*, 80, 533–539.

Higgins, L. A., Jones K. M., and Wayne, M. L. 2005. Quantitative genetics of natural variation of behavior in *Drosophila melanogaster*: the possible role of the social environment on creating persistent patterns of group activity. *Evolution*, 59, 1529–1539.

Hill, E. M., Ross, L. T., and Low, B. S. 1997. The role of future unpredictability in human risk taking. *Human Nature*, 8, 287–325.

Hogan, J. A., and Bolhuis, J. J., eds. 1994. *Causal Mechanisms of Behavioural Development*. Cambridge: Cambridge University Press.

Huntingford, F. A. 1982. Do interspecific and intraspecific aggression vary in relation to predation pressure in sticklebacks? *Animal Behaviour*, 30, 909–916.

Johnson, J. C., and Sih, A. 2005. Precopulatory sexual cannibalism in fishing spiders (*Dolomedes triton*): a role for behavioral syndromes. *Behavioral Ecology and Sociobiology*, 58, 390–396.

———. 2007. Fear, food, sex and parental care: a syndrome of boldness in the fishing spider, *Dolomedes triton*. *Animal Behaviour*, 74, 1131–1138.

Kagan, J. 1994. *Galen's Prophecy: Temperament in Human Nature*. New York: Basic Books.

Kortet, R., and Hedrick, A. 2007. A behavioural syndrome in the field cricket *Gryllus integer*: intrasexual aggression is correlated with activity in a novel environment. *Biological Journal of the Linnean Society*, 91, 475–482.

Kortet, R., Rantala, M. J., and Hedrick, A. 2007. Boldness in antipredator behaviour and immune defence in field crickets. *Evolutionary Ecology Research*, 9, 185–197.

Logue, D. M., Mishra, S., McCaffrey, D., Ball, D., and Cade, W. H. 2009. A behavioral syndrome linking courtship behavior toward males and females predicts reproductive success from a single mating in the hissing cockroach, *Gromphadorhina portentosa*. *Behavioral Ecology*, 20, 781–788.

Mather, J. A. 1995. Cognition in cephalopods. *Advances in the Study of Behavior*, 24, 316–353.

———. 2006. Behaviour development: a cephalopod perspective. *International Journal of Comparative Psychology*, 19, 98–115.

Mather, M. A., and Anderson, R. C. 1993. Personalities of octopuses (*Octopus rubescens*). *Journal of Comparative Psychology*, 107, 336–340.

———. 1999. Exploration, play, and habituation in octopuses (*Octopus dofleini*). *Journal of Comparative Psychology*, 113, 333–338.

———. 2007. Ethics and invertebrates: a cephalopod perspective. *Diseases of Aquatic Organisms*, 75, 119–129.

Maupin, J. L., and Riechert, S. E. 2001. Superfluous killing in spiders: a consequence of adaptation to food-limited environments? *Behavioral Ecology*, 12, 569–576.

Mery, F., Belay, A. T., So, A. K. C., Sokolowski, M. B., and Kawecki, T. J. 2007. Natural polymorphism affecting learning and memory in *Drosophila*. *Proceedings of the National Academy of Sciences U S A*, 104, 13051–13055.

Nemiroff, L., and Despland, E. 2007. Consistent individual differences in the foraging behaviour of forest tent caterpillars (*Malacosoma disstria*). *Canadian Journal of Zoology*, 85, 1117–1124.

Pechenik, J. A. 2000. *Biology of the Invertebrates*. 2nd ed. New York: McGraw-Hill.

Phares, E. J., and Chaplin, W. F. 1997. *Introduction to Personality*. 4th ed. New York: Longman.

Pruitt, J. N., and Riechert, S. E. 2012. The ecological consequences of temperament in spiders. *Cunent Zoology*, 58, 589–596.

Pruitt, J. N., Riechert, S. E., and Jones, T. C. 2008. Behavioural syndromes and their fitness consequences in a socially polymorphic spider, *Anelosimus studiosus*. *Animal Behaviour*, 76, 871–879.

Reaney, L. T., and Backwell, P. R. Y. 2007. Risk-taking behavior predicts aggression and mating success in a fiddler crab. *Behavioral Ecology*, 18, 521–525.

Retana, J., and Cerda, X. 1991. Behavioural variability and development of *Cataglyphis cursor* ant workers (Hymenoptera, Formicidae). *Ethology*, 89, 275–286.

Riechert, S. E. 1993. Investigation of potential gene flow limitation of behavioral adaptation in an arid lands spider. *Behavioral Ecology and Sociobiology*, 32, 355–363.

Riechert, S. E., and Hall, R. F. 2000. Local population success in heterogeneous habitats: reciprocal transplant experiments completed on a desert spider. *Journal of Evolutionary Biology*, 13, 541–550.

Riechert, S. E., and Hedrick, A. V. 1990. Levels of predation and genetically based antipredator behaviour in the spider *Agelenopsis aperta*. *Animal Behaviour*, 40, 679–687.

———. 1993. A test for correlations among fitness-linked behavioural traits in the spider *Agelenopsis aperta* (*Araneae, Agelenidae*). *Animal Behaviour*, 46, 669–675.

Rutherford, P. L., Baker, R. L., and Forbes, M. R. 2007. Do larval damselflies make adaptive choices when exposed to both parasites and predators? *Ethology*, 113, 1073–1080.

Scharrer, B. 1987. Insects as models in neuroendocrine research. *Annual Review of Entomology*, 32, 1–16.

Sih, A., Bell, A. M., Johnson, J. C., and Ziemba, R. E. 2004. Behavioral syndromes: an integrative overview. *Quarterly Review of Biology*, 79, 241–277.

Sih, A., and Watters, J. V. 2005. The mix matters: behavioural types and group dynamics in water striders. *Behaviour*, 142, 1417–1431.

Sinn, D. L., Apiolaza, L. A., and Moltschaniwskyj, N. A. 2006. Heritability and fitness-related consequences of squid personality traits. *Journal of Evolutionary Biology*, 19, 1437–1447.

Sinn, D. L., Gosling, S. D., and Moltschaniwskyj, N. A. 2008. Development of shy/bold behaviour in squid: context-specific phenotypes associated with developmental plasticity. *Animal Behaviour*, 75, 433–442.

Sinn, D. L., and Moltschaniwskyj, N. A. 2005. Personality traits in dumpling squid (*Euprymna tasmanica*): context-specific traits and their correlation with biological characteristics. *Journal of Comparative Psychology*, 119, 99–110.

Sinn, D. L., Moltschaniwskyj, N. A., Wapstra, E., and Dall, R. X. 2010. All behavioral syndromes invariant? Spatiotemporal variation in shy/bold behavior in squid. *Behavioral Ecology and Sociology*, 64, 693–702.

Sinn, D. L., Perrin, N. A., Mather, J. A., and Anderson, R. C. 2001. Early temperamental traits in an octopus (*Octopus bimaculoides*). *Journal of Comparative Psychology*, 115, 351–364.

Smith, B. R., and Blumstein, D. T. 2008. Fitness consequences of personality: a meta-analysis. *Behavioral Ecology*, 19, 448–455.

Smith, J. M., and Riechert, S. E. 1984. A conflicting-tendency model of spider agonistic behavior—hybrid-pure population line comparisons. *Animal Behaviour*, 32, 564–578.

Sokolowski, M. B. 2001. *Drosophila*: genetics meets behaviour. *Nature Reviews Genetics*, 2, 879–890.

Starks, P. T., and Fefferman, N. H. 2006. *Polistes* nest founding behavior: a model for the selective maintenance of alternative behavioral phenotypes. *Annales Zoologici Fennici*, 43, 456–467.

Stearns, S. 1992. *The Evolution of Life Histories*. Oxford: Oxford University Press.

Stephens, D. W., and Krebs, J. R. 1986. *Foraging Theory*. Princeton, NJ: Princeton University Press.

Taborsky, B. 2006. The influence of juvenile and adult environments on life-history trajectories. *Proceedings of the Royal Society of London B*, 273, 741–750.

Turner, A. M., Turner, S. E., and Lappi, H. M. 2006. Learning, memory, and predator avoidance by freshwater snails: effects of experience on predator recognition and defensive strategy. *Animal Behaviour*, 72, 1443–1450.

Van Oers, K., de Jong, G., Van Noordwijk, A. J., Kempenaers, B., and Drent, P. J. 2005. Contribution of genetics to the study of animal personalities: a review of case studies. *Behaviour*, 142, 1191–1212.

Von Uexküll, J. (1934) 1957. A stroll through the worlds of animals and men: a picture book of invisible worlds. In: *Instinctive Behaviour: The Development of a Modern Concept* (C. H. Schiller, ed.), pp. 5–80. New York: International Universities Press.

Weinstein, P., and Maelzer, D. A. 1997. Leadership behaviour in sawfly larvae *Perga dorsalis* (Hymenoptera: Pergidae). *Oikos*, 79, 450–455.

West-Eberhard, M. J. 2003. *Developmental Plasticity and Evolution*. Oxford: Oxford University Press.

Wolf, M., Van Doorn, S., Leimar, O., and Weissing, F. J. 2007. Life-history trade-offs favour the evolution of animal personalities. *Nature*, 447, 581–585.

Zuk, M., and Kolluru, G. R. 1998. Exploitation of sexual signals by predators and parasitoids. *Quarterly Review of Biology*, 73, 415–438.

2

Evolutionary Perspectives on

Personality in Stickleback Fish

ALISON M. BELL, SUSAN A. FOSTER, AND MATTHEW WUND

Introduction

Consistent individual differences in behavior within populations, also known as personality, are attracting the attention of many evolutionary biologists (e.g., Sih et al. 2004a; Sih et al. 2004b; A. M. Bell 2007; Réale et al. 2007). One of the reasons why personality is interesting from an evolutionary perspective is that behaving in a consistent way across a variety of contexts or through time seems maladaptive. At some point, a mismatch between an individual's behavior and the optimal strategy for a given situation may be expected (Sih et al. 2004b). For example, an individual that is consistently uninhibited and bold will at some point incur an elevated risk of predation; the optimal individual should be bold only when it makes sense to be bold, and adjust its behavior when the situation changes. However, personality implies that individuals are limited in their behavioral plasticity; the individuals that are, for example, particularly active in the absence of predation risk are also those that are most active in the presence of predation risk (Sih et al. 2003).

But personality does not simply refer to the fact that individuals behave consistently. The other key characteristic of personality is that individuals *differ* in their behavior; some individuals are consistently bolder than others, for example. When individual differences in behavior are heritable and linked to fitness, theory predicts that natural selection will eliminate behavioral variation, rather than maintain it within populations (D. S. Wilson 1998). Therefore the second puzzle about personality is how we explain the maintenance of individual variation in behaviors that are related to fitness (Dingemanse and Réale 2005).

A third issue that personality raises for evolutionary biologists is that a suite of correlated behaviors that together form a behavioral syndrome (e.g., boldness toward predators, aggressiveness toward conspecifics, and exploratory behavior) will evolve together if the behaviors are genetically correlated with each other. For the purposes of this review, we use the terms "personality" and "behavioral syndrome" interchangeably, as both refer to individual differences in behavior that are correlated through time or across situations. When behaviors are genetically correlated with each other, selection on one behavior will produce a correlated response to selection on other behaviors. As a result, a genetically correlated suite of traits can prevent the independent optimization of different traits (Lynch and Walsh 1998). What, then, maintains correlations among behaviors?

Threespine stickleback fish, *Gasterosteus aculeatus*, have proven especially good subjects for tackling these challenging evolutionary questions about personality. At the most basic level, a large number of studies have already shown that consistent individual differences in behavior are common in sticklebacks (Bakker 1986; Huntingford and Giles 1987; Godin and Crossman 1994; Walling et al. 2004; A. M. Bell 2005; Brydges et al. 2008a). However, we are also starting to appreciate that not all stickleback behave consistently through time or across situations (A. M. Bell 2005; A. M. Bell and Sih 2007; Dingemanse et al. 2007). By comparing groups of stickleback that either do or do not behave consistently (A. M. Bell 2005; Dingemanse et al. 2007), we have an opportunity to understand the selective factors that can favor within-individual consistency and ultimately to experimentally test for selection (A. M. Bell and Sih 2007).

As a complement to studies of interindividual differences in personality, the substantial diversity among freshwater stickleback populations has proven useful for testing the hypothesis that behavioral correlations act as constraints. According to the "constraint" hypothesis, behaviors should be similarly correlated with each other even in different populations inhabiting different selective environments (A. M. Bell 2005). If, on the other hand, behavioral correlations differ among populations, the behaviors may be free to vary independently, or the correlation itself might be favored by selection in certain environments.

Finally, the unusual evolutionary history of stickleback allows us to compare syndromes in what we believe to be a living ancestor to derivative members of a freshwater radiation that exhibits high levels of adaptive parallelism. For example, we can ask whether behavioral syndromes that arise in divergent freshwater populations are the products of the winnowing of ancestral variation by selection, whether derived syndromes are produced *de novo* in isolated freshwater populations, or whether ancestral syndromes are held together by genetic correlations through the evolutionary process. We can also evaluate whether behavioral syndromes are composites of plastic or fixed traits, and whether genetic accommodation (*sensu* West-Eberhard

2003) has been in involved in the production of novel syndromes in derived populations. Although these questions cannot all be answered in this chapter, we can offer insights into the ways these issues can be addressed with this adaptive radiation, into the processes leading to the assembly of individual personalities, and into their functions.

The adaptive radiation of stickleback fish

The threespine stickleback is a small teleost fish that breeds at 3–8 cm standard length and has a holarctic distribution. The species includes three fundamental life history types: marine, anadromous (sea run), and freshwater (reviewed in Wootton 1976; Baker 1994). Freshwater populations of threespine stickleback are the descendants of marine or anadromous stickleback (both hereafter referred to as oceanic). Despite a wide geographic distribution, oceanic stickleback exhibit remarkably uniform morphology (Walker and Bell 2000), which has changed little over the past 7 million– 12 million years (M. A. Bell and Foster 1994). The large size of oceanic populations, interconnected by high levels of gene flow as indicated by molecular data (Withler and McPhail 1985; Taylor and McPhail 1999; Cresko 2000; Taylor and McPhail 2000), suggests that little genetic drift is likely to have occurred since postglacial colonization began years ago. This means that modern oceanic populations can be used to infer phenotypes possessed by stickleback that gave rise to the postglacial freshwater radiation with unusual certainty (Foster and Bell 1994; Walker and Bell 2000; Colosimo et al. 2005; Shaw et al. 2007).

In a pattern similar to other freshwater adaptive radiations (Foster and Bell 1994; Lee and Bell 1999; Schluter 2000; West-Eberhard 2003), oceanic stickleback have undoubtedly invaded freshwater habitats repeatedly over their long evolutionary history (M. A. Bell 1994). The most recent wave of extensive colonization occurred in the early Holocene as glaciers retreated (Taylor and McPhail 1999; Raeymaekers et al. 2005; Malhi et al. 2006). Whereas the large oceanic populations are likely to have changed little over this time, their freshwater derivative populations have diversified to an extraordinary degree. For example, freshwater populations have undergone rapid differentiation in behavior in response to divergent environmental conditions such as habitat type (e.g., streams vs. ponds; Girvan and Braithwaite 1998; Braithwaite and Girvan 2003; Alvarez and Bell 2007), predation pressure (Huntingford et al. 1994), and lake type (e.g., shallow relatively eutrophic lakes versus deep comparatively oligotropic lakes; Foster 1995; McKinnon et al. 2004). This diversity offers an opportunity to perform widespread comparative studies to understand the proximate and ultimate factors that generate behavioral diversity.

Freshwater populations of stickleback exposed to similar environments have independently evolved similar traits, a phenomenon termed parallel evolution (M. A. Bell and Foster 1994; McPhail 1994; Lee and Bell 1999; Schluter 2000; Arendt and Reznick 2008). When the component populations and species consistently differ in phenotype between habitat types, and the similarities among populations of the same habitat type have evolved independently (parallel evolution), comparative studies can offer very strong inference of adaptive value, or function of particular traits or trait complexes (Foster and Bell 1994; Foster 1999; Lee and Bell 1999; Schluter 2000; West-Eberhard 2003). For example, if we identify an ecological factor or a proximate mechanism that we suspect is driving a phenotypic difference between two populations, we can test this hypothesis by comparing additional populations. If the same factor is associated with differences among multiple, independent populations, we infer that the hypothesized factor is, indeed, driving the phenotypic difference and that the variation is not simply a consequence of other factors such as genetic drift (figure 2.1). Therefore the radiation is a naturally replicated experiment and hence offers initial tests of evolutionary hypotheses.

A brief history of stickleback behavioral research

Behavioral research on the threespine stickleback was initiated by Niko Tinbergen, who, in a quest for a subject of behavioral study for students in his university classes, thought of the stickleback, so common near his home that it had captured his attention as a child. The research of Tinbergen and coworkers provided fundamental insights into the behavior of this small fish, and contributed in no small part to the insights that earned him a Nobel Prize, along with the two other prominent ethologists, Konrad Lorenz and Karl von Frisch. However, ethological research at the time was focused upon understanding consistent, predictable pathways that would translate an external stimulus into a specific, consistent, and adaptive behavior (Burkhardt 2005). Thus, although the insights of the classical ethologists laid the foundations for subsequent research on the behavior of stickleback, they missed the remarkable diversity of behavior within this adaptive radiation (M. A. Bell and Foster 1994).

The first hint that there existed behavioral differentiation among stickleback populations came from a comparative study by Wilz (1973) in which he described persistent differences in courtship between a European and a

Figure 2.1. The replicated evolution of behavior in stickleback provides a natural experiment for testing evolutionary hypotheses. This idealized map of Scotland shows the movement of stickleback from the ocean (on left) into freshwater rivers following glacial retreat. Stickleback inhabiting water bodies with abundant predators (dark fish) independently evolved increased levels of antipredator behavior compared with stickleback inhabiting water bodies without predators (light fish).

North American population. His description of these differences suggests he was comparing a European limnetic population with a North American benthic population. Concurrently, Hagen and Gilbertson (1973a; 1973b) were characterizing morphological differences among North American populations linked to differences in predation. Later, Huntingford and colleagues (Giles and Huntingford 1984) showed population differences in aggressive and antipredator behaviors associated with differences in predation in British populations, and McPhail and Lindsey (1970) argued that freshwater populations of threespine sticklebacks were descendants of oceanic populations, and that they had evolved often independently and in parallel during the freshwater radiation of the species. At this point, the stage was set for the development of the stickleback radiation as a model for evolutionary study that would ultimately benefit from the confluence of elegant evolutionary inference from the fossil record (M. A. Bell 1994; Hunt et al. 2008), extensive research on the characteristics of a diversity of freshwater populations (M. A. Bell and Foster 1994; Östlund-Nilsson et al. 2007), and the development of genomic tools (Kingsley and Peichel 2007). Together, these developments have shown the value of the radiation for understanding evolution from the four levels of study: causation, development, function, and evolution, as articulated by Tinbergen (Tinbergen 1951; Hinde and Tinbergen 1958).

While stickleback are renowned for their behavioral variation among populations (Foster and Endler 1999), they also possess remarkable behavioral variation between individuals, a fact first documented by Huntingford (1976). She found that individual male subadult stickleback differed in how they responded to a pike predator: while some individuals approached the pike, other individuals hid in the weeds. Remarkably, months later, when those same males were reproductively mature, the individuals that had been particularly bold toward a pike as subadults were also particularly aggressive toward conspecifics. In other words, how "bold" an individual was as a subadult was a reliable predictor of the individual's aggressive behavior at a completely different life stage (after sexual maturity). This study showed that interindividual variation among male stickleback was correlated across different functional contexts and stable over a relatively long period of time (but see A. M. Bell and Stamps 2004). In the next section, we show how relating variation within and among populations offers insights into the boldness-aggressiveness behavioral syndrome.

Predation and the boldness-aggressiveness behavioral syndrome

Stickleback provide a primary food source for a diverse assemblage of predators in postglacial freshwater lakes, and presumably also in the ocean. Predatory vertebrates, as well as some insects, can impose strong selection on stickleback populations and consequently select for particular armoring and body shape attributes (reviewed in Reimchen 1994). Stickleback exhibit tremendous variation in the number of lateral plates (the rigid, bony plates along the side of the body). Following the invasion of freshwater, most populations lose the posterior lateral plates that typically characterize their oceanic counterparts; this evolutionary reduction in armor can be rapid (M. A. Bell et al. 2004). The causes might include relaxed pressure from predators that typically pierce the skin of their prey, reduced overall predation, decreased ionic strength water, and linkage of the main loci regulating plate development to a locus that influences parasite resistance (Colosimo et al. 2005; Miller et al. 2007). Selective predation also strongly influences retention of the anterior lateral plates, which connect the pelvic girdle and associated spines to the dorsal spines. In combination, these defenses offer significant protection from gape-limited predators (Hoogland et al. 1957; Hagen and Gilbertson 1973b; M. A. Bell and Haglund 1978). In addition, both the absence of piscine predators and low ionic strength water have been associated with the loss of the anterior armor complex (M. A. Bell and Ortí 1994).

Analogous to its impact on the diversity of defensive morphology, variation in predation pressure is a strong organizing determinant of behavioral variation among stickleback populations. Stickleback from areas where there are high levels of predation tend to show heightened sensitivity to predators and perform more predator-inspection behavior than their counterparts in relatively low predation environments (Huntingford and Coulter 1989; Huntingford et al. 1994; Walling et al. 2003; 2004; Messler et al. 2007). This influence extends to traits beyond those directly related to predator detection and avoidance. Learning and memory, as well as activity (Brydges et al. 2008a; 2008b) and aggressiveness, vary across populations differing in predation pressure. Indeed, echoing the aforementioned pattern at the within-population level, boldness toward predators (here, predator inspection) and aggressiveness toward conspecifics positively covary across populations (figure 2.2; A.M. Bell, unpublished data; Giles and Huntingford 1984).

Clearly then, predation has a strong selective influence on stickleback populations—affecting both behavior and morphology. A relatively unexplored but intriguing concept is how these phenotypic axes might be integrated as a result of correlational selection, or even pleiotropic control (Pigliucci 2003). At the phenotypic level, rigid, bony armor might help prevent a toothed predator from piercing the body, but it can also interfere with swimming ability, thereby reducing the chance of eluding capture by a predator (Huntingford 1981; Andraso and Barron 1995; Grand 2000; Krause et al. 2000; Mikolajewski and Johansson 2004). Studying the possible tradeoff between morphological and behavioral defenses offers an opportunity to understand the overall phenotypic integration among morphology, physiology, and behavior (M. A. Bell and Foster 1994; Foster 1995; DeWitt et al. 1999).

OCEANIC STICKLEBACK—THE ANCESTRAL SYNDROME
One of the main advantages of studying stickleback traits from an evolutionary perspective is that, in many cases, the ancestral state of a trait is

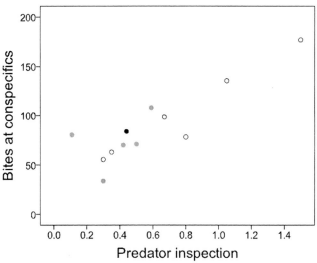

Figure 2.2. Average differences between populations in "boldness" (here, predator inspection, along the x-axis) and "aggressiveness" (biting at conspecifics, along the y-axis). Each data point represents a mean based on 20 individuals per population. "High risk" populations, where there are abundant predatory fishes (pike, brown trout, eels), are in open circles, while populations with few predatory fishes are in gray. Note that the oceanic population (in black) is intermediate. On average, populations that are more aggressive are also more bold ($r = 0.881$, $P < 0.0001$, $n = 12$).

exhibited by extant oceanic fish. Examining the antipredator responses of oceanic stickleback can shed important insights into the course of evolution that leads to differences in behavior between high and low predation freshwater populations. Oceanic stickleback are exposed to a wide variety of predators, whereas many derived, freshwater populations show substantially reduced predator assemblages. Thus, relaxed selection pressure from predators could have profound implications for the evolution of behavior (Coss 1999; Blumstein and Daniel 2002; Blumstein et al. 2004; Blumstein 2006; Messler et al. 2007). Surprisingly, despite substantial differences in predation regime, Messler et al. (2007) found that latencies to recover from simulated predator attacks, an indicator of fearfulness (Wibe et al. 2001; Quinn and Creswell 2005), were similar between oceanic stickleback and an armor-reduced, freshwater population devoid of all predatory fish. Boldness in response to the presence of a predator is considered an important aspect of the shy-bold behavioral syndrome, along with other behaviors such as overall activity, willingness to explore novel environments, and foraging under novel or risky circumstances (Lopez et al. 2005; Quinn and Creswell

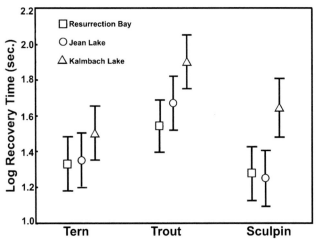

Figure 2.3. Variation in log-transformed recovery times of wild-caught fish from three Alaskan populations of stickleback responding to three model predators (from Messler et al. 2007). Resurrection Bay is a marine population, Jean Lake is a freshwater population lacking predatory fish, and Kalmbach Lake is a freshwater population to which rainbow trout have been introduced. Means (95% CIs) are shown. Results of analysis of variance were as follows: predator, $F_{2,257} = 10.3$, P < 0.001; population, $F_{2,257} = 11.3$, P < 0.001; interaction, $F_{4,272} = 0.76$, P = 0.55.

2005; A. D. M. Wilson and Godin 2009). Antipredator behavior has also been linked to aggression toward conspecifics in the boldness-aggressiveness syndrome (Huntingford 1976; A. M. Bell 2005). We hypothesize that oceanic fish can "afford" to be as bold as their freshwater conspecifics because they rely on their robust armor for defense against predators. By another measure (time spent frozen following an attack), and in a different set of oceanic and freshwater populations, the strongest fear responses were exhibited by wild-caught fish from a population to which trout were introduced over the past several decades (Kalmbach Lake) (figure 2.3; Messler et al. 2007). This result suggests that the reintroduction of predators to a freshwater population can lead to a rapid increase in fright responses. We suspect that relative boldness is ancestral in stickleback, and predator introductions to freshwater population may promote (via either selection or learning) heightened fear responses. Experiments are currently underway to determine whether latency to recover is indeed correlated with other measures of boldness within each of these populations (e.g., exploratory behavior, willingness to feed in novel circumstances), and whether the presence of behavioral syndromes varies among populations that encounter different types of predators, as has been found elsewhere (Dingemanse et al. 2007; A. M. Bell 2005; 2007).

INDIVIDUAL VARIATION

A great deal of the behavioral variation among stickleback populations is associated with variation in predation pressure. Because many of the observed evolutionary responses are adaptive, behavioral variation among populations is relatively easy to explain by natural selection. Explaining variation *within* populations, however, is more difficult.

Variation within populations of stickleback has been described along several behavioral axes. Aggressiveness is particularly well characterized in this regard (Bakker 1994a; Rowland 1984). Intraspecific aggression in stickleback occurs during competition over access to resources, including food and territories, and is manifested as biting, chasing, and attacking conspecifics. These behaviors are disadvantageous because in addition to energetic costs (Thorpe et al. 1995), aggression can result in injury (Neat et al. 1998) and exposure to predators while fighting (Diaz-Uriarte 1999). The territorial aggression of male stickleback is especially well studied (Rowland 1994), but juveniles and females also exhibit aggression (Bakker 1986).

Individual stickleback also vary in their antipredator behavior and can be generally classified along a shy-bold continuum according to their willingness to take risks in the presence of a predator. Predator inspection is one of the most obvious forms of risk-taking behaviors and has been widely studied in stickleback and other small fishes (Pitcher et al. 1986; Murphy and Pitcher 1991; Brown and Warburton 1999; Smith and Belk 2001). Despite the obvious danger involved in performing this behavior (Dugatkin and Godin 1992; Milinski et al. 1997), it is thought that predator inspection can provide reliable information about predation risk via both olfactory and visual cues, or can eliminate the element of surprise, and therefore can be advantageous.

Individual differences in risk-taking behavior are also observed when individuals balance the benefits of feeding against the costs of potential predation. "Risk-prone" individuals are more willing to assume the risk of predation in order to get food than are others, even when differences in size, sex, or hunger level are controlled for (Ibrahim and Huntingford 1989; Lima and Dill 1990; Dugatkin and Godin 1992; Godin and Crossman 1994; Jonsson et al. 1996; Godin and Clark 1997; Krause et al. 1998; Aeschlimann et al. 2000; Grand 2000; A. M. Bell 2004). Therefore, differences in the willingness to forage in the presence of a predator can reflect intrinsic differences in the propensity to engage in risk-taking behaviors.

In a classic study, Huntingford (1976) showed that individual stickleback that were particularly bold toward predators were also exceptionally aggressive toward other stickleback. Recent studies in some populations have confirmed Huntingford's result: interindividual variation in boldness is related to interindividual variation in aggressiveness. However, in other populations, how an individual behaves in the presence of a predator is not related to how that individual behaves during competitive interactions with conspecifics (A. M. Bell 2005; A. M. Bell and Sih 2007; Dingemanse et al. 2007). In other words, there are behavioral syndromes in some populations of stickleback but not in all of them.

Predation pressure appears to be the major determinant of whether boldness and aggressiveness covary: individuals from populations with a history of strong selection by predators behave predictably across contexts, while individuals from relatively safe environments do not (Dingemanse et al. 2004; A. M. Bell 2005; A. M. Bell and Sih 2007). Dingemanse et al. (2007) examined boldness-aggressiveness syndromes in juvenile stickleback from twelve populations and found that aggression, activity, and exploratory behavior were correlated, but only in populations that coexist with predatory fish. This is consistent with the adaptive hypothesis that behavioral syndromes exist in environments where correlations among particular suites of behaviors are selectively favored (A. M. Bell 2005).

The importance of predation in determining personality in stickleback was confirmed in an experimental test. Bell and Sih (2007) measured the behavior of individually marked stickleback from a population that did not have a history of co-occurring with major predators. As predicted, boldness (here, willingness to forage under predation risk) was not correlated with aggressiveness among the individuals at the start of the experiment. Then, they allowed rainbow trout to consume half the stickleback population, and remeasured the behavior of the survivors. Among the survivors, boldness was correlated with aggressiveness. Thus, exposure to predation risk generated the boldness-aggressiveness behavioral syndrome.

The predation-induced behavioral syndrome could have been produced via one of two mechanisms, or both. On the one hand, the correlation could have been the result of correlational selection if predators selectively consumed bold and nonaggressive, or shy and aggressive behavioral types. Not surprisingly, bolder individuals were less likely to survive predation by trout—that is, there was directional selection against boldness, but correlational selection was not detected, although that could be due to lack of statistical power. Alternatively, behavioral plasticity might have been the cause of the predation-induced behavioral syndrome if "mismatched" behavioral types shifted their behavior toward the correlation (figure 2.4). Indeed, the survivors of the experiment changed their behavior after exposure to predation risk, so the animals' behavior was clearly plastic. An exciting new insight from this study is that the boldness-aggressiveness syndrome in response to predation pressure is probably a product of both selected and plastic responses, and a promising avenue for further investigation is to quantify the relative importance of selection and plasticity in generating behavioral syndromes.

WHY ARE BOLDNESS AND AGGRESSIVENESS CORRELATED UNDER HIGH RISK?

Regardless of the mechanism producing the correlation, predation clearly emerges as an important ultimate determinant of personality in stickleback. An outstanding question, however, is why behavioral consistency is favored in high predation populations. What is it about predation pressure that causes behavioral reactions to a predator to covary with aggressiveness toward conspecifics? There are several possible explanations. Recent work suggests that life history tradeoffs can favor the evolution of consistent individual differences in behavior, or personality (Stamps 2007; Wolf et al. 2007). The reasoning is that different suites of behaviors should be associated with different life history strategies. For example, individuals investing in future reproduction should be generally risk averse because their residual reproductive value is high. Individuals investing in current reproduction, on the other hand, should be more willing to take risks generally. That is, behavioral correlations are a product of tradeoffs between life history traits. Thus, behavioral syndromes are to be expected when tradeoffs between life history traits are strong. If predators force a tradeoff between life history traits, then this theory could explain why boldness and aggressiveness are correlated in high predation populations of stickleback but not in low predation populations.

An alternative hypothesis to explain why boldness and aggressiveness are correlated when predation pressure is high draws from theory that shows that personality can be favored when the environment is unpredictable or noisy (Sih 1992; McElreath and Strimling 2006). If the environment varies stochastically, individuals might do better if they maintain a consistent behavioral strategy rather than if they try in vain to track noisy cues. According to this reasoning, and assuming that predation risk is unpredictable and

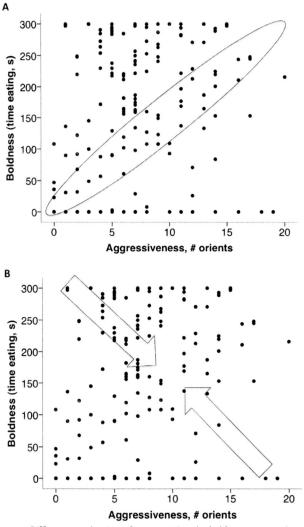

Figure 2.4. Different mechanisms for generating the boldness-aggressiveness behavioral syndrome in response to predation. These data show the relationship between boldness and aggressiveness prior to exposure to trout (from Bell and Sih, 2007). Note that there is a broad distribution of behavioral types within this population prior to exposure to predators. A) Selection might have caused the behavioral correlation via nonrandom mortality of shy and aggressive, or bold and nonaggressive individuals, such that individuals "along" the correlation (circled) survived. B) Alternatively, individuals might have plastically shifted their behavior, indicated with arrows.

cues indicating risk are noisy, individuals should behave consistently and not bother to try to track environmental cues in high predation populations— but not in low predation populations.

Another explanation for why boldness and aggressiveness are correlated in high predation populations is that different strategies work equally well for coping with predation risk: some individuals rely on predator inspection in order to obtain reliable cues about predation pressure while other individuals rely on the safety of the group. Presumably, high levels of aggression are incompatible with effective shoaling behavior (Magurran and Seghers 1994), so timid individuals should also be nonaggressive. Bold predator inspectors, on the other hand, do not have to pay the costs of aggression in a shoal because they use another tactic (predator inspection) to avoid predation. Therefore, a mix of behavioral strategies could be maintained in high predation populations but not in low predation populations. However, this hypothesis predicts that there is a range of behavioral types in high predation populations but not in low ones, which is not consistently observed (A. M. Bell and Stamps 2004).

ASSEMBLING THE SHY-BOLD AXIS

Several lines of evidence have shown that individual differences in behavioral tendencies such as aggressiveness and boldness have a heritable component in stickleback. For example, aggressiveness responds to selection (Bakker 1986; 1994b), and common garden studies have also shown that phenotypic differences between populations in antipredator responses are maintained in the lab, although maternal or paternal effects are rarely excluded (A. M. Bell 2005). However, there is also obviously an important role for the environment. Boldness is influenced by an individual's immediate social context (Webster 2007) and conditions experienced over the course of ontogeny (Tulley and Huntingford 1987a; Huntingford and Wright 1989; Huntingford et al. 1994). In a variety of fish species, variation in antipredator behavior has been shown to have a significant heritable component, yet laboratory rearing also indicates an important role for experience in many of these cases (reviewed in Kelley and Magurran 2003). Emerging evidence from several Alaskan stickleback populations indicates a particularly strong role for experience in the development of population-level differences in antipredator behavior. In an examination of three populations that differ in predation regime, Messler et al. (2007) found that populations differed in their responses to models of predatory fish. However, these population-level differences disappeared when fish from these populations were reared in the laboratory, suggesting a strong role for learning in the development

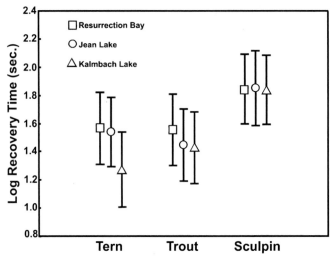

Figure 2.5. Population differences in recovery times disappear when fish are reared in the lab. Variation in log-transformed recovery times of laboratory reared fish from three Alaskan populations of stickleback responding to three model predators. Resurrection Bay is a marine population, Jean Lake is a freshwater population lacking predatory fish, and Kalmbach Lake is a freshwater population to which rainbow trout have been introduced. Means (95% CIs) are shown. Results of analysis of variance were as follows: predator, $F_{2,118} = 8.652$, $P < 0.001$; population, $F_{2,118} = 0.959$, $P = 0.386$; interaction, $F_{4,118} = 0.465$, $P = 0.761$.

of antipredator behavior in these populations (compare figures 2.3 and 2.5; Wund and Foster, unpublished data). Laboratory-reared fish did respond to simulated attacks, but interindividual variation was substantial, whereas interpopulation variation was undetectable.

An especially interesting environmental effect comes via paternal care. During the reproductive season, male stickleback defend nesting territories. Males attract females to spawn in their nests and defend the breeding territory from intruders and predators. After spawning, the female leaves the male's territory and the male is solely responsible for the care of the eggs. During the ~9-day incubation period, the male "fans" (oxygenates) the eggs, removes rotten eggs and debris, and defends the territory. A breeding male stickleback tends his newly hatched offspring for ~10 days. In some populations, fathers chase and catch fry that stray and spit them back into the nest. The fry are not injured during these interactions, but Tulley and Huntingford (1987b), following a suggestion by Benzie (1965), showed that these interactions help prepare the stickleback fry for later encounters with predators. Stickleback reared by their father avoided a model pike predator more than stickleback reared without paternal care (Tulley and Huntingford 1987b). Interestingly, the behavioral difference between father-reared and orphan stickleback was apparent only among stickleback from a population where predators are abundant. The fact that stickleback from high predation localities learn faster about predators suggests an inherited predisposition to respond to experience.

In several species, the shy-bold continuum and the proactive-reactive axis have been associated with individual differences in stress responsiveness (Koolhaas et al. 1999). Considering the important role of predators in shaping behavioral variation within and among stickleback populations, and the fact that predators are natural stressors, stress responsiveness is a promising neuroendocrine mechanism underlying behavioral variation. For example, how an individual reacts to a variety of different dangerous situations—for example, a fight with risk of injury, or an encounter with a predator with risk of death—might be influenced by the individual's stress-responsiveness. If the stress response is generalized across behavioral contexts, and individual stickleback differ in stress responsiveness, as has been shown for other animals (Koolhaas et al. 1999), then variation in stress responsiveness could be the underlying root of the covariance of behavioral responses in stickleback.

To evaluate this possibility, Bell et al. (2007) measured the turnover of monoamines in the brain as well as whole-body concentrations of cortisol in stickleback at different intervals following an encounter either with a potential competitor or with a predator. If differences in stress responsiveness underlie the covariance between behaviors, then we would expect the fish that reacted in a particularly bold way to exhibit a neuroendocrine profile similar to the fish that reacted in a particularly aggressive way toward a conspecific. However, while both fighting with conspecifics and confrontation with a predator produce a cortisol response, boldness toward a predator (here, inspection of a live pike) and aggressiveness were associated with different patterns of monoamine turnover in the brain. Individuals that were particularly "bold" in the face of predation had higher concentrations of serotonin in the brain, but aggressiveness was negatively associated with serotonin turnover (A. M. Bell et al. 2007). These results suggest that at least with respect to serotonin, the two behaviors are mechanistically independent.

The benthic-limnetic axis

A second axis of ecological variation that is studied as much as the boldness-aggressiveness axis, but from different perspectives, is the benthic-limnetic continuum in stickleback. In this case, the majority of our information comes from comparisons among populations, rather than within them. This axis of differentiation has been driven by differences in trophic conditions among lakes. Deep, relatively oligotrophic lakes favor slender bodies adapted for foraging in open water, and trophic structures specialized for feeding on plankton. In contrast, shallow, relatively eutrophic lakes favor deeper bodies suitable for maneuvering in complex habitats and trophic structures specialized for feeding on large, bottom-dwelling invertebrates (figure 2.6; Lavin and McPhail 1985; 1986; Walker 1997). In benthic populations, cannibalistic foraging groups attack nests defended by male stickleback and consume offspring within them. These groups appear to detect nests visually as a consequence of activity at the nest, and thus have favored drab male courtship coloration and inconspicuous courtship behavior. In contrast, limnetic males are characterized by bright courtship coloration and exhibit more conspicuous courtship. Aggression between neighboring males disappears quickly after territory establishment in benthic populations, presumably

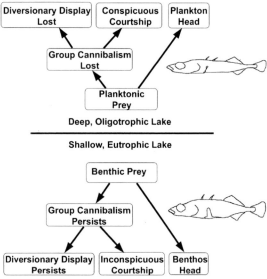

Figure 2.6. Differences between benthic and limnetic ecotypes of threespine stickleback. Benthic populations of threespine stickleback possess a deeper body, larger mouth, and shorter, more widely spaced gill rakers than do limnetic populations. They also retain group cannibalism, diversionary display behavior, and inconspicuous courtship behavior typical of ancestral, oceanic populations, whereas extreme limnetic ecotypes have lost the first two behavior patterns, and exhibit much more conspicuous courtship.

because it attracts cannibalistic groups, whereas it is maintained in limnetic populations (Foster 1994; 1995; Hulslander 2003; Scotti and Foster 2007; Foster et al. 2008; Robert 2008). Thus, these two ecotypes have diverged remarkably consistently in trophic morphology and body shape, foraging, cannibalistic and courtship behavior, and the intensity of territorial aggression between neighbors during the reproductive cycle.

The consistency of these patterns across populations suggests that different suites of correlated behavior patterns are favored in these two environments. As in the case of behavioral syndromes driven by predation intensity, behavioral syndromes characterizing the extremes of the benthic-limnetic continuum could reflect underlying genetic differentiation, plasticity, or the evolution of norms of reaction in response to differing environments (Shaw et al. 2007). Research on the evolution of correlated suites of behavioral traits along the benthic-limnetic continuum has focused on understanding the ancestral condition through examination of the oceanic form, and on understanding how norms of reaction for ancestral traits have evolved in the many independently derived benthic and limnetic lacustrine populations of stickleback (Shaw et al. 2007; Wund et al. 2008). Moreover, the coexistence of benthic and limnetic types within lakes offers an unusual opportunity to understand the selective factors that can maintain behavioral variation within the same lake (Schluter 1996; Foster et al. 1998; Albert and Schluter 2004; Vines and Schluter 2006) and ultimately, to understand the contributions of behavioral syndromes to the speciation process.

INSIGHTS FROM THE FIELD

Both field and laboratory data suggest that oceanic populations of threespine stickleback exhibit phenotypic plasticity for traits that have differentiated along the benthic-limnetic axis. Observations of behavior in two different years in the oceanic Anchor River population in Alaska (Foster et al. 1998) demonstrated that, as in most oceanic populations, benthic foraging groups are common and courtship primarily entails dorsal pricking. In one year when settlement was light in the tidal pools at this site, however, effective foraging groups could not form and males increased the frequency of zigzagging in their courtship, while retaining the less conspicuous dorsal pricking. This shift in behavior was our first indication that plasticity was maintained in ancestral populations by environmental variation, and that the pattern of plasticity paralleled the divergence between the benthic and limnetic ecotypes (Shaw et al. 2007). Laboratory and field assessments of nuptial signal intensity similarly suggest that the pattern of plasticity in the ancestor parallels the divergence between these ecotypes (Hulslander

2003). Thus, the pattern of ancestral plasticity could have guided the evolution of phenotypic norms of reaction for courtship behavior and nuptial signal expression, because the pattern of plasticity determines the phenotypes exposed to selection (West-Eberhard 2003) and could therefore have influenced the evolution of the benthic-limnetic behavioral variation. Our field observations also demonstrate that behavioral phenotypes that exhibit high levels of plasticity in the ancestral condition can be retained despite lack of use for as long as 15,000–20,000 years. In one limnetic population (Lynne Lake, AK), increased productivity has initiated a cascade of effects that is causing the population to regain ancestral characteristics like those of benthic populations (Chock 2008). In this lake, increased productivity has resulted in larger average adult body size in the population and a shift to benthic foraging. Remarkably, the diversionary display, thought to have been lost from this and other limnetic populations (Foster 1994; 1995), has re-emerged—elicited by foraging groups near male nests. In contrast, males in this population have retained the conspicuous zigzag dance as a consistent part of their courtship behavior, despite the increased risk posed by foraging groups. Thus, the Lynne Lake population appears to have lost ancestral plasticity for courtship behavior, causing it to retain a phenotype unlikely to be adaptive under the new conditions to which the population is exposed (Shaw et al. 2007). Our expectation is that selection by cannibalistic foraging groups will, over time, favor less conspicuous courtship—a hypothesis we will test in future years (Chock 2008).

INSIGHTS FROM LABORATORY-REARED ECOTYPES

Comparison of laboratory-reared stickleback from one benthic (Willow Lake, AK) and one limnetic (Lynne Lake, AK) population demonstrates that different elements of these syndromes are likely to have different underpinnings. A comparison of wild-caught stickleback from these populations with their laboratory-reared counterparts demonstrated that the frequency of the zigzag component of courtship in naive laboratory-reared males and wild-caught fish were closely matched. In contrast, relatively aggressive elements of courtship were more common in wild-caught fish of both ecotypes than in comparable laboratory-reared fish. Thus, the difference between the two study populations in the tendency to perform the zigzag element of the courtship display seems clearly to be influenced by genetic differentiation, whereas learning/experience has a greater impact on differences in aggressive components of courtship behavior in both ecotypes (Shaw et al. 2007). In contrast, differences in the tendency of courting males to behave aggressively toward intruding males in particular situations does appear to reflect underlying genetic differentiation between the two populations, suggesting that aggressive elements of behavior, at least, can respond to selection to different degrees depending upon context (Shaw et al. 2007; Robert 2008). These results are based on one population of each ecotype and thus may not be reflective of an overall pattern of influence of genes and learning/environment across a broader spectrum of benthic and limnetic populations. The generality of this pattern is currently being evaluated and will offer unusual insight into the consistency of genetic and plastic changes that result in similar patterns of expression across the populations comprising individual ecotypes. A possibility not yet explored is that there exist genetic differences in the ability to learn that contribute to these patterns, or to differences between ancestral and derivative freshwater ecotypes.

Taken together, the fascinating insights here are, first, that behavioral types can include both plastic and genetically fixed elements, an insight that echoes the plastic and selected responses to predation risk discussed earlier. Second, the way in which behavioral syndromes are assembled following invasion of a new environment, or local environmental change, is likely to be determined by the extent and nature of plastic contributions to behavior types experiencing the change in environment. By definition, plastic elements can change rapidly, whereas those that do not exhibit plasticity will be slower to respond, if they do so at all. Plasticity that does exist has the potential to direct future evolution by determining which trait values are exposed to selection (West Eberhard 2003; Shaw et al. 2007). Third, plastic elements of ancestral syndromes, such as the diversionary display in Lynne Lake, can sometimes re-emerge, even after thousands of years without use, if the environment shifts toward ancestral conditions (Chock 2008). Overall, the assembly of behavioral syndromes is likely to involve complex processes—as diverse as the syndromes themselves. We should, however, be able to understand these processes with proper comparative and experimental studies of stickleback populations that include more attention to individual variation in behavior.

WITHIN-POPULATION INSIGHTS

We know comparatively little about within-population behavioral differences that parallel differences between benthic and limnetic populations, although there are reasons to expect that they exist. Within two populations, one in British Columbia and one in Alaska, fish foraging on plankton versus benthic macroinvertebrates differ morphologically in the directions that would be predicted from allopatric ecotypes (Cresko and Baker 1996; Robinson 2000). Differences in behavior are likely to exist as well, at least in the Alaskan population. In Benka Lake, AK, benthic foragers nest and feed in shallow bays, while limnetic foragers do so where the lake drops steeply to deep water and planktivorous fish can feed in open water. Thus, differences in habitat use could promote assortative mating, thus enhancing the possibility that strong, possibly genetically correlated behavioral syndromes could evolve. Further research in this lake could also provide an outstanding addition to the hierarchical study of behavioral syndrome evolution related to the benthic-limnetic ecological axis in stickleback (see below).

The differences we see in male responses to foraging groups between benthic and limnetic populations are likely to be especially interesting and tractable in terms of understanding proximate causation. The difference in responses is likely to be mediated, in part at least, through the stress response system. As in the case of the shy-bold continuum, individual differences in stress responsiveness (Koolhaas et al. 1999; A. M. Bell et al. 2007) may contribute to both individual and population differences in responses to cannibalistic groups—a possibility we are currently exploring. Males in cannibalistic populations may also be particularly vulnerable to extreme stress imposed upon them by interactions with foraging groups. Some males, faced with repeated interactions, appear to be forced into over-reactivity, responding to encounters by tearing up nests and by becoming excessively aggressive. This vulnerability seems to affect all components of aggressive and reproductive behavior. At present we cannot discriminate the relative roles of individual personality and experience, but the shift in behavior is dramatic and likely to be influenced by personality and responsiveness to stress.

Conclusions

The truly unusual aspect of the stickleback adaptive radiation lies in the existence of a living ancestor: modern populations are likely to closely resemble, in both phenotypic and genetic terms, those that gave rise to an extensive adaptive radiation. Given the importance of oceanic stickleback for establishing character polarity in the radiation, and the nature of genetic and plastic changes that have given rise to the high levels of parallelism, the paucity of research on these populations is remarkable. We do know that ancestral populations are large and genetically diverse, and this genetic diversity has provided the fuel for adaptive radiation, being repeatedly and differentially winnowed by selection in each of the derived freshwater populations (Withler and McPhail 1985; Cresko et al. 2004; Colosimo et al. 2005). These oceanic populations offer a remarkable opportunity to understand the antecedents of derivative behavioral syndromes, from both phenotypic and genomic perspectives.

Thus, personalities and their causes can be examined in the ancestor and then can be used to ask whether the assembly of behavioral syndromes in derivative freshwater populations involves winnowing of genetically correlated variation within the ancestor, or whether syndromes in daughter populations are assembled *de novo* from uncorrelated, genetically based variation and/or plasticity in the ancestor, or a combination of the above. One could then continue to the next logical step to look at variation in personalities within derivative freshwater populations, again with an eye to understanding the way in which these syndromes have evolved.

As mentioned above, growing genomic resources for stickleback, including a whole-genome sequence, are allowing to address long-standing questions about the evolution of behavior (Kingsley and Peichel 2007). In addition to identifying quantitative trait loci (and ultimately the genes themselves) contributing to differences in behavior between populations (Peichel et al. 2001; Cresko et al. 2004; Shapiro et al. 2004; Colosimo et al. 2005), developing tools for the analysis of gene expression are especially promising for understanding genotype × environment interactions (Fitzpatrick et al. 2005). For example, we will have an opportunity to ask about shared genetic control underlying natural variation in ecologically relevant behaviors. By comparing gene expression in differentiated populations, we can begin to understand how genetic and environmental variation interact to construct behavioral syndromes. Moreover, these approaches should lend insight into the possibility that common genetic control has guided or constrained differentiation among freshwater populations.

In addition to its role in our realization that personality is not restricted to higher primates (Huntingford 1976), the unusual evolutionary history of the stickleback has provided a unique opportunity to gain insight into the evolution of personality. In particular, their propensity for rapid evolution has shown that ecology is a major driving force behind personality in this species—whether personality occurs and what form it takes is strongly influenced by the selective factors in the environment. Studies of stickleback have also spurred a new wave of research beyond simply documenting patterns of behavioral correlations. We can now ask more sophisticated questions about the proximate and ultimate mechanisms that shape variation in behavioral correlations.

In conclusion, stickleback have already taught us a great deal about personality in animals. Some emerging themes to come out of studies of variation along both the shy-bold, aggressive-nonaggressive and the benthic-limnetic axis are that in most cases, behavioral covariation is a result of both an evolutionary response to selection and phenotypic plasticity within the lifetime. Along the same lines, stickleback have taught us that it is not enough to simply study inherited behavioral variation without also considering how the expression of variation depends on environmental conditions such as the immediate social environment, prior experience, rearing conditions, and experience with parents. Given the rapidly emerging tools to study the evolution of behavior from a more mechanistic perspective, it is likely that stickleback will remain at the forefront of research on animal personality for many years to come.

References

Aeschlimann, P., Haberli, M., and Milinski, M. 2000. Threat-sensitive feeding strategy of immature sticklebacks (*Gasterosteus aculeatus*) in response to recent experimental infection with the cestode *Schistocephalus solidus*. *Behavioral Ecology and Sociobiology*, 49, 1–7.

Albert, A. Y. K., and Schluter, D. 2004. Reproductive character displacement of male stickleback mate preference: reinforcement or direct selection? *Evolution*, 58, 1099–1107.

Alvarez, D., and Bell, A. M. 2007. Stream sticklebacks are more bold than pond sticklebacks. *Behavioural Processes*, 76, 215–217.

Andraso, G. M., and Barron, J. N. 1995. Evidence for a trade-off between defensive morphology and startle response performance in the brook stickleback (*Culaea inconstans*). *Canadian Journal of Zoology*, 73, 1147–1153.

Arendt, J., and Reznick, D. 2008. Convergence and parallelism reconsidered: what have we learned about the genetics of adaptation? *Trends in Ecology & Evolution*, 23, 26–32.

Baker, J. A. 1994. Life history variation in female threespine stickleback. In: *The Evolutionary Biology of the Threespine Stickleback* (Bell, M. A., and Foster, S. A., eds.), pp. 144–187. Oxford: Oxford University Press.

Bakker, T. C. M. 1986. Aggressiveness in sticklebacks (*Gasterosteus aculeatus*): a behavior-genetic study. *Behaviour*, 98, 1–144.

———. 1994a. Evolution of aggressive behaviour in the threespine stickleback. In: *The Evolutionary Biology of the Threespine Stickleback* (Bell, M. A., and Foster, S. A., eds.), pp. 345–379. Oxford: Oxford.

———. 1994b. Genetic correlations and the control of behavior, exemplified by aggressiveness in sticklebacks. *Advances in the Study of Behavior*, 23, 135–171.

Bell, A. M. 2004. An endocrine disrupter increases growth and risky behavior in threespined stickleback (*Gasterosteus aculeatus*). *Hormones and Behavior*, 45, 108–114.

———. 2005. Differences between individuals and populations of threespined stickleback. *Journal of Evolutionary Biology*, 18, 464–473.

———. 2007. Future directions in behavioral syndromes research. *Proceedings of the Royal Society of London B*, 274, 755–761.

Bell, A. M., Backstrom, T., Huntingford, F. A., Pottinger, T. G., and Winberg, S. 2007. Variable behavioral and neuroendocrine responses to ecologically relevant challenges in sticklebacks. *Physiology and Behavior*, 91, 15–25.

Bell, A. M., and Sih, A. 2007. Exposure to predation generates personality in threespined sticklebacks. *Ecology Letters*, 10, 828–834.

Bell, A. M., and Stamps, J. A. 2004. The development of behavioural differences between individuals and populations of stickleback. *Animal Behaviour*, 68, 1339–1348.

Bell, M. A. 1994. Paleobiology and evolution of threespine stickleback. In: *The Evolutionary Biology of the Threespine Stickleback* (Bell, M. A., and Foster, S. A., eds.), pp. 438–471. Oxford: Oxford University Press.

Bell, M. A., Aguirre, W. E., and Buck, N. J. 2004. Twelve years of contemporary armor evolution in a threespine stickleback population. *Evolution*, 58, 814–824.

Bell, M. A., and Foster, S. A. 1994. Introduction to the evolutionary biology of the threespine stickleback. In: *The Evolutionary Biology of the Threespine Stickleback* (Bell, M. A., and Foster, S. A., eds.), pp. 1–27. Oxford: Oxford University Press.

Bell, M. A., and Haglund, T. R. 1978. Selective predation of threespine sticklebacks *Gasterosteus aculeatus* by garter snakes. *Evolution*, 32, 304–319.

Bell, M. A., and Ortí, G. 1994. Pelvic reduction in threespine stickleback from Cook Inlet lakes: geographical distribution and intrapopulation variation. *Copeia*, 1994, 314–325.

Benzie, V. 1965. Some aspects of the antipredator responses of two species of stickleback. PhD diss., Oxford University.

Blumstein, D. T. 2006. The multipredator hypothesis and the evolutionary persistence of antipredator behavior. *Ethology*, 112, 209–217.

Blumstein, D. T., and Daniel, J. C. 2002. Isolation from mammalian predators differentially affects two congeners. *Behavioral Ecology*, 13, 657–663.

Blumstein, D. T., Daniel, J. C., and Springett, B. P. 2004. A test of the multipredator hypothesis: rapid loss of antipredator behavior after 130 years of isolation. *Ethology*, 110, 919–934.

Braithwaite, V. A., and Girvan, J. R. 2003. Use of water flow direction to provide spatial information in a small-scale orientation task. *Journal of Fish Biology*, 63, 74–83.

Brown, C., and Warburton, K. 1999. Differences in timidity and escape responses between predator-naive and predator-sympatric rainbowfish populations. *Ethology*, 105, 491–502.

Brydges, N. M., Colegrave, N., Heathcote, R. J. P., and Braithwaite, V. A. 2008a. Habitat stability and predation pressure affect temperament behaviours in populations of three-spined sticklebacks. *Journal of Animal Ecology*, 77, 229–235.

Brydges, N. M., Heathcote, R. J. P., and Braithwaite, V. A. 2008b. Habitat stability and predation pressure influence learning and memory in populations of three-spined sticklebacks. *Animal Behaviour*, 75, 935–942.

Burkhardt, R. W. 2005. *Patterns of Behavior: Konrad Lorenz, Niko Tinbergen, and the Founding of Ethology*. Chicago: University of Chicago Press.

Chock, R. Y. 2008. Re-emergence of ancestral plasticity and the loss of a rare limnetic phenotype in an Alaskan population of threespine stickleback. Master's thesis, Clark University.

Colosimo, P. F., Hosemann, K. E., Balabhadra, S., Guadalupe Villarreal, J., Dickson, M., Grimwood, J., Schmutz, J., Myers, R. M., Schluter, D., and Kingsley, D. M. 2005. Widespread parallel evolution in sticklebacks by repeated fixation of ectodysplasin alleles. *Science*, 307, 1928–1933.

Coss, R. G. 1999. Effects of relaxed natural selection on the evolution of behavior. In:

Geographic Variation in Behavior: Perspectives on Evolutionary Mechanisms (Foster, S. A., and Endler, J. A., eds.), pp. 180–208. New York: Oxford University Press.

Cresko, W. A. 2000. The ecology and geography of speciation: a case study using an adaptive radiation of threespine stickleback in Alaska. PhD diss., Clark University, Worcester, Massachusetts.

Cresko, W. A., Amores, A., Wilson, C., Murphy, J., Currey, M., Phillips, P., Bell, M. A., Kimmel, C. B., and Postlethwait, J. H. 2004. Parallel genetic basis for repeated evolution of armor loss in Alaskan threespine stickleback populations. *Proceedings of the National Academy of Sciences U S A*, 101, 6050–6055.

Cresko, W. A., and Baker, J.A. 1996. Two morphotypes of lacustrine threespine stickleback, *Gasterosteus aculeatus* in Benka Lake, Alaska. *Environmental Biology of Fishes*, 45, 343–350.

DeWitt, T. J., Sih, A., and Hucko, J. A. 1999. Trait compensation and cospecialization in a freshwater snail: size, shape, and antipredator behaviour. *Animal Behaviour*, 58, 397–407.

Diaz-Uriarte, R. 1999. Antipredator behaviour changes following an aggressive encounter in the lizard *Tropidurus hispidus*. *Proceedings of the Royal Society of London B*, 266, 2457–2464.

Dingemanse, N. J., Both, C., Drent, P. J., and Tinbergen, J. M. 2004. Fitness consequences of avian personalities in a fluctuating environment. *Proceedings of the Royal Society of London B*, 271, 847–852.

Dingemanse, N. J., and Réale, D. 2005. Natural selection and animal personality. *Behaviour*, 142, 1159–1184.

Dingemanse, N. J., Thomas, D. K., Wright, J., Kazem, A. J. N., Koese, B., Hickling, R., and Dawnay, N. 2007. Behavioural syndromes differ predictably between twelve populations of three-spined stickleback. *Journal of Animal Ecology*, 76, 1128–1138.

Dugatkin, L. A., and Godin, J.-G. J. 1992. Predator inspection shoaling and foraging under predation hazard in the Trinidadian guppy, *Poecilia reticulata*. *Environmental Biology of Fishes*, 34, 265–276.

Fitzpatrick, M. J., Ben-Shahar, Y., Smid, H. M., Vet, L. E. M., Robinson, G. E., and Sokolowski, M. B. 2005. Candidate genes for behavioural ecology. *Trends in Ecology and Evolution*, 20, 96–104.

Foster, S. A. 1994. Evolution of the reproductive behaviour of threespine stickleback. In: *The Evolutionary Biology of the Threespine Stickleback* (Bell, M. A., and Foster, S. A., eds.), pp. 381–398. Oxford: Oxford University Press.

———. 1995. Understanding the evolution of behavior in threespine stickleback: the value of geographic variation. *Behaviour*, 132, 1107–1129.

———. 1999. The geography of behavior: an evolutionary perspective. *Trends in Ecology & Evolution*, 14, 190–195.

Foster, S. A., and Bell, M. A. 1994. Evolutionary inference: the value of viewing evolution through stickleback-tinted glasses. In: *The Evolutionary Biology of the Threespine Stickleback* (Bell, M. A., and Foster, S. A., eds.), pp. 472–486. Oxford: Oxford University Press.

Foster, S. A., and Endler, J. A. 1999. *Geographic Variation in Behavior: Perspectives on Evolutionary Mechanisms*. New York: Oxford University Press.

Foster, S. A., Scott, R. J., and Cresko, W. A. 1998. Parallel hierarchical variation and speciation. *Proceedings of the Royal Society of London B*, 353, 207–218.

Foster, S. A., Shaw, K. A., Robert, K. L., and Baker, J. A. 2008. Benthic, limnetic, and oceanic threespine stickleback: profiles of reproductive behaviour. *Behaviour*, 145, 485–508.

Giles, N., and Huntingford, F. A. 1984. Predation risk and interpopulation variation in antipredator behaviour in the threespined stickleback. *Animal Behaviour*, 32, 264–275.

Girvan, J. R., and Braithwaite, V. A. 1998. Population differences in spatial learning in three-spined sticklebacks. *Proceedings of the Royal Society of London B*, 265, 913–918.

Godin, J.-G. J., and Clark, K. A. V. 1997. Risk-taking in stickleback fishes faced with different predatory threats. *Ecoscience*, 4, 246–251.

Godin, J.-G. J., and Crossman, S. L. 1994. Hunger-dependent predator inspection and foraging behaviours in the threespine stickleback (*Gasterosteus aculeatus*) under predation risk. *Behavioral Ecology and Sociobiology*, 34, 359–366.

Grand, T. C. 2000. Risk-taking by threespine stickleback (*Gasterosteus aculeatus*) pelvic phenotypes: does morphology predict behaviour? *Behaviour*, 137, 889–906.

Hagen, D. W., and Gilbertson, L. G. 1973a. Selective predation and the intensity of selection acting upon the lateral plates of threespine sticklebacks. *Heredity*, 30, 273–287.

Hagen, D. W., and Gilbertson, L. G. 1973b. The genetics of plate morphs in freshwater threespine sticklebacks. *Heredity*, 31, 75–84.

Hinde, R. A., and Tinbergen, N. 1958. The comparative study of behavior. In: *Behavior and Evolution* (Roe, A., and Simpson, G. G., eds.), pp. 251–268. New Haven: Yale University Press.

Hoogland, R., Morris, D., and Tinbergen, N. 1957. The spines of sticklebacks (*Gasterosteus* and *Pygosteus*) as a means of defence against predators (*Perca* and *Esox*). *Behaviour*, 10, 205–236.

Hulslander, C. L. 2003. The evolution of the male threespine stickleback color signal. PhD diss., Clark University, Worcester, Massachusetts.

Hunt, G., Bell, M. A., and Travis, M. P. 2008. Evolution toward a new adaptive optimum: phenotypic evolution in a fossil stickleback lineage. *Evolution*, 62, 700–710.

Huntingford, F. A. 1976. The relationship between antipredator behaviour and aggression among conspecifics in the three-spined stickleback. *Animal Behaviour*, 24, 245–260.

———. 1981. Further evidence for an association between lateral scute number and aggressiveness in the threespine stickleback. *Copeia*, 3, 717–720.

Huntingford, F. A., and Coulter, R. M. 1989. Habituation of predator inspection in the three-spined stickleback *Gasterosteus aculeatus*. *Journal of Fish Biology*, 35, 153–154.

Huntingford, F. A., and Giles, N. 1987. Individual variation in antipredatory responses in the three-spined stickleback. *Ethology*, 74, 205–210.

Huntingford, F. A., and Wright, P. J. 1989. How sticklebacks learn to avoid dangerous feeding patches. *Behavioural Processes*, 19, 181–189.

Huntingford, F. A., Wright, P. J., and Tierney, J. F. 1994. Adaptive variation and antipredator behaviour in threespine stickleback. In: *The Evolutionary Biology of the Threespine Stickleback* (Bell, M. A., and Foster, S. A., eds.), pp. 277–295. Oxford: Oxford University Press.

Ibrahim, A. A., and Huntingford, F. A. 1989. Laboratory and field studies of the effect of predation risk on foraging in three-spined sticklebacks (*Gasterosteus aculeatus*). *Behaviour*, 109, 46–57.

Jonsson, E., Johnsson, J. I., and Bjornsson, B. T. 1996. Growth hormone increases predation exposure of rainbow trout. *Proceedings of the Royal Society of London B*, 263, 647–651.

Kelley, J. L., and Magurran, A. E. 2003. Learned predator recognition and antipredator responses in fishes. *Fish and Fisheries*, 4, 216–226.

Kingsley, D. M., and Peichel, C. L. 2007. The molecular genetics of evolutionary change in sticklebacks. In: *Biology of the Three-Spined Stickleback* (Östlund-Nilsson, S., Mayer, I., and Huntingford, F. A., eds.), pp. 271–284. Boca Raton: CRC Press.

Koolhaas, J. M., Korte, S. M., De Boer, S. F., Van Der Vegt, B. J., Van Reenen, C. G., Hopster, H., De Jong, I. C., Ruis, M. A. W., and Blokhuis, H. J. 1999. Coping styles in animals: current status in behavior and stress-physiology. *Neuroscience and Biobehavioral Reviews*, 23, 925–935.

Krause, J., Hoare, D. J., Croft, D., Lawrence, J., Ward, A., Ruxton, G. D., Godin, J.-G. J., and James, R. 2000. Fish shoal composition: mechanisms and constraints. *Proceedings of the Royal Society of London B*, 267, 2011–2017.

Krause, J., Loader, S. P., McDermont, J., and Ruxton, G. D. 1998. Refuge use by fish as a function of body length–related metabolic expenditure and predation risks. *Proceedings of the Royal Society of London B*, 265, 2373–2379.

Lavin, P. A., and McPhail, J. D. 1985. The evolution of freshwater diversity in the threespine stickleback (*Gasterosteus aculeatus*): site-specific differentiation of trophic morphology. *Canadian Journal of Zoology*, 63, 2632–2638.

———. 1986. Adaptive divergence of trophic phenotype among freshwater populations of threespine stickleback (*Gasterosteus aculeatus*). *Canadian Journal of Fisheries and Aquatic Science*, 43, 2455–2463.

Lee, T., and Bell, M. A. 1999. Causes and consequences of recent freshwater invasions by saltwater animals. *Trends in Ecology & Evolution*, 14, 284–288.

Lima, S. L., and Dill, L. M. 1990. Behavioral decisions made under the risk of predation: a review and prospectus. *Canadian Journal of Zoology*, 68, 619–640.

Lopez, P., Hawlena, D., Polo, V., Amo, L., and Martin, J. 2005. Sources of individual shy-bold variations in antipredator behaviour of Iberian male rock lizards. *Animal Behaviour*, 69, 1–9.

Lynch, M., and Walsh, B. 1998. *Genetics and Analysis of Quantitative Traits*. Sunderland, MA: Sinauer Associates.

Magurran, A. E., and Seghers, B. H. 1994. Predator inspection behaviour covaries with schooling tendency amongst wild guppy, *Poecilia reticulata*, populations in Trinidad. *Behaviour*, 128, 121–134.

Malhi, R. S., Rhett, G., and Bell, A. M. 2006. Mitochondrial DNA evidence of an early Holocene population expansion of threespine sticklebacks from Scotland. *Molecular Phylogenetics and Evolution*, 40, 148–154.

McElreath, R., and Strimling, P. 2006. How noisy information and individual asymmetries can make "personality" an adaptation: a simple model. *Animal Behaviour*, 72, 1135–1139.

McKinnon, J. S., Mori, S., Blackman, B. K., David, L., Kingsley, D. M., Jamieson, L., Chou, J., and Schluter, D. 2004. Evidence for ecology's role in speciation. *Nature*, 429, 294–298.

McPhail, J. D. 1994. Speciation and the evolution of reproductive isolation in the sticklebacks (*Gasterosteus*) of southwestern British Columbia. In: *The Evolutionary Biology of the Threespine Stickleback* (Bell, M. A., and Foster, S. A., eds.), pp. 399–437. Oxford: Oxford University Press.

McPhail, J. D., and Lindsey, C. C. 1970. Freshwater fishes of northwestern Canada and Alaska. *Bulletin of the Fisheries Research Board of Canada*, 173, 1–381.

Messler, A., Wund, M. A., Baker, J. A., and Foster, S. A. 2007. The effects of relaxed and reversed selection by predators on the antipredator behavior of the threespine stickleback, *Gasterosteus aculeatus*. *Ethology*, 113, 953–963.

Mikolajewski, D. J., and Johansson, F. 2004. Morphological and behavioral defenses in dragonfly larvae: trait compensation and cospecialization. *Behavioral Ecology*, 15, 614–620.

Milinski, M., Luthi, J. H., Eggler, R., and Parker, G. A. 1997. Cooperation under predation risk: experiments on costs and benefits. *Proceedings of the Royal Society of London B*, 264, 831–837.

Miller, M. R., Dunham, J. P., Amores, A., Cresko, W. A., and Johnson, E. A. 2007. Rapid and cost-effective polymorphism identification and genotyping using restriction site–associated DNA (RAD) markers. *Genome Research*, 17, 240–248.

Murphy, K. E., and Pitcher, T. J. 1991. Individual behavioral strategies associated with predator inspection in minnow shoals. *Ethology*, 88, 307–319.

Neat, F. C., Taylor, A. C., and Huntingford, F. A. 1998. Proximate costs of fighting in male cichlid fish: the role of injuries and energy metabolism. *Animal Behaviour*, 55, 875–882.

Östlund-Nilsson, S., Mayer, I., and Huntingford, F. A. 2007. Biology of three-spined stickleback. In: *Marine Biology* (Lutz, P. L., ed.), p. xx. Boca Raton, FL: CRC Press.

Peichel, C. L., Nereng, K. S., Ohgi, K. A., Cole, B. L. E., Colosimo, P. F., Buerkle, C. A., Schluter, D., and Kingsley, D. M. 2001. The genetic architecture of divergence between threespine stickleback species. *Nature*, 414, 901–905.

Pigliucci, M., and Murrent, C. J. 2003. Perspective: genetic assimilation and a possible evolutionary paradox: can macroevolution sometimes be so fast as to pass us by? *Evolution*, 57, 1455–1464.

Pitcher, T. J., Green, D. A., and Magurran, A. E. 1986. Dicing with death: predator inspection behavior in minnow (*Phoxinus phoxinus*) shoals. *Journal of Fish Biology*, 28, 439–448.

Quinn, J.L., and Creswell, W. 2005. Escape response delays in wintering redshank, *Tringa tetanus*, flocks: perceptual limits and economic decisions. *Animal Behaviour*, 69, 1285–1292.

Raeymaekers, J., Maes, G., Audenaert, E., and Volckaert, F. 2005. Detecting Holocene divergence in the anadromous-freshwater three-spined stickleback (*Gasterosteus aculeatus*) system. *Molecular Ecology*, 14, 1001–1014.

Réale, D., Reader, S. M., Sol, D., McDougall, P. T., and Dingemanse, N. J. 2007. Integrating animal temperament within ecology and evolution. *Biological Reviews*, 82, 291–318.

Reimchen, T. E. 1994. Predators and morphological evolution in threespine stickleback. In: *The Evolutionary Biology of the Threespine Stickleback* (Bell, M. A., and Foster, S. A., eds.), pp. 240–273. Oxford: Oxford University Press.

Robert, K. L. T. 2008. The role of male nuptial coloration in aggression of threespine stickleback: a two-population, context-dependent analysis. Master's thesis, Clark University.

Robinson, B. W. 2000. Trade-offs in habitat-specific foraging efficiency and the nascent adaptive divergence of sticklebacks in lakes. *Behaviour*, 137, 865–888.

Rowland, W. J. 1984. The relationship among coloration, aggression, and courtship of male three-spined sticklebacks. *Canadian Journal of Zoology*, 62, 999–1004.

Rowland, W. J. 1994. Proximate determinants of stickleback behavior: an evolutionary perspective. In: *The Evolutionary Biology of the Threespine Stickleback* (Bell, M. A., and Foster, S. A., eds.). Oxford: Oxford University Press.

Schluter, D. 1996. Ecological speciation in postglacial fishes. *Philosophical Transactions of the Royal Society of London B*, 351, 807–814.

———. 2000. Ecological character displacement in adaptive radiation. *American Naturalist*, 156, S4–S16.

Scotti, M. L., and Foster, S. A. 2007. Phenotypic plasticity and the ecotypic differentiation of aggressive behavior in threespine stickleback. *Ethology*, 113, 190–198.

Shapiro, M. D., Marks, M. E., Peichel, C. L., Blackman, B. K., Nereng, K. S., Jonsson, B., Schluter, D., and Kingsley, D. M. 2004. Genetic and developmental basis of evolutionary pelvic reduction in threespine sticklebacks. *Nature*, 428, 717–723.

Shaw, K. A., Scotti, M. L., and Foster, S. A. 2007. Ancestral plasticity and the evolutionary diversification of courtship behaviour in threespine sticklebacks. *Animal Behaviour*, 73, 415–472.

Sih, A. 1992. Prey uncertainty and the balancing of antipredator and feeding needs. *American Naturalist*, 139, 1052–1069.

Sih, A., Bell, A. M., and Johnson, J. C. 2004a. Behavioral syndromes: an ecological and evolutionary overview. *Trends in Ecology and Evolution*, 19, 372–378.

Sih, A., Bell, A. M., Johnson, J. C., and Ziemba, R. 2004b. Behavioral syndromes: an integrative overview. *Quarterly Review of Biology*, 79, 241–277.

Sih, A., Kats, L. B., and Maurer, E. F. 2003. Behavioral correlations across situations and the evolution of antipredator behaviour in a sunfish-salamander system. *Animal Behaviour*, 65, 29–44.

Smith, M. E., and Belk, M. C. 2001. Risk assessment in western mosquitofish (*Gambusia affinis*): do multiple cues have additive effects? *Behavioral Ecology and Sociobiology*, 51, 101–107.

Stamps, J. A. 2007. Growth-mortality tradeoffs and "personality traits" in animals. *Ecology Letters*, 10, 355–363.

Taylor, E. B., and McPhail, J. D. 1999. Evolutionary history of an adaptive radiation in species pairs of threespine sticklebacks (*Gasterosteus*): insights from mitochondrial DNA. *Biological Journal of the Linnean Society*, 66, 271–291.

———. 2000. Historical contingency and ecological determinism interact to prime speciation in sticklebacks, *Gasterosteus aculeatus*. *Proceedings of the Royal Society of London B*, 267, 2375–2384.

Thorpe, K. E., Taylor, A. C., and Huntingford, F. A. 1995. How costly is fighting? Physiological effects of sustained exercise and fighting in swimming crabs, *Necora puber*. *Animal Behaviour*, 50, 1657–1666.

Tinbergen, N. 1951. *The Study of Instinct*. Oxford: Oxford University Press.

Tulley, J. J., and Huntingford, F. A. 1987a. Age, experience and the development of adaptive variation in antipredator responses in threespined sticklebacks. *Ethology*, 75, 285–290.

Tulley, J. J., and Huntingford, F. A. 1987b. Paternal care and the development of adaptive variation in antipredator responses in sticklebacks. *Animal Behaviour*, 35, 1570–1572.

Vines, T.H., and Schluter, D. 2006. Strong assortative mating between allopatric sticklebacks as a by-product of adaptation to different environments. *Proceedings of the Royal Society of London B*, 273, 911–916.

Walker, J. A. 1997. Ecological morphology of lacustrine threespine stickleback *Gasterosteus aculeatus* L. (Gasterosteidae) body shape. *Biological Journal of the Linnean Society*, 61, 3–50.

Walker, J. A., and Bell, M. A. 2000. Net evolutionary trajectories of body shape evolution within a microgeographic radiation of threespine stickleback (*Gasterosteus aculeatus*). *Journal of Zoology*, 252, 293–302.

Walling, C. A., Dawnay, N., Kazem, A. J. N., Hickling, R., and Wright, J. 2003. Do competing males cooperate? Familiarity and its effect on cooperation during predator inspection in male three-spined sticklebacks (*Gasterosteus aculeatus*). *Journal of Fish Biology*, 63, 243–244.

Walling, C. A., Dawnay, N., Kazem, A. J. N., and Wright, J. 2004. Predator inspection behaviour in three-spined sticklebacks (*Gasterosteus aculeatus*): body size, local predation pressure, and cooperation. *Behavioral Ecology and Sociobiology*, 56, 164–170.

Webster, M. M. 2007. Boldness is influenced by social context in threespine sticklebacks. *Behaviour*, 144, 351–371.

West-Eberhard, M. J. 2003. *Developmental Plasticity and Evolution*. Oxford: Oxford University Press.

Wibe, A. E., Nordtug, T., and Jenssen, B. M. 2001. Effects of bis(tributyltin)oxide on antipredator behavior in threespine stickleback *Gasterosteus aculeatus* L. *Chemosphere*, 44, 475–481.

Wilson, D. S. 1998. Adaptive individual differences within single populations. *Philosophical Transactions of the Royal Society of London B*, 353, 199–205.

Wilson, A. D. M., and Godin, J.-G. J. 2009. Boldness and behavioral syndromes in bluegill sunfish, *Lepomis macrochirus*. *Behavioral Ecology*, 20, 231–237.

Wilz, K. J. 1973. Quantitative differences in the courtship of two populations of three-spined sticklebacks, *Gasterosteus aculeatus*. *Zeitschrift für Tierpsychologie*, 33, 141–146.

Withler, R. E., and McPhail, J. D. 1985. Genetic variability in freshwater and anadromous sticklebacks (*Gasterosteus aculeatus*) of southern British Columbia. *Canadian Jorunal of Zoology*, 63, 528–533.

Wolf, M., Van Doorn, G. S., Leimar, O., and Weissing, F. J. 2007. Life history tradeoffs favour the evolution of personality. *Nature*, 447, 581–585.

Wootton, R. J. 1976. *The Biology of the Sticklebacks*. London: Academic Press.

Wund, M. A., Baker, J. A., Clancy, B., Golub, J. L., and Foster, S. A. 2008. A test of the "flexible stem" model of evolution: ancestral plasticity, genetic accommodation, and morphological divergence in the threespine stickleback radiation. *American Naturalist*, 174, 449–462.

3

Avian Personality

KEES VAN OERS AND MARC NAGUIB

Introduction

Birds are widely distributed and highly diversified, and also generally more conspicuous and observable in natural environments than many other vertebrates. Birds also exhibit complex behavior and social processes. Because of these attributes, birds are key model organisms that have allowed behavioral biologists to address a wide range of ecological and evolutionary questions (Konishi et al. 1989; Danchin et al. 2008; Davies et al. 2012). Moreover, owing to the contributions from a diverse group of ornithologists ranging from amateur bird-watchers to professional scientists, the knowledge of bird behavior under natural conditions is more extensive than for any other vertebrate taxa. Many avian species are diurnal, conspicuous, and resilient, and also permit relatively invasive investigations. In consequence, they are well suited for experimental field research using a wide range of methods such as manipulation of breeding conditions (Tinbergen and Boerlijst 1990), bioacoustic analyses (Marler and Slabbekoorn 2004; Catchpole and Slater 2008), capture and marking procedures (Lebreton et al. 1992), and analyses of energy expenditure and of endocrine and immune function (Wingfield 2005). Birds are also excellent study organisms for experimental investigations of behavioral adaptations to changes in environmental conditions, a topic that has become of particular interest in recent years (Visser 2008). Furthermore, captive bird populations proved to be extremely valuable in studies on the genetics of behavior (Berthold and Querner 1981; Jones and Hocking 1999; Price 2002; Drent et al. 2003; Van Oers et al. 2005a), the neurological correlates of behavior (Konishi 1985; Gil et al. 2006), physiological mechanisms underlying behavior and its variation (Wingfield 2005; Soma et al. 2008), and developmental influences on behavior (Naguib and Nemitz 2007), as well as in studies on sexual selection (Andersson 1994). Even in more applied fields such as conservation, animal welfare, and animal husbandry, birds have been shown to be good models for studying behavior, both in the field and in captivity (Sutherland 1998; Dawkins 1999; Rodenburg and Turner 2012).

Considering all of the above, it is not surprising, therefore, that the majority of studies on animal personality have been conducted on birds. The study of personality traits in birds can be framed into an ecological context more easily than in other taxa, allowing analyses of ecological and evolutionary aspects of personality. Birds of resident species can be followed individually, often throughout their lives, their behavior can be measured both under standardized conditions in captivity and in their natural environment, and fitness can be quantified. The opportunity to conduct behavioral tests in the laboratory also allows researchers to effectively document consistent differences in individual behavior within and across contexts (Dingemanse et al. 2002; Van Oers et al. 2005b; Martins et al. 2007; Schuett and Dall 2009).

The aim of this chapter is to provide a broad overview of studies on avian personality, including those that address genetic variation linked to personality (Van Oers et al. 2005a; Fidler et al. 2007; Gil and Faure 2007; Bell and Aubin-Horth 2010; Van Oers and Muller 2010), the behavioral and fitness consequences of personality in natural populations (e.g., Dingemanse et al. 2004; Dingemanse and Wolf 2010; Wolf and Weissing 2010), mathematical models that investigate possible scenarios for the evolution of personality (Wolf et al. 2007; Amy et al. 2010; Dingemanse and Wolf 2010; Houston 2010), and the physiological substrate of personality (e.g., Carere et al. 2005a; Kralj-Fiser et al. 2007; Fucikova et al. 2009; Coppens et al. 2010; Baugh et al. 2012). We also provide a historical background of personality research in birds and discuss recent studies from a historical perspective.

Historical overview of avian personality research

Darwin wrote extensively on animal behavior (Darwin 1872) and, after him, scientists working on animal behavior could be separated into two different fields, namely, comparative psychologists and ethologists. There was one camp of ethologists such as Douglas Spalding (1841–1877), who studied imprinting in chicks (Gray 1967), and another camp of mainly animal psychologists such as Lloyd Morgan (1852–1936), who wrote a book, very influential at the time, introducing the field of comparative psychology (Lloyd Morgan 1923). Wallace Craig, Oskar Heinroth, and Charles Otis Whitman contributed to the emergence of ethology as a major separate biological discipline in the study of behavior. Although both comparative psychology and ethology address basic principles of animal behavior, these disciplines have been separated for a long time. Interestingly, however, the study of individuality in the behavior of birds, and their personality, has found its way in research both by ethologists and by comparative psychologists, and also has emerged in publications by semiprofessional bird banders. A crucial factor for recognizing individual differences in behavior is spending sufficient time observing the individuals. Yet, a larger part of ornithological research up until the first part of the twentieth century was done in museums so that few data were available on behavioral differences among individuals (Armstrong 1947).

The first published use of the word *personality* to describe characteristics of individual birds was probably by Talbot in 1922. Talbot (1922) described inter- and intraspecific differences in the motivation to fly through a hole that was situated at the entrance of a gathering cage, when birds were caught for banding. In the *Manual of Bird Banders*, Lincoln and Baldwin (1929) even stated that documenting bird personality, defined as the individual peculiarities in appearance, habits, and manners (Baechle 1947), is one of the main achievements of bird banding. A nice example that recognizing individual personality differences requires spending sufficient time with these individuals is provided by Gwendolen (Len) Howard, a musician who kept birds in captivity as hobby. Most of her birds were tame, or at least habituated to human presence, and in a book called *Birds as Individuals*, Howard (1953) describes in great detail the lives of great tits, black birds, robins, and other birds in her garden or inside her house. Most remarkable about this is that she recognized the birds not only by their individual plumage characteristics, but also by their *"characteristic mannerisms and poses and their facial expressions."* She states that great tits especially were easily recognized since *"their whole bearing and personality was too individual for confusion to arise when I had them at close quarters"* (Howard 1953).

While the early descriptions of personality-like behavior in birds were largely anecdotal, it was probably Lack (1947) who made the first scientifically based description of individual differences in aggressiveness in robins, when he tested several males and their reaction to a model. Some males were hardly interested in the model, while others vigorously attacked it. Another series of papers related to individual differences in behavior came from Burtt (1967), a psychologist who combined his interest in bird banding with the scientific study of personality. While banding over 17,000 blackbirds for the Federal Wildlife Service, he studied bird behavior for several years from a psychological perspective (Thayer and Austin 1992). In his book *The Psychology of Birds*, he made the connection between personality traits, such as extroversion, dominance, and emotional stability, and the behavior of individual birds (Burtt 1967). Together with Giltz, he investigated, for instance, the importance of personality in the interpretation of bird behavior (Burtt and Giltz 1969b). In connection with a banding program at a decoy trap, the pair measured the personality traits "complacency" and "aggression" in common grackles (*Quiscalus quiscula*), red-winged blackbirds (*Agelaius phoeniceus*), brown-headed cowbirds (*Molothrus ater*), and starlings (*Sturnus vulgaris*). They defined what they called *the complacency-agitation continuum* as the way an individual bird behaved in a small cage, directly after banding. Complacent birds moved more compared with agitated birds. Aggression was assessed while birds were held in a hand; birds were threatened with a finger and the tendency to bite was measured on a scale from zero to ten (Burtt and Giltz 1969a). Both traits showed within- and between-species variation and were found to be highly repeatable. They found cowbirds to be the most complacent and starlings the most agitated species, while grackles and cowbirds were the most aggressive, and starlings and red-winged blackbirds the least aggressive species (Burtt and Giltz 1969a; 1969b). They also found that grackles had the greatest tendency to re-enter the trap, and that they were the most resident species compared with the other species (Burtt and Giltz 1973).

As suggested by this brief historical overview, recognizing consistent individual differences in bird behavior has a long history, dating back to the roots of ethology. Scientific research focusing on such differences, however, has flourished mainly in the last decade (Table 1), after individual differences in behavior were framed into the well-defined concepts of personality, behavioral syndromes, and coping styles.

Recent advances in avian personality research
BEHAVIORAL CONSISTENCY AND BEHAVIORAL TESTS

Many recent studies on behavioral consistency in birds have been inspired by the work of Verbeek, Drent, and coworkers in the early 1990s on great tits (Verbeek et al. 1994; 1996; 1999). This work was based on earlier studies on the individual consistency in the response toward changes in the environment in mice (e.g., Van Oortmerssen and Bakker 1981; Benus et al. 1987) and studies on foraging and exploratory behavior in great tits (*Parus major*) (e.g., Krebs et al. 1972; Krebs and Perrins 1978). Individual differences in exploratory behavior had been noticed in several studies on foraging behavior in great tits (Partridge 1976; Kacelnik et al. 1981), so the expansion of research with a focus on such behavioral differences could be built on a substantial body of knowledge of this topic. The behavioral tests developed by Verbeek and coworkers to investigate novel object and novel environment exploration (Verbeek et al. 1994) are now used as standard tests in most studies on great tits (e.g., Dingemanse et al. 2002; Van Oers et al. 2008; Hollander et al. 2008; Titulaer et al. 2012) and also on other bird species (e.g., Fox and Millam 2007; Martins et al. 2007; Kurvers et al. 2009; Schuett and Dall 2009). Since the behavioral response to a novel object or novel environment might be species-specific, some caution is required when comparing studies of different species. A test that might be very meaningful in the life-history of one species might be less suitable for another one, thus resulting in between-species differences in the consistency of a trait (Mettke-Hofmann et al. 2005). Hence, although exploratory behavior is now considered one of the major animal personality traits (Réale et al. 2007; Sih and Bell 2008), species-specific tests to validate this and other behaviors are very important. Related to exploratory behaviors are boldness or risk-taking behaviors (Réale et al. 2007; but see chapter 6). Whereas the exploration of a novel object is likely to reflect a mixture of curiosity and fearfulness, risk-taking behaviors are often related to predation risk and foraging (Van Oers et al. 2004b). Therefore, these behaviors are more closely linked to stress responses (Martins et al. 2007). Social relationships (Stowe et al. 2006; Stowe and Kotrschal 2007) and the behavior of unrelated flock mates (Marchetti and Drent 2000; Van Oers et al. 2005b; Kurvers et al. 2009) might also be important factors. The behavior of a group of birds as a whole can, therefore, be affected by the mix of personality types present within the group (Kurvers et al. 2009; Schuett and Dall 2009; Amy et al. 2010).

Apart from exploratory behavior, several other traits have been investigated in birds as personality traits (see table 3.1 for some examples). One trait that has been widely used in bird research is agonistic behavior, which has been shown to be individually consistent and to have a genetic correlate in mice (Benus 1988) and fish (Bakker 1994). Aggression toward a conspecific is a crucial part of the life-history of social animals such as birds. Especially for social dominance (Verbeek et al. 1996; Dingemanse and De Goede 2004) and in territorial behavior (Duckworth 2006b; Garamszegi et al. 2009; Amy et al. 2012), agonistic behavior is an important personality trait determining the outcome of an interaction and subsequent access to resources. In great tits, Verbeek and coworkers (1996) investigated whether individual differences in exploratory behavior were related to agonistic behavior and how this, in turn, related to dominance. In an experimental set-up, two males were placed in a cage, with an opaque partition between them. After removing the partition, the authors noted which of the two males attacked first. They found that this measure of aggressive behavior was consistent over time and that fast explorers started more fights than slow explorers, independent of sex and morphological traits. Individuals that initiated a fight were also more likely to win that fight.

Table 3.1. A noncomprehensive species overview with studies on the assessment of avian personality traits.

Author	Year	Species		Personality traits
		Anseriformes		
Kurvers et al.	2009	Barnacle goose	(*Branta bernicla*)	Boldness
Kralj-Fiser et al.	2007	Greylag goose	(*Anser anser*)	Reaction to handling
		Ciconiiformes		
Blas et al.	2007	European white stork	(*Ciconia ciconia*)	Reaction to handling
		Falconiformes		
Costantini et al.	2005	European kestrel	(*Falco tinnunculus*)	Feeding habits
		Galliformes		
Richard et al.	2008	Japanese quail	(*Coturnix japonica*)	Boldness
Uitdehaag et al.	2008	Jungle fowl	(*Gallus gallus*)	Boldness
Faure	1980	Jungle fowl	(*Gallus gallus*)	Exploration
		Passeriformes		
Guillette et al.	2009	Black-capped chickadee	(*Poecile atricapillus*)	Exploration
Arnold et al.	2007	Blue tit	(*Cyanistes caeruleus*)	Boldness
Burtt and Giltz	1973	Brown-headed cowbird	(*Molothrus ater*)	Aggressiveness
		Common grackle Red-winged blackbird Starling	(*Quiscalus quiscala*) (*Agelaius phoeniceus*) (*Sturnus vulgaris*)	Complacency
Harvey and Freeberg	2007	Carolina chickadees	(*Poecile carolinensis*)	Aggressiveness Sociability
Quinn and Cresswell	2005	Chaffinch	(*Fringilla coelebs*)	Risk-taking
Garamszegi et al.	2009	Collared flycatcher	(*Ficedula albicollis*)	Boldness Aggressiveness Risk-taking
Stöwe and Kotrschal	2007	Common raven	(*Corvus corax*)	Boldness
Mettke-Hofmann et al.	2005	Garden warbler	(*Sylvia borin*)	Exploration
		Sardinian warbler	(*Sylvia melanocephala*)	Boldness
Verbeek et al.	1994	Great tit	(*Parus major*)	Exploration Boldness
Verbeek et al.	1996	Great tit	(*Parus major*)	Aggressiveness
Van Oers et al.	2004b	Great tit	(*Parus major*)	Risk-taking
Hollander et al.	2008	Great tit	(*Parus major*)	Exploration
Quinn et al.	2009	Great tit	(*Parus major*)	Exploration
Fucikova et al.	2009	Great tit	(*Parus major*)	Reaction to handling
Fox et al.	2009	Mountain chickadee	(*Poecile gambeli*)	Exploration Boldness
Marchetti and Zehtindjiev	2009	Sedge warbler	(*Acrocephalus schoenobaenus*)	Exploration
Minderman et al.	2009	Starling	(*Sturnus vulgaris*)	Exploration
Duckworth	2006a	Western bluebird	(*Sialia mexicana*)	Aggressiveness

(continued)

Table 3.1. (*continued*)

Author	Year	Species		Personality traits
Beauchamp	2000	Zebra finch	(*Taenopygia guttata*)	Exploration
Martins et al.	2007	Zebra finch	(*Taenopygia guttata*)	Exploration Boldness
Schuett and Dall	2009	Zebra finch	(*Taenopygia guttata*)	Exploration
		Psittaciformes		
Fox and Millam	2004	Orange-winged Amazon	(*Amazona amazonica*)	Neophobia
Mettke-Hofmann et al.	2004	10 species		Boldness
		Sphenisciformes		
Ellenberg et al.	2009	Yellow-eyed penguin	(*Megadyptes antipodes*)	Reaction to handling
		Struthioniformes		
De Azevedo and Young	2006	Greater Rhea	(*Rhea americana*)	Boldness

Agonistic behavior in birds is linked to the ability to obtain and maintain a breeding territory (Stamps and Krishnan 1997; Naguib 2005). In studies on agonistic behavior in relation to territorial competitiveness (Duckworth 2006a) and parental behavior (Duckworth 2006b) in Western bluebirds (*Sialia mexicana*) it was shown, for instance, that agonistic behavior is repeatable and costly (Duckworth 2006b). Moreover, Duckworth (2006a) showed experimentally that more aggressive males compete more effectively for territories in areas with a higher density of nest boxes. As a consequence, aggressive and nonaggressive males occurred in breeding habitats that differed in the strength of selection on morphological traits. These results show that aggression can affect selection on a local scale by determining individual settlement patterns, thereby also providing opportunity for correlational selection.

Tests of agonistic behavior are, by definition, conducted in a social context, and the consistency in behavior could also be dependent on the context (Van Oers et al. 2005b). To test whether males of black-capped chickadees (*Poecile atricapillus*) behaved consistently with different mates, they were paired to a female and several aspects of their behavior were observed (Harvey and Freeberg 2007). After being paired with a new mate, the males showed behavior that was consistent with the behavior previously shown with their former mate, indicating that agonistic behavior is consistent also when the social context has been changed (Harvey and Freeberg 2007).

A crucial issue in assessing personality traits is the identification of contexts that are most suitable for revealing consistent differences in behavior among individuals. Exploration tests in artificial environments, such as a novel room or cage, and novel object tests, provide standard situations. The artificial nature of such a context has the advantage that behavior can be tested in the absence of a specific resource value, which may confound any measurement of intrinsic behavioral traits. Resources vary in their value to different individuals so that in tests conducted in natural situations it is difficult to separate individual differences in behavior from differences in resource values. Individuals may behave differently not as a result of personality but as a result of differences in resource values. Conducting tests in the natural habitat could potentially reveal a more meaningful variation in behavior compared with the standard tests done in captivity, and therefore yield results with high ecological and evolutionary relevance. Yet, when testing birds for personality traits during resource defense in the field, one needs to be careful in developing tests so that the recorded behavior is not primarily related to the resource value rather than to their personality traits. Demonstrating consistencies in behavior in contexts that are independent of resources crucial for reproduction, thus remains a powerful tool in research aimed at unraveling the evolutionary significance of personality traits.

Causes of personality variation: genes and physiology

GENETICS

The majority of our knowledge of the genetic background of traits such as exploratory behavior, aggressiveness, and fearfulness comes from studies on domestic birds. Selection experiments on chicken, turkey, ducks, and Japanese quail have shown that the observed variation in open field, pair-wise, and novel object tests has a substantial genetic basis (e.g., Brown and Nestor 1973; Francois et al. 1999; Jones and Hocking 1999; Arnaud et al. 2008). Although the presence of a genetic component is an important factor to consider in studies on evolutionary change of personality traits, genetic studies of animal personality traits in natural populations are scarce (Bell and Aubin-Horth 2010). Most quantitative genetic and molecular analyses of bird personality traits have been conducted with the great tit (quantitative: Van Oers et al. 2005a; Quinn et al. 2009; molecular: Fidler et al. 2007; Van Bers et al. 2010; Korsten et al. 2010). In this species it is apparent that personality traits typically have moderate levels of additive genetic variation (Dingemanse et al. 2002; Drent et al. 2003; Van Oers et al. 2004b; Quinn et al. 2009) and of genetic dominance (Van Oers et al. 2004c), and that traits are genetically correlated (Van Oers et al. 2004a). An important gene has been identified for great tit personality: the dopamine receptor D4 gene

(Fidler et al. 2007) explains between 5% and 8% of the phenotypic variation in some but not all European great tit populations (Korsten et al. 2010). This and other genes are thereby likely not only to affect behavioral responses early in life (i.e., temperament traits), but also to affect learning and plasticity later in life.

One difficult question is to assess the extent to which the underlying genetic and physiological mechanisms are a constraint on the evolution of behavioral traits vs. the extent to which they are the product of variation and selection (Bell 2005; Dingemanse et al. 2007). An important issue here is how the word *constraint* is used. In a biological context, the word can refer to a factor that impedes but does not necessarily prevent evolution in particular directions, or it can indicate that specific evolutionary trajectories are unavailable to selection (Roff and Fairbairn 2007). Therefore, determining the role of genetic constraints in the evolutionary change of personality will require not only more detailed ecological studies, but also better knowledge than we currently possess of the relevant genes and their importance in natural populations (Blows and Hoffmann 2005; Van Oers et al. 2005a). A more comprehensive overview of the current knowledge of the genetic basis of personality variation in captive and wild animals is provided in chapter 6.

HORMONAL CORRELATES

Many vertebrate species are characterized by the flexibility with which they cope with stressors, and birds provide good examples of this. When birds are under conditions of low environmental predictability, they experience stress (Wiepkema 1992). Behavioral and physiological efforts to master the situation (coping strategies) can be important determinants of health and disease both in humans and in animals (Koolhaas et al. 1999). Moreover, individuals often show different physiological responses in these situations, also referred to as coping styles. Individuals appear to cope with stressors in a predominantly sympathetic or parasympathetic way. Individual variation in agonistic behavior can be considered as an example of more general variation in coping with environmental challenges, where highly aggressive individuals adopt a proactive coping style and submissive individuals adopt a more passive or reactive style (Koolhaas et al. 2007). Therefore, most physiological research relevant to avian personality involves stress physiology and coping (see chapter 12). In mammals, it has been suggested that individual differences in stress responses reflect variation in personality or coping styles (Korte et al. 2005). Passive copers are expected to have higher hypothalamic-pituitary-adrenal (HPA) responses to stressors, a lower sympathetic adrenomedullary reactivity, a higher humoral immunity, but also a higher vulnerability to stress-induced illnesses compared with active copers (Ellis et al. 2006).

In birds, fast exploration, high risk-taking, and high responsiveness to a novel object appear to be associated with high corticosterone levels during the test (Martins et al. 2007; Richard et al. 2008). Birds also show individual consistency in the secretion of stress hormones. Kralj-Fisher and coworkers (2007) have shown in greylag geese that corticosterone levels measured in fecal samples collected after repeated handling episodes were consistent within individuals. This repeatability was especially pronounced when individuals were feeding in low competition areas; in high competition areas individuals were less consistent. Whether these individual differences in stress responses reflect genetic personality differences remains unclear. Yet, some support for this possibility has been found in other avian species where selection experiments have shown that the stress response has an additive genetic background (Brown and Nestor 1973; Edens and Siegel 1975; Satterlee and Johnson 1988). Moreover, great tit genetic lines selected for different personalities (Drent et al. 2003) differed in the level of stress hormone secretion after being in a social contest with a conspecific (Carere et al. 2003). Birds of the less aggressive and more cautious line (slow explorers) showed a trend for a higher response compared with birds of the more aggressive and bolder line (fast explorers), which could be taken to suggest a physiological basis of different coping strategies in birds (Carere et al. 2003). Alternatively, personality types might differ in the perception of the stressor, causing indirect differences in stress response to a standard stressor (Sapolsky 1994). Most likely both mechanisms play a role. However, the extent to which one or the other is more important in shaping variation in personality needs to be studied in more detail. Studies in great tits have shown that exploratory behavior is related to stress responses in both juvenile (Fucikova et al. 2009) and adult birds (Carere and Van Oers 2004). Adult birds that explore a novel environment more quickly (fast explorers) show a lower stress response after being handled, but nestlings that become fast explorers in adulthood show an increased response after being socially isolated as nestlings. This indicates that differences in personality traits measured just after independence might act as a predictor of individual variation in the stress response in adulthood, and as a result of variation in responsiveness as juveniles. A more direct way of measuring the link between variation in personality and stress is through stress hormones. In birds, corticosterone is the major glucocorticosteroid hormone secreted in response to stress, and many studies have shown that individual birds differ in their hormonal response to stress. This response varies not only with intrinsic factors such as sex and age, but also in relation to behavioral types, and it is modified by factors such as body condition (Schwabl 1995; Silverin 1998; Cockrem and Silverin 2002; Cockrem 2007). These modifications may allow the adjustment of physiological and behavioral responses to adverse environmental circumstances and help explain individual differences in responses.

Glucocorticosteroid responses are seen as an evolutionary mechanism that maintains physiological homeostasis within an adaptive range. Only recently, however, have studies raised the question of how to explain the consistent individual differences in these responses that are unrelated to age or size (Cockrem 2007). It will be interesting to develop this and determine whether individual variation in short-term elevation of circulating glucocorticosteroids also explains fitness differences, and whether fitness consequences vary for different behavioral types. A first step in that direction was made by the work of Blas et al. (2007) on European white storks (*Ciconia ciconia*). This study showed that the magnitude of the adrenocortical response to a standardized perturbation during development is negatively correlated to survival and recruitment (Blas et al. 2007). The next step would be to investigate whether these fitness consequences differ for different behavioral types. Blas et al. (2007) discuss this possibility, by arguing that the success of proactive vs. reactive types varies as a function of population density and predictability of food resources. If this is associated with differences in glucocorticosteroid responses, it could provide an explanation for varying success of the behavioral types. Direct measurements and experimental changes, however, are needed to test this hypothesis. More on this subject can be found in chapter 12, on neuroendocrine and autonomic correlates of personalities.

Developmental aspects of personality
PLASTICITY

Although it has been shown that the genetic and physiological structure of personality is relatively rigid (Koolhaas et al. 1999; Van Oers et al. 2005a), personality nevertheless has substantial phenotypic plasticity and can be affected by environmental factors, specifically when they act during early development (Carere et al. 2005b; Arnold et al. 2007; Krause et al. 2009; Stamps and Groothuis 2010a; 2010b; Naguib et al. 2011). Such environmental factors include changes in the context in which a trait is expressed, for example, and the social factors that will change the adaptive value of a certain behavioral response (e.g., Schuett and Dall 2009). More subtle factors include maternal effects acting during embryonic development, such as the amount of hormones transferred to the egg by the female (Schwabl 1993; Gil 2008; chapter 11), circulating hormones that influence habituation and sensitization to stressful events, and learning how to cope with certain stressors (Kant et al. 1985; Stam et al. 2000). Avian personalities are therefore open to change, though within certain boundaries, and traits such as shyness or boldness are therefore highly consistent within an individual. Individuals that are classified as being one or the other personality type, however, may still fluctuate in their behavior (Carere et al. 2005a), and repeatability of behavioral traits is typically moderate (Bell et al. 2009). Phenotypic correlations between personality traits may be strong in some environments but weak in others. For example, in great tits, both males and females of genetic lines selected for fast exploratory behavior return more quickly to the feeding table after a startle (risk-taking behavior). Lines selected for risk-taking behavior also differ in their exploratory tendency, such that a positive genetic correlation of 0.84 exists between the two traits (Van Oers et al. 2004a). In a follow-up study investigating the context generality of this correlation, fast- or slow-exploring great tits were tested for their risk-taking behavior in a nonsocial context followed by a social context (Van Oers et al. 2005b). Van Oers and coworkers found that the relation between exploratory behavior and risk-taking behavior depended on the social context. Females in general returned later in the social test, while the reaction of males to the presence of a companion was dependent on their behavioral type. Slow males came back sooner with faster companions and fast males did not react to the companion (Van Oers et al. 2005b). Similar results were found in a study on zebra finches (Schuett and Dall 2009): males and females differed in how consistently they behaved across social and nonsocial contexts. Schuett and Dall (2009) also tested whether males and females differed in their influence as companions, and found that individuals of both sexes influenced each other's exploratory behavior in a similar way within the social context: the more exploratory the companion, the more exploratory the focal individual. In great tits, birds of different personality type differ in their foraging strategy (Marchetti and Drent 2000). Moreover, the presence and strategy of a conspecific can affect the foraging strategies of individuals differently, depending on their personality. As a result, birds of different exploration types differ in their tendency to copy a tutor's foraging decision (Marchetti and Drent 2000; see below).

MATERNAL HORMONES

Since birds lay eggs, embryos are separated from the mother before they start to develop. Females are, however, able to influence embryonic development by transferring nutrients and hormones to the yolk and albumen of their eggs (Gil 2008). Although it is not completely clear how females can actively vary the amounts of nutrients and hormones, these substances can exert short- and long-term effects on the offspring (Schwabl 1993; Eising et al. 2001; Groothuis et al. 2005; Eising et al. 2006; chapter 11). It is possible that maternal hormones in the egg may influence personality variation by modulating genetic differences or physiologically programming the offspring in certain ways. These maternal effects, however, may have a genetic basis, as individual differences in the deposition of maternal hormones in eggs have been shown to be, in part, genetically determined. Genetic variation in yolk hormones has been found in lines of domesticated species (Daisley et al. 2005; Gil and Faure 2007). In lines selected for fast and slow exploratory behavior in great tits, Groothuis et al. (2008) found that females of the slow line deposited lower amounts of testosterone in the yolk of their eggs compared with females from the fast line. This was especially true for the eggs laid early in the laying sequence (see chapter 11). This could indicate that females of different personality types have different strategies in rearing offspring. In this view, fast females would aim at rearing all offspring, whereas slow females invest less in those offspring that have lower chance of surviving (Tobler and Sandell 2007). Not only can maternal hormones influence behavioral differences, they can also affect the persistence of certain behaviors. In an experiment, the behavior of birds that came from testosterone-injected (T-treated) and control eggs were compared for their behavior toward a novel food source (Tobler and Sandell 2007). Birds from T-treated eggs did not differ in their latencies to approach and eat novel food during their first encounter. However, testosterone treatment affected subsequent encounters with the novel food source. Owing to habituation, latencies decreased in both groups over a period of 5 days, but considerably more so in T-offspring (Tobler and Sandell 2007). Whether this is different for offspring hatched from eggs at the beginning or end of the laying sequence is still unknown.

EARLY DEVELOPMENT

Environmental factors acting during ontogeny have been shown to significantly affect a whole range of morphological (Tinbergen and Boerlijst 1990), physiological (Kitaysky et al. 2001; Naguib et al. 2004; Verhulst et al. 2006), behavioral (DeKogel and Prijs 1996; Nowicki et al. 1998; Naguib and Nemitz 2007; Krause et al. 2009), and life history traits (DeKogel 1997; Naguib et al. 2006) and therefore may also play a role in personality development. It is well established that conditions experienced during early development are important in shaping the behavior of an animal against the background of the reaction norm (Mason 1979; Metcalfe and Monaghan 2001). Such factors can involve a wide range of stressors, such as low-quality food (Krause et al. 2009), brood size (Neuenschwander et al. 2003), and parasites (Tschirren and Richner 2006). Yet, relatively little is known about how conditions experienced during early development affect personality (Stamps and Groothuis 2010a; 2010b). One trait, used as a proxy for personality, boldness in response to novel objects, has been studied by Fox and Millam (2004), who investigated the effects of rearing condition on the reaction to novel objects in orange-winged Amazon parrots (*Amazona amazonica*). Hand-reared, parent-reared, and parent-reared/human-handled birds, which were handled five times a week for 20 minutes, were tested for their latency to feed in the presence of five different novel objects between 4.5 and 6 months after hatching. At an age of 12 months the response to a novel object in their home cage was measured. They found that handling birds did not influence the reaction to novelty, but whether an individual was reared by their own parents or hand-reared did. They concluded that the development of neophobia in orange-winged Amazon parrots may be related to novelty the chicks experience during early life (Fox and Millam 2004). Moreover, we recently showed in great tits that personality is affected by the sex ratio in the nest. Birds that grew up in female-biased nests became faster explorers than birds that grew up in male-biased nests. These findings that early social interactions shape personality in animals but future research is required to unravel the precise behavioral mechanism leading to this effect.

Another important influence on behavior during early development is the quantity, quality, or composition of food (e.g., see Birkhead et al. 1999). Variation in food characteristics can be caused by parental choice, sibling competition, or habitat quality but also by differences in the parents' ability to search and find suitable food. Experiments on birds investigating the influence of food during ontogeny can be conducted relatively easily in altricial species compared with precocial species. Carere et al. (2005b), for example, manipulated the early rearing condition in two great tit lines selected for fast and slow exploratory behavior by a food rationing protocol. Birds from both fast and slow exploration lines, but also control chicks, decreased their growth rate and increased their begging behavior compared with unmanipulated chicks within the same nests. This resulted in slow chicks becoming much faster than their parents, but without any changes in aggressiveness. In contrast, fast chicks had exploration scores similar to their parents but an increased level of aggressiveness. As a consequence, there was no apparent line difference in exploration behavior at independence. The effect, however, turned out to be partly temporary: although the offspring of the slow line were still relatively fast six months later, birds of the fast line became even faster, restoring the line differences again. A side effect of the treatment on the experimental birds was that control chicks also begged more. To rule out these effects of sibling competition, the authors conducted a second experiment with experimental and control siblings in separate nests. Here, only the food-rationed chicks became faster in exploration, indicating that the shift in the controls in the within-nests design was indeed due to enhanced sibling competition. Krause et al. (2009) also showed that the feeding conditions experienced during early development affect exploration behavior in zebra finches. They showed that female zebra finches raised as nestlings under low-quality food conditions were more explorative in a novel environment than females that had been raised under high-quality food conditions. The same individuals were also more sensitive to short periods of food deprivation by losing more weight than those raised under high-quality food, underlining the link between behavioral differences in personality with physiological differences in responses to a stressor.

Aside from the amount of food, its quality may also be very important for the development of behavioral consistency. Essential amino-acids are known to be limiting factors during development (Murphy and Pearcy 1993) and are also relatively scarce in the bulk food of many passerines (Izhaki 1998; Ramsay and Houston 2003). Several tit species therefore supply their nestlings with a high proportion of spiders early during development (Ramsay and Houston 2003). Spiders contain a relatively high amount of taurine. To investigate whether taurine has a developmental effect on personality variation in blue tits (*Cyanistes caeruleus*), Arnold and coworkers (2007) conducted a feeding experiment in which they supplemented taurine to blue tit nestlings during the period of maximum growth. Juveniles that had received additional taurine as nestlings were significantly bolder when investigating novel objects, and also were more successful at a spatial learning task than controls. They concluded that prey selection is a mechanism by which parents can alter the behavioral phenotype of their offspring. Further experiments in natural settings should be conducted to see whether parents indeed use this mechanism to prepare their offspring for the future. Taken together, these findings suggest that personality traits may well be shaped by the conditions experienced during early development, a topic that is worth exploring more in the future.

A question arises as to the extent to which individuals differing in personality also differ in learning abilities or learning strategies. Individuals that differ in environmental exploration are likely to also differ in the way they acquire, process, recall, and use environmental information (Guillette et al. 2010; Cole and Quinn 2012; Amy et al. 2012; Titulaer et al. 2012). Animals have to learn to find food, possibly from successful conspecifics, and to remember food locations. Marchetti and Drent (2000) found that, in great tits, slow and fast explorers differ in their routines to revisit known feeding sites and in using information they obtain by observing conspecifics when foraging. Birds were first trained to find food at a specified feeder, which during the experiment was then left without food. When tested alone, fast explorers kept on visiting the now unrewarded feeder they had previously learned to visit. Slow explorers, in contrast, were quicker in finding new food locations and did not revisit the previously rewarded feeder as often. In other words, fast birds were less flexible in finding food locations and expressed more behavioral routines than did slow explorers. Interestingly, when subjects had the opportunity to observe a tutor bird to explore a specific food source (colored feeder), then slow and fast explorers behaved in opposite ways. Fast explorers were faster in copying the tutor's behavior (and in feeding from the feeder indicated by the tutor) whereas slow explorers did not learn to explore the rewarded feeder from the tutor. In other words, slow explorers were better in finding new food sources on their own, whereas fast explorers did better in exploring new food sources when given the opportunity to learn from others. Such differences in foraging strategies may explain in part why slow and fast explorers perform differently under natural conditions depending on food availability. Alternatively, this could be mediated by differential susceptibility to stress, where differential stress levels may alter the way information is processed. In a recent experiment, Titulaer et al. (2012) further showed that personality affected learning only in difficult tasks, but in a sex-specific way. Such effects presumably are related to selection acting differently on the sexes with respect to behavioral strategies in terms of space use, foraging, and social behavior. Such relations between cognition and personality have also been addressed in a number of other recent studies (Guillette et al. 2010; Cole and Quinn 2012; Amy et al. 2012).

Animal husbandry and welfare

Obviously, research on personality is not restricted to natural contexts. Personality has been a key issue in studies on animal welfare and husbandry (Wemelsfelder et al. 2000; Bolhuis et al. 2005; Würbel 2009; see chapter 14) and is of emerging relevance in the development of nature conservation programs that deal, for instance, with habitat defragmentation and reintroduction of animals to new areas (see chapter 13). Although researchers with more applied interests have often studied behavioral consistency, their concepts and terminology are often different from those used by behavioral ecologists. Studies investigating personality in quail and chicken, for example, often use the terms fear and fearfulness (Boissy 1995), where fearfulness is measured as the emergence into a larger compartment, the reaction to novel food or objects, or the response to a predator (Miller et al. 2005; 2006).

One of the central problems in applied bird ethology is feather pecking, a common behavior in laying hens, with substantial economic and welfare implications. Feather pecking has a genetic component but is also affected by various social and housing factors (Van Krimpen et al. 2005; Van de Weerd and Elson 2006; Rodenburg et al. 2008). Research has shown that individuals are consistent in this behavior, reflecting a potential personality trait. Understanding the causes of feather pecking is thus of high applied significance as it will contribute to develop rearing strategies and selection processes that will reduce this problem. With the new regulations of housing laying hens in larger groups, feather pecking needs to be monitored carefully, and identifying behavioral and genetic correlates of this behavior will help to find optimal solutions that balance welfare and economic interests.

Personality differences also might be of great importance in reintroduction of individuals reared in captivity into wild populations or in transferring wild individuals to new locations. Before reintroduction, individuals of many species, for example, have to learn to avoid predators. Training individuals in these capacities is therefore a crucial factor for increasing the probability of postrelease survival (Box 1991). The ability to learn to avoid certain dangers might, however, be dependent on the personality type of an individual. That this can be the case is nicely shown in a study on greater rheas (*Rhea americana*) (De Azevedo and Young 2006). Captive individuals were tested for their response to several novel objects, before and after being trained to avoid predators. Birds were less bold after training compared with before training, and the responses to the novel object before the training sessions were a good predictor of how the bird would react during training. Bolder birds behaved more calmly than shy birds. Similar results were obtained studying the natural antipredator behavior of chaffinches (*Fringilla coelebs*): calm individuals were better able to assess the risk of a hawk flying over compared with very active individuals. They showed greater behavioral plasticity in high-risk versus low-risk situations, while hyperactive individuals in general showed more fleeing behavior (Quinn and Cresswell 2005), possibly because of differences in the risk perception between the activity types (Butlers et al. 2006). As dispersal, territorial, and foraging strategies can be linked to personality (Dingemanse et al. 2003; Amy et al. 2010; see below; chapter 7), including information about personality in decisions made about when to release individuals may affect the success of a reintroduction project (Bremner-Harrison et al. 2004; Merton 2006).

Field studies and fitness correlates

The growing interest in research on animal personality in part has been driven by the ecological and evolutionary significance of personality— that is, that personality matters in the dynamics of wild populations. Even though the origin and nature of individual variation in itself is interesting, the ecological and evolutionary framework adds another reason for interest, as it shows that personality affects selection while at the same time being under selection itself. Ecologically relevant personality correlates include nest defense behavior (Hollander et al. 2008), song (Garamszegi et al. 2008; Amy et al. 2010; Naguib et al. 2010), social dominance (Dingemanse and De Goede 2004), feeding behavior (Costantini et al. 2005), as well as dispersal and mating behavior (Dingemanse et al. 2003; Van Oers et al. 2008).

Dingemanse et al. (2003) showed, for instance, that natal dispersal distance (i.e., the distance between place of origin and the place of breeding) was linked to personality in great tits. In their study population of great tits with known personalities (assessed under standard laboratory conditions using a novel environment test), they found that the personality of the father mattered. Offspring from males that were fast explorers in the novel environment test dispersed farther from their original nest box than did slow explorers. Furthermore, immigrants into the population had higher exploration scores than resident birds. Such effects presumably resulted from genetic differences and environmental effects acting during development. Indeed, as personality has a considerable heritable component (Van Oers et al. 2005a), personality-related differences in dispersal strategies may affect the genetic variance of a population, depending on the extent to which residents and immigrants succeed in breeding.

Dingemanse et al. (2004) found in great tits that personalities had differential fitness effects, that these effects were different for males and females, and that they were reversed in different years depending on food availability. Studying local survival across three years, they found that slow males did better in the two years with limited food availability in winter than they did in the year with high winter food availability. Females over the same period were affected in the opposite way. These results suggest that fast males, being more aggressive and dominant over females (Krebs and Perrins 1978; Drent 1987; Verbeek et al. 1996), do better in poor years than females, as the limiting resource is the food over which they compete. In good years, the higher winter survival results in higher competition among males for territories so that slower and less aggressive males do relatively less well than females. Indeed, Amy et al (2010) showed that territorial behavior is affected by personality, with faster males being more aggressive at the site of intrusions while slower males follow a different strategy by singing more from the distance and exploring more the boundaries of their territory. Dingemanse et al. (2004) also found that in the poor year, pairs of individuals with similar personality produced more local recruits than pairs in which the male and female had different personalities; these effects, however, may have been caused by differential survival or differential dispersal (Dingemanse et al. 2003). Along this line, Both et al. (2005) reported that pairs with similar personality have a higher reproductive success, measured as offspring condition. These findings may explain the production of more recruits by pairs with similar extreme personality, as shown by a different study for the same population (Dingemanse et al. 2004), as fledgling condition has been shown to predict survival (Tinbergen and Boerlijst 1990). While the nature of one's own and the partner's personality appears to have fitness consequences measured as offspring condition, recruits, and survival, personality also can also affect more immediate reproductive decisions. Monogamous animals are known to produce extra-pair offspring in addition to the offspring produced with their social partner (Petrie and Kempenaers 1998; Griffith et al. 2002). Decisions about extra-pair matings most commonly have been linked to the expression of sexually selected traits of the social mate and the extra-pair mate (Kempenaers et al. 1992; Hasselquist et al. 1996). Apparently, decisions about extra-pair matings also depend on within-pair personality differences in great tits, as pairs with more extreme similar personality (slow-slow and fast-fast pairs) are more likely to raise extra-pair offspring (Van Oers et al. 2008). In other words, females are more likely to engage in extra-pair copulations when their social mate is similar in personality compared with her own personality. Such disassortative decision making in reproduction may ensure high offspring variability and also contributes to maintaining high variation in personality within a population, in a similar way as has been argued for links between mate preferences for partners with different immune characteristics (Wedekind et al. 1995; Milinski 2006).

Conclusions

Birds are excellent model organisms for studying the causes and consequences of personality in descriptive and experimental ways in the field and in the laboratory. Integrating information obtained from studies under controlled experimental lab conditions with experimental studies in the field and analyses of fitness-relevant traits under natural conditions has generated multifaceted insights into principles of personality. So far, much of the field research addressing evolutionary and ecological questions has been conducted on great tits, which remain the species in which the study of causation and consequences of personality under natural conditions has been best integrated. With research on personality now gradually expanding to other avian species, the opportunity to obtain comparative data sets allowing us to link basic species differences in life history to our current understanding on personality is becoming available.

References

Amy, M., Sprau, P., de Goede, P., and Naguib, M. 2010. Effects of personality on territory defence in communication networks: a playback experiment with radio-tagged great tits. *Proceedings of the Royal Society of London B*, 227, 3685–3692.

Amy, M., Van Oers, K., and Naguib, M. 2012. Worms under cover: relationships between peformance in learning tasks and personality in great tits (*Parus major*). *Animal Cognition*, 15, 763–770.

Andersson, M. 1994. *Sexual Selection*. Princeton, NJ: Princeton University Press.

Armstrong, A. E. 1947. *Bird Display and Behaviour*. London: Lindsay Drummond.

Arnaud, I., Mignon-Grasteau, S., Larzul, C., Guy, G., Faure, J. M., and Guemene, D. 2008. Behavioural and physiological fear responses in ducks: genetic cross effects. *Animal*, 2, 1518–1525.

Arnold, K. E., Ramsay, S. L., Donaldson, C., and Adam, A. 2007. Parental prey selection affects risk-taking behaviour and spatial learning in avian offspring. *Proceedings of the Royal Society of London B*, 274, 2563–2569.

Baechle, J. W. 1947. Bird banding on a college campus. *American Biology Teacher*, 9, 208–213.

Bakker, T. C. M. 1994. Genetic correlations and the control of behavior, exemplified by aggressiveness in sticklebacks. *Advances in the Study of Behavior*, 23, 135–171.

Baugh, A. T., Schaper, S., Hau, M., Cockrem, J. F., de Goede, P., and Van Oers, K. 2012. Corticosterone responses differ between lines of great tits (*Parus major*) selected for divergent personalities. *General and Comparative Endocrinology*, 175, 488–494.

Beauchamp, G. 2000. Individual differences in activity and exploration influence leadership in pairs of foraging zebra finches. *Behaviour*, 137, 301–314.

Bell, A. M. 2005. Behavioural differences between individuals and two populations of stickleback (*Gasterosteus aculeatus*). *Journal of Evolutionary Biology*, 18, 464–473.

Bell, A. M., and Aubin-Horth, N. 2010. What can whole genome expression data tell us about the ecology and evolution of personality? *Philosophical Transactions of the Royal Society of London B*, 365, 4001–4012.

Bell, A. M., Hankison, S. J., and Laskowski, K. L. 2009. The repeatability of behaviour: a meta-analysis. *Animal Behaviour*, 77, 771–783.

Benus, R. F. 1988. Aggression and coping: differences in behavioural strategies between aggressive and nonaggressive male mice. PhD diss., University of Groningen, the Netherlands.

Benus, R. F., Koolhaas, J. M., and Van Oortmerssen, G. A. 1987. Individual differences in behavioural reaction to a changing environment in mice and rats. *Behaviour*, 100, 105–122.

Berthold, P., and Querner, U. 1981. Genetic basis of migratory behavior in European warblers. *Science*, 212, 77–79.

Birkhead, T. R., Fletcher, F., and Pellatt, E. J. 1999. Nestling diet, secondary sexual traits, and fitness in the zebra finch. *Proceedings of the Royal Society of London B*, 266, 385–390.

Blas, J., Bortolotti, G. R., Tella, J. L., Baos, R., and Marchant, T. A. 2007. Stress response during development predicts fitness in a wild, long-lived vertebrate. *Proceedings of the National Academy of Sciences U S A*, 104, 8880–8884.

Blows, M. W., and Hoffmann, A. A. 2005. A reassessment of genetic limits to evolutionary change. *Ecology*, 86, 1371–1384.

Boissy, A. 1995. Fear and fearfulness in animals. *Quarterly Review of Biology*, 70, 165–191.

Bolhuis, J. E., Schouten, W. G. P., Schrama, J. W., and Wiegant, V. M. 2005. Behavioural development of pigs with different coping characteristics in barren and substrate-enriched housing conditions. *Applied Animal Behaviour Science*, 93, 213–228.

Both, C., Dingemanse, N. J., Drent, P. J., and Tinbergen, J. M. 2005. Pairs of extreme avian personalities have highest reproductive success. *Behavioral Ecology*, 74, 667–674.

Box, H. O. 1991. Training for life after release: simian primates as examples. *Symposia of the Zoological Society of London*, 62, 111–123.

Bremner-Harrison, S., Prodohl, P. A., and Elwood, R. W. 2004. Behavioural trait assessment as a release criterion: boldness predicts early death in a reintroduction programme of captive-bred swift fox (*Vulpes velox*). *Animal Conservation*, 7, 313–320.

Brown, K. I., and Nestor, K. E. 1973. Some physiological responses of turkeys selected for high and low adrenal response to cold stress. *Poultry Science*, 52, 1948–1954.

Burtt, H. E. 1967. *The Psychology of Birds*. New York: Macmillan.

Burtt, H. E., and Giltz, M. 1969a. A statistical analysis of blackbird aggressiveness. *Ohio Journal of Science*, 69, 58–62.

———. 1969b. Measurement of complacency in blackbirds. *Ohio Journal of Science*, 69, 109–114.

———. 1973. Personality as a variable in the behaviour of birds. *Ohio Journal of Science*, 73, 65–82.

Butlers, S. J., Whittingham, M. J., Quinn, J. L., and Cresswell, W. 2006. Time in captivity, individual differences, and foraging behaviour in wild-caught chaffinches. *Behaviour*, 143, 535–548.

Carere, C., Drent, P. J., Koolhaas, J. M., and Groothuis, T. G. G. 2005a. Personalities in great tits (*Parus major*): stability and consistency. *Animal Behaviour*, 70, 795–805.

———. 2005b. Epigenetic effects on personality traits: early food provisioning and sibling competition. *Behaviour*, 142, 1329–1355.

Carere, C., Groothuis, T. G. G., Mostl, E., Daan, S., and Koolhaas, J. M. 2003. Fecal corticosteroids in a territorial bird selected for different personalities: daily rhythm and the response to social stress. *Hormones and Behavior*, 43, 540–548.

Carere, C., and Van Oers, K. 2004. Shy and bold great tits (*Parus major*): body temperature and breath rate in response to handling stress. *Physiology and Behavior*, 82, 905–912.

Catchpole, C. K., and Slater, P. J. B. 2008. *Bird Song: Biological Themes and Variation*. New York: Cambridge University Press.

Cockrem, J. F. 2007. Stress, corticosterone responses, and avian personalities. *Journal of Ornithology*, 148, S169-S178.

Cockrem, J. F., and Silverin, B. 2002. Variation within and between birds in corticosterone responses of great tits (*Parus major*). *General and Comparative Endocrinology*, 125, 197–206.

Cole, E. F., and Quinn, J. L. 2012. Personality and problem-solving performance explain competitive ability in the wild. *Proceedings of the Royal Society of London B*, 279, 1168–1175.

Coppens, C. M., de Boer, S. F., and Koolhaas, J. M. 2010. Coping styles and behavioural flexibility: towards underlying mechanisms. *Philosophical Transactions of the Royal Society of London B*, 365, 4021–4028.

Costantini, D., Casagrande, S., Di Lieto, G., Fanfani, A., and Dell'Omo, G. 2005. Consistent differences in feeding habits between neighbouring breeding kestrels. *Behaviour*, 142, 1403–1415.

Daisley, J. N., Bromundt, V., Mostl, E., and Kotrschal, K. 2005. Enhanced yolk testosterone influences behavioral phenotype independent of sex in Japanese quail chicks *Coturnix japonica*. *Hormones and Behavior*, 47, 185–194.

Danchin, E., Giraldeau, L. A., and Cézilly, F. 2008. *Behavioural Ecology*. New York: Oxford University Press.

Darwin, C. 1872. *The Expression of the Emotions in Man and Animals*. London: John Murray.

Davies, N. B., Krebs, J. R., and West, S. A. 2012. *An Introduction to Behavioural Ecology*. 4th ed. Oxford: Wiley-Blackwell.

Dawkins, M. S. 1999. The role of behaviour in the assessment of poultry welfare. *World's Poultry Science Journal*, 55, 295–303.

De Azevedo, C. S., and Young, R. J. 2006. Shyness and boldness in greater rheas *Rhea americana* Linnaeus (Rheiformes, Rheidae): the effects of antipredator training on the personality of the birds. *Rivista Brasileira de Zoologia*, 23, 202–210.

DeKogel, C. H. 1997. Long-term effects of brood size manipulation on morphological development and sex-specific mortality of offspring. *Journal of Animal Ecology*, 66, 167–178.

DeKogel, C. H., and Prijs, H. J. 1996. Effects of brood size manipulations on sexual attractiveness of offspring in the zebra finch. *Animal Behaviour*, 51, 699–708.

Dingemanse, N. J., Both, C., Drent, P. J., and Tinbergen, J. M. 2004. Fitness consequences of avian personalities in a fluctuating environment. *Proceedings of the Royal Society of London B*, 271, 847–852.

Dingemanse, N. J., Both, C., Drent, P. J., Van Oers, K., and Van Noordwijk, A. J. 2002. Repeatability and heritability of exploratory behaviour in wild great tits. *Animal Behaviour*, 64, 929–937.

Dingemanse, N. J., Both, C., Van Noordwijk, A. J., Rutten, A. L., and Drent, P. J. 2003. Natal dispersal and personalities in great tits (*Parus major*). *Proceedings of the Royal Society of London B*, 270, 741–747.

Dingemanse, N. J., and De Goede, P. 2004. The relation between dominance and exploratory behaviour is context-dependent in wild great tits. *Behavioral Ecology*, 15, 1023–1030.

Dingemanse, N. J., Wright, J., Kazem, A. J. N., Thomas, D. K., Hickling, R., and Dawnay, N. 2007. Behavioural syndromes differ predictably between 12 populations of three-spined stickleback. *Journal of Animal Ecology*, 76, 1128–1138.

Dingemanse, N. J., and Wolf, M. 2010. Recent models for adaptive personality differences: a review. *Philosophical Transactions of the Royal Society of London B*, 365, 3947–3958.

Drent, P. J. 1987. The importance of nest boxes for territory settlement, survival, and density of the great tit. *Ardea*, 75, 59–71.

Drent, P. J., Van Oers, K., and Van Noordwijk, A. J. 2003. Realized heritability of personalities in the great tit (*Parus major*). *Proceedings of the Royal Society of London B*, 270, 45–51.

Duckworth, R. A. 2006a. Aggressive behaviour affects selection on morphology by influencing settlement patterns in a passerine bird. *Proceedings of the Royal Society of London B*, 273, 1789–1795.

———. 2006b. Behavioral correlations across breeding contexts provide a mechanism for a cost of aggression. *Behavioral Ecology*, 17, 1011–1019.

Edens, F. W., and Siegel, H. S. 1975. Adrenal responses in high and low ACTH response lines of chickens during acute heat stress. *General and Comparative Endocrinology*, 25, 64–73.

Eising, C. M., Eikenaar, C., and Groothuis, T. G. G. 2001. Maternal androgens in black-headed gull (*Larus ridibundus*) eggs: consequences for chick development. *Proceedings of the Royal Society of London B*, 268, 839–846.

Eising, C. M., Muller, W., and Groothuis, T. G. G. 2006. Avian mothers create different phenotypes by hormone deposition in their eggs. *Biology Letters*, 2, 20–22.

Ellenberg, U., Mattern, T., and Seddon, P. J. 2009. Habituation potential of yellow-eyed penguins depends on sex, character, and previous experience with humans. *Animal Behaviour*, 77, 289–296.

Ellis, B. J., Jackson, J. J., and Boyce, W. T. 2006. The stress response systems: universality and adaptive individual differences. *Developmental Review*, 26, 175–212.

Faure, J. M. 1980. Selection for open-field behavior in the chicken—the relationship between open-field and social behavior. *Applied Animal Ethology*, 6, 385–385.

Fidler, A. E., Van Oers, K., Drent, P. J., Kuhn, S., Mueller, J. C., and Kempenaers, B. 2007. *Drd4* gene polymorphisms are associated with personality variation in a passerine bird. *Proceedings of the Royal Society of London B*, 274, 1685–1691.

Fox, R. A., Ladage, L. D., Roth, T. C., and Pravosudov, V. V. 2009. Behavioural profile predicts dominance status in mountain chickadees, *Poecile gambeli. Animal Behaviour*, 77, 1441–1448.

Fox, R. A., and Millam, J. R. 2004. The effect of early environment on neophobia in orange-winged Amazon parrots (*Amazona amazonica*). *Applied Animal Behaviour Science*, 89, 117–129.

———. 2007. Novelty and individual differences influence neophobia in orange-winged Amazon parrots (*Amazona amazonica*). *Applied Animal Behaviour Science*, 104, 107–115.

Francois, N., Mills, A. D., and Faure, J. M. 1999. Interindividual distances during open-field tests in Japanese quail (*Coturnix japonica*) selected for high or low levels of social reinstatement behaviour. *Behavioural Processes*, 47, 73–80.

Fucikova, E., Drent, P. J., Smiths, N., and Van Oers, K. 2009. Handling stress as a measurement of personality in great tit nestlings (*Parus major*). *Ethology*, 115, 366–374.

Garamszegi, L. Z., Eens, M., and Janos, T. 2009. Behavioural syndromes and trappability in free-living collared flycatchers, *Ficedula albicollis*. *Animal Behaviour*, 77, 803–812.

Garamszegi, L. Z., Eens, M., and Török, J. 2008. Birds reveal their personality when singing. *PLoS ONE*, 3, e2647, 1–7.

Gil, D. 2008. Hormones in avian eggs: physiology, ecology, and behavior. *Advances in the Study of Behavior*, 38, 337–398.

Gil, D., and Faure, J. M. 2007. Correlated response in yolk testosterone levels following divergent genetic selection for social behaviour in Japanese quail. *Journal of Experimental Zoology Part A—Ecological Genetics and Physiology*, 307A, 91–94.

Gil, D., Naguib, M., Riebel, K., Rutstein, A., and Gahr, M. 2006. Early condition, song learning, and the volume of song brain nuclei in the zebra finch (*Taeniopygia guttata*). *Journal of Neurobiology*, 66, 1602–1612.

Gray, P. H. 1967. Spalding and his influence on research in developmental behavior. *Journal of the History of the Behavioral Sciences*, 3, 168–179.

Griffith, S. C., Owens, I. P. F., and Thuman, K. A. 2002. Extra-pair paternity in birds: a review of interspecific variation and adaptive function. *Molecular Ecology*, 11, 2195–2212.

Groothuis, T. G. G., Carere, C., Lipar, J., Drent, P. J., and Schwabl, H. 2008. Selection on personality in a songbird affects maternal hormone levels tuned to its effect on timing of reproduction. *Biology Letters*, 4, 465–467.

Groothuis, T. G. G., Muller, W., von Engelhardt, N., Carere, C., and Eising, C. M. 2005. Maternal hormones as a tool to adjust offspring phenotype in avian species. *Neuroscience and Biobehavioral Reviews*, 29, 329–352.

Guillette, L. M., Reddon, A. R., Hoeschele, M., and Sturdy, C. B. 2010. Sometimes slower is better: slow-exploring birds are more sensitive to changes in a vocal discrimination task. *Proceedings of the Royal Society of London B*, 278, 767–773.

Guillette, L. M., Reddon, A. R., Hurd, P. L., and Sturdy, C. B. 2009. Exploration of a novel space is associated with individual differences in learning speed in black-capped chickadees, *Poecile atricapillus*. *Behavioural Processes*, 82, 265–270.

Harvey, E. M., and Freeberg, T. M. 2007. Behavioral consistency in a changed social context in Carolina chickadees. *Journal of General Psychology*, 134, 229–245.

Hasselquist, D., Bensch, S., and Von Schantz, T. 1996. Correlation between male song repertoire, extra-pair paternity, and offspring survival in the great reed warbler. *Nature*, 381, 229–232.

Hollander, F. A., Van Overveld, T., Tokka, I., and Matthysen, E. 2008. Personality and nest defence in the great tit (*Parus major*). *Ethology*, 114, 405–412.

Houston, A. I. 2010. Evolutionary models of metabolism, behaviour, and personality. *Philosophical Transactions of the Royal Society of London B*, 365, 3969–3975.

Howard, L. 1953. *Birds as Individuals*. London: Readers Union/Collins.

Izhaki, I. 1998. Essential amino acid composition of fleshy fruits versus maintenance requirements of passerine birds. *Journal of Chemical Ecology*, 24, 1333–1345.

Jones, R. B., and Hocking, P. M. 1999. Genetic selection for poultry behaviour: big bad wolf or friend in need? *Animal Welfare*, 8, 343–359.

Kacelnik, A., Houston, A. I., and Krebs, J. R. 1981. Optimal foraging and territorial defense in the great tit (*Parus major*). *Behavioral Ecology and Sociobiology*, 8, 35–40.

Kant, G. J., Egglestone, T., Landman-Roberts, L., Kenion, C. C., Driver, G. C., and Meyerhoff, J. L. 1985. Habituation to repeated stress is stressor specific. *Pharmacology Biochemistry and Behavior*, 22, 631–634.

Kempenaers, B., Verheyen, G. R., Vandenbroeck, M., Burke, T., Van Broeckhoven, C., and Dhondt, A. A. 1992. Extra-pair paternity results from female preference for high-quality males in the blue tit. *Nature*, 357, 494–496.

Kitaysky, A. S., Kitaiskaia, E. V., Wingfield, J. C., and Piatt, J. F. 2001. Dietary restriction causes chronic elevation of corticosterone and enhances stress response in red-legged kittiwake chicks. *Journal of Comparative Physiology B—Biochemical Systemic and Environmental Physiology*, 171, 701–709.

Konishi, M. 1985. Birdsong: from behavior to neuron. *Annual Review of Neuroscience*, 8, 125–170.

Konishi, M., Emlen, S. T., Ricklefs, R. E., and Wingfield, J. C. 1989. Contributions of bird studies to biology. *Science*, 246, 465–472.

Koolhaas, J. M., De Boer, S. F., Buwalda, B., and Van Reenen, K. 2007. Individual variation in coping with stress: a multidimensional approach of ultimate and proximate mechanisms. *Brain Behavior and Evolution*, 70, 218–226.

Koolhaas, J. M., Korte, S. M., De Boer, S. F., Van der Vegt, B. J., Van Reenen, C. G., Hopster, H., De Jong, I. C., Ruis, M. A. W., and Blokhuis, H. J. 1999. Coping styles in animals: current status in behavior and stress-physiology. *Neuroscience and Biobehavioral Reviews*, 23, 925–935.

Korsten, P., Mueller, J. C., Hermannstädter, C., Bouwman, K. M., Dingemanse, N. J., Drent, P. J., Liedvogel, M., Matthysen, E., Van Oers, K., Van Overveld, T., Patrick, S. C., Quinn, J. L., Sheldon, B. C., Tinbergen, J. M., and Kempenaers, B. 2010. Association between DRD4 gene polymorphism and personality variation in great tits: a test across four wild populations. *Molecular Ecology*, 19, 832–843.

Korte, S. M., Koolhaas, J. M., Wingfield, J. C., and McEwen, B. S. 2005. The Darwinian concept of stress: benefits of allostasis and costs of allostatic load and the trade-offs in health and disease. *Neuroscience and Biobehavioral Reviews*, 29, 3–38.

Kralj-Fiser, S., Scheiber, I. B. R., Blejec, A., Moestl, E., and Kotrschal, K. 2007. Individualities in a flock of free-roaming greylag geese: behavioral and physiological consistency over time and across situations. *Hormones and Behavior*, 51, 239–248.

Krause, E. T., Honarmand, M., Wetzel, J., and Naguib, M. 2009. Early fasting is long-lasting: differences in early nutritional conditions reappear under stressful conditions in adult zebra finches. *PLoS ONE*, 4, e5015–e5020.

Krebs, J. R., MacRoberts, M. H., and Cullen, J. M. 1972. Flocking and feeding in the great tit *Parus major*—an experimental study. *Ibis*, 114, 507–530.

Krebs, J. R., and Perrins, C. M. 1978. Behaviour and population regulation in the great tit (*Parus major*). In: *Population Control by Social Behaviour* (Ebling, F. J., and Stoddard, D. M., eds.), pp. 23–47. London: Institute of Biology.

Kurvers, R., Eijkelenkamp, B., Van Oers, K., Van Lith, H. A., Van Wieren, S., Ydenberg, R., and Prins, A. J. A. 2009. Personality differences explain leadership in Barnacle geese. *Animal Behaviour*, 78, 447–453.

Lack, D. 1947. *The Life of the Robin*. London: Witherby.

Lebreton, J. D., Burnham, K. P., Clobert, J., and Anderson, D. R. 1992. Modeling survival and testing biological hypotheses using marked animals—a unified approach with case studies. *Ecological Monographs*, 62, 67–118.

Lincoln, F. C., and Baldwin, S. P. 1929. *Manual for Bird Banders*. Washington, DC: US Department of Agriculture.

Lloyd Morgan, C. 1923. *Emergent Evolution*. London: Williams and Norgate.

Marchetti, C., and Drent, P. J. 2000. Individual differences in the use of social information in foraging by captive great tits. *Animal Behaviour*, 60, 131–140.

Marchetti, C., and Zehtindjiev, P. 2009. Migratory orientation of sedge warblers (*Acrocephalus schoenobaenus*) in relation to eating and exploratory behaviour. *Behavioural Processes*, 82, 293–300.

Marler, P., and Slabbekoorn, H. 2004. *Nature's Music: The Science of Birdsong*. San Diego, CA: Elsevier.

Martins, T. L. F., Roberts, M. L., Giblin, I., Huxham, R., and Evans, M. R. 2007. Speed of exploration and risk-taking behavior are linked to corticosterone titres in zebra finches. *Hormones and Behavior*, 52, 445–453.

Mason, W. 1979. Ontogeny of social behavior. In: *Handbook of Behavioral Neurobiology: Social Behavior and Communication* (Marler, P., and Vandenbergh, J. G., eds.), vol. 3, pp. 1–28. New York: Plenum Press.

Merton, D. V. 2006. The Kakapo: some highlights and lessons from five decades of applied conservation. *Journal of Ornithology*, 147, 4.

Metcalfe, N. B., and Monaghan, P. 2001. Compensation for a bad start: grow now, pay later? *Trends in Ecology and Evolution*, 16, 254–260.

Mettke-Hofmann, C., Ebert, C., Schmidt, T., Steiger, S., and Stieb, S. 2005. Personality traits in resident and migratory warbler species. *Behaviour*, 142, 1357–1375.

Mettke-Hofmann, C., Wink, M., Winkler, H., and Leisler, B. 2004. Exploration of environmental changes relates to lifestyle. *Behavioral Ecology*, 16, 247–254.

Milinski, M. 2006. The major histocompatibility complex, sexual selection, and mate choice. *Annual Review of Ecology and Systematics*, 37, 159–186.

Miller, K. A., Garner, J. P., and Mench, J. A. 2005. The test-retest reliability of four behavioural tests of fearfulness for quail: a critical evaluation. *Applied Animal Behaviour Science*, 92, 113–127.

———. 2006. Is fearfulness a trait that can be measured with behavioural tests? A validation of four fear tests for Japanese quail. *Animal Behaviour*, 71, 1323–1334.

Minderman, J., Reid, J. M., Evans, P. G. H., and Whittingham, M. J. 2009. Personality traits in wild starlings: exploration behavior and environmental sensitivity. *Behavioral Ecology*, 20, 830–837.

Murphy, M. E., and Pearcy, S. D. 1993. Dietary amino-acid complementation as a foraging strategy for wild birds. *Physiology and Behavior*, 53, 689–698.

Naguib, M. 2005. Singing interactions in song birds: implications for social relations, territoriality, and territorial settlement. In: *Communication Networks* (McGregor, P. K., ed.), pp. 300–319. Cambridge: Cambridge University Press.

Naguib, M., Floercke, C., and Van Oers, K. 2011. Effects of social conditions during early development on stress response and personality traits in great tits (*Parus major*). *Developmental Psychobiology*, 53, 592–600.

Naguib,. M., Kazek, A., Schaper, S., Van Oers, K., and Visser, M. E. 2010. Singing activity reveals personality traits in great tits. *Ethology*, 116, 763–769.

Naguib, M., and Nemitz, A. 2007. Living with the past: nutritional stress in juvenile males has immediate effects on their plumage ornaments and on adult attractiveness in zebra finches. *PLoS ONE*, 2, e901–e910.

Naguib, M., Nemitz, A., and Gil, D. 2006. Maternal developmental stress reduces reproductive success of female offspring in zebra finches. *Proceedings of the Royal Society of London B*, 273, 1901–1905.

Naguib, M., Riebel, K., Marzal, A., and Gil, D. 2004. Nestling immunocompetence and testosterone covary with brood size in a songbird. *Proceedings of the Royal Society of London B*, 271, 833–838.

Neuenschwander, S., Brinkhof, M. W. G., Kolliker, M., and Richner, H. 2003. Brood size, sibling competition, and the cost of begging in great tits (*Parus major*). *Behavioral Ecology*, 14, 457–462.

Nowicki, S., Peters, S., and Podos, J. 1998. Song learning, early nutrition, and sexual selection in songbirds. *American Zoologist*, 38, 179–190.

Partridge, L. 1976. Individual differences in feeding efficiencies and feeding preferences of captive great tits. *Animal Behaviour*, 24, 230–240.

Petrie, M., and Kempenaers, B. 1998. Extra-pair paternity in birds: explaining variation between species and populations. *Trends in Ecology and Evolution*, 13, 52–58.

Price, T. D. 2002. Domesticated birds as a model for the genetics of speciation by sexual selection. *Genetica*, 116, 311–327.

Quinn, J. L., and Cresswell, W. 2005. Personality, antipredation behaviour, and behavioural plasticity in the chaffinch *Fringilla coelebs*. *Behaviour*, 142, 1377–1402.

Quinn, J. L., Patrick, S. C., Bouwhuis, S., Wilkin, T. A., and Sheldon, B. C. 2009. Heterogeneous selection on a heritable temperament trait in a variable environment. *Journal of Animal Ecology*, 78, 1203–1215.

Ramsay, S. L., and Houston, D. C. 2003. Amino acid composition of some woodland arthropods and its implications for breeding tits and other passerines. *Ibis*, 145, 227–232.

Réale, D., Reader, S. M., Sol, D., McDougall, P. T., and Dingemanse, N. J. 2007. Integrating temperament in ecology and evolutionary biology. *Biological Reviews*, 82, 291–318.

Richard, S., Wacrenier-Cere, N., Hazard, D., Saint-Dizier, H., Arnould, C., and Faure, J. M. 2008. Behavioural and endocrine fear responses in Japanese quail upon presentation of a novel object in the home cage. *Behavioural Processes*, 77, 313–319.

Rodenburg, T. B., Komen, H., Ellen, E. D., Uitdehaag, K. A., and Van Arendonk, J. A. M. 2008. Selection method and early-life history affect behavioural development, feather pecking, and cannibalism in laying hens: a review. *Applied Animal Behaviour Science*, 110, 217–228.

Rodenburg, T. B., and Turner, S. P. 2012. The role of breeding and genetics in the welfare of farm animals. *Animal Frontier*, 2, 16–21.

Roff, D. A., and Fairbairn, D. J. 2007. The evolution of trade-offs: where are we? *Journal of Evolutionary Biology*, 20, 433–447.

Sapolsky, R. M. 1994. Individual differences and the stress response. *Seminars in the Neurosciences*, 6, 261–269.

Satterlee, D. G., and Johnson, W. A. 1988. Selection of Japanese quail for contrasting blood corticosterone response to immobilization. *Poultry Science*, 67, 25–32.

Schuett, W., and Dall, S. R. X. 2009. Sex differences, social context, and personality in zebra finches, *Taeniopygia guttata*. *Animal Behaviour*, 77, 1041–1050.

Schwabl, H. 1993. Yolk is a source of maternal testosterone for developing birds. *Proceedings of the National Academy of Sciences U S A*, 90, 11446–11450.

———. 1995. Individual variation of the acute adrenocortical response to stress in the white-throated sparrow. *Zoology-Analysis of Complex Systems*, 99, 113–120.

Sih, A., and Bell, A. M. 2008. Insights for behavioral ecology from behavioral syndromes. *Advances in the Study of Behavior*, 38, 227–281.

Silverin, B. 1998. Stress responses in birds. *Poultry and Avian Biology Reviews*, 9, 153–168.

Soma, K. K., Scotti, M. A. L., Newman, A. E. M., Charlier, T. D., and Demas, G. E. 2008. Novel mechanisms for neuroendocrine regulation of aggression. *Frontiers in Neuroendocrinology*, 29, 476–489.

Stam, R., Bruijnzeel, A. W., and Wiegant, V. M. 2000. Long-lasting stress sensitisation. *European Journal of Pharmacology*, 405, 217–224.

Stamps, J. A., and Groothuis, T. G. G. 2010a. Developmental perspectives on personality: implications for ecological and evolutionary studies of individual differences. *Philosophical Transactions of the Royal Society of London B*, 365, 4029–4041.

———. 2010b. The development of animal personality: relevance, concepts, and perspectives. *Biological Reviews*, 85, 301–325.

Stamps, J. A., and Krishnan, V. V. 1997. Functions of fights in territory establishment. *American Naturalist*, 150, 393–405.

Stowe, M., Bugnyar, T., Loretto, M. C., Schloegl, C., Range, F., and Kotrschal, K. 2006. Novel object exploration in ravens (*Corvus corax*): effects of social relationships. *Behavioural Processes*, 73, 68–75.

Stowe, M., and Kotrschal, K. 2007. Behavioural phenotypes may determine whether social context facilitates or delays novel object exploration in ravens (*Corvus corax*). *Journal of Ornithology*, 148, S179-S184.

Sutherland, W. J. 1998. The importance of behavioural studies in conservation biology. *Animal Behaviour*, 56, 801–809.

Talbot, L. R. 1922. Bird-banding at Thomasville, Georgia, in 1922. *Auk*, 39, 334–350.

Thayer, P. W., and Austin, J. T. 1992. Harold E. Burtt (1890–1991): obituary. *American Psychologist*, 47, 1677.

Tinbergen, J. M., and Boerlijst, M. C. 1990. Nestling weight and survival in individual great tits (*Parus major*). *Journal of Animal Ecology*, 59, 1113–1127.

Titulaer, M., Van Oers, K., and Naguib, M. 2012. Personality affects learning performance in difficult tasks in a sex-dependent way. *Animal Behaviour*, 83, 723–730.

Tobler, M., and Sandell, M. I. 2007. Yolk testosterone modulates persistence of neophobic responses in adult zebra finches, *Taeniopygia guttata*. *Hormones and Behavior*, 52, 640–645.

Tschirren, B., and Richner, H. 2006. Parasites shape the optimal investment in immunity. *Proceedings of the Royal Society of London B*, 273, 1773–1777.

Uitdehaag, K., Kornen, H., Rodenburg, T. B., Kemp, B., and Van Arendonk, J. 2008. The novel object test as predictor of feather damage in cage-housed Rhode Island Red and White Leghorn laying hens. *Applied Animal Behaviour Science*, 109, 292–305.

Van Bers, N. E. M., Van Oers, K., Kerstens, H. H. D., Dibbits, B. W., Crooijmans, R. P. M. A., Visser, M. E., and Groenen, M. A. M. 2010. Genome-wide SNP detection in the great tit *Parus major* using high throughput sequencing. *Molecular Ecology*, 19, 89–99.

Van de Weerd, H. A., and Elson, A. 2006. Rearing factors that influence the propensity for injurious feather pecking in laying hens. *World's Poultry Science Journal*, 62, 654–664.

Van Krimpen, M. M., Kwakkel, R. P., Reuvekamp, B. F. J., Van der Peet-Schwering, C. M. C., Den Hartog, L. A., and Verstegen, M. W. A. 2005. Impact of feeding management on feather pecking in laying hens. *World's Poultry Science Journal*, 61, 663–685.

Van Oers, K., De Jong, G., Drent, P. J., and Van Noordwijk, A. J. 2004a. Genetic correlations of avian personality traits: correlated response to artificial selection. *Behavior Genetics*, 34, 611–619.

Van Oers, K., De Jong, G., Van Noordwijk, A. J., Kempenaers, B., and Drent, P. J. 2005a. Contribution of genetics to the study of animal personalities: a review of case studies. *Behaviour*, 142, 1185–1206.

Van Oers, K., Drent, P. J., De Goede, P., and Van Noordwijk, A. J. 2004b. Realized heritability and repeatability of risk-taking behaviour in relation to avian personalities. *Proceedings of the Royal Society of London B*, 271, 65–73.

———. 2004c. Additive and nonadditive genetic variation in avian personality traits. *Heredity*, 93, 496–503.

Van Oers, K., Drent, P. J., Dingemanse, N. J., and Kempenaers, B. 2008. Personality is associated with extra-pair paternity in great tits (*Parus major*). *Animal Behaviour*, 76, 555–563.

Van Oers, K., Klunder, M., and Drent, P. J. 2005b. Context dependence of avian personalities: risk-taking behavior in a social and a nonsocial situation. *Behavioral Ecology*, 16, 716–723.

Van Oers, K., and Mueller, J. C. 2010. Evolutionary genomics of animal personality. *Philosophical Transactions of the Royal Society of London B*, 365, 3991–4000.

Van Oortmerssen, G. A., and Bakker, T. C. M. 1981. Artificial selection for short and long attack latencies in wild *Mus musculus domesticus*. *Behavior Genetics*, 11, 115–126.

Verbeek, M. E. M., Boon, A., and Drent, P. J. 1996. Exploration, aggressive behaviour, and dominance in pair-wise confrontations of juvenile male great tits. *Behaviour*, 133, 945–963.

Verbeek, M. E. M., De Goede, P., Drent, P. J., and Wiepkema, P. R. 1999. Individual behavioural characteristics and dominance in aviary groups of great tits. *Behaviour*, 136, 23–48.

Verbeek, M. E. M., Drent, P. J., and Wiepkema, P. R. 1994. Consistent individual differences in early exploratory behaviour of male great tits. *Animal Behaviour*, 48, 1113–1121.

Verhulst, S., Holveck, M. J., and Riebel, K. 2006. Long-term effects of manipulated natal brood size on metabolic rate in zebra finches. *Biology Letters*, 2, 478–480.

Visser, M. E. 2008. Keeping up with a warming world: assessing the rate of adaptation to climate change. *Proceedings of the Royal Society of London B*, 275, 649–659.

Wedekind, C., Seebeck, T., Bettens, F., and Paepke, A. J. 1995. MHC-dependent mate preferences in humans. *Proceedings of the Royal Society of London B*, 260, 245–249.

Wemelsfelder, F., Hunter, E. A., Mendl, M. T., and Lawrence, A. B. 2000. The spontaneous qualitative assessment of behavioural expressions in pigs: first explorations of a novel methodology for integrative animal welfare measurement. *Applied Animal Behaviour Science*, 67, 193–215.

Wiepkema, P. R. 1992. Stressing farm animals. *Tijdschrift Voor Diergeneeskunde*, 117, 141–147.

Wingfield, J. C. 2005. Historical contributions of research on birds to behavioral neuroendocrinology. *Hormones and Behavior*, 48, 395–402.

Wolf, M., Van Doorn, G. S., Leimar, O., and Weissing, F. J. 2007. Life-history trade-offs favour the evolution of animal personalities. *Nature*, 447, 581–584.

Wolf, M., and Weissing, F. J. 2010. An explanatory framework for adaptive personality differences. *Philosophical Transactions of the Royal Society of London B*, 365, 3959–3968.

Würbel, H. 2009. Ethology applied to animal ethics. *Applied Animal Behaviour Science*, 118, 118–127.

4

Differential Behavioral Ecology

The Structure, Life History, and

Evolution of Primate Personality

ALEXANDER WEISS AND MARK J. ADAMS

Introduction

Behavioral variation is highly relevant to evolution. For example, behavioral differences in courtship can lead to reproductive isolation and speciation, a process referred to as ethological isolation (Dobzhansky 1970). Thus, the study of behavioral variation can yield important clues about other kinds of diversity. While the study of human behavioral variation in the form of personality research is a mature field with a distinguished record of accomplishments, there has only recently been renewed interest in studying individual differences in nonhuman animal behavior. Research in this area has appeared under various guises, including temperament (Réale et al. 2007), behavioral syndromes (Sih et al. 2004), and personality (Gosling 2001). While these frameworks all concern individual differences in behavior, they differ in several critical ways. Notably, compared with animal personality research based in human differential psychology (i.e., the study of all interindividual differences in psychological function; Galton 1883), animal personality research arising from behavioral ecology prominently features evolutionary theory as a unifying theme. Other differences include basic assumptions, central research questions, methods, and measurement (Nettle and Penke 2010). Much like behavioral differences that act as barriers to reproduction between individuals within a species, these differences between research programs may lead to a less "fruitful" evolutionary science of personality (see also chapter 5, and Nettle and Penke 2010).

The realization by scientists that nonhuman primates are individually recognizable and behaviorally distinct probably began in early laboratory colonies and preceded the formal study of their personality (Crawford 1938; Yerkes 1939). Later contributions came from researchers such as Itani (1957), who considered introversion and extraversion in Japanese macaques, and Goodall (1986), who described the personalities of the chimpanzees she studied at Gombe. Rather than being simple anthropomorphism (a charge that Yerkes anticipated), the assumption that primate personality should resemble our own has a strong phylogenetic basis. Without evidence to the contrary, we assume that closely related species will be more similar to each other than to more distantly related species (Darwin 1859). Thus, how nonhuman primates differ behaviorally should resemble how we differ.

Our own research on nonhuman primate personality led us to believe that the approaches of human differential psychology can benefit research on personality conducted by behavioral ecologists. This revelation came about because nonhuman primates are ideal subjects for studies that link approaches from differential psychology and behavioral ecology. For example, compared with other animals, our phylogenetic affinity with nonhuman primates, and especially the great apes, makes us more able to "read" a primate's personality and thus provide reliable and valid ratings (e.g., see Uher and Asendorpf 2008). This, and the fact that nonhuman primates' social lives and behaviors are complex and distinct, readily lends these species to more comprehensive measures that can educe a whole personality structure. In addition, while practical and ethical barriers prevent studying humans with some of the behavioral ecologists' experimental manipulations (e.g., see Réale et al. 2007) or conducting "field" research, such barriers in studying nonhuman primates are few. Finally, this also means that tools for assessment adapted from human studies will capture much of the variety present in nonhuman primates.

Results from nonhuman primate personality research can often be directly extrapolated to humans. This point is underscored by a study on the genetic correlation of human personality and subjective well-being (Weiss et al. 2008), which replicated the findings of an earlier chimpanzee study (Weiss et al. 2002). Also, recent studies on the relationship between human personality and immune functioning (O'Cleirigh et al. 2007; Ironson and Hayward, 2008; Ironson et al. 2008) were anticipated by studies in rhesus macaques (Capitanio et al. 1999; Maninger et al. 2003).

We hope to show that, while each of these research programs asks interesting questions, to understand the *nature* of personality—that is, its ultimate origins in humans and other species—requires a combination of approaches. On the basis of studies of personality ratings in humans and in nonhuman primates, we believe that nonhuman personality research in other animals would benefit by incorporating approaches from differential psychology research. These approaches include more broadly defining and measuring traits, conducting epidemiological studies in natural communities, looking at personality profiles, and examining the covariance among traits or personality structure.

The study of personality in behavioral ecology
TRADE-OFFS AND LIFE-HISTORY STRATEGIES

Behavioral ecologists who study animal personality follow the approach outlined by Tinbergen ([1963] 2005); they formulate and test hypotheses concerning ultimate causation or how a particular behavior is adaptive. This research program also posits that existing behaviors represent an optimal trade-off between fitness costs (e.g., energy expended) and benefits (e.g., reproductive outcomes). The study of optimization in behavioral ecology has typically been at the individual level. For example, one might test whether the pattern of foraging within a species is such that individuals switch to another patch when the amount of energy from food they gain from one patch is exceeded by the cost of continued foraging in that patch (MacArthur and Pianka, 1966). Studies of personality variation by behavioral ecologists have brought this focus on trade-offs to the study of intra-specific behavioral variation.

Behavioral ecology also concerns the study of life-history strategies (Pianka 1970). Life-history strategies comprise a suite of traits that are co-adapted for particular environments. These traits reflect trade-offs between reproductive and somatic effort. Unstable environments lead to species that are r-selected, that is, characterized by high reproductive effort (Pianka 1970). These r-selected species (e.g., rabbits) have short generation times, a lower probability of offspring survival, and small body size. On the other hand, stable environments lead to species that are characterized by high somatic effort or K-selected (Pianka 1970). Individuals within K-selected species (e.g., elephants) possess traits indicative of high somatic effort, namely larger body size, high parental investment, and greater longevity.

Dall, Houston, and McNamara (2004) conceptualized personality as alternative behavioral strategies and Wolf, Van Doorn, Leimar, and Weissing (2007) placed it within the context of life-history evolution. An individual's optimal behavior depends both on its own condition and behavioral history as well as on the behaviors of others in the population. For example, an animal with poor body condition and, therefore, a low potential for reproductive success, would do better to invest in foraging, even if this increases predation risk. Likewise, the advantage of behaviors such as aggression depends both on how others are behaving and on population density.

Normally, aggression leads an individual to outcompete its neighbors for resources, but at low population densities with moderate competition, high aggression might be counterproductive, leading to lower fitness (Dingemanse et al. 2004; Sih et al. 2004).

Personality may also become co-adapted with life-history traits, such as the trade-off between growth and mortality (Stamps 2007; Biro and Stamps 2008). As growth rate is known to vary consistently between individuals in many different species, individuals that require an increased food intake benefit from a personality style suitable for high foraging rates—for example, by being bold, aggressive, or explorative. The maintenance of variation in both productivity and behavioral traits require and reinforce each other.

This approach to the study of personality and life-history strategies should be distinguished from that typically conducted in humans or other primates. In the study of humans and nonhuman primates, the focus is typically on how individual differences in some personality dimensions are related to individual differences in life-history outcomes. In the nonprimate personality literature, researchers are less interested in species-typical behaviors than in how personality is expressed, the relation between personality and fitness, how combinations of personality traits may be adaptations for particular social and ecological niches, how the presence or absence of variation can be explained by ecological factors, and the maintenance of heritable genetic variation (e.g., Bergmüller and Taborsky 2010; Dingemanse and Wolf 2010; Dochtermann and Roff 2010; Réale et al. 2010; Van Oers and Mueller 2010; Wolf and Weissing 2010). On a methodological level, the phylogenetically informed inferences and arguments from theory made by workers coming from the tradition of human psychology (Gosling and John 1999; Gosling and Graybeal 2007; Penke et al. 2007b; Gosling 2008) are distinct from the heavy lifting of measuring the strength and mode of natural selection practiced by behavioral ecologists (Dingemanse and Réale 2005). While both traditions are engaged with evolutionary theory, empirical analyses from evolutionary biology are featured more frequently and more prominently in behavioral ecology (Dingemanse and Réale 2005; Penke et al. 2007c) than in differential psychology (see, however, Eaves et al. 1990).

THE PERSISTENCE OF HERITABLE VARIATION

Personality variation in humans and other animals is, in part, genetic. The heritability of personality traits has been estimated for a number of animal species (Van Oers et al. 2005). Both behavioral ecologists and differential psychologists have adopted a number of methods for estimating heritability, although a method uniquely available to studies of nonprimate animals is realized heritability via selection experiments. Behavioral ecology research has also examined the genetic variance of personality in wild populations and has gone beyond estimating heritability by attempting to understand how genetic variation in personality traits is maintained within specific populations. If personality traits in primates show similar levels of heritability, then differential psychologists should share a concern for how this variation is maintained.

A review of the literature suggested that approximately 50% of the variance of all five human personality domains was heritable with little to no variance being accounted for by the shared family environment (Bouchard and Loehlin 2001). There have also been several recent studies on the genetics of personality in nonhuman primates. One study that examined the heritability of personality in 145 zoo chimpanzees found that, of the six personality dimensions identified by personality ratings (King and Figueredo 1997), only Dominance was significantly heritable ($h^2 = .62$; Weiss et al. 2000). A later study of rhesus macaques examined the heritability of seven factors derived from reactions to anxiety tests (Williamson et al. 2003). They found that individual differences in the Movement ($h^2 = 1.00$), Distress cues ($h^2 = .58$), Early independence ($h^2 = .83$), Explore familiar environment ($h^2 = .47$), and Explore novelty ($h^2 = 1.00$) factors were heritable; neither the Distress vocalizations nor Delayed independence factors were heritable. In a study of vervet monkeys, Fairbanks et al. (2004) found that Social Impulsivity was heritable ($h^2 = .35$). Fairbanks and her colleagues also found evidence for the heritability of the two facets of Social Impulsivity: Impulsive Approach ($h^2 = .25$) and Aggression ($h^2 = .61$). Finally, Rogers and colleagues (2008) found evidence that two measures of behavioral inhibition (likely related to Neuroticism) in juvenile rhesus macaques were influenced by additive genetic effects (freezing: $h^2 = .38$; vigilance: $h^2 = .91$).

Molecular genetic research in nonhuman primates supports the findings of quantitative geneticists. Bethea et al. (2004) found evidence linking the serotonin transporter linked polymorphic region (5HTTLPR) and reactions to several tests of anxious behavior or Neuroticism: monkeys that had two short forms of this allele (s/s) displayed greater anxiety than monkeys who had two long forms of the allele (l/l) or who were heterozygous (s/l). A molecular genetic study of a heritable phenotype related to Neuroticism (novelty seeking) found that latency to approach a novel object was lower in vervet monkeys that had the five- as opposed to six-repeat polymorphism of the dopamine D4 receptor gene (Bailey et al. 2007). Interestingly, experimental studies of genotyped rhesus macaques suggest that the effects of genetic polymorphisms on behavioral indicators of personality differ as a function of early environmental stressors (Champoux et al. 2002; Newman et al. 2005; Kraemer et al. 2008).

How this additive genetic variation is maintained remains a puzzle, for under directional or stabilizing selection one would expect the heritable variance in any given trait to be negligible (Barton and Turelli 1989; Falconer and Mackay 1996). This has led to considerable debate. Some theorists have suggested that variation in human psychological traits is maintained by neutral selection (Tooby and Cosmides 1990). The neutral theory states that most mutations have neither beneficial nor deleterious fitness consequences (Kimura 1983; 1986), leading to a balance between the input of new variation by mutation and its removal by drift (Lynch and Hill 1986). Others have taken an adaptationist stance, and see individual differences in humans and other animals as arising from frequency-dependent selection, sexual selection, or any number of mechanisms by which individual differences emerge because there are multiple niches or because what is "optimal" depends on the social and nonsocial context (see Penke et al. 2007a; 2007b; 2007c).

Exploring these hypotheses necessitates adopting additional methods from behavioral ecology to examine microevolutionary trends such as fluctuations in selection pressure across time (Dingemanse et al. 2004) or across environments (Dingemanse et al. 2007). Unlike research on primates, these ecologically informed studies have measured selection on temperament (Réale and Festa-Bianchet 2003; Bell and Sih 2007) rather than relating personality to fitness in only the broadest terms.

PERSONALITY MEASURES

Behavioral ecologists who study personality typically use similar methods. While there are notable exceptions (e.g., Réale and Festa-Bianchet 2003), one methodological feature common to much of this research is the use of small numbers of subjects that are easily (and safely) subjected to behavioral tests or experimental manipulations. Another common methodological feature is a preference for behaviorally based measures of personality, which has its roots in the traditional ethological emphasis on observation and the aversion to seeming anthropomorphism (Tinbergen 1951). Finally, these studies typically focus on one or two strictly defined traits such as boldness, aggressiveness, or exploratory behavior. The definition of these traits does not constrain how they are measured, which might include behavioral or physiological (hormonal, serotonergic, etc.) indicators. This framework does not assume that intercorrelations among traits are the product of underlying latent constructs or personality factors, and leaves this as a question to be investigated (Réale et al. 2007). Instead, the approach typically taken by many behavioral ecologists is to measure several behaviors thought to indicate a particular personality trait and test whether they are correlated (Stamps 2007). In one example of this approach, Réale and colleagues (2000) measured boldness (in terms of trappability) and docility (behavior during handling) in female sheep, found that they were weakly correlated, and concluded that they represented two separate traits. Sometimes a single trait is examined, as in Daniewski and Jezierski's (2003) study of exploratory behavior in rabbits using an open field test. In short, the behavioral differences currently studied by ecologists were never meant to be comprehensive. Instead, these studies rely on hypothesis-driven observations and experiments that attempt to assess how a few narrowly defined traits are related to ecological or social variables. It will be necessary to add further traits to fully define most of the intraspecific behavioral variation present, particularly in species with more complex social lives.

The study of personality in differential psychology
HISTORICAL BACKGROUND
The study of human—and to some extent nonhuman primate—personality has a variety of origins. While early biographers such as Plutarch (46 AD–120 AD) ascribed personality characteristics to their subjects, the study of human personality as a science has a shorter history. Research in human personality originated from Francis Galton's (1822–1911) study of human individual differences (Galton 1883), which also led to the development of modern psychometrics, statistics, behavior genetics, and cognitive abilities research.

The insight that personality descriptors may be found in natural languages and can be used to construct personality inventories came later (Allport and Odbert 1936), and is still one of the most popular ways of studying human personality. These data are most commonly analyzed using factor or principal components analysis (Gorsuch 1983), which involves examining covariances among variables to determine whether groups of variables cluster together—that is, are indicative of one or more latent underlying constructs (factors or components). Each factor or component can be described with respect to the amount of variance it accounts for (its eigenvalue) and the degree to which each variable is related to the factors or components (loadings).

DEBATES ABOUT PERSONALITY TRAITS
The study of human personality via ratings was and still is controversial. The most serious charge by critics was that personality traits did a poor job of

predicting behavior and were inconsistent across situations (Mischel 1968). In fact, like early (and some contemporary) ethologists and behavioral ecologists, Mischel favored using behavioral measures. This challenge was later answered by research showing that personality measures were related to behavior if behavior was aggregated over time (Epstein 1979). Personality researchers also showed that self-ratings and observer ratings were correlated (McCrae et al. 2004), that personality was mostly stable in adulthood (Costa and McCrae 2002), and that personality predicted a broad range of important outcomes (Ozer and Benet-Martínez 2006; Roberts et al. 2007).

Other debates, many of which are still ongoing, revolved around the number of personality dimensions needed to explain human personality differences (Costa and McCrae 1992a; 1992b; Eysenck 1992a; 1992b; Lee et al. 2005). For the purposes of this chapter, we will focus our discussion on the Five-Factor Model (FFM; Digman 1990), as it is the dominant model in human personality research, which contends that five normally distributed dimensions or domains—Neuroticism, Extraversion, Openness to Experience, Agreeableness, and Conscientiousness—describe human personality differences.

While it is not possible to comprehensively summarize each of the five personality domains, it is useful to highlight some of their cardinal characteristics as described by Costa and McCrae (1992c). Individuals on the low end of Neuroticism are emotionally stable and well adjusted, and exhibit low levels of negative affect; individuals on the high end tend to be emotionally unstable, have problems with adjustment, and have high levels of negative affect. Individuals high in Extraversion are more sociable, assertive, and active, and experience more positive affect; whereas those low in Extraversion do not seek out others' company and are independent and less active, and experience less positive affect. Individuals scoring higher in Openness to Experience tend to be curious, value new experiences and feelings, and are unconventional; whereas individuals scoring low tend to be less curious, prefer the familiar to the novel, and are more conventional in their outlooks and behavior. Individuals higher in Agreeableness tend to be helpful, trusting, and more inclined to cooperate rather than to compete. On the other hand, individuals lower in Agreeableness tend to be less inclined to help others, suspicious and cynical, and more competitive. Finally, individuals ranking high in Conscientiousness are reliable, organized, directed, and self-disciplined; whereas those lower in Conscientiousness are often less reliable, disorganized, directionless, and lacking in self-discipline.

Although laymen would be inclined to see one end of each personality domain as a desirable state of affairs, when considered within the framework of evolution, it is easy to imagine the advantages and disadvantages presented by various levels of each of these five domains. In fact, Nettle (2006) has outlined possible benefits and costs to the high and low end of each domain, which we briefly detail. While high Neuroticism could offer benefits such as increased vigilance, it might also carry costs including poorer interpersonal relationships, which are critical in social species. Extraversion, while possibly leading to benefits such as mating success, might also cost in terms of harm from risk-taking behaviors. High Openness to Experience, while being related to benefits such as increased creativity, carries the possible costs of susceptibility to unusual or even harmful beliefs. High Agreeableness may benefit individuals by making them more valued partners in coalitions, though it may also lead to individuals being at greater risk from social cheaters. Finally, while Conscientiousness may benefit individuals by helping them meet long-term goals, it can potentially cost them the benefits of more immediate fitness gains.

This framework offers fascinating possibilities for considering the evolutionary basis of heritable variation in the FFM. Recent research that couches personality differences within evolutionary theory has made progress in understanding the ultimate causes of personality. Moreover, prior and ongoing research in seemingly unrelated areas has also yielded insights relevant to personality evolution.

IS THE FIVE-FACTOR MODEL RELATED TO LIFE-HISTORY STRATEGIES?

One reason the differential psychology approach may appeal to behavioral ecologists is that the measures derived using this approach are likely related to evolutionarily relevant characteristics, such as life-history strategies. Figueredo and colleagues (2004; 2007) emphasized that, although life-history strategy is typically used to differentiate species, it is also a dimension on which individuals within a species differ. In other words, while humans are generally suited to a long-term (K-selected) strategy, there is still considerable variation. To test whether human personality dimensions were related to how K-selected the individuals were, Figueredo et al. (2004; 2007) conducted two analyses. In the first they showed that factor analysis of either the phenotypic or genetic covariance among questionnaire items concerning altruism, reproductive effort, parenting effort, and other aspects of life history revealed a single factor, which they named K. Figueredo et al. also conducted a second factor analysis on the covariance among K and the five human personality factors and found a single higher-order factor. These findings are certainly suggestive. However, these studies are limited in that they are cross-sectional and do not rule out the possibility that correlations may have arisen because personality and outcome measures were assessed using the same method, that is, questionnaires (see Campbell and Fiske 1959, for a detailed discussion).

Behavioral ecologists who are accustomed to objectively measured life-history variables will be understandably skeptical of these cross-sectional questionnaire-based results. However, prospective epidemiological studies linking personality to real-world health outcomes also suggest a relationship between human personality dimensions and life-history strategy. This is especially true for Conscientiousness, which, of the five human dimensions, is most consistently related to a wide variety of health outcomes (Bogg and Roberts 2004; Roberts et al. 2007). One set of findings demonstrates that, across studies of samples differing widely in health, age, socioeconomic status, culture, cognitive ability, and other characteristics, people who are high in Conscientiousness live longer (Kern and Friedman 2008). Conscientiousness is also related to slower disease progression; research on HIV-infected persons has shown that higher Conscientiousness was related to reductions in viral load and increases in CD4 counts (O'Cleirigh et al. 2007; Ironson et al. 2008). Moreover, recent research has shown that higher Conscientiousness is associated with a reduced risk of Alzheimer's disease and cognitive decline, suggesting slower senescence (Wilson et al. 2007).

The fact that Conscientiousness is a particularly robust and strong predictor of mortality and other health outcomes is interesting. More interesting still is that, despite the fact that Conscientiousness leads to health behaviors (Bogg and Roberts 2004), these possible mediating variables explain only part of the relationship between Conscientiousness and mortality (Martin et al. 2007). Thus, the protective effect of Conscientiousness may be due to Conscientiousness reflecting a more long-term or K-selected orientation, including facets of competence, order, dutifulness, achievement striving, self-discipline, and deliberation (Costa and McCrae 1992c). This suggests that Conscientiousness may be particularly well-positioned to capture individual differences in life history.

If Conscientiousness were the only personality domain related to these variables, it would be problematic. This is because, to date, Conscientiousness dimensions have been identified in only chimpanzees and humans (Gosling and John 1999). It is therefore good that other personality domains are also related to health outcomes. Agreeableness comprises facets such as altruism and compliance (Costa and McCrae 1992c), which are related to the K-selected end of the life-history strategy spectrum (Pianka 1970). Epidemiological research has generally found that Agreeableness is related to reduced cardiovascular disease risk (see Whiteman et al. 2000, for a review) and longevity (e.g., Weiss and Costa 2005). However, these findings have not been replicated in all studies (e.g., Iwasa et al. 2008).

Neuroticism, too, may reflect some aspects of life-history strategy, especially in terms of short-term orientation, for example, the impulsiveness facet. However, the relationship between Neuroticism and mortality is markedly less clear: some studies show that it is a risk factor (e.g., Wilson et al. 2004), others have found that it is protective (e.g., Weiss and Costa 2005), and still others find no significant relationship (e.g., Iwasa et al. 2008).

The reduced fidelity among studies examining the relationship between personality domains other than Conscientiousness and mortality should give one pause; it suggests that the association between personality and life-history strategy is probably more complex than suggested by the results of questionnaire-based studies (e.g., Figueredo et al. 2004; 2007). There are several possible explanations for why personality dimensions potentially related to life history are not related to health outcomes. Among these explanations are that effect sizes are smaller, and, as such, it might be difficult to replicate these findings. However, failures to find this effect have occurred in some studies (Almada et al. 1991) that used very large samples. This also does not explain why in some studies of Neuroticism and mortality, the effects are significant, but in the opposite direction (e.g., Weiss and Costa 2005). A final possibility is that contextual or environmental factors, such as population density or resource availability, may attenuate or modify the way in which human personality domains are related to these outcomes.

How differential psychology can enrich studies of personality in behavioral ecology

As pointed out by Gosling (see chapter 5 in this volume), Nettle and Penke (2010), and Capitanio (2011), researchers can gain insight into human personality from studies of animal personality. We echo this sentiment, but add that animal personality studies, especially those conducted by behavioral ecologists, can benefit from human personality research. In particular, we emphasize the benefits of adopting a multivariate approach to measurement, asking questions about the impact of multiple dimensions, and examining personality structure.

IMPROVING MEASURES

As we previously indicated, there are marked differences in how personality is measured in these disciplines. While neither approach possesses an absolute advantage, we feel that behavioral ecologists could benefit from using the same tools developed by differential psychologists. To do so we will first highlight the advantages and disadvantages of behavioral measures. We will then offer several recommendations.

ADVANTAGES OF BEHAVIORAL MEASURES: UNIFORMITY

As noted previously, behavioral ecology research on personality typically relies on single "item" measures of one or two aspects of personality. There are multiple advantages to this approach. First, these measures are easily replicable within and across species, thus facilitating their use by multiple investigators. Second, unlike rating scales, individuals within a sample can be quickly assessed. Third, unlike using rating scales, where raters need to know individuals for a considerable period of time, responses to behavioral tests can be assessed by naive individuals who have had little or no experience with the individual animals. Finally, these measures benefit those wishing to conduct meta-analyses, as the uniformity in personality measures makes comparing studies straightforward.

DISADVANTAGES OF BEHAVIORAL MEASURES: THE LACK OF CONTEXT

As with all methods, there are drawbacks to this approach. One drawback is that this method assumes that a behavior measured in one species has the same meaning in other species. While this assumption might very well hold, it is impossible to know unless there are multiple measures of multiple personality domains (Cronbach and Meehl 1955; Uher 2008a; 2008b). To illustrate the problem, consider the bold-shy dimension often studied by behavioral ecologists (e.g., Réale et al. 2007). Behavioral ecologists usually define this dimension as an individual's reaction to risky yet non-novel situations and have drawn broad conclusions concerning the evolutionary importance of individual differences in the bold-timid dimension on the basis of this definition. However, an examination of a related measure in studies of zoo chimpanzees (King and Figueredo 1997) and orangutans (Weiss et al. 2006) suggests that this assumption does not always hold. The two studies used comparable questionnaires, which differed only in that the questionnaire used to assess orangutan personality included five additional items (for more details, see Weiss et al. 2006). The item *timid* is common to both questionnaires and defined as "Subject lacks self-confidence, is easily alarmed and is hesitant to venture into new social or non-social situations." In chimpanzees, this item is a clear marker of the Dominance factor, having a loading of -.81. By examining the other items that load on Dominance, we see that individuals described as *timid* are also, for example, less *persis-*

tent and more *dependent* (see table 1 in King and Figueredo 1997). On the other hand, in orangutans *timid* is a clear marker of the Neuroticism factor, its loading being .70. Orangutans described as timid are more likely to be *excitable* and *impulsive*, but less likely to be *cool* (see table 3 in Weiss et al. 2006). Clearly, without the context of other items, the meaning of *timid* would be misunderstood in chimpanzees, orangutans, or both. Worse for the study of nonhuman personality is that, without this context, researchers may be led down blind alleys and comparisons of results across studies may be misleading.

RECOMMENDATIONS

It is not our intention to be scolds, and, given their advantages, we think it would be foolish for behavioral ecologists to abandon their measures. Instead we advocate that, where feasible, researchers use multiple types of measures (e.g., ratings, experimental tests, behavioral observations) of multiple personality constructs (e.g., Neuroticism, Extraversion, Agreeableness). This method, first advanced by Campbell and Fiske (1959), later expanded upon by others (see, e.g., Widaman 1985; Shadish 1992) and applied in a study of great ape personality (Uher and Asendorpf 2008), is based on the premise that traits may be correlated because they reflect the same underlying trait (e.g., Neuroticism) or method (e.g., an experimental paradigm). Thus, by measuring multiple constructs using multiple methods, one can better understand and more reliably measure the traits of interest. Once we understand the species-specific contexts in which a trait such as *timid* is expressed, we can begin to place it within the behavioral and social ecology of the species. For instance, under what ecological conditions or social organizations does boldness versus timidity become a facet related to social dominance rather than anxiety?

ADOPTING MULTIDIMENSIONAL AND MULTIVARIATE APPROACHES

In addition to our recommendations concerning measures, we feel that behavioral ecology could reap great rewards and insights by studying multiple traits. To support this suggestion, we will highlight research in psychology suggesting that individual traits or dimensions do not always act alone. More fundamentally, we will next highlight how studies of personality structure may benefit behavioral ecology.

PERSONALITY PROFILES IN DIFFERENTIAL PSYCHOLOGY

Until relatively recently, like behavioral ecology research in personality, human personality research has focused on examining a single dimension at a time. However, sometimes it may be useful to study or understand the "whole person" (or animal in this case). That is, one wishes to examine how the combination of an individual's traits may be related to important outcomes. At present, the study of personality profiles has been focused on mental as opposed to physical health outcomes, including subjective well-being and happiness, affective disorders such as depression, and personality disorders.

The study of subjective well-being or happiness has shown major advances since the advent of interest in the area (Diener et al. 1999; 2006). While there is also growing evidence from large longitudinal panel studies that subjective well-being can be changed by major life events, for example, marriage (Lucas et al. 2003), a consistent finding across several studies has been that subjective well-being is reliably predicted by all five human personality dimensions, especially low Neuroticism and high Extraversion (De-Neve and Cooper 1998; Steel et al. 2008).

Some may wish to dismiss the study of happiness and related personality profiles as merely reflecting whether fortune (genetic or environmental) smiles upon an individual. However, a recent review of the literature suggests that this may be premature, as subjective well-being actually predicts numerous positive outcomes, such as more successful marriages and higher income (Lyubomirsky et al. 2005) as well as longevity (Danner et al. 2001). As such, one could hypothesize that subjective well-being and associated personality profiles predict outcomes reflecting adjustment to one's environment.

As suggested by its relationship with mortality (Kinder et al. 2008), major depression may have high fitness costs or be related to life-history strategy. Prospective studies have highlighted the depression risk conferred by high Neuroticism and low Extraversion over long periods and that this relationship is mediated by common genetic factors (Kendler et al. 2006). Even though high Neuroticism and low Extraversion are risk factors for depression, many individuals exhibiting either or both of these risk factors do not become depressed. To study how other personality dimensions moderate this risk, Weiss, Sutin, et al. (2009) studied the relationship between the five personality dimensions and 28 of the personality styles (see figure 4.1) and depression in older adults. Their results indicated that, while high Neuroticism and low Conscientiousness were risk factors, they were predictors of depression only in the context of the individual's standing on other personality dimensions.

Personality disorders refer to underlying and persistent sets of behaviors, cognitions, and affective dispositions that can lead individuals and those close to them to experience many difficulties and general impairment (American Psychiatric Association 2004). Unlike personality measures, personality disorders have traditionally been operationalized as categories or types. This conceptualization is captured in the major classification manuals used by psychiatry, such as the fourth edition text revision of the *Diagnostic and Statistical Manual* (*DSM-IV-TR*; American Psychiatric Association 2004). For example, individuals exhibiting antisocial personality disorder may encounter repeated legal difficulties because of their violation of social norms and rules. The *DSM-IV-TR* criteria used to classify individuals as having antisocial personality disorder are:

A. A pervasive pattern of disregard for and violation of the rights of others occurring since age 15, as indicated by at least three of the following:
   (1) Failure to conform to social norms and repeated lawbreaking.
   (2) Deceitfulness.
   (3) Impulsivity or failure to plan ahead.
   (4) Irritability and aggressiveness.
   (5) Reckless disregard for safety of self or others.
   (6) Consistent irresponsibility.
   (7) Lack of remorse.
B. Age at least 18 years.
C. Evidence of Conduct Disorder with onset before age 15.

It is beyond the present chapter's scope to detail the problems with this typological approach. However, these problems have led to a growing unease with its use and advocacy for a dimensional approach based on the FFM (Widiger and Frances 2002).

The research supporting the move toward a dimensional approach suggests that personality disorders are, in fact, reflections of maladaptive combinations of extremely high or low standings on the five personality dimensions. A recent meta-analysis on the relationship between the five human personality dimensions and ten DSM-IV-TR personality disorders revealed that high Neuroticism and low Agreeableness were consistent across most disorders (Saulsman and Page 2004). This same meta-analysis also found that, for the most part, personality disorders can be distinguished by standings on the other dimensions. For example, antisocial personality disorder is also characterized by low Conscientiousness, whereas avoidant personality disorder is characterized by low Extraversion.

The accumulated research suggests that personality profiles are critical to understanding potentially fitness-related outcomes in humans, such as health and adjustment. Although it might be difficult or impossible to study phenomena as complex as personality or affective disorders in many nonhuman species, including primates, incorporating personality profiles could expand the scope of nonhuman personality research and may yield numerous insights. In the case of the role of different environmental or ecological contexts in maintaining genetic personality variance (e.g., Dingemanse et al. 2004), personality dimensions other than the one of interest in a particular study may be an additional contextual background in which the personality dimension of interest operates. This would provide yet another means by which heritable variation may be maintained through selection on correlated traits (Lande and Arnold 1983; Merilä et al. 2001). A hypothetical example is a species with a positive genetic correlation between exploratory behavior and boldness (e.g., great tits; Van Oers et al. 2004). In an environment with a high population density but also significant predation, these two traits could be selected in the opposite direction. While a fast-exploring individual would be able to outcompete conspecifics for access to food, its high boldness would raise its risk to predation. Access to food and mates is usually compared with rank, which in primates is captured by personality dimensions such as Dominance. For a primate, the social networks through which influence operates can be equally effective for controlling group dynamics, so high Extraversion and Agreeableness would be beneficial.

PERSONALITY STRUCTURE

Another way in which a multivariate approach can advance the study of nonhuman personality research is via the elucidation of personality, that is, the way in which traits cluster together as revealed by their loadings on different factors. Instead of focusing on the relationship of each individual's standing on one or more personality traits to life-history strategy or context and how heritable personality variation is maintained, the study of personality structure would permit researchers to examine personality at higher levels of organization, including species.

THE UNIVERSALITY OF PERSONALITY STRUCTURE IN HUMANS

Human personality research strongly indicates that while variation on any number of personality traits is the norm, the nature of the covariance among these traits is a human universal. The majority of this evidence is based on cross-cultural studies of human personality in which a well-validated measure developed in one culture is administered to members of another culture to see whether a similar factor structure emerges. The most ambitious study to date examined ratings on a well-validated measure of the FFM, the Revised NEO Personality Inventory (NEO-PI-R; Costa and McCrae 1992c). The participants were college students who were members of 50 Western and non-Western cultures spanning six continents and were asked to rate other members of the culture whom they knew well (McCrae et al. 2005). Overall, McCrae and his colleagues found evidence that the five factors found in American samples were reliably replicated in all of the cultures, with poorer replicability being mostly attributable to problems with data quality and smaller sample sizes.

THE UNIVERSALITY OF PERSONALITY STRUCTURE IN CHIMPANZEES
Evidence showing that personality structure may also be stable across different populations of nonhuman primates in different settings is accumulating. King and Figueredo (1997) obtained ratings from zoo keepers and volunteers on 100 zoo-housed chimpanzees. The 43-item questionnaire was based on human personality questionnaires and attempted to sample items from all five human domains; factor analysis of the ratings revealed a broad chimpanzee-specific factor labeled Dominance as it reflected dominance and competitiveness and five factors similar, though not identical, to the five human factors.

Three studies using identical or comparable questionnaires have examined whether the factor structure derived by King and Figueredo (1997) replicated in different samples. The first study used a sample comprised of ratings on 43 chimpanzees living in a naturalistic sanctuary in the Republic of the Congo rated on a French-language translation of the questionnaire and 74 zoo chimpanzees rated on the same questionnaire (King et al. 2005). The second study (Weiss et al. 2007) involved 175 chimpanzees housed at the Yerkes National Primate Research Center rated using the same questionnaire as the original 100 chimpanzees. The third study was based on ratings of 146 chimpanzees housed in zoos, research centers, and a sanctuary in Japan (Weiss, Inoue-Murayama, et al. 2009). The chimpanzees in this last study were rated on a translated version of the questionnaire and their factor structure was compared with that of the original 100 chimpanzees. In all three studies, high congruence coefficients (Haven and ten Berge 1977) suggested that Dominance, Extraversion, Conscientiousness, and Agreeableness clearly replicated and that the overall structure of chimpanzee

Style of Anger Control

Vertical Axis: Neuroticism (=45 T)

Horizontal Axis: Agreeableness (=4 T)

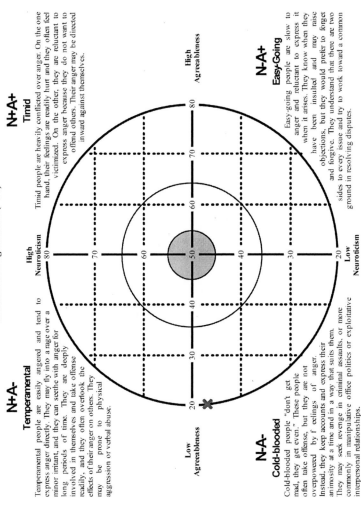

N+A−
Temperamental

Temperamental people are easily angered and tend to express anger directly. They may fly into a rage over a minor irritant, and they can seethe with anger for long periods of time. They are deeply involved in themselves and take offense readily, and they often overlook the effects of their anger on others. They may be prone to physical aggression or verbal abuse.

N+A+
Timid

Timid people are heavily conflicted over anger. On the one hand, their feelings are readily hurt and they often feel victimized. On the other, they are reluctant to express anger because they do not want to offend others. Their anger may be directed inward against themselves.

N−A−
Cold-blooded

Cold-blooded people "don't get mad, they get even." These people often take offense, but they are not overpowered by feelings of anger. Instead, they keep accounts and express their animosity at a time and in a way that suits them. They may seek revenge in criminal assaults, or more commonly in manipulative office politics or exploitative interpersonal relationships.

N−A+
Easy-Going

Easy-going people are slow to anger and reluctant to express it when it arises. They know when they have been insulted and may raise objections, but they would prefer to forget and forgive. They understand that there are two sides to every issue and try to work toward a common ground in resolving disputes.

Figure 4.1. Style of Anger Control graph for "Madeline G" from Costa and Piedmont (2003). Revised NEO Personality Inventory (NEO-PI-R) factor scores are reported as T-scores, which have a mean of 50 and standard deviation of 10. Individuals whose scores place them outside the shaded circle can be characterized in terms of one of the four styles. Madeline G's scores classify her Style of Anger Control as "Cold-blooded."

personality was similar to that described by King and Figueredo (1997). On the other hand, considerably lower congruence coefficients indicated that neither Neuroticism nor Openness was replicated in these studies. However, loadings on these two factors were not at odds with definitions of Neuroticism or Openness, and a follow-up analysis revealed that these factors also did not replicate in a second sample of zoo chimpanzees. These findings suggest that the failure to replicate the Neuroticism and Openness dimensions is probably attributable to properties of the questionnaire and not to their absence in other samples (see King et al. 2005 for more details).

TOP-DOWN VERSUS BOTTOM-UP APPROACHES

Saucier and Goldberg (2001) have argued that the apparent universality of personality structure is an artifact reflecting the top-down or etic approach used by McCrae and his colleagues (2005)—that is, the structure is imposed by virtue of the items within the questionnaire. Saucier and Goldberg favor bottom-up or emic approaches, such as the lexical approach, which involve assessing individuals within cultures using personality-descriptive adjectives derived from within those cultures. A similar criticism could, of course, be raised to question the seeming universality of chimpanzee personality structure or rating nonhuman primates and other animals on questionnaires based on the FFM (Uher 2008a; 2008b). Uher's alternative approach is also a bottom-up approach in that she recommends that researchers base measures or questionnaires of nonhuman personality on a species' naturally occurring behaviors.

However, there is evidence that these criticisms of the top-down approach may be overstated and that use of the same questionnaire items will not obscure species differences. For example, while the questionnaire King and Figueredo (1997) used to study chimpanzees was based on the FFM, the structure that emerged in chimpanzees differed from the human structure: it included a sixth factor, Dominance, and the other factors were not identical to their human analogues. Similarly, when a slightly expanded version of this questionnaire was used to assess orangutan personality (Weiss et al. 2006), the structure differed from that of chimpanzees and humans: Conscientiousness and Openness items loaded on a single factor named Intellect, and Dominance was more narrowly defined.

Studies using a personality questionnaire designed by Stevenson-Hinde and Zunz (1978) have also revealed differences. A study of rhesus macaque personality found four factors: Sociable, Equable, Confident, and Excitable (Capitanio 1999). On the other hand, a study of gorillas found four seemingly different factors: Extroverted, Dominant, Fearful, and Understanding (Gold and Maple 1994). Of course, in the latter example, different names may obscure similarities between the rhesus macaque and gorilla personality structures. However, a visual inspection of the loadings suggests that these factors differ (Capitanio 1999, table II; Gold and Maple 1994, table 1). To test how similar these factors were, we conducted an orthogonal targeted Procrustes rotation on the published results of these studies; this type of factor rotation involves rotating the structure derived from one sample so that it maximally resembles that of another structure (McCrae et al. 1996). For the present analysis we rotated the gorilla factor structure to the rhesus macaque factor structure using only the loadings of the 12 items that were in common between the two studies. Also, because exact factor loadings were not provided for the gorilla data, we defined loadings in both samples as being equal to 0, 1, or -1 depending on whether the item was not associated with the factor, positively associated with the factor, or negatively associated with the factor, respectively. The congruence coefficients suggest that rhesus structure was not replicated in gorillas (.73), that only the Equable factor replicated (.86), and that the loadings of only five items (*equable, understanding, sociable, playful*, and *curious*) replicated (see table 4.1). In short, this analysis suggested that the factor structures of gorillas and rhesus macaques differed.

WHAT CAN BE LEARNED FROM A STUDY OF STRUCTURE?

Once a sufficient number of personality structures are determined using comparable instruments, researchers can begin comparing them in a phylogenetic context to infer the historical patterns (when, where, and why) of personality evolution (Gosling and Graybeal 2007). For example, while the broad personality domains found in the great apes are similar, we are still left to puzzle out the ultimate causes of structural differences—for example, why orangutans share no domain equivalent to Conscientiousness or why humans lack a Dominance domain. Presumably, these structural differences will be related to adaptations by each species to their ecological and social environments. A challenge in phylogenetic inference about personality structures is that personality dimensions are not species-typical traits (as a new anatomical feature or behavior might be), but rather a feature that differs among individuals. Orangutans lack a distinct Conscientious domain but this does not mean they are unconscientious. The maintenance of variation of personality within species can also be informed by the determination of structure. If dimensions of a species' personality structure are genetically correlated, then knowing the whole structure (not just one or

Table 4.1. Gorilla Personality Factor Structure Rotated to Rhesus Macaque Factor Structure

Factor Item	Equable	Confidence	Excitable	Sociable	Congruence
Equable	.98	.00	.03	.19	.98
Understanding	.98	.00	.03	.19	.98
Slow	.19	.01	−.22	−.96	.19
Confident	−.01	.55	−.81	.19	.55
Aggressive	.01	.83	.54	−.11	.83
Effective	.01	.83	.54	−.11	.83
Excitable	.01	.83	.54	−.11	.54
Active	−.19	−.01	.22	.96	.22
Subordinate	.01	−.55	.81	−.19	.81
Sociable	−.19	−.01	.22	.96	.96
Playful	−.19	−.01	.22	.96	.96
Curious	−.19	−.01	.22	.96	.96
Congruence	.86	.78	.58	.76	.73

Note. Gorilla structure derived from Table 1 in Gold and Maple (1994). Macaque structure derived from Table II in Capitano (1999). Congruence coefficients ≥ .85 indicate replication of factors, item loadings, or the structure.

two dimensions) would be necessary to understand how personality traits respond to correlated selection and relate to life-history trade-offs.

Conclusions

We are pleased to see the recent growth and development of research in nonhuman animal personality research. However, we think that the relatively minor schism between behavioral ecologists and differential psychologists, if allowed to grow wider, can potentially threaten progress in this endeavor (see also Nettle and Penke 2010; chapter 5 in this volume). Such an event would not be without precedent in the study of animal behavior (see, e.g., the early rift, now mostly healed, between ethology and comparative psychology). As in biological evolution, we think that when research programs become overly specialized and rigid (i.e., lack diversity), they are threatened with extinction, or, at the very least, becoming a degenerating research program (Lakatos 1976).

While these potential consequences are dire, preventing this form of ethological isolation may be relatively simple. The present chapter and volume represent two such efforts. In addition, we suggest that behavioral ecologists and differential psychologists try to incorporate each other's methods, publish in each other's journals, and attend each other's conferences. Also, when appropriate, graduating PhDs in behavioral ecology should consider seeking placements in differential psychology labs. In short, researchers studying personality could opt to follow their own ways and end up as the ring-tailed lemurs of animal behavior research, whiling away on isolated scientific islands. Alternatively, these researchers could choose to increase their behavioral diversity and spread over a wide range of environments, much like rhesus macaques, or even humans.

References

Allport, G. W., and Odbert, H. S. 1936. Trait-names, a psycho-lexical study. *Psychological Monographs*, 47, 1–171.

Almada, S. J., Zonderman, A. B., Shekelle, R. B., Dyer, A. R., Daviglus, M. L., Costa, P. T., Jr., Stamler, J. 1991. Neuroticism and cynicism and risk of death in middle-aged men: the Western Electric Study. *Psychosomatic Medicine*, 53, 165–175.

American Psychiatric Association. 2004. *Diagnostic and Statistical Manual of Mental Disorders* 4th ed., rev. Washington, DC: American Psychiatric Association.

Bailey, J. N., Breidenthal, S. E., Jorgensen, M. J., McCracken, J. T., and Fairbanks, L. A. 2007. The association of DRD4 and novelty seeking is found in a nonhuman primate model. *Psychiatric Genetics*, 17, 23–27.

Barton, N. H., and Turelli, M. 1989. Evolutionary quantitative genetics: how little do we know? *Annual Review of Genetics*, 23, 337–370.

Bell, A. M., and Sih, A. 2007. Exposure to predation generates personality in threespined sticklebacks (*Gasterosteus aculeatus*). *Ecology Letters*, 10, 828–834.

Bergmüller, R., and Taborsky, M. 2010. Animal personality due to social niche specialisation. *Trends in Ecology and Evolution*, 25, 504–511.

Bethea, C. L., Streicher, J. M., Coleman, K., Pau, F. K.-Y., Moessner, R., and Cameron, J. L. 2004. Anxious behavior and fenfluramine-induced prolactin secretion in young rhesus macaques with different alleles of the serotonin reuptake transporter polymorphism (5HTTLPR). *Behavior Genetics*, 34, 295–307.

Biro, P. A., and Stamps, J. A. 2008. Are animal personality traits linked to life-history productivity? *Trends in Ecology and Evolution*, 23, 361–368.

Bogg, T., and Roberts, B. W. 2004. Conscientiousness and health-related behaviors: a meta-analysis of the leading behavioral contributors to mortality. *Psychological Bulletin*, 130, 887–919.

Bouchard, T. J., Jr., and Loehlin, J. C. 2001. Genes, evolution, and personality. *Behavior Genetics*, 31, 243–273.

Campbell, D. T., and Fiske, D. W. 1959. Convergent and discriminant validation by the multitrait-multimethod matrix. *Psychological Bulletin*, 56, 81–105.

Capitanio, J. P. 1999. Personality dimensions in adult male rhesus macaques: prediction of behaviors across time and situation. *American Journal of Primatology*, 47, 299–320.

———. 2011. Nonhuman primate personality and immunity: mechanisms of health and disease. In: *Personality and Temperament in Nonhuman Primates* (Weiss, A., King, J. E., and Murray, L. E., eds.), pp. 233–256. New York: Springer.

Capitanio, J. P., Mendoza, S. P., and Baroncelli, S. 1999. The relationship of personality dimensions in adult male rhesus macaques to progression of simian immunodeficiency virus disease. *Brain Behavior and Immunity*, 13, 138–154.

Champoux, M., Bennett, A., Shannon, C., Higley, J. D., Lesch, K. P., and Suomi, S. J. 2002. Serotonin transporter gene polymorphism, differential early rearing, and behavior in rhesus monkey neonates. *Molecular Psychiatry*, 7, 1058–1063.

Costa, P. T., Jr., and McCrae, R. R. 1992a. Four ways five factors are basic. *Personality and Individual Differences*, 13, 653–665.

———. 1992b. Reply to Eysenck. *Personality and Individual Differences*, 13, 861–865.

———. 1992c. *Revised NEO Personality Inventory (NEO-PI-R) and NEO Five-Factor Inventory (NEO-FFI) Professional Manual*. Odessa, FL: Psychological Assessment Resources.

———. 2002. Looking backward: changes in the mean levels of personality traits from 80 to 12. In: *Advances in Personality Science* (Cervone, D., and Mischel, W., eds.), pp. 219–237. New York: Guilford.

Costa, P. T., Jr., and Piedmont, R. L. 2003. Multivariate assessment: NEO-PI-R profiles of Madeline G. In: *Paradigms of Personality Assessment* (Wiggins, J. S., ed.), pp. 262–280. New York: Guilford.

Crawford, M. P. 1938. A behavior rating scale for young chimpanzees. *Journal of Comparative Psychology*, 26, 79–91.

Cronbach, L. J., and Meehl, P. E. 1955. Construct validity in psychological tests. *Psychological Bulletin*, 52, 281–302.

Dall, S. R. X., Houston, A. I., and McNamara, J. M. 2004. The behavioural ecology of personality: consistent individual differences from an adaptive perspective. *Ecology Letters*, 7, 734–739.

Daniewski, W., and Jezierski, T. 2003. Effectiveness of divergent selection for open-field activity in rabbits and correlated response for body weight and fertility. *Behavior Genetics*, 33, 337–345.

Danner, D. D., Snowdon, D. A., and Friesen, W. V. 2001. Positive emotions in early life and longevity: findings from the nun study. *Journal of Personality and Social Psychology*, 80, 804–813.

Darwin, C. 1859. *On the Origin of Species*. Cambridge, MA: Harvard University Press.

DeNeve, K. M., and Cooper, H. 1998. The happy personality: a meta-analysis of 137 personality traits and subjective well-being. *Psychological Bulletin*, 124, 197–229.

Diener, E., Lucas, R. E., and Scollon, C. N. 2006. Beyond the hedonic treadmill: revising the adaptation theory of well-being. *American Psychologist*, 61, 305–314.

Diener, E., Suh, E. M., Lucas, R. E., and Smith, H. L. 1999. Subjective well-being: three decades of progress. *Psychological Bulletin*, 125, 276–302.

Digman, J. M. 1990. Personality structure: emergence of the Five-Factor Model. *Annual Review of Psychology*, 41, 417–440.

Dingemanse, N. J., Both, C., Drent, P. J., and Tinbergen, J. M. 2004. Fitness consequences of avian personalities in a fluctuating environment. *Proceedings of the Royal Society of London B*, 271, 847–852.

Dingemanse, N. J., and Réale, D. 2005. Natural selection and animal personality. *Behaviour*, 142, 1159–1184.

Dingemanse, N. J., and Wolf, M. 2010. Recent models for adaptive personality differences: a review. *Philosophical Transactions of the Royal Society of London B*, 365, 3947–3958.

Dingemanse, N. J., Wright, J., Kazem, A. J. N., Thomas, D. K., Hickling, R., and Dawnay, N. 2007. Behavioural syndromes differ predictably between 12 populations of three-spined stickleback. *Journal of Animal Ecology*, 76, 1128–1138.

Dobzhansky, T. 1970. *Genetics of the Evolutionary Process*. New York: Columbia University Press.

Dochtermann, N. A., and Roff, D. A. 2010. Applying a quantitative genetics framework to behavioural syndrome research. *Philosophical Transactions of the Royal Society of London B*, 365, 4013–4020.

Eaves, L. J., Martin, N. G., Heath, A. C., Hewitt, J. K., and Neale, M. C. 1990. Personality and reproductive fitness. *Behavior Genetics*, 20, 563–568.

Epstein, S. 1979. The stability of behavior, I: on predicting most of the people much of the time. *Journal of Personality and Social Psychology*, 37, 1097–1126.

Eysenck, H. J. 1992a. Four ways five factors are not basic. *Personality and Individual Differences*, 13, 667–673.

———. 1992b. A reply to Costa and McCrae: P or A and C—the role of theory. *Personality and Individual Differences*, 13, 867–868.

Fairbanks, L. A., Newman, T. K., Bailey, J. N., Jorgensen, M. J., Breidenthal, S. E., Ophoff, R. A., Comuzzie, A. G., Martin, L. J., Rogers, J. 2004. Genetic contributions to social impulsivity and aggressiveness in vervet monkeys. *Biological Psychiatry*, 55, 642–647.

Falconer, D. S., and Mackay, T. F. C. 1996. *Introduction to Quantitative Genetics*. 4th ed. Essex, England: Longman.

Figueredo, A. J., Vásquez, G., Brumbach, B. H., and Schneider, S. M. R. 2004. The heritability of life history strategy: the K-factor, covitality, and personality. *Social Biology*, 51, 121–143.

———. 2007. The K-factor, covitality, and personality: a psychometric test of life history theory. *Human Nature*, 18, 47–73.

Galton, F. 1883. *Inquiries into Human Faculty and Its Development*. London: Macmillan.

Gold, K. C., and Maple, T. L. 1994. Personality assessment in the gorilla and its utility as a management tool. *Zoo Biology*, 13, 509–522.

Goodall, J. 1986. *The Chimpanzees of Gombe: Patterns of Behavior*. Cambridge, MA: Belknap Press of Harvard University.

Gorsuch, R. L. 1983. *Factor Analysis*. 2nd ed. Hillsdale, NJ: Lawrence Erlbaum.

Gosling, S. D. 2001. From mice to men: what can we learn about personality from animal research? *Psychological Bulletin*, 127, 45–86.

———. 2008. Personality in nonhuman animals. *Social and Personality Psychology Compass*, 2, 985–1001.

Gosling, S. D., and Graybeal, A. 2007. Tree thinking: a new paradigm for integrating comparative data in psychology. *Journal of General Psychology*, 134, 259–277.

Gosling, S. D., and John, O. P. 1999. Personality dimensions in nonhuman animals: a cross-species review. *Current Directions in Psychological Science*, 8, 69–75.

Haven, S., and ten Berge, J. M. F. 1977. *Tucker's Coefficient of Congruence as a Measure of Factorial Invariance: an Empirical Study* (Heymans Bulletins, HB 77–290-EX). Groningen, the Netherlands: University of Groningen.

Ironson, G., and Hayward, H. 2008. Do positive psychosocial factors predict disease progression in HIV-1? a review of the evidence. *Psychosomatic Medicine*, 70, 546–554.

Ironson, G., O'Cleirigh, C., Weiss, A., Schneiderman, N., and Costa, P. T., Jr. 2008. Personality and HIV disease progression: the role of NEO-PI-R openness, extraversion, and profiles of engagement. *Psychosomatic Medicine*, 70, 245–253.

Itani, J. 1957. Personality of Japanese monkeys. *Iden*, 11, 29–33.

Iwasa, H., Masui, Y., Gondo, Y., Inagaki, H., Kawaai, C., and Suzuki, T. 2008. Personality and all-cause mortality among older adults dwelling in a Japanese community: a five-year population-based prospective cohort study. *American Journal of Geriatric Psychiatry*, 16, 399–405.

Kendler, K. S., Gatz, M., Gardner, C. O., and Pedersen, N. L. 2006. Personality and major depression: a Swedish longitudinal, population-based twin study. *Archives of General Psychiatry*, 63, 1113–1120.

Kern, M. L., and Friedman, H. S. 2008. Do conscientious individuals live longer? a quantitative review. *Health Psychology*, 27, 505–512.

Kimura, M. 1983. *The Neutral Theory of Molecular Evolution*. Cambridge: Cambridge University Press.

———. 1986. DNA and the neutral theory. *Philosophical Transactions of the Royal Society of London B*, 312, 343–354.

Kinder, L. S., Bradley, K. A., Katon, W. J., Ludman, E., McDonell, M. B., and Bryson, C. L. 2008. Depression, posttraumatic stress disorder, and mortality. *Psychosomatic Medicine*, 70, 20–26.

King, J. E., and Figueredo, A. J. 1997. The Five-Factor Model plus Dominance in chimpanzee personality. *Journal of Research in Personality*, 31, 257–271.

King, J. E., Weiss, A., and Farmer, K. H. 2005. A chimpanzee (*Pan troglodytes*) analogue of cross-national generalization of personality structure: zoological parks and an African sanctuary. *Journal of Personality*, 73, 389–410.

Kraemer, G. W., Moore, C. F., Newman, T. K., Barr, C. S., and Schneider, M. L. 2008. Moderate-level fetal alcohol exposure and serotonin transporter gene promoter polymorphism affect neonatal temperament and limbic-hypothalamic-pituitary-adrenal axis regulation in monkeys. *Biological Psychiatry*, 63, 317–324.

Lakatos, I. 1976. *Proofs and Refutations: The Logic of Mathematical Discovery*. Cambridge: Cambridge University Press.

Lande, R., and Arnold, S. J. 1983. The measurement of selection on correlated characters. *Evolution*, 37, 1210–1226.

Lee, K., Ogunfowora, B., and Ashton, M. C. 2005. Personality traits beyond the Big Five: are they within the HEXACO space? *Journal of Personality*, 73, 1437–1463.

Lucas, R. E., Clark, A. E., Georgellis, Y., and Diener, E. 2003. Reexamining adaptation and the set point model of happiness: reactions to changes in marital status. *Journal of Personality and Social Psychology*, 84, 527–539.

Lynch, M., and Hill, W. G. 1986. Phenotypic evolution by neutral mutation. *Evolution*, 40, 915–935.

Lyubomirsky, S., King, L., and Diener, E. 2005. The benefits of frequent positive affect: does happiness lead to success? *Psychological Bulletin*, 131, 803–855.

MacArthur, R. H., and Pianka, E. R. 1966. On optimal use of a patchy environment. *American Naturalist*, 100, 603–609.

Maninger, N., Capitanio, J. P., Mendoza, S. P., and Mason, W. A. 2003. Personality influences tetanus-specific antibody response in adult male rhesus macaques after removal from natal group and housing relocation. *American Journal of Primatology*, 61, 73–83.

Martin, L. R., Friedman, H. S., and Schwartz, J. E. 2007. Personality and mortality risk across the life span: the importance of conscientiousness as a biopsychosocial attribute. *Health Psychology*, 26, 428–436.

McCrae, R. R., Costa, P. T., Jr., Martin, T. A., Oryol, V. E., Rukavishnikov, A. A., Senin, I. G., Hrebickova M., Urbanek T. 2004. Consensual validation of personality traits across cultures. *Journal of Research in Personality*, 38, 179–201.

McCrae, R. R., Terracciano, A., and 78 members of the Personality Profiles of Cultures Project. 2005. Universal features of personality traits from the observer's perspective: data from 50 cultures. *Journal of Personality and Social Psychology*, 88, 547–561.

McCrae, R. R., Zonderman, A. B., Bond, M. H., Costa, P. T., Jr., and Paunonen, S. V. 1996. Evaluating replicability of factors in the Revised NEO Personality Inventory: confirmatory factor analysis versus Procrustes rotation. *Journal of Personality and Social Psychology*, 70, 552–566.

Merilä, J., Sheldon, B. C., and Kruuk, L. E. B. 2001. Explaining stasis: microevolutionary studies in natural populations. *Genetica*, 112, 199–222.

Mischel, W. 1968. *Personality and Assessment*. New York: Wiley.

Nettle, D. 2006. The evolution of personality variation in humans and other animals. *American Psychologist*, 61, 622–631.

Nettle, D., and Penke, L. 2010. Personality: bridging the literatures from human psychology and behavioural ecology. *Philosophical Transactions of the Royal Society of London B*, 365, 4043–4050.

Newman, T. K., Syagailo, Y. V., Barr, C. S., Wendland, J. R., Champoux, M., Graessle, M., Suomib, S. J., Higleya, J. D., Leschc, K.-P. 2005. Monoamine oxidase A gene promoter variation and rearing experience influences aggressive behavior in rhesus monkeys. *Biological Psychiatry*, 57, 167–172.

O'Cleirigh, C., Ironson, G., Weiss, A., and Costa, P. T., Jr. 2007. Conscientiousness predicts disease progression (CD4 number and viral load) in people living with HIV. *Health Psychology*, 26, 473–480.

Ozer, D. J., and Benet-Martínez, V. 2006. Personality and the prediction of consequential outcomes. *Annual Review of Psychology*, 57, 401–421.

Penke, L., Denissen, J. J. A., and Miller, G. F. 2007a. Evolution, genes, and interdisciplinary personality research. *European Journal of Personality*, 21, 639–665.

———. 2007b. The evolutionary genetics of personality. *European Journal of Personality*, 21, 549–587.

———. 2007c. Open peer commentary. *European Journal of Personality*, 21, 589–637.

Pianka, E. R. 1970. On r- and K-selection. *American Naturalist*, 104, 592–597.

Réale, D., and Festa-Bianchet, M. 2003. Predator-induced natural selection on temperament in bighorn ewes. *Animal Behaviour*, 65, 463–470.

Réale, D., Gallant, B. Y., Leblanc, M., and Festa-Bianchet, M. 2000. Consistency of temperament in bighorn ewes and correlates with behaviour and life history. *Animal Behaviour*, 60, 589–597.

Réale, D., Garant, D., Humphries, M. M., Bergeron, P., Careau, V., and Montiglio, P. O. 2010. Personality and the emergence of the pace-of-life syndrome concept at the population level. *Philosophical Transactions of the Royal Society of London B*, 365, 4051–4063.

Réale, D., Reader, S. M., Sol, D., McDougall, P. T., and Dingemanse, N. J. 2007. Integrating animal temperament within ecology and evolution. *Biological Reviews*, 82, 291–318.

Roberts, B. W., Kuncel, N. R., Shiner, R., Caspi, A., and Goldberg, L. R. 2007. The power of personality: the comparative validity of personality traits, socioeconomic status, and cognitive ability for predicting important life outcomes. *Perspectives on Psychological Science*, 2, 313–345.

Rogers, J., Shelton, S. E., Shelledy, W., Garcia, R., and Kalin, N. H. 2008. Genetic influences on behavioral inhibition and anxiety in juvenile rhesus macaques. *Genes, Brain, and Behavior*, 7, 463–469.

Saucier, G., and Goldberg, L. R. 2001. Lexical studies of indigenous personality factors: premises, products, and prospects. *Journal of Personality*, 69, 847–879.

Saulsman, L. M., and Page, A. C. 2004. The Five-Factor Model and personality disorder empirical literature: a meta-analytic review. *Clinical Psychology Review*, 23, 1055–1085.

Shadish, W. R. 1992. Critical multiplism: a research strategy and its attendant tactics. *New Directions for Program Evaluation*, 60, 13–57.

Sih, A., Bell, A. M., Johnson, J. C., and Ziemba, R. E. 2004. Behavioral syndromes: an integrative overview. *Quarterly Review of Biology*, 79, 241–277.

Stamps, J. A. 2007. Growth-mortality tradeoffs and "personality traits" in animals. *Ecology Letters*, 10, 355–363.

Steel, P., Schmidt, J., and Shultz, J. 2008. Refining the relationship between personality and subjective well-being. *Psychological Bulletin*, 134, 138–161.

Stevenson-Hinde, J., and Zunz, M. 1978. Subjective assessment of individual rhesus monkeys. *Primates*, 19, 473–482.

Tinbergen, N. 1951. *The Study of Instinct*. Oxford: Oxford University Press.

———. [1963] 2005. On aims and methods of ethology. *Animal Biology*, 55, 297–321. Reprinted from *Zeitschrift fur Tierpsychologie*, 20, 410.

Tooby, J., and Cosmides, L. 1990. On the universality of human nature and the uniqueness of the individual: the role of genetics and adaptation. *Journal of Personality*, 58, 17–67.

Uher, J. 2008a. Comparative personality research: methodological approaches. *European Journal of Personality*, 22, 475–496.

———. 2008b. Three methodological core issues of comparative personality research. *European Journal of Personality*, 22, 475–496.

Uher, J., and Asendorpf, J. B. 2008. Personality assessment in the great apes: comparing ecologically valid behavior measures, behavior ratings, and adjective ratings. *Journal of Research in Personality*, 42, 821–838.

Van Oers, K., de Jong, G., Drent, P. J., and Van Noordwijk, A. J. 2004. A genetic analysis of avian personality traits: correlated response to artificial selection. *Behavior Genetics*, 34, 611–619.

Van Oers, K., de Jong, G., Van Noordwijk, A. J., Kempenaers, B., and Drent, P. J. 2005. Contribution of genetics to animal personalities: a review of case studies. *Behaviour*, 142, 1191–1212.

Van Oers, K., and Mueller, J. C. 2010. Evolutionary genomics of animal personality. *Philosophical Transactions of the Royal Society of London B*, 365, 3991–4000.

Weiss, A., Bates, T. C., and Luciano, M. 2008. Happiness is a personal(ity) thing—the genetics of personality and well-being in a representative sample. *Psychological Science*, 19, 205–210.

Weiss, A., and Costa, P. T., Jr. 2005. Domain and facet personality predictors of all-cause mortality among Medicare patients aged 65 to 100. *Psychosomatic Medicine*, 67, 724–733.

Weiss, A., Inoue-Murayama, M., Hong, K.-W., Inoue, E., Udono, S., Ochiai, T., Matsuzawa T, Hirata, S., King, J. E. 2009. Assessing chimpanzee personality and subjective well-being in Japan. *American Journal of Primatology*, 71, 283–292.

Weiss, A., King, J. E., and Enns, R. M. 2002. Subjective well-being is heritable and genetically correlated with dominance in chimpanzees (*Pan troglodytes*). *Journal of Personality and Social Psychology*, 83, 1141–1149.

Weiss, A., King, J. E., and Figueredo, A. J. 2000. The heritability of personality factors in chimpanzees (*Pan troglodytes*). *Behavior Genetics*, 30, 213–221.

Weiss, A., King, J. E., and Hopkins, W. D. 2007. A cross-setting study of chimpanzee (*Pan troglodytes*) personality structure and development: zoological parks and Yerkes National Primate Research Center. *American Journal of Primatology*, 69, 1264–1277.

Weiss, A., King, J. E., and Perkins, L. 2006. Personality and subjective well-being in orangutans (*Pongo pygmaeus* and *Pongo abelii*). *Journal of Personality and Social Psychology*, 90, 501–511.

Weiss, A., Sutin, A. R., Duberstein, P. R., Friedman, B., Bagby, R. M., and Costa, P.T., Jr. 2009. The personality domains and styles of the Five-Factor Model are related to incident depression in Medicare recipients aged 65 to 100. *American Journal of Geriatric Psychiatry*, 17, 591–601.

Whiteman, M. C., Deary, I. J., and Fowkes, F. G. R. 2000. Personality and health: cardiovascular disease. In: *Advances in Personality Psychology* (vol. 1, Hampson, S., ed.), pp. 157–198. New York: Psychology Press.

Widaman, K. F. 1985. Hierarchically nested covariance structure models for multitrait-multimethod data. *Applied Psychological Measurement*, 9, 1–26.

Widiger, T. A., and Frances, A. J. 2002. Toward a dimensional model for the personality disorders. In: *Personality Disorders and the Five-Factor Model of Personality* (2nd ed., Costa, P. T., Jr., and Widiger, T. A., eds.), pp. 23–44. Washington, DC: American Psychological Association.

Williamson, D. E., Coleman, K., Bacanu, S.-A., Devlin, B. J., Rogers, J., Ryan, N. D., Cameron, J. L. 2003. Heritability of fearful-anxious endophenotypes in infant rhesus macaques: a preliminary report. *Biological Psychiatry*, 53, 284–291.

Wilson, R. S., de Leon, C. F. M., Bienias, J. L., Evans, D. A., and Bennett, D. A. 2004. Personality and mortality in old age. *Journals of Gerontology Series B—Psychological Sciences and Social Sciences*, 59, P110-P116.

Wilson, R. S., Schneider, J. A., Arnold, S. E., Bienias, J. L., and Bennett, D. A. 2007. Conscientiousness and the incidence of Alzheimer disease and mild cognitive impairment. *Archives of General Psychiatry*, 64, 1204–1212.

Wolf, M., Van Doorn, S., Leimar, O., and Weissing, F. J. 2007. Life-history trade-offs favour the evolution of animal personalities. *Nature*, 447, 581–585.

Wolf, M., and Weissing, F. J. 2010. An explanatory framework for adaptive personality differences. *Philosophical Transactions of the Royal Society of London B*, 365, 3959–3968.

Yerkes, R. M. 1939. The life history and personality of the chimpanzee. *American Naturalist*, 73, 97–112.

5

Personalities in a Comparative Perspective:

What Do Human Psychologists Glean from

Animal Personality Studies?

SAMUEL D. GOSLING AND PRANJAL H. MEHTA

Introduction

Comparative research has long played a central role in many areas of psychology, including learning, sensation and perception, memory, and psychopathology (Domjan and Purdy 1995). As we and others have argued before, comparative research also has an important contribution to make to personality psychology (Gosling 2001; Nettle and Penke 2010; chapter 4, this volume). Indeed, with advances in genomics, neuroscience, and phylogenetics, the potential contributions to be made by cross-species research are now greater than ever. And with continued progress in the measurement of personality in animals and in identifying cross-species generalities in personality traits, the assessment of personality in animals also stands on increasingly solid ground.

Pet owners, zoo keepers, and many practitioners and scientists who work with animals have long been aware of individual differences in animals and have happily referred to these differences in terms of personalities (Hebb 1946; Stevenson-Hinde et al. 1980). However, many researchers, even those who acknowledge the existence of individual differences in animals, feel it is anthropomorphic to apply to nonhumans a term that includes the word "person" in it. As a result, terms such as *temperament* and *behavioral syndrome* or *style* are often used instead of *personality* (see introduction, this volume).

In human research, temperament is considered a construct closely related to personality, and has been defined as the inherited, early-appearing tendencies that continue throughout life and serve as the foundation for personality. This definition is not adopted uniformly by animal or even human researchers (McCrae et al. 2000). However, a similar definition has gained some acceptance among nonhuman primate researchers; specifically,

Clarke and Boinski (1995) use the term *temperament* to refer to behavioral styles or tendencies that show continuity over time and can be identified in early infancy, and which are reflected in the degree and nature of responsivity to novel or stressful stimuli.

The term *behavioral syndrome* has gained recent popularity in the field of behavioral ecology (Sih et al. 2004). Behavioral syndromes are defined as suites of correlated behaviors expressed either within a given behavioral context or across different contexts. This definition very closely matches the concept of personality in humans.

Weinstein, Capitanio, and Gosling (2008) argued in favor of using the term *personality* for three reasons. "First, it is confusing to create new terms without a compelling conceptual reason to do so. Second, using the term *personality* facilitates connections with the enormous existing research on personality in humans" (p. 330). Third, Weinstein et al. argue, "it is not useful, as some have suggested, to adopt the term *temperament* for nonhumans because this entails a priori assumptions (e.g., about traits being inherited and appearing early) that may or may not be appropriate; for example, it is increasingly clear that individual differences in adult animal behavior are a function of both biological tendencies and experience, as is the case with humans" (p. 330).

On the basis of these arguments, we use the term *personality* and define it rather broadly as those characteristics of individuals that describe and account for consistent patterns of feeling, cognition, and behaving.

Review of animal personality literature

CONCERNS ABOUT THE EXISTENCE OF PERSONALITY IN ANIMALS

As noted above, many people, even some working in the sciences, have been reluctant to concede that personality exists in nonhuman animals. Their concerns range from philosophical arguments regarding the uniqueness of humans to methodological concerns about the perils of anthropomorphism (Gosling 2001). To address concerns about the existence of personality in animals, Gosling, Lilienfeld, and Marino (2003; see also Gosling and Vazire 2002) proposed borrowing three criteria that emerged from the debate concerning the existence of personality in humans (Kenrick and Funder 1988): (1) assessments by independent observers must agree with one another, (2) those assessments must predict behaviors or other real-world outcomes, and (3) observer ratings must be shown to reflect genuine attributes of the individuals rated, not just the observers' implicit theories about how personality traits covary.

Criterion 1: Independent assessments must agree. If individual differences in personality exist and can be detected, then independent observers should agree about the relative standing of individuals on personality traits (Gosling et al. 2003a, b). Studies of humans rating other humans typically elicit interobserver agreement correlations in the region of .50 (e.g., Funder et al. 1995), supporting the idea that humans agree with their ratings of one another and providing a standard by which judgments of animals can be evaluated.

There is now a substantial corpus of research showing that observers agree strongly in their ratings of animals. Gosling (2001) summarized the findings from 21 rating studies of animal personality; the mean interobserver agreement correlation was .52, matching the magnitude of consensus correlations typically obtained in human research. Gosling's estimate is probably somewhat inflated because the studies reviewed by him were generally interested in creating reliable measures and not in estimating reliability *per se*; as a result the studies probably did not retain unreliable measures. However, even studies that retained all the measures show that strong levels of interobserver agreement are typically obtained; for example, in a study of personality ratings of hyenas, pairwise interjudge agreement correlations averaged .38 across the 44 traits examined (Gosling 1998).

Criterion 2: Assessments must predict behaviors and real-world outcomes. For personality traits to be of any use, ultimately they must predict behaviors and real-world outcomes. Though few animal studies exist in which personality measures have been tested (Gosling 2001), the evidence for concurrent and predictive validity is strong. Personality traits have been shown to predict specific behaviors (Pederson et al. 2005), occupational success (Maejima et al. 2007), and health outcomes (Capitanio et al. 1999). For example, Capitanio et al. (1999) found that four to six weeks after experimental inoculation with the simian immunodeficiency virus, highly sociable animals had lower viral load than did less sociable individuals.

Criterion 3: Ratings must reflect attributes of targets, not observers' implicit personality theories. Several studies of personality structure in animals have identified a number of broad dimensions, which often resemble dimensions found in studies of humans (Gosling and John 1999). These findings could be taken as evidence that animals have human-like personalities. However, it is possible that observers are not detecting the true structure of personality traits in animals, but are instead simply "filling in the blanks" using their knowledge of human personality structure. Although many animal studies of personality structure are based on personality ratings (e.g., "curiosity"), some others are based on behavioral tests (e.g., response to novel objects) and carefully recorded ethological observations (e.g., time spent exploring environment). Unlike the ratings-based factors, such behavior-based factors cannot be explained solely in terms of observers "filling in the blanks" on the basis of the semantic similarity of the traits. Moreover, in cases where cross-study comparisons can be made, the factors obtained from behavioral codings often resemble factors obtained from observer ratings, suggesting that the two methods assess the same underlying constructs (Gosling and John 1999; Rouff et al. 2005; Konečná et al. 2008). Together, the findings suggest that the structure of personality ratings is based, at least in part, on real attributes of the individuals being rated.

Further evidence that humans are not inappropriately projecting human traits onto animals was provided by a study that simultaneously evaluated projection of personality onto dogs and humans (Kwan et al. 2008). Results indicated that humans were no more likely to inappropriately project their own traits onto dogs than they were onto humans. Overall, the research literature suggests that when held to the standards met by studies of personality in humans, there is substantial evidence for the existence of personality in animals.

PERSONALITY RATINGS IN HUMANS AND ANIMALS
Animal personality can be measured in different ways but until recently no research had explicitly compared assessments of humans and animals in a single design (see also chapter 4). So Gosling et al. (2003a) directly compared side-by-side the accuracy of personality ratings of dogs to the accuracy of personality ratings of humans. Parallel procedures and instruments were used to compare personality judgments of 78 dogs and their owners in terms of three accuracy criteria: internal consistency, consensus, and correspondence. On all three criteria, judgments of dogs were as accurate as judgments of humans, again suggesting that personality differences do exist and demonstrating that personality traits can be measured in animals.

Rating versus coding. Broadly speaking, the two main methods for measuring personality in individual animals are coding of an animal's overt behaviors, and subjective ratings of broad traits by knowledgeable observers (e.g., human caretakers). These two methods reflect different resolutions to the supposed trade-off between quantifying personality in terms of objective behaviors and using humans to record and collate information more subjectively (Gosling 2001; Uher 2008; Uher and Asendorpf 2008). Behavioral coding methods are used to gather data from test situations designed to elicit personality-relevant responses (e.g., exposure to a novel stimulus) or from observations of naturalistic behavior (Freeman et al. 2011). Ratings are used to gather data from either of the above contexts (test situations and naturalistic behavior) as well as to draw on the accumulated experience of humans who know the animals well. To quantify personality traits with the rating method, humans are generally asked to rate each of the animals on a number of personality traits. There are typically multiple observers who complete ratings, and occasionally the ratings are made at several points in time (Gosling 2001; Weiss et al. 2006).

Behavior codings have been used more widely than ratings in animal-personality studies; a review by Gosling (2001) found that 74% of animal-personality studies had used behavior codings to assess personality, and only 34% had used trait ratings. However, direct comparisons of the two methods suggest that rating methods may be better than behavior coding methods for capturing personality traits because rating methods are generally more reliable, are not as subjective as is sometimes assumed, and are much more practical (Vazire et al. 2007). In one study of chimpanzees, the mean intraclass correlation (a measure of reliability) was .42 for behavior codings but was .61 for trait ratings, suggesting that trait ratings are well suited for detecting consistencies in animals' behaviors, the very foundation of personality (Vazire et al. 2007; see chapter 4). Behavior codings, in contrast, are often difficult to measure reliably, particularly when observations are made across different times of day or under varying conditions. It should be noted that some animal behavior laboratories have been able to achieve substantial reliability in behavior coding through rigorous training of coders. Indeed, many of these laboratories require that novice coders achieve a criterion level of agreement with experienced coders before participation in an actual study. But even when behaviors can be coded reliably, they may reflect other characteristics of the environment (e.g., situational influences) and not personality. Thus, the extant data suggest not that behavior codings are poor measures of *behavior*, but that they may be poor measures of *personality*. Behavior-coding methods may be better suited for experimental manipulations, where researchers are concerned with detecting the effects of situational variables on behavior (Vazire et al. 2007).

PERSONALITY DIMENSIONS ACROSS SPECIES

Empirical research on animal personality essentially comprises studies of traits—behavioral regularities that are relatively consistent across time and contexts. Commonly studied personality traits include exploration, boldness, fearfulness, aggression, general activity, emotionality, confidence, and timidity.

To get a better idea of which traits emerge in structural analyses of personality, Gosling and John (1999) reviewed 19 factor-analytic studies across

12 nonhuman species. They used the most widely accepted structure of human personality—the Five-Factor Model (John and Srivastava 1999)—along with the additional factors of Dominance and Activity to organize their findings. The Five-Factor Model is a hierarchical model with five broad factors, representing personality traits at the broadest level of abstraction. These five bipolar factors are commonly labeled Extraversion, Neuroticism, Agreeableness, Conscientiousness, and Openness (see chapter 4). The dimensions of Extraversion, Neuroticism, and Agreeableness showed considerable generality across the 12 species included in their review. Of the 19 studies reviewed, 17 identified a factor closely related to Extraversion, capturing dimensions ranging from Surgency in chimpanzees, Sociability in pigs, dogs, and rhesus monkeys, Energy in cats and dogs, and Vivacity in donkeys, to a dimension contrasting Bold approach vs. Avoidance in octopuses. Of course, the way these personality dimensions are manifested depends on the species; whereas the human scoring low on Extraversion stays at home on Saturday night, or tries to blend into a corner at a large party, the octopus scoring low on Boldness stays in its protective den during feedings and attempts to hide itself.

Factors related to Neuroticism appeared in 15 of the 19 studies, capturing dimensions such as Fearfulness, Emotional Reactivity, Excitability, and low Nerve Stability. Factors related to Agreeableness appeared in 14 studies, with Affability, Affection, and Social Closeness representing the high pole, and Aggression, Hostility, and Fighting representing the low pole. Factors related to Openness were identified in all but four of the 12 species; the two major components defining this dimension were Curiosity-Exploration and Playfulness. Dominance emerged as a clear separate factor in seven of the 19 studies, and a separate Activity dimension was identified in two of the studies.

Chimpanzees were the only nonhuman species with a separate Conscientiousness factor, which was defined more narrowly than in humans but included the lack of attention and goal-directedness and erratic, unpredictable, and disorganized behavior typical of the low pole. The existence of a separate Conscientiousness factor in only humans and their closest relative suggests that the trait evolved relatively recently in the evolution of Homininae. The finding is consistent with the fact that both humans and chimpanzees have relatively developed frontal cortices, the area of the brain associated with higher executive function such as making plans and controlling impulses (Beer et al. 2004). However, if one adopts a relatively broad definition of Conscientiousness (e.g., being reliable, organized, goal-directed, and self-disciplined; see chapter 4), then even some distantly related organisms (e.g., honey bees) may exhibit some aspects of this trait.

The benefits of animal personality studies for research on humans

Researchers examining basic questions about human personality must overcome a number of methodological hurdles. For example, in the field of personality and health, human researchers rely heavily on longitudinal studies, which can take decades to complete and are costly and logistically challenging because of the long lifespan and slow time course of disease progression. With reliance on human studies alone, it could take several decades or more to identify personality-health relationships to understand the factors that influence these relationships and to delineate the mechanisms by which these relationships come about. Another challenge for human research is that it is difficult or impossible to experimentally manipulate biological and environmental variables that may affect personality (e.g., rearing experience, hormone levels) because of ethical and practical considerations (Gosling 2001; Vazire and Gosling 2003; Mehta and Gosling 2006). As a result, even the most comprehensive human studies are subject to important limitations. Equivalent animal studies can be conducted with more efficiency and in more detail, and thus have the potential to complement research on human personality (Mehta and Gosling 2008).

Animal research should not replace human research, but studies in animals have already enriched our understanding of human personality and it appears that they will continue to do so (e.g., Nettle and Penke 2010). Animal studies are particularly well suited for identifying mediators and moderators in personality processes (Mehta and Gosling 2008; chapter 15). Mediators are factors that stand between personality traits and various outcomes (e.g., health outcomes such as longevity and disease progression). These can include behavioral factors that may be influenced by personality, such as aggression and exploration, as well as biological factors, such as endocrine function, gene expression, brain activity, and immune parameters. Moderators are factors that affect the direction or magnitude of links between personality and other variables. For example, certain environmental variables (e.g., rearing environment, social status, stress exposure) could moderate relationships between personality traits and health outcomes; that is, a specific trait could have a beneficial impact on health under certain environmental conditions, but the same trait could negatively influence or have no influence on health under other conditions. In particular, we suggest that there are four major benefits of animal studies for understanding personality, which we illustrate below with reference to elucidating links between personality and health.

Benefit 1: Greater experimental control. Animal studies permit experimental manipulations that are not possible in humans. These include the manipulation of biological (e.g., hormones, neurotransmitters, genes) and environmental factors (e.g., rearing conditions) that may influence personality. For example, research in nonhuman primates has shown that experimental manipulation of an animal's early life experience can affect both personality and immunity (cf. Capitanio 2011). These studies show that prenatal stress and maternal separation lead to the development of a behaviorally inhibited/low sociability personality, lower cellular immune function, and increased natural killer (NK) cell function. In contrast, nursery rearing leads to the development of a high-emotionality personality, enhanced cellular immune function, and decreased NK function (Capitanio 2011). These findings suggest that early life experiences might also influence the sociability-immunity or neuroticism-immunity relationships in humans. Animal research also permits selective breeding that can result in production of large numbers of animals with desired temperament characteristics, as illustrated in the Mouse Phenome Database (www.jax.org/phenome). These large populations can be used to study how personality traits such as neuroticism or sociability affect immunity, disease, and mortality or how specific personality traits interact with particular drugs or environmental conditions.

Benefit 2: Greater ability to measure physiological parameters. Animal studies afford greater opportunities to measure physiological parameters—such as markers of immune function, gene expression, and neuroendocrine function—that may affect personality traits (see chapters 10, 12, and 15, this volume). For example, a recent study by Vegas et al. (2006) exposed tumor-bearing and non–tumor-bearing mice to social stress. The researchers measured corticosterone responses to a social stressor and behavioral coping strategies during the stressor. Immune function was measured one hour and three days after the stress, and tumor development was measured 15 days after the stressor. The researchers found that social stress led to increased corticosterone levels and tumor development. However, mice characterized by a defensive and avoidant coping strategy during the social stressor showed the most severe tumor development. These findings suggest that social stress exacerbates tumor development, but that this effect depends on an individual's coping strategy in response to the social stressor. Such animal models can inform human research because avoidant and defensive coping styles also predict negative health outcomes in humans (Sapolsky 1998; Segerstrom and Miller 2004; chapter 15, this volume).

Benefit 3: Greater opportunities for naturalistic observation. The observational opportunities afforded by animal research are far greater than those available in human research; relative to humans, animals can be observed for greater periods, in more detail, and in more contexts. These observational data allow researchers to study relationships among personality, the environment, and health in far greater depth than is possible in humans. To illustrate, consider a research program by Capitanio, which for over a decade has been accruing personality data on over 175 rhesus monkeys (see Weinstein et al. 2008, for a description of this research program). Members of Capitanio's group assessed their personalities at 5–10 years of age, identifying a four-factor structure, which was later replicated with confirmatory factor analysis in a separate subsample (Capitanio 1999; Capitanio and Widaman 2005). Animals were tested in a variety of social and nonsocial situations, and behavioral and physiological measures were obtained in these situations for up to several years following the initial personality assessments; personality was found to predict various measures of social behavior and emotionality, plasma cortisol concentrations, tetanus- and herpesvirus-specific antibody responses, heart rate, and central nervous system functioning (Capitanio 2002; 2011). Most recently (Capitanio et al. 2008), this group has demonstrated that Sociability scores moderate the response to a social stressor, and influence expression of genes associated with innate immune responses. This latter finding shows that the effect of sociability on immunity depends on the social environment, which could inform the study of sociability and health in humans. Another example of naturalistic observations is Sapolsky's research program (e.g., Sapolsky 1998), which is based on naturalistic studies of personality, stress, and health in male olive baboons.

Benefit 4: Reduced time and cost of longitudinal studies. The shorter lifespan of many animal species makes it possible to conduct longitudinal studies that yield important insights in a timely manner and at a fraction of the cost of equally comprehensive human studies. Thus, animal research is well suited for study of such topics as the causes of personality development and the onset and progression of disease over the lifespan (see chapter 15). For example, Cavigelli et al. (2006) used a longitudinal design to track female rats prone to spontaneous mammary and pituitary tumors. They assessed variation in exploratory temperament during infancy and hormonal function at various periods during the life span. They found that individuals with low-exploratory infant temperament died earlier than individuals with high-exploratory infant temperament (see chapter 15). Moreover, there was evidence for differences in prolactin, estrogen, and progesterone functioning among low-exploratory and high-exploratory individuals. Taken together, the results demonstrate that infant nonexploratory temperament may be linked to tumor development and lifespan via neuroendocrine mechanisms. This animal model can inform human research because studies in humans have also identified an inhibited versus exploratory infant temperament style and have shown that it predicts outcomes relevant to stress and health (e.g., emotional reactivity; Schwartz et al. 2003).

What contributions can be made
by research on animal personality?

Drawing on the methodological benefits of animal studies outlined above, researchers have begun to make contributions to three broad domains, which are summarized below.

NEW WAYS TO ADDRESS LONGSTANDING QUESTIONS

Animal studies will never replace research on humans altogether but they do offer a number of powerful methods and unique opportunities for augmenting traditional human research. So far, much of the animal personality work in psychology has used animal models to understand the biological and environmental bases of personality (e.g., Ray et al. 2006; Willis-Owen and Flint 2007) and to examine how personality is related to various outcome measures, ranging from disease progression to the occurrence of specific behaviors (e.g., Capitanio et al. 1999; Pederson et al. 2005). The studies have addressed questions that have long interested personality psychologists but have been difficult to tackle using human studies alone. These include examining the degree to which individuals behave consistently across situations, examining the interplay of biological and environmental factors on personality development, and exploring the impact of personality on health. Some examples of these applications and others are described below.

Malloy and colleagues (2005) used mice to examine a set of three basic questions at the heart of traditional personality research: (1) To what extent do individuals behave consistently across interaction partners? ("actor effects"; Kenny 1994: e.g., is A generally aggressive?); (2) To what extent do individuals consistently elicit behaviors from others? ("partner effects"; e.g., does B generally elicit aggression?); and (3) To what extent do social behaviors reflect factors idiosyncratic to that particular combination of individuals? ("relationship effects"; is A uniquely aggressive toward B?). The best way to address these questions is to measure the behavior of individuals as they engage in numerous social interactions. Malloy et al. (2005) ran just such a study, but instead of humans they used 80 mice, organized into 10 groups of 8. Using a round robin design, each mouse was observed interacting with every other individual in its group and then observed again a week later, yielding 56 dyadic interactions per group. Across an array of social behaviors, Malloy et al. estimated actor, partner, and relationship effects, all of which showed evidence for cross-situational consistency. Although such a study could have been done in humans, it would have required a huge logistical investment and it would have been much more difficult to control extraneous factors for the duration of the research (also see Uher et al. 2008).

Animal studies afford numerous new options for investigating personality development and change. For example, Weinstein et al. (2008) have highlighted the opportunities afforded by animal research for examining the effects of development during the prenatal period. Personality is an emergent property of brain activity so, Weinstein et al. argue, it makes sense to examine the principal time in an organism's life when brain development proceeds most rapidly. Logistical difficulties in studying this developmental period reduce the number of human studies that can be conducted. Nonhuman primate studies can play a vital role in research on this period because it is possible to experimentally manipulate conditions and to obtain samples (both behavioral and physiological) regularly, and to follow animals longitudinally in a time frame that is considerably accelerated compared with that for humans. Schneider and colleagues induced stress in prenatal individuals by exposing pregnant rhesus monkeys to randomly distributed noise bursts over a 10-minute period, five days per week for a few weeks (Schneider 1992). Compared with control animals, prenatally stressed animals showed impaired neuromotor development and attentional deficits at birth (Schneider 1992) and a suite of personality differences assessed in the subsequent months and years (Schneider 1992; Clarke and Schneider 1993; 1997; Clarke et al. 1996).

Cross-fostering designs are another powerful way to investigate biological and environmental influences on personality development. For example, Suomi's (1987) cross-fostering study on rhesus monkeys suggested that infants' responses to separation from foster mothers is best predicted by their inherited levels of reactivity, not their foster mother's reactivity or caretaking style (also see Benus and Röndigs 1997; Suomi 1999; Drent et al. 2003). Of course, cross-fostering studies can also highlight the effects of environmental factors; studies of rats and rhesus monkeys have emphasized the role of maternal care, rather than genetics, in the transmittal of some traits (Francis et al. 1999; Maestripieri 2005; see chapter 11, this volume).

Psychologists have long been aware of the contribution to be made by animals to studies of heritability (Hall 1941; Scott 1953). For example, Scott and Fuller (1965) used standard cross-breeding techniques in five breeds of dog to examine the genetic components of behavioral traits, such as timidity and aggressiveness. In species with short life spans, researchers can selectively breed animals and then monitor the development of the offspring. For example, Van Oers et al. (2003) bred great tits to demonstrate that risk-taking behavior had a genetic component. Because great tits reach sexual maturity relatively early (compared with humans) and because breeding periods are brief, the researchers were able to test birds' risk-taking tendencies, then selectively breed birds with high and low levels of risk-taking behavior for two generations. Results indicated that the trait has a heritability of about 19% in great tits (see also chapter 6, this volume).

Staying within the realm of genetics, researchers have used animal models (in this case, dogs) to examine the specific genomic processes that could account for interbreed differences in behavior (Niimi et al. 1999). Animal personality research provides avenues of genetic research that are more cost-effective than research on humans and can offer some important advantages (Gershenfeld et al. 1997; Gershenfeld and Paul 1998). For example, a great deal of quantitative and molecular genetic research has already been done on some model species (Plomin and Crabbe 2000), so large amounts of genomic information are now available (see Van Oers and Mueller 2010). In addition, modern gene-mapping technologies are allowing researchers to search for genes or multiple-gene systems (e.g., quantitative trait loci [QTL]) for complex traits (e.g., Flint et al. 1995; Talbot et al. 1999). Transgenic methods and cloning techniques (e.g., Wilmut et al. 1997; Wakayama et al. 1998) will also provide new opportunities for animal research on genetic influences on personality.

NEW QUESTIONS

As the topic of animal personality has spread to new disciplines, it has raised novel questions that are of interest to human personality researchers but which they may not have thought to ask. For instance, researchers in behavioral ecology are primarily interested in learning about the ecological and evolutionary implications of consistent individual differences in behavior (e.g., Bell and Stamps 2004; Carere and Eens 2005; Dingemanse and Réale 2005; McElreath and Strimling 2006; see chapters 7, 8, and 9, this volume).

The existence of personality traits across a broad array of species raises the possibility of using phylogenetic methods to explore the evolutionary origins of traits (Gosling and Graybeal 2007). Combining evidence from across numerous species with maps of phylogenetic relationships provides a means for estimating when a trait evolved. Such maps can be used to chart the emergence of physical traits such as *wings* or behavioral traits such as *curiosity* or *pair bonding*. Just as we can date the evolution of wings to the common ancestor of birds and again independently to the point at which winged mammals diverged from nonwinged mammals, we can do the same for behavioral traits. Although the same trait can evolve independently on multiple occasions (via convergent evolution, as in the case of wings in birds and bats), cross-species similarities are more typically due to homology (i.e., inheritance from a common ancestor). For example, pair bonding is a trait humans share with some other mammals as well as other nonmammalian vertebrates (e.g., some fish and birds). In their recent phylogenetic analysis of pair bonding, Fraley and colleagues (2005) have shown how the present-day distribution of this trait across species almost certainly reflects several independent origins plus trait retention in descendant lineages. Their analyses direct attention to likely ancestral candidates, paving the way for more focused phylogenetic analyses hone in on the conditions under which this trait emerged. Similarly, it was the discovery of a separate Conscientiousness dimension in humans and chimpanzees (but not in orangutans or gorillas) that permits phylogenetically oriented researchers to date the emergence of Conscientiousness to the period after the common ancestor to chimpanzees and humans diverged from the other apes (i.e., 7 million–10 million years ago). Such analyses demonstrate the intellectual leaps that can be made by integrating research on animals with research on humans within a common evolutionary framework. More generally, knowledge of the evolutionary history of a trait illuminates the likely adaptive forces that shaped the trait and its probable function.

Behavioral ecologists and others have also been asking foundational questions about the adaptive function of personality traits (Sih et al. 2004; Nettle 2006; see chapters 7 and 8, this volume). As Sih and his collaborators have noted, traits presumably reflect underlying genetic or physiological mechanisms that constrain the flexibility of individuals' behaviors. This constraint in behavioral flexibility can generate tradeoffs, such that a certain personality characteristic may prove advantageous for an animal in one situation but not another. For example, highly aggressive individuals may be successful in defending their territories but these same individuals may act too aggressively toward potential mates. Thus, differing environments (e.g., those with scarce vs. abundant territorial resources) may have supported individual differences in behavior during the evolution of a species. Evidence supporting this idea is provided by a Dutch research group that has been conducting long-term studies of personality in a natural population of great tits (e.g., Dingemanse et al. 2004; Groothuis and Carere 2005).

One intriguing new opportunity for animal studies to contribute to our knowledge of human psychology is in the area of Functional Genomics (FG), which holds the promise of revealing how information encoded in a gene gets translated into behavior (Gosling and Mollaghan 2006). FG can be defined as the study of genes, their resulting proteins, and the role played by the proteins in the body's biochemical process. Genes are essentially information bytes consisting of DNA, which undergo a two-step process to encode unique proteins. The two steps are transcription (i.e., the synthesis of a messenger RNA [mRNA] copy from a sequence of DNA) and translation (the process in which the genetic code carried by mRNA directs the synthesis of proteins from amino acids). Understanding the process by which proteins are encoded is important because it represents the link between an individual's genetic code and its behavior. The most useful genes for understanding behavior will be those that play a role in the neurotransmitter and neuroendocrine systems because they are known to regulate social behavior (Hofmann 2003).

Functional genomicists are interested in measuring how the encoding of proteins is affected by various social and environmental influences. This is done by measuring the mRNA created in response to environmental conditions. For example, a functional genomicist might be interested in looking at the output of mRNA in cichlids (a taxa of fish) as a function of threatening stimuli (e.g., the presence of a predator). By looking at gene expression in the brain we can examine how the cellular processes underlying gene-behavior links vary as a function of environmental changes. The cellular processes can be revealed by monitoring the activity of genes in specific brain areas while an organism is undergoing environmentally induced changes.

Currently, invasive procedures, such as decapitation, are required for most genomic analyses of brain processes, ruling out human studies. However, if social psychologists can find animal analogs of the situations in which they are interested, they may soon be able to harness the power of functional-genomic analyses. New genomic techniques now make it possible to examine gene expression (by measuring the abundance of mRNA) in the brains of animal models (Reinke and White 2002; Robinson 2004). Thus, FG studies hold great promise for scientists interested in social behavior because they have the potential to show how genes affect specific brain mechanisms associated with social behaviors. In addition, FG studies have the potential to identify sources of individual differences at a molecular level. Although still in its infancy, this research holds great promise for understanding the most fundamental biological processes underlying social behavior.

Researchers in various applied fields, such as applied ethology, focus on practical issues, such as predicting working dog performance (e.g., Svartberg 2005; Maejima et al. 2007), and applications in animal welfare and management (e.g., McDougall et al. 2006; Watters and Meehan 2007). A number of organizations have adapted the principles of human personnel selection to identify individual animals suited to various tasks (Jones and Gosling 2005). For example, patrol or detection dogs must work in environments that are unusual, unpredictable, and noisy, requiring individuals low on fearfulness. In one study of drug-detection dogs, personality trait scores obtained after two weeks of training predicted the dogs' detection success assessed after four months of training (Maejima et al. 2007). Specifically, 93.3% of dogs scoring high on desire for work (assessed at 2 weeks) passed the final detection test, whereas only 53.3% of dogs scoring low on desire for work passed the final detection test. Thus, with the use of this single dimension approximately half the dogs (47%) that would later be rejected could be removed from the program at two weeks instead of four months.

Personality assessments of domestic animals have also been developed to help potential owners identify a pet that matches their needs (Coren 1998) and to assist with adoption decisions at animal shelters; in one study behavioral responses of animal-shelter dogs to an unfamiliar person entering their kennel correlated 0.64 with ratings of excitability subsequently made by their new owners after adoption (Ledger and Baxter 1996). Such information is very useful in setting realistic owner expectations about a new dog and in matching dogs to suitable homes, both of which improve the rate of successful adoptions.

Principles for cross-species comparisons in personality psychology

The sections above illustrated the benefits of taking a comparative perspective in personality research. In this section, we discuss some basic principles for making such cross-species comparisons (Mehta and Gosling 2006; Gosling and Harley 2009).

One challenge facing any comparative researcher is determining the degree to which apparently similar traits really are tapping the same underlying trait. How can it be determined that what appears to be boldness in squid or trout or chimpanzees is in any way similar to boldness in humans? After all, there are very few literal similarities in how the same trait could be expressed across species. There are numerous cases of easy errors of interpretation; for example, the chimpanzee facial display in which the lips are retracted so that clenched teeth are exposed reflects fear, not happiness, as might be assumed by the expression's apparent similarity to a human smile (Hayes 1994).

To solve the challenge of finding meaningful points of cross-species equivalence, cross-species researchers can draw from the lessons developed by cross-cultural researchers. In a sense, a comparative researcher asking whether the apparently sociable behavior of a rhesus monkey reflects the sociability that we know in humans is analogous to the cross-cultural emotions researcher asking whether the apparently angry expression of a hitherto isolated group of humans reflects the anger that we know in our own culture. The solution to determining cross-cultural equivalence of anger expressions is examining what comes before and after the expressions, and where possible, looking for commonalities in underlying physiology. Thus, if this expression that resembles anger comes after an event that might reasonably elicit anger and results in actions that might reasonably follow anger and displays the physiological signature of anger, then the researcher can be reasonably confident that the facial expression does indeed reflect anger as we know it. Likewise, an animal researcher can examine the apparently sociable behavior in the context of what comes before and after the behavior and, where possible, examine whether it shares physiological, biological, and genetic commonalities with human sociability. In essence, following this procedure is what many animal researchers already do implicitly, which explains why researchers with experience of a species do not make mistakes about behavior (e.g., mistaking a chimpanzee's fear grin for a smile). Nonetheless, this procedure offers a set of steps that researchers can take when they encounter unfamiliar species or when they want to establish cross-species equivalences empirically. As an example of an empirical approach to establishing cross-species equivalence, Flint (2004) reviewed evidence that the set of genes that influences emotionality in rodents also influences neuroticism in humans. These similarities were offered as evidence for cross-species equivalence in human neuroticism and rodent emotionality.

Conclusions

Despite several decades in which the study of nonhuman animals was largely ignored by investigators of human personality psychology, comparative research in personality is making a comeback. The past five years in particular have shown incredible growth in comparative personality research, as researchers capitalize on its advantages for addressing questions that are difficult or impossible to address with humans or animals alone (Gosling 2008). And with rapid advances in neuroscience, genomics, and personality measurement, the potential translational contributions of cross-species research will only continue to grow. This chapter illustrated the benefits of a comparative approach for personality research, but we anticipate that in the coming years cross-species research in personality will continue to influence many other diverse fields, ranging from genetics, neuroscience, and psychology to behavioral ecology, primatology, and veterinary medicine.

References

Beer, J. S., Shimamura, A. P., and Knight, R. T. 2004. Frontal lobe contributions to executive control of cognitive and social behavior. In: *The Cognitive Neurosciences III* (Gazzaniga, M. S., ed.), pp. 1091–1104. Cambridge, MA: MIT Press.

Bell, A. M., and Stamps, J. A. 2004. Development of behavioural differences between individuals and populations of sticklebacks, *Gasterosteus aculeatus. Animal Behaviour*, 68, 1339–1348.

Benus, R. F., and Röndigs, M. 1997. The influence of the postnatal maternal environment in accounting for differences in aggression and behavioural strategies in *Mus domesticus. Behaviour*, 134, 623–641.

Capitanio, J. P. 1999. Personality dimensions in adult male rhesus macaques: prediction of behaviors across time and situation. *American Journal of Primatology*, 47, 299–320.

———. 2002. Sociability and responses to video playbacks in adult male rhesus monkeys (*Macaca mulatta*). *Primates*, 43, 169–177.

———. 2011. Nonhuman primate personality and immunity: mechanisms of health and disease. In: *Personality and Temperament in Nonhuman Primates* (Weiss, A., King, J. E., and Murray, L., eds.), pp. 233–256. New York: Springer.

Capitanio, J. P., Abel, K., Mendoza, S. P., Blozis, S. A., McChesney, M. B., Cole, S. W., and Mason, W. A. 2008. Personality and serotonin transporter genotype interact with social context to affect immunity and viral set-point in simian immunodeficiency virus disease. *Brain, Behavior, and Immunity*, 22, 676–689.

Capitanio, J. P., Mendoza, S. P., and Baroncelli, S. 1999. The relationship of personality dimensions in adult male rhesus macaques to progression of simian immunodeficiency virus disease. *Brain, Behavior, and Immunity*, 13, 138–154.

Capitanio, J. P., and Widaman, K. F. 2005. A confirmatory factor analysis of personality structure in adult male rhesus monkeys (*Macaca mulatta*). *American Journal of Primatology*, 65, 289–294.

Carere, C., and Eens, M. 2005. Unravelling animal personalities: how and why individuals consistently differ. *Behaviour*, 142, 1149–1157.

Cavigelli, S., Yee, J., and McClintock, M. 2006. Infant temperament predicts life span in female rats that develop spontaneous tumors. *Hormones and Behavior*, 50, 454–462.

Clarke, A. S., and Boinski, S. 1995. Temperament in nonhuman primates. *American Journal of Primatology*, 37, 103–125.

Clarke, A. S., and Schneider, M. L. 1993. Prenatal stress has long-term effects on behavioral responses to stress in juvenile rhesus monkeys. *Developmental Psychobiology*, 26, 293–304.

———. 1997. Effects of prenatal stress on behavior in adolescent rhesus monkeys. *Annals of the New York Academy of Sciences USA*, 807, 490–491.

Clarke, A. S., Soto, A., Bergholz, T., and Schneider, M. L. 1996. Maternal gestational stress alters adaptive and social behavior in adolescent rhesus monkey offspring. *Infant Behavior and Development*, 19, 451–461.

Coren, S. 1998. *Why We Love the Dogs We Do: How to Find the Dog That Matches Your Personality*. New York: Free Press.

Dingemanse, N. J., Both, C., Drent, P. J., and Tinbergen, J. M. 2004. Fitness consequences of avian personalities in a fluctuating environment. Proceedings of the Royal Society of London B, 271, 847–852.

Dingemanse, N. J., and Réale, D. 2005. Natural selection and animal personality. *Behaviour*, 142, 1159–1184.

Domjan, M., and Purdy, J. 1995. Animal research in psychology: more than meets the eye of the general psychology student. *American Psychologist*, 50, 496–503.

Drent, P. J., Van Oers, K., and Van Noordwijk, A. J. 2003. Realized heritability of personalities in the great tit (*Parus major*). *Proceedings of the Royal Society of London B*, 270, 45–51.

Flint, J. 2004. The genetic basis of neuroticism. *Neuroscience and Biobehavioral Reviews*, 28, 307–316.

Flint, J., Corley, R., Defries, J. C., Fulker, D. W., Gray, J. A., Miller, S., and Collins, A. C. 1995. A simple genetic basis for a complex psychological trait in laboratory mice. *Science*, 269, 1432–1435.

Fraley, R. C., Brumbaugh, C. C., and Marks, M. J. 2005. The evolution and function of adult attachment: a comparative and phylogenetic analysis. *Journal of Personality and Social Psychology*, 89, 731–746.

Francis, D., Diorio, J., Liu, D., and Meaney, M. J. 1999. Nongenomic transmission across generations of maternal behavior and stress responses in the rat. *Science*, 286, 1155–1158.

Freeman, H., Gosling, S. D., and Schapiro, S. J. 2011. Comparison of methods for assessing personality in nonhuman primates. In: *Personality and Temperament in Nonhuman Primates* (Weiss, A., King, J. E., and Murray, L. E., eds.), pp. 17–40. New York: Springer.

Funder, D., Kolar, D., and Blackman, M. 1995. Agreement among judges of personality: interpersonal relations, similarity, and acquaintanceship. *Journal of Personality and Social Psychology*, 69, 656–672.

Gershenfeld, H. K., Neumann, P. E., Mathis, C., Crawley, J. N., Li, X., and Paul, S. M. 1997. Mapping quantitative trait loci for open-field behavior in mice. *Behavior Genetics*, 27, 201–210.

Gershenfeld, H. K., and Paul, S. M. 1998. Towards a genetics of anxious temperament: from mice to men. *Acta Psychiatrica Scandinavica*, 98, 56–65.

Gosling, S. D. 1998. Personality dimensions in spotted hyenas (*Crocuta crocuta*). *Journal of Comparative Psychology*, 112, 107–118.

————. 2001. From mice to men: what can we learn about personality from animal research? *Psychological Bulletin*, 127, 45–86.

————. 2008. Personality in nonhuman animals. *Social and Personality Psychology Compass*, 2, 985–1001.

Gosling, S. D., and Graybeal, A. 2007. Tree thinking: a new paradigm for integrating comparative data in psychology. *Journal of General Psychology*, 134, 259–277.

Gosling, S. D., and Harley, B. A. 2009. Animal models of personality and cross-species comparisons. In: *Cambridge Handbook of Personality Psychology* (Corr, P., and Matthews, G., eds.), pp. 275–286. Cambridge: Cambridge University Press.

Gosling, S. D., and John, O. P. 1999. Personality dimensions in nonhuman animals: a cross-species review. *Current Directions in Psychological Science*, 8, 69–75.

Gosling, S. D., Kwan, V. S. Y., and John, O. P. 2003a. A dog's got personality: a cross-species comparative approach to evaluating personality judgments. *Journal of Personality and Social Psychology*, 85, 1161–1169.

Gosling, S. D., Lilienfeld, S. O., and Marino, L. 2003b. Personality. In: *Primate Psychology* (Maestripieri, D., ed.), pp. 254–288. Cambridge, MA: Harvard University Press.

Gosling, S. D., and Mollaghan, D. M. 2006. Animal research in social psychology: a bridge to functional genomics and other unique research opportunities. In: *Bridging Social Psychology: Benefits of Transdisciplinary Approaches* (Van Lange, P. A. M., ed.), pp. 123–128. Mahwah, NJ: Erlbaum.

Gosling, S. D., and Vazire, S. 2002. Are we barking up the right tree? Evaluating a comparative approach to personality. *Journal of Research in Personality*, 36, 607–614.

Groothuis, T. G. G., and Carere, C. 2005. Avian personalities: characterization and epigenesis. *Neuroscience and Biobehavioral Reviews*, 29, 137–150.

Hall, C. S. 1941. Temperament: a survey of animal studies. *Psychological Bulletin*, 38, 909–943.

Hayes, N. 1994. *Principles of Comparative Psychology*. New York: Psychology Press.

Hebb, D. O. 1946. Emotions in man and animals: an analysis of the intuitive process of recognition. *Psychological Review*, 53, 88–106.

Hofmann, H. A. 2003. Functional genomics of neural and behavioral plasticity. *Journal of Neurobiology*, 54, 272–282.

John, O. P., and Srivastava, S. 1999. The Big Five trait taxonomy: history, measurement, and theoretical perspectives. In: *Handbook of Personality: Theory and Research* (2nd ed., Pervin, L. A., and John, O. P., eds.), pp. 102–138. New York: Guilford.

Jones, A. C., and Gosling, S. D. 2005. Temperament and personality in dogs (*Canis familiaris*): a review and evaluation of past research. *Applied Animal Behaviour Science*, 95, 1–53.

Kenny, D. A. 1994. *Interpersonal Perception: A Social Relations Analysis*. New York: Guilford Press.

Kenrick, D. T., and Funder, D. C. 1988. Profiting from controversy: lessons from the person-situation debate. *American Psychologist*, 43, 23–34.

Konečná, M., Lhota, S., Weiss, A., Urbánek, T., Adamová, T., and Pluháček, J. 2008. Personality in free-ranging Hanuman langur (*Semnopithecus entellus*) males: subjective ratings and recorded behavior. *Journal of Comparative Psychology*, 122, 379–389.

Kwan, V. S. Y., Gosling, S. D., and John, O. P. 2008. Anthropomorphism as a special case of social perception: a cross-species social relations model analysis of humans and dogs. *Social Cognition*, 26, 129–142.

Ledger, R. A., and Baxter, M. 1996. A validated test to assess the temperament of dogs. In: *Proceedings of the 30th International Congress of the ISAE* (Duncan, I. J. H., Widowski, T. M., and Haley, D. B., eds.). Guelph, Canada: Campbell Centre for the Study of Animal Welfare, University of Guelph.

Maejima, M., Inoue-Murayama, M., Tonosaki, K., Matsuura, N., Kato, S., Saito, Y., Weiss, A., Murayama, Y., and Ito, S. 2007. Traits and genotypes may predict the successful training of drug detection dogs. *Applied Animal Behaviour Science*, 107, 287–298.

Maestripieri, D. 2005. Early experience affects the intergenerational transmission of infant abuse in rhesus monkeys. *Proceedings of the National Academy of Sciences U S A*, 102, 9726–9729.

Malloy, T., Barcelos, S., Arruda, E., DeRosa, M., and Fonseca, C. 2005. Individual differences and cross-situational consistency of dyadic social behavior. *Journal of Personality and Social Psychology*, 89, 643–654.

McCrae, R. R., Costa, P. T., Jr., Ostendorf, F., Angleitner, A., Hrebickova, M., Avia, M. D., Sanz, J., Sanchez-Bernardos, M. L., Kusdil, M. E., Woodfield, R., Saunders, P. R., and Smith, P. B. 2000. Nature over nurture: temperament, personality, and life span development. *Journal of Personality and Social Psychology*, 78, 173–186.

McDougall, P. T., Réale, D., Sol, D., and Reader, S. M. 2006. Wildlife conservation and animal temperament: causes and consequences of evolutionary change for captive, reintroduced, and wild populations. *Animal Conservation*, 9, 39–48.

McElreath, R., and Strimling, P. 2006. How noisy information and individual asymmetries can make "personality" an adaptation: a simple model. *Animal Behaviour*, 72, 1135–1139.

Mehta, P. H., and Gosling, S. D. 2006. How can animal studies contribute to research on the biological bases of personality? In: *The Biological Bases of Personality and Individual Differences* (Canli, T., ed.), pp. 427–448. New York: Guilford Press.

———. 2008. Bridging human and animal research: a comparative approach to studies of personality and health. *Brain, Behavior, and Immunity*, 22, 651–661.

Nettle, D. 2006. The evolution of personality variation in humans and other animals. *American Psychologist*, 61, 622–631.

Nettle, D., and Penke, L. 2010. Personality: bridging the literatures from human psychology and behavioural ecology. *Philosophical Transactions of the Royal Society of London B*, 365, 4043–4050.

Niimi, Y., Inoue-Murayama, M., Murayama, Y., Ito, S., and Iwasaki, T. 1999. Allelic variation of the D4 dopamine receptor polymorphic region in two dog breeds, Golden Retriever and Shiba. *Journal of Veterinary Medical Science*, 61, 1281–1286.

Pederson, A. K., King, J. E., and Landau, V. I. 2005. Chimpanzee (*Pan troglodytes*) personality predicts behavior. *Journal of Research in Personality*, 39, 534–549.

Plomin, R., and Crabbe, J. 2000. DNA. *Psychological Bulletin*, 126, 806–828.

Ray, J., Hansen, S., and Waters, N. 2006. Links between temperamental dimensions and brain monoamines in the rat. *Behavioral Neuroscience*, 120, 85–92.

Reinke, V., and White, K. P. 2002. Developmental genomic approaches in model organisms. *Annual Review of Genomics and Human Genetics*, 3, 153–178.

Robinson, G. E. 2004. Genomics: beyond nature and nurture. *Science*, 304, 397–399.

Rouff, J. H., Sussman, R. W., and Stube, M. J. 2005. Personality traits in captive lion-tailed macaques (*Macaca silenus*). *American Journal of Primatology*, 67, 177–198.

Sapolsky, R. M. 1998. *Why Zebras Don't Get Ulcers*. New York: W. H. Freeman.

Schneider, M. L. 1992. Prenatal stress exposure alters postnatal behavioral expression under conditions of novelty challenge in rhesus monkey infants. *Developmental Psychobiology*, 25, 529–540.

Schwartz, C., Wright, C., Shin, L., Kagan, J., and Rauch, S. 2003. Inhibited and uninhibited infants "grown up": adult amygdalar response to novelty. *Science*, 300, 1952–1953.

Scott, J. P. 1953. New directions in the genetic study of personality and intelligence. *Eugenical News*, 38, 97–101.

Scott, J. P., and Fuller, J. L. 1965. *Genetics and the Social Behavior of the Dog*. Chicago: University of Chicago Press.

Segerstrom, S., and Miller, G. 2004. Psychological stress and the human immune system: a meta-analytic study of 30 years of inquiry. *Psychological Bulletin*, 130, 601–630.

Sih, A., Bell, A. M., Johnson, J. C., and Ziemba, R. E. 2004. Behavioral syndromes: an integrative overview. *Quarterly Review of Biology*, 79, 241–277.

Stevenson-Hinde, J., Stillwell-Barnes, R., and Zunz, M. 1980. Individual differences in young rhesus monkeys: consistency and change. *Primates*, 21, 498–509.

Suomi, S. J. 1987. Genetic and maternal contributions to individual differences in rhesus monkey biobehavioral development. In: *Perinatal Development: A Psychobiological Perspective* (Krasnegor, N., Blass, E., Hofer, M., and Smotherman, W., eds.), pp. 397–420. New York: Academic Press.

———. 1999. Behavioral inhibition and impulsive aggressiveness: insights from studies with rhesus monkeys. In: *Child Psychology: A Handbook of Contemporary Issues* (Balter, L., and Tamis-Lemonda, C. S., eds.), pp. 510–525. Philadelphia: Taylor and Francis.

Svartberg, K. 2005. A comparison of behaviour in test and in everyday life: evidence of three consistent boldness-related personality traits in dogs. Applied Animal Behaviour Science, 91, 103–128.

Talbot, C. J., Nicod, A., Cherny, S. S., Fulker, D. W., Collins, A. C., and Flint, J. 1999. High-resolution mapping of quantitative trait loci in outbred mice. *Nature Genetics*, 21, 305–308.

Uher, J. 2008. Comparative personality research: methodological approaches. *European Journal of Personality*, 22, 427–455.

Uher, J., and Asendorpf, J.B. 2008. Personality assessment in the great apes: comparing ecologically valid behavior measures, behavior ratings, and adjective ratings. *Journal of Research in Personality*, 42, 821–838.

Uher, J., Asendorpf, J. B., and Call, J. 2008. Personality in the behaviour of great apes: temporal stability, cross-situational consistency, and coherence in response. *Animal Behaviour*, 75, 99–112.

Van Oers, K., Drent, P. J., de Goede, P., and Van Noordwijk, A. J. 2003. Realized heritability and repeatability of risk-taking behavior in relation to avian personalities. Proceedings of the Royal Society of London B, 271, 65–73.

Van Oers, K., and Mueller, J. C. 2010. Evolutionary genomics of animal personality. *Philosophical Transactions of the Royal Society of London B*, 365, 3991–4000.

Vazire, S., and Gosling, S. D. 2003. Bridging psychology and biology with animal research. *American Psychologist*, 5, 407–408.

Vazire, S., Gosling, S. D., Dickey, A. S., and Schapiro, S. J. 2007. Measuring personality in nonhuman animals. In: *Handbook of Research Methods in Personality Psychology* (Robins, R.W., Fraley, R. C., and Krueger, R., eds.), pp. 190–208. New York: Guilford.

Vegas, O., Fano, E., Brain, P., Alonso, A., and Azpiroz, A. 2006. Social stress, coping strategies, and tumor development in male mice: behavioral, neuroendocrine, and immunological implications. *Psychoneuroendocrinology*, 31, 69–79.

Wakayama, T., Perry, A. C. F., Zuccotti, M., Johnson, K. R., and Yanagimachi, R. 1998. Full-term development of mice enucleated oocytes injected with cumulus cell nuclei. *Nature*, 394, 369–374.

Watters, J. V., and Meehan, C. L. 2007. Different strokes: can managing behavioral types increase postrelease success? *Applied Animal Behaviour Science*, 102, 364–379.

Weinstein, T. A. R., Capitanio, J. P., and Gosling, S. D. 2008. Personality in animals. In: *Handbook of Personality Theory and Research* (John, O. P., Robins, R. W., and Pervin, L. A., eds.), pp. 328–350. New York: Guilford.

Weiss, A., King, J. E., and Perkins, L. 2006. Personality and subjective well-being in orangutans (*Pongo pygaeus* and *Pongo abelii*). *Journal of Personality and Social Psychology*, 90, 501–511.

Willis-Owen, S. A. G., and Flint, J. 2007. Identifying the genetic determinants of emotionality in humans: insights from rodents. *Neuroscience and Biobehavioral Reviews*, 31, 115–124.

Wilmut, I., Schnieke, A. E., McWhir, J., Kind, A. J., and Campbell, K. H. S. 1997. Viable offspring derived from fetal and adult mammalian cells. *Nature*, 385, 810–813.

II

GENETICS, ECOLOGY, AND EVOLUTION

OF ANIMAL PERSONALITIES

6

Quantitative and Molecular Genetics

of Animal Personality

KEES VAN OERS AND DAVID L. SINN

Introduction

Correlated suites of behavior are collectively known as *animal personality* (Gosling 2001), *behavioral syndromes* (Sih et al. 2004), *behavioral strategies* (Benus et al. 1990), or *behavioral profiles* (Rodgers et al. 1997). Each of these terms, to some extent, describes an emergent phenomenon of the total biases in behavioral reactions an individual expresses compared with other individuals within the same population or species. While there is some debate on terminology (e.g., Réale et al. 2007; Gosling 2008; introduction, this volume), we will use the term *animal personality* throughout this chapter.

Recent evidence suggests an important role for within-population variation in animal personality for our understanding of ecology and evolution, since behavioral variation often covaries with several important indicators of fitness, such as growth, reproduction, and survival (Dingemanse and Réale 2005; Smith and Blumstein 2008; see chapters 3, 7, and 13). In line with several other fields in the life sciences (Pigliucci 2007; Wingfield et al. 2008), a re-emerging theme or emphasis of many studies of animal personality is how selection on correlated suites of traits (as opposed to a singular focus on any one trait in an isolated context) may influence and be influenced by evolutionary and population dynamics (see Van Oers and Mueller 2010). To take one example, predators selecting on individual prey with certain behavioral characteristics can result in particular personality types (i.e., covariance between aggression and boldness) in some populations but not in others, which in threespined stickleback fish (*Gasterosteus aculeatus*) has been demonstrated in a descriptive way (Dingemanse et al. 2007; Alvarez and Bell 2007; Brydges et al. 2008) as well as in an experimental way (Bell and Sih 2007; see chapter 2). This effect of correlated selection on evolutionary and population dynamics can also be found in human-induced selection on fisheries: fast-growing individuals, which are also the boldest foragers, are most likely to be caught in fishing nets, causing the phenotypic and genotypic behavioral and growth structure of natural fish populations to be altered (Biro and Post 2008). In another example, consistent behavioral differences influence natural selection by influencing settlement patterns in Western bluebirds (*Sialia mexicana*), where males aggressively compete for territories. Males with higher levels of aggressiveness are more successful in preferred habitats with multiple nesting possibilities per territory (Duckworth 2006b). The physical habitat characteristics of these areas, however, also induce correlated selection on morphology: in preferred areas there was positive selection on longer tarsi and tails where agility was favored, while in less preferred habitats (where agility was also not as important) there was no morphological selection on these physical characteristics (Duckworth 2006a). In short, variation in animal personality represents an important behavioral mechanism that links life-history characteristics to variation in fitness across a wide range of taxa (Biro et al. 2003; Réale et al. 2007; Biro and Stamps 2008).

Evolution by natural selection presents an elementary paradox: if selection increases individuals' fitness, why do less fit genotypes and phenotypes still persist? When selection on a trait is weak or absent, genetic drift is expected to drive alleles to fixation. Additionally, when selection is either directional or disruptive, particular alleles are favored and as a result of this, additive genetic variation is expected to decrease or disappear (Fisher 1930). For decades, evolutionary biologists have sought to understand the evolutionary forces that influence genetic variation within and among natural populations (Lewontin 1974; Houle 1992; Barton and Keightley 2002; Mitchell-Olds et al. 2007). Traditionally, genetic variation was thought to originate within populations through migration of individuals from nearby populations and/or genetic mutations. For example, genetic variation will be relatively high for mildly deleterious mutations when migration or mutation rates are high (Mitchell-Olds et al. 2007), or if selection on genetic variation is neutral (Ohta 1992; Ohta 2002); for many quantitative traits these mutation rates can be high enough to make a substantial contribution to molecular genetic variation (Lynch and Walsh 1998). Recent work suggests that genetic variation also originates from genetic accommodation of developmentally induced phenotypic variation (West-Eberhard 2003; Badyaev 2005). Therefore, environmentally induced heritable changes in gene expression and function often cannot be explained by changes in DNA sequence alone. These epigenetic changes are based on a set of molecular processes that can alter the activity of particular genes by methylation of cytosine, remodelling of DNA chromatin structure, or regulatory processes mediated by small RNA molecules (Bossdorf et al. 2008).

The maintenance of genetic variation can occur through, for example, trade-offs between components of fitness, antagonistic genetic effects, maternal effects, and genotype-by-age (GxA) and genotype-by-environment (GxE) interactions (Kruuk et al. 2008). It is now also well established that genetic and environmental effects can interact in such a way that the effect of any given environment on the phenotype is dependent on an individual's genotype (GxE interaction; Barton and Turelli 1989). GxE means that the phenotypic response to changing environments will differ among individuals such that phenotypic plasticity may itself be thought of as a heritable trait (Via et al. 1995). The question of whether a phenotypic trait shows GxE is equivalent to asking whether plasticity in the trait has a heritable basis of variation (Via and Lande 1985; Kruuk et al. 2008).

Another theme emerging from many studies of life-history evolution is that several other different types of selection exist, which actively maintain polymorphisms within a population, and are collectively referred to as balancing selection. Examples of these processes include overdominance or heterozygous advantage (when a heterozygote has higher fitness than either homozygote), frequency-dependent selection, and temporal or spatial variation in selection (Roff 1997). Furthermore, the inheritance of behavioral variation may not be only a simple function of the additive effect of both parent's genotypes. Instead, the biological evolution of behavior most likely includes many nonadditive genetic effects (Crnokrak and Roff 1995; Meffert et al. 2002), indirect genetic effects (Wolf et al. 1999; Wilson et al. 2009), stress-induced genetic accommodation (Badyaev 2005), cultural inheritance (Boyd and Richerson 1985), learning effects (Gibbons et al. 2005), structural interactions with other life history and physiological characters (Stirling et al. 2002), and environmental variation that covaries (negatively or positively) with fitness (Mitchell-Olds et al. 2007). All these factors may also contribute to reduce or prevent the erosion of additive genetic variation.

Animal personality research can be conducted from a genetic or a phenotypic perspective and these two approaches are complementary (Van Oers and Sinn 2011). Although a phenotypic approach can predict which optimal combinations of correlated suites of traits results in stable population-level phenotypic equilibria (i.e., a mix of personality phenotypes), it cannot, in general, predict the evolutionary trajectory taken by evolution to reach that optimum (Roff 2007). While genetic approaches have proven to be important to answer questions about the adaptive significance and the evolution of life-history traits, the genetic basis of behavioral traits in studies within an eco-logical or evolutionary context has still largely been neglected (Sokolowski 2001; Boake et al. 2002; Higgins et al. 2005). Reasons for this may be the lack of genetically tractable natural systems (Wolf 2001) or the behavioral ecolo-gists' focus on field studies, in which genetic analyses can be extremely dif-ficult (Boake et al. 2002). Additionally, abiotic sources of noise (such as tem-perature) or physiological processes (such as age and stress) can also generate experimental noise in behavior assays (Boake 1994; Higgins et al. 2005). Nev-ertheless, most behavioral traits are expected to be at least partially heritable (Lynch and Walsh 1998) and to influence life-history characteristics (Tur-kheimer 1998; Stirling et al. 2002); accordingly, behavioral ecologists have generally assumed that there is heritable variation available for the evolution of behavior by natural selection (e.g., the phenotypic gambit: Owens 2006). However, this assumption may not always be tenable. Phenotypic correlations (r_p), which form the basis of most studies of animal personality, may often be different from genetic correlations (r_g), since r_g also depends on the extent of age- and environment-specific expressed additive genetic variation, which can often result in large deviations of r_g from r_p (Kruuk et al. 2008). Similar to other complex traits, the genetic architecture of personality traits is likely to consist of many small effect factors (pleiotropy) and a range of gene-gene and gene-environment interactions. Since the expected response to natural selec-tion on any trait depends on its underlying genetic structure (Dobzhansky et al. 1977), quantitative and molecular genetics are integral components of our understanding of studies on the fitness, natural selection, and evolution of animal personality traits. These complex aspects of personality genetics are shared by many quantitative traits, and therefore we believe the field of ani-mal personality genetics is well poised to make significant contributions to a greater understanding of evolutionary biology (Van Oers and Mueller 2010). While it is probably fair to say that the genetic analysis of animal personality traits, especially in wild populations, is still in its infancy, relatively descrip-tive questions yet to be adequately addressed—such as "How much genetic variation is there for a personality trait in wild populations?" and "What is the genetic architecture of a personality profile?"—will prove fruitful not only to our understanding of the evolution of personality but also to a larger under-standing of the processes involved in evolution by natural selection.

Currently, most information available on the structure of inheritance of personality traits comes from humans (e.g., Bouchard 2004; Savitz and Ramesar 2004), rodents (e.g., Sluyter et al. 1996), and domesticated species (e.g., Burrow 1997). Although genetic studies of human personality have been immensely valuable in demonstrating genetic influences on personality traits, interpreting patterns of genetic variation in personality in humans in evolutionary terms remains difficult (Bouchard 1994; Nettle 2005; 2006). Animal models, on the other hand, have proven more useful in understanding the underlying genetic mechanisms of behavioral traits (e.g., Wehner et al. 2001). Unfortunately, however, most genetic studies on personality traits in animals to date have been on populations that were bred in captivity over long periods (laboratory animals and domesticated species; table 6.1). Although studying laboratory-model organisms offers tremendous advantages in terms of control, replication, and convenience, laboratory studies also provide novel, stable, uniform, benign environments where selection is unlikely to operate as it would in wild populations (Merilä and Sheldon 2001). In cases where laboratory populations have been maintained for long enough (sometimes only for several generations), and with sufficient competition, genetic adaptation to this novel environment can occur (e.g., Hoffman et al. 2001; Blanchet et al. 2008). Laboratory estimates of heritability therefore may not be good predictors of heritability in natural populations, since heritability estimates depend on the environmental conditions experienced during ontogeny (Riska et al. 1989; but see Dingemanse et al. 2002; Drent et al. 2003). In these cases, selection caused by laboratory environments may alter patterns of genetic inheritance relative to wild populations (Kruuk et al. 2008). Note that this does not mean that there is no selection in captive populations, that artificial selection acts fundamentally differently from natural selection, or that one cannot measure selection in a captive population. Selection processes in natural populations might, however, be important for generating and maintaining specific patterns of genetic variation in particular personality traits, while captive populations would presumably show captive-specific personality genetic profiles. Studies of artificial selection therefore may give important insight on the selection process and its outcomes, but not necessarily on the strength of natural selection and corresponding genetic response given a particular set of population genetic and environmental conditions. Variation in fitness outcomes in captive populations therefore might not be very valuable for translation to wild ones. Hence, studies of fitness and the genetic factors influencing it are best carried out in environments where selection occurs, that is, in the wild.

In a previous review on the genetics of animal personality, Van Oers et al. (2005a) concluded that (a) qualitatively, significant heritabilities have now been documented for several personality traits in different animal taxa, (b) the genetic architecture of some animal personality traits appears to include more than just additive genetic variance (e.g, dominance variation and epistasis in bird personality traits: Van Oers et al. 2004b; 2005a), (c) it is

Table 6.1. List of studies used in meta-analysis of animal personality traits. Strain represents the evolutionary history of the subjects for a given study (D = domesticated, W = wild. WO = wild population. WO = offspring from wild-caught parents, Z = zoo animals, and LM = laboratory monkeys). Asterisks in "Trait name used in analysis" indicate traits that were not used in the second set of analyses (i.e., they were not used to address our first question of interest but not our second).

Author	Year	Taxon	Strain	No. of Reported Estimates	No. of Traits	Trait Name Given	Trait Name Used in Analysis	h^2 Estimate	No. of Subjects
Adalsteinsson	1977	Equidae	D	1	1	Spirit and willingness to run	Disposition	0.85	77
Agyemang et al.	1982	Ruminatia	D	3	1	Good disposition	Disposition	0.07	5601
						Average disposition	Disposition	0.06	5601
						Poor disposition	Disposition	−0.01	5601
Aitchison et al.	1972	Ruminatia	D	4	1	Disposition	Disposition	0.15	11106
						Disposition	Disposition	0.11	11106
						Disposition	Disposition	0.1	11106
						Disposition	Disposition	0.09	11106
Arnason	1979	Equidae	D	1	1	Spirit and willingness to run	Disposition	0.14	1208
Arnason	1984	Equidae	D	2	2	Spirit and willingness to run	Disposition	0.21	1603
						Ease of handling	Disposition	0.36	1603
Bakker	1994	Teleostei	WO	5	2	Aggression	Aggression	0.37	128
						Aggression	Aggression	0.25	171
						Aggression	Aggression	0.23	250
						Aggression	Aggression	0.31	296
						Dominance	Dominance	0.34	88
Beckman et al.	2007	Ruminatia	D	6	1	Docility	Boldness	0.34	49459
						Docility	Boldness	0.37	49459
						Docility	Boldness	0.31	49459
						Docility	Boldness	0.31	49459
						Docility	Boldness	0.38	49459
						Docility	Boldness	0.29	49459
Beilharz et al.	1966	Ruminatia	D	1	1	Dominance	Dominance	0.4	113

Author	Year	Family	Code			Measure	Category	Value	N
Beilharz and Cox	1967	Suidae	D	2	1	Open field test	Exploration	0.16	1853
						Open field test	Exploration	0.46	1853
Bell	2005	Teleostei	OW	6	3	Activity	Activity	0.048	29
						Activity	Activity	0.156	42
						Aggression	Aggression	0.011	29
						Aggression	Aggression	0.14	42
						Boldness	Boldness	0.036	29
						Boldness	Boldness	0.002	42
Boenigk et al.	2005	Canidae	D	1	1	Temperament	Disposition	0.07	5608
Boenigk et al.	2006	Canidae	D	1	1	Temperament	Disposition	0.11	4113
Boissy et al.	2005	Ruminatia	D	10	10	Locomotion	Activity	0.34	1347
						High bleats (hbleat)	Activity	0.48	1347
						Low bleats (lbleat)	Activity	0.28	1347
						Sniffing in an unfamiliar environment	Activity	0.36	1347
						Vigilant postures	*	0.24	1347
						Attempts to escape isolation	*	0.14	1347
						Proximity to bucket	*	0.16	1347
						Capacity to approach novel human	Boldness	0.32	1347
						Tolerance to approach by novel human	Boldness	0.23	1347
						Tolerance to human approach when partly restrained	Boldness	0.23	1347
Burrow	2001	Ruminatia	D	1	1	Flight speed	Disposition	0.42	1871

(continued)

Table 6.1. (continued)

Author	Year	Taxon	Strain	No. of Reported Estimates	No. of Traits	Trait Name Given	Trait Name Used in Analysis	h² Estimate	No. of Subjects
Burrow and Corbet	2000	Ruminatia	D	3	1	Flight speed	Disposition	0.35	851
						Flight speed	Disposition	0.5	4518
						Flight speed	Disposition	0.48	4518
Burrow et al.	1988	Ruminatia	D	2	1	Flight speed	Disposition	0.54	561
						Flight speed	Disposition	0.26	558
Colvin and Gatehouse	1993	Lepidoptera	OW	1	1	Total flight time	Activity	0.39	120
Craig et al.	1965	Phasianidae	D	2	1	Dominance	Dominance	0.16	632
						Dominance	Dominance	0.28	620
Detelleux et al.	1999	Ruminatia	D	1	1	Temperament	Disposition	0.11	50905
Dickson et al	1970	Ruminatia	D	4	2	Docility	Boldness	0.52	1017
						Docility	Boldness	0.47	1017
						Dominance	Dominance	0.15	1017
						Dominance	Dominance	0	1017
Dingemanse et al.	2002	Passeriformes	W	5	1	Exploratory behaviour	Exploration	0.22	42
						Exploratory behaviour	Exploration	0.31	59
						Exploratory behaviour	Exploration	0.34	63
						Exploratory behaviour	Exploration	0.61	50
						Exploratory behaviour	Exploration	0.37	33
Drent et al.	2003	Passeriformes	W	3	1	Exploratory behaviour	Exploration	0.54	414
						Exploratory behaviour	Exploration	0.247	414
						Exploratory behaviour	Exploration	0.331	414

Reference	Year	Taxon	W/D/LM	n	Trait	Category	Value	N
Duckworth and Badyaev	2007	Passeriformes	W	1	Aggression	Aggression	0.45	231
Dusek	1980	Equidae	D	1	Spirit and willingness to run	Disposition	0.07	579
Erf et al.	1992	Ruminatia	D	2	Disposition	Disposition	0.1	5814
					Trouble-free workability	Disposition	0.11	5353
Fairbanks et al.	2004	Platyrrhini	LM	3	Social impulsivity	Sociability	0.35	352
					Aggression	Aggression	0.61	352
					Approach behavior	Boldness	0.25	352
Figueiredo et al.	2005	Ruminatia	D	1	Docility	Boldness	0.165	5754
Fordyce and Goddard	1984	Ruminatia	D	2	Docility	Boldness	0	889
					Docility	Boldness	0.09	889
Fordyce et al.	1982	Ruminatia	D	3	Docility in crush test	Boldness	0.25	957
					Docility in race test	Boldness	0.17	957
					Docility in ball test	Boldness	0.67	957
Foster et al.	1988	Ruminatia	D	1	Disposition	Disposition	0.08	43428
Francis	1984	Teleostei	D	2	High dominance	Dominance	0.12	322
					Low dominance	Dominance	0.6	317
Gauly et al.	2001	Ruminatia	D	8	Docility	Boldness	0.13	206
					Docility	Boldness	0.61	206
					Docility	Boldness	0.11	206
					Docility	Boldness	0.18	206
					Docility	Boldness	0.17	259
					Docility	Boldness	0.55	259
					Docility	Boldness	0.35	259
					Docility	Boldness	0.52	259

(continued)

Table 6.1. (continued)

Author	Year	Taxon	Strain	No. of Reported Estimates	No. of Traits	Trait Name Given	Trait Name Used in Analysis	h^2 Estimate	No. of Subjects
Gauly et al.	2002	Ruminatia	D	6	1	Docility	Boldness	0.06	271
						Docility	Boldness	0.1	258
						Docility	Boldness	0.06	249
						Docility	Boldness	0.17	219
						Docility	Boldness	0.29	204
						Docility	Boldness	0.27	192
Gerken and Petersen	1992	Phasianidae	D	13	6	Dustbathing activity	Activity	0.28	2365
						Dustbathing activity	Activity	0.32	2333
						Dustbathing activity	Activity	0.22	2591
						Locomotion	Activity	0.2	900
						Locomotion	Activity	0.17	900
						Locomotion	Activity	0.1	900
						Aggression	Aggression	0.18	939
						Reaction toward novelty	Exploration	0.05	2889
						Reaction toward novelty	Exploration	0.33	2590
						Reaction toward novelty	Exploration	0.29	2692
						Latency to emerge from a box	Boldness	0.11	632
						Latency to emerge from a box	Boldness	0.1	632
						Tonic immobility	Boldness	0.56	632

Gervai and Csányi	1985	Teleostei	D	7	6	Exploration of novelty	Exploration	0.97	358
						Defense in novel environment	Exploration	0.96	358
						Timidity	Boldness	0.92	358
						Territoriality	*	0.12	358
						Activity 1	Activity	0.55	358
						Activity 2	Activity	0.89	358
						Emotionality	*	0.88	358
Goddard and Beilharz	1983	Canidae	D	10	10	Nervousness	*	0.58	249
						Suspicion	*	0.1	249
						Concentration	*	0.28	249
						Willingness	*	0.22	249
						Distraction	*	0.08	249
						Dog distraction	*	0.27	249
						Noise distraction	*	0	249
						Sound shy	*	0.14	249
						Hearing sensitivity	*	0	249
						Body sensitivity	*	0.33	249
Guhl et al.	1960	Phasianidae	D	2	1	Aggression	Aggression	0.18	326
						Aggression	Aggression	0.22	326
Han and Gatehouse	1993	Endopterygota	OW	1	1	Duration of flight	Activity	0.27	31
Hayes	1999	Ruminatia	D	1	1	Nervousness	*	0.05	232064
Hearnshaw and Morris	1984	Ruminatia	D	1	1	Docility	Boldness	0.44	529
Hemsworth et al.	1990	Suidae	D	1	1	Fear of humans	Boldness	0.38	334
Humphries et al.	2005	Hymenoptera	D	1	1	Activity	Activity	0.2	402
Isles et al.	2004	Rodentia	D	3	2	Reward impulsivity	*	0.16	31
						Reward impulsivity	*	0.16	31
						Activity	Activity	0.75	31

(continued)

Table 6.1. (continued)

Author	Year	Taxon	Strain	No. of Reported Estimates	No. of Traits	Trait Name Given	h² Estimate	Trait Name Used in Analysis	No. of Subjects
Kenttamies et al.	2006	Canidae	D	3	3	Confidence in feeding in presence of human	0.22	Boldness	1569
						Aggression	0	Aggression	1569
						Ease of capture	0.07	*	1569
Lawstuen et al.	1988	Ruminatia	D	1	1	Docility	0.12	Boldness	9546
Le Neindre et al.	1995	Ruminatia	D	2	1	Docility	0.22	Boldness	906
						Docility	0.18	Boldness	906
Lindberg et al.	2004	Canidae	D	3	3	Excitement	0.49	*	920
						Willingness to search and retrieve	0.28	*	920
						Independence	0.16	*	920
Mackenzie et al.	1985	Canidae	D	1	1	Willingness to chase and attack decoy and use olfaction	0.51	*	575
Morris et al.	1994	Ruminatia	D	4	1	Calving docility	0.09	*	5973
						Docility in handling yards	0.22	Boldness	765
						Docility in handling yards	0.32	Boldness	250
						Docility in handling yards	0.23	Boldness	653
Mourao et al.	1998	Ruminatia	D	3	1	Docility in crush test	0.06	Boldness	273
						Docility in crush test	0.15	Boldness	459
						Docility in crush test	0.27	Boldness	182

Author	Year	Group	Code	N₁	Behavior	Temperament	Value	N
O'Bleness et al.	1960	Ruminatia	D	1	Temperament	Disposition	0.4	842
Oikawa et al.	1989	Ruminatia	D	1	Docility during judging	Boldness	0.27	474
Oki et al.	2007	Equidae	D	1	Response to stressful human inspection	*	0.23	4452
Phocas et al.	2006	Ruminatia	D	1	Docility in handling	Boldness	0.18	2781
Réale et al.	2000	Ruminatia	W	1	Boldness	Boldness	0.21	35
Rogers et al.	2008	Platyrrhini	LM	5	Freezing duration	Boldness	0.38	285
					Orient to intruder	Boldness	0.91	285
					Duration of locomotion	Activity	0	285
					Hostility toward intruder	Boldness	0	285
					Frequency of cooing	*	0	285
Ruefenacht et al.	2002	Canidae	D	1	Overall behavior in a standardized test with regard to people and environment	Disposition	0.18	3497
Saetre et al.	2006	Canidae	D	2	Boldness	Boldness	0.25	5964
					Boldness	Boldness	0.27	4589
Samore et al.	1997	Equidae	D	2	Irritable/passive	*	0.06	3902
					Docility	Boldness	0.02	3902
Sato	1981	Ruminatia	D	2	Nervousness while weighing	*	0.45	191
					Nervousness while weighing	*	0.67	191

(continued)

Table 6.1. (*continued*)

Author	Year	Taxon	Strain	No. of Reported Estimates	No. of Traits	Trait Name Given	Trait Name Used in Analysis	h^2 Estimate	No. of Subjects
Sharma and Khanna	1980	Ruminatia	D	1	1	Disposition	Disposition	0.19	319
Silva et al.	2003	Ruminatia	D	1	1	Flight distance	Disposition	0.13	8800
Sinn et al.	2006	Cephalopoda	OW	6	3	Boldness in threat context	Boldness	0.21	147
						Activity in threat context	Activity	0.67	147
						Reactivity in threat context	*	0.89	147
						Boldness in feeding context	Boldness	0.08	147
						Activity in feeding context	Activity	0.05	147
						Reactivity in feeding context	*	0	147
Strandberg et al.	2005	Canidae	D	5	5	Boldness	Boldness	0.27	5959
						Playfulness	*	0.26	5959
						Chase proneness	*	0.18	5959
						Curiosity/ fearlessness	Boldness	0.31	5959
						Aggression	Aggression	0.2	5959
Thompson et al.	1981	Ruminatia	D	1	1	Disposition	Disposition	0.07	8977
Van Oers et al.	2004	Passeriformes	W	2	1	Risk taking	Boldness	0.19	73
						Risk taking	Boldness	0.32	92
Van Vleck	1964	Ruminatia	D	2	1	Nervousness	Disposition	0.16	1400
						Nervousness	Disposition	0	4080

Author	Year	Family	Type	N1	Trait	N2	Category	Value	N
Varo	1965	Equidae	D	1	Temperament	1	Disposition	0.23	5996
Visscher and Goddard	1995	Ruminatia	D	2	Temperament	1	Disposition	0.22	14596
Weiss et al.	2002	Platyrrhini	Z	2	Temperament	2	Disposition	0.25	4695
					Subjective well-being		*	0.4	128
Weiss et al.	2000	Platyrrhini	Z	6	Dominance	6	Dominance	0.66	128
					Dominance		Dominance	0.63	145
					Surgency		*	0	145
					Dependability		*	0.21	145
					Agreeableness		*	0.03	145
					Emotional stability		*	0.08	145
					Openness		*	0	145
Wickham	1979	Ruminatia	D	4	Docility during milking	1	Boldness	0.11	7847
					Docility during milking		Boldness	0.12	7421
					Docility during milking		Boldness	0.11	6377
					Docility during milking		Boldness	0.09	6213
Wilsson and Sundgren	1997	Canidae	D	2	Temperament	1	Disposition	0.15	1310
					Temperament		Disposition	0.1	797
Wilsson and Sundgren	1998	Canidae	D	3	Approach behavior to novel human	3	Exploration	0.21	554
					Fearfulness of large ball		Exploration	0.27	554
					Activity		Activity	0.53	554
Wright et al.	2003	Teleostei	D	1	Shoaling tendency	1	Sociability	0.23	80

unlikely that animal personality traits have genetic inheritance patterns that are independent of one another, and (d) structural pleiotropy, in which developmental constraints prevent additive genetic covariance from changing sign across environments (De Jong 1990), is a strong candidate to account for the structural architecture of animal personality. Since this review there have been a number of significant conceptual, theoretical, and analytic developments within the field, and our aim in this chapter is to (1) review the current standing of the field of quantitative and molecular genetic studies on animal personality, (2) quantify, through a statistical meta-analysis, patterns of heritability of animal personality traits in both captive and wild populations, and (3) highlight and discuss some of the advances in molecular genetics the field has experienced since 2005 (see also Van Oers and Mueller 2010). We use this information to then provide what we believe are fruitful directions for future research.

Current status of the field
ADDITIVE GENETIC EFFECTS

Variation in behavior in wild populations of animals is ubiquitous, and most behavioral traits are expected to be at least partly heritable and to influence life-history traits, thereby being the target of selection. Therefore, the identification of a heritable influence (heritability) is an essential starting point for evolutionary research on behavioral traits (Boake et al. 2002). Heritability is defined as the relative amount of genetic variation (V_G) in relation to the phenotypic variation (V_P) (Falconer and Mackay 1996). Genetic variation can further be subdivided into several components. The additive genetic component is a fraction of genetic variation in which separate alleles contribute a fixed value to the metric mean of a quantitative trait. In other words, additive genetic variation implies fixed additive contributions among a (most likely) large number of alleles. The presence of additive genetic variation is the *sine qua non* for predicting evolutionary change and for calculating heritability and genetic correlations. This means that although the presence of an additive genetic component is not necessarily a prerequisite for a personality trait, it is a prerequisite when one is interested in the evolutionary background or the genetic architecture of personality traits. In human personality, a heritable component has been found in almost every personality trait (Penke et al. 2007) and psychologists sometimes even refer to it as a law (Turkheimer 1998).

Two different types of heritabilities can be distinguished. The first, broad-sense heritability, is an estimate of the proportion of variance due to additive (V_A) and nonadditive (i.e., dominance variance V_D and interaction variance V_I) genetic variance, while the second, narrow-sense heritability, is an estimate of the proportion of phenotypic variance in a trait due to only additive genetic effects. A special case of a narrow-sense heritability is realized heritability: the change in mean phenotype as a consequence of selection as a fraction of applied selection (Lynch and Walsh 1998). In other words, realized heritability is an estimate of additive genetic variation based on the actual genetic change across generations. Generally speaking, narrow-sense heritability is normally estimated by using phenotypic data from parents and offspring predicting the selection response; broad-sense heritability is normally estimated by variance components analysis of groups of full- or half-sibs. Estimates of narrow- and broad-sense heritability are often broadly consistent with one another (Lynch and Walsh 1998), but the latter can often overestimate narrow-sense heritability and is often influenced by dominance genetic variation, which often exists for many fitness-related traits (Crnokrak and Roff 1995).

Currently, most genetic studies on animal personality traits focus on establishing at least some sort of genetic basis for their phenotypic traits of interest (see table 6.1 for a list of studies), and there is now evidence that there is a significant genetic influence, including additive genetic effects, on many animal personality traits (see Van Oers et al. 2005a; and meta-analysis, this chapter). However, most studies also normally estimate heritability at only one point in time for a single population, and therefore often neglect the fact that heritabilities might vary. Importantly, the expression of genetic variation is often dependent on the quality or predictability of the environment (Hoffmann and Merilä 1999; Wilson et al. 2006; Dingemanse et al. 2009). In this case, differences in heritabilities may be caused by covariation between gene expression and environmental conditions.

In a recent meta-analysis of genetic parameters estimated across heterogeneous environments in wild populations, Charmantier and Garant (2005) reported an emergent trend of higher heritabilities under more favorable conditions, which was statistically significant for morphological but not for life-history traits, indicating that environmental conditions can have important consequences for predicted responses to selection. Comparisons across environments both in the laboratory and in the field indicate that there can often be significant covariance between the expression of genetic variance and measures of environmental quality; however, there appear to be no strong generalizations as to how heritabilities are likely to change in direction with regards to environmental conditions (Hoffmann and Parsons 1991; Weigensberg and Roff 1996). Studies on the response of heritabilities of personality traits to changing environments are surely needed (but see Dingemanse et al. 2009).

Another way in which heritabilities can vary is when genes are differentially expressed, dependent on an animal's age or gender. For many traits population-level phenotypic variation for traits and their underlying gene expression in individuals changes with age. Constancy of estimates of genetic contributions to traits with age cannot be assumed—in many cases, for example, heritability in numerous morphological traits is known to vary over ontogeny (Réale et al. 1999; Badyaev and Martin 2000; Uller et al. 2002). For many morphological traits, estimates of additive genetic variation differ also between the sexes, suggesting significant sexual differences in genetic architecture (Coltman et al. 2005; Kruuk et al. 2008). Sex differences in genetic architecture of fitness-related traits appear to be linked to mating systems, and may be most pronounced in highly polygynous species with high degrees of sexual dimorphism. Note that basic patterns of sex-specific additive genetic variation are central to understanding antagonistic pleiotropic effects and, therefore, the maintenance of variation in personality. For example, it is conceivable that some alleles would be associated with enhanced male fighting success but reduced female fecundity owing to their antagonistic effects on levels of hormones in each sex (e.g., testosterone; Kruuk et al. 2008).

Although it is now generally accepted that animal personality traits have at least some genetic basis (but see Sinn et al. 2006), there is still a basic lack of understanding of environmental, age-, and sex-specific effects on contributions of genetic variation to animal personality traits. Also unclear is how these parameters may have differential effects on some personality traits, but not others, within an animal system (Weiss et al. 2000; Sinn et al. 2006). Information on environmental qualities and age- and sex-specific effects on the relative difference in the estimates of genetic variation between different animal personality traits could give information on the selection pressures and fitness consequences that act or have been acting on these traits (Kimura 1958). Currently, most animal personality studies report the heritability for only a single trait, and are hampered by small sample sizes/lack of statistical power. It is unlikely that most animals express behavioral variation along a single trait axis. Further work is needed on differences in genetic influences on different personality traits (see Réale et al. 2007 for a proposed framework) through time and across environments using longer-term studies of known-pedigreed individuals. Recently developed animal models (see below) should allow for increased power in at least some study systems.

Nonadditive genetic effects

Because genes not only may act additively, but also may interact, it is important to include in the genetic value a measure of the amount of interaction. Nonadditive genetic variance is defined as the component of phenotypic variance in a trait that cannot be predicted from the combined additive effects of a genotype's nuclear alleles (Mazer and Damuth 2001). Nonadditive genetic variation can be broken down into dominance and epistatic variation (Crnokrak and Roff 1995; Meffert et al. 2002).

When one is looking at a single locus, interaction between alleles that results in phenotypic expression that is not purely additive is referred to as the dominance deviation; in other words, alleles at a single locus interact to produce a genotype that would not be predicted on the basis of the average effects of those alleles working alone (i.e., heterozygotes express the phenotype of the dominant allele). Depending on the study, several methods for studying dominance variation are possible, but what they all have in common is that researchers require certain information about the pedigree of the study population. This can be obtained either in a controlled experimental breeding design with crosses between lines or populations (Hayman 1960) or in a natural population with a structured pedigree (Kruuk 2004). Significant dominant genetic effects on a trait can result in inflated estimates of broad-sense heritability (h^2), and comparisons of estimates of h^2 using different experimental designs (i.e., parent-offspring analysis versus full-sib analysis) can often help to evaluate whether there is significant dominance variation acting on a trait (Mousseau and Roff 1987). From an evolutionary standpoint, significant nonadditive genetic variation in a trait (dominance and epistasis) may contribute to the maintenance of additive genetic variation for some traits in small populations, either after selective "bottlenecks" or through metapopulation dynamics (e.g., Naciri-Graven and Goudet 2003). Nonadditive genetic variation is probably converted to additive genetic variation owing to chance changes in the genetic background of small populations under directional selection, which increases the frequency of rare partially recessive alleles (Willis and Orr 1993), which are common in many animal species due to inbreeding depression for many components of fitness (Charlesworth and Charlesworth 1987). In essence, as the genetic background of individuals in the population changes, the relationships between dominant and epistatic gene interactions also change, and can alter the availability of additive genetic variation for response to selection.

To date, the only study that has examined genetic dominance in animal personality traits was conducted on selected lines of great tits. Van Oers et al. (2004b) carried out a quantitative genetic analysis of two personality traits (exploration and boldness) and the combination of these two traits (early exploratory behavior). This study was carried out on the lines resulting from a two-directional artificial selection experiment on early exploratory behavior (EEB) of great tits (*Parus major*) originating from a wild population (Drent et al. 2003). In analyses using the original lines, reciprocal F_1, and reciprocal first backcross generations they found that dominance variation may range between 0.43 (exploration) and 1.2 (boldness) times the additive genetic component (Van Oers et al. 2004b). Another example of nonadditive genetic effects is given by epistatic effects. Interaction between genes at different loci that act on the same characteristic is referred to as epistasis. Dominance variation can be defined as nonadditivity of allelic effects within loci; epistasis here refers to the nonadditivity of allelic effects between loci. Similar methodology used for dominance effects is also used for detecting epistasis, but to our knowledge, no animal personality studies have yet attempted to estimate epistatic effects.

Psychologists do not yet fully understand the role of nonadditive genetic influences on personality traits (McCrae 2007) and these effects have been largely neglected in animal personality research (Van Oers et al. 2005a). Recent studies in psychology have shown that nonadditive components (mainly genetic dominance) can encompass a significant part of the phenotypic variance (Van Kampen 1999; Keller et al. 2005; Rettew et al. 2008), having an effect similar to that of additive components, but even up to 50% of the total phenotypic variation for a General Personality Factor, recently proposed to underlie all personality traits in humans (Rushton et al. 2008). In general, little is known about how genotypic influences on animal personality traits are influenced by the genetic background of individuals (gene x gene interactions); research in this area could provide important genetic and molecular clues to help explain the maintenance of genetic variation of animal personality traits, even under directional selection.

Genetic correlations and animal personality

As heritability quantifies the amount of variation that can be attributed to additive genetic variation, genetic correlations are a measure of the degree to which traits have genes in common or to which genes are linked due to linkage disequilibrium (Roff 1996). Genetic correlations between traits can constrain evolutionary change of single traits, since during selection on one trait genetic correlations influence the selection response of the other (Gromko 1995). However, a genetic correlation will not act as a constraint if the effects of a gene on two traits are themselves independent. If consistent individual differences are adaptive (Wilson 1998; Buss and Greiling 1999), the coherence between different personality traits could be a product of natural selection as well (Wolf et al. 2007). The existence of trade-offs between different components of fitness is fundamental to the concept of correlated suites of traits and, therefore, of personality. Prerequisites for trade-offs to act as evolutionary constraints from a quantitative genetic standpoint are that (a) the two traits have a genetic basis, and (b) either there is antagonistic pleiotropy where the same or some genes are involved in both traits or there is a genetic correlation between the traits (Lande 1982; Kruuk et al. 2008). The word constraint as used here should not be confused with an absolute evolutionary constraint (Roff and Fairbairn 2007). Additive genetic effects and genetic correlations are *states* in an evolutionary trajectory, not necessarily absolute constraints that might hamper evolution of a trait (see Bell 2005; Dingemanse et al. 2007; Roff and Fairbairn 2007). Genetic covariance will limit evolution of possible trait combinations only when the genetic correlation between two traits is -1 and constant, since independent selection will not have any effect on the separate traits (Roff and Fairbairn 2007). Since this scenario is very unlikely in traits involved in personality, a particular genetic structure for a personality profile most likely is not an absolute constraint for an adaptive evolutionary response, but instead may represent a consequence of past evolutionary processes.

Since personality traits have moderate genetic heritability and are likely to have complex genetic correlations with other traits (e.g., Merilä and Sheldon 1999; 2000), the potential for evolutionary change may be limited by a lack of genetic covariance in the multidimensional direction favored by selection (Blows and Hoffmann 2005; Moretz et al. 2007). To date, however, there is a surprising dearth of studies that report genetic correlations of animal personality traits in variable environments in wild populations of animals, while genetic correlations are known to vary and even switch signs over environments for morphological and life-history traits (Sgro and Hoffmann 2004). The calculation of genetic correlations for behavioral traits is often complicated by the notoriously high standard errors (see Van Oers et al. 2004a). In short-term studies of wild populations this is problematic, but with information on multiple generations and pedigree files (see below), the calculation of meaningful genetic correlations is not impossible. Simpler methods such as full- or half-sib mean correlations have in some occasions proven to be valuable as well (Astles et al. 2006).

Estimates of genetic correlations are fundamental for understanding the evolution of behavioral constructs like personality or behavioral syndromes. The functional architecture of personality traits has been debated in various approaches to human personality research, but all human personality approaches share commonality in that they report an underlying genetic structure that causes the coherence of personality traits (Bouchard and Loehlin 2001). In the case of pleiotropy, individual genes have effects on several traits. The effects of a gene on two traits might themselves be independent, or structurally linked (De Jong 1990). Independence of the effect of a gene on different traits is usually assumed in quantitative genetics, rather than structural pleiotropy. Linkage disequilibrium exists when traits are affected by different sets of genes, but a selective force generates and preserves particular combinations of alleles at a particular locus (Price and Langen 1992; Falconer and Mackay 1996; Lynch and Walsh 1998).

Several animal personality studies have now reported moderate to high genetic correlation values between various personality traits ranging from 0.4 to 0.9 (Van Oers et al. 2005a; Moretz et al. 2007). What effect these correlations have on behavioral evolution is currently a matter of discussion. Moretz et al. (2007) point out that the evolutionary influence of strong genetic correlations may be small, since over time strong genetic correlations might decrease differences among individuals in a population. Therefore, unless correlational selection is acting on the two traits simultaneously or there are temporal fluctuations in environmental conditions (Roff 1996), the originally strong phenotypic or genetic correlations are likely to disappear. In other words, selection against particular combinations of traits will cause other combinations to be more frequent, and this will eventually decouple the correlation. Hence, unless correlational selection is strong and chronic (Sinervo and Svensson 2002), linkage disequilibria built up by correlational selection are expected to weaken rapidly and therefore will not constrain evolution of the separate traits (Bulmer 1989; Falconer and Mackay 1996). Mutations will most likely counteract these effects since they will more likely be of the rare type. Moreover, selection in both human and animal personality traits has been found to fluctuate over time and space (Dingemanse and Réale 2005; Penke et al. 2007), possibly causing genetic correlations to fluctuate (see above) and therefore preventing simultaneous erosion of genetic variation in the two traits. On the other hand, weak, but continual, genetic correlations may have a profound evolutionary impact when considered over long periods, especially for quantitative behavioral traits (Moretz et al. 2007). Here, strong genetic linkage between some of the genes involved may produce only weak phenotypic correlations, which over long periods have profound effects on the direction of evolutionary change. The next step, therefore, in animal personality research is to begin measuring genetic correlations in various environments and to track these correlations over several generations to understand how they may change owing to variation in selection pressures and direction.

A second indication for the selection for sets of behavioral traits was found by Dingemanse et al. (2004), who showed differences in selection pressures on explorative behavior for males and females and different selection pressures over three different years in a study on exploration in a natural population. Considering the differences in selection pressure together with the prerequisites of correlational selection, the genetic correlations found will be built up and maintained by correlated selection only if variation in natural selection on one trait covaries with variation in selection on another trait. This seems to be unlikely, and therefore structural pleiotropy seems to be one of the potential explanations for the observed genetic correlations. This does not exclude differences in correlational selection from being a major factor for explaining differences among populations. Phenotypic and genetic correlation among traits in a population can also be the result of selection on sets of traits, when certain combinations of traits are more fit than others. Nevertheless, the absence of phenotypic correlations in a population does not automatically imply that the existence of a correlation in other populations is caused by correlational selection. For example, in a study by Bell and Sih (2007), no correlation between boldness and aggression was found in a population of stickleback fish before a predation event. In a controlled predation experiment, predation was found to act on only one trait, namely boldness, and although there was no selection on the combination of the two traits, the surviving population showed a positive phenotypic correlation between the two traits, boldness and aggression (Bell and Sih 2007). Hence, the existence of genetic correlations is not proof of an absolute constraint or lack of a potential response to selection, and the presence of correlated selection does not imply independent evolution of personality traits.

A question that remains unanswered is which behaviors are correlated across which contexts and how stable these genetic correlations are (Sih et al. 2004). A study on sticklebacks found genetic correlations for some populations but not for others (Bell 2005), which suggests that the syndromes that cause traits to be phenotypically correlated are not necessarily difficult to uncouple. In two studies on the same species, the existence of these syndromes was coupled to predatory selection pressures, and predator-induced selection was hypothesized to cause these syndromes to emerge (Bell and Sih 2007; Dingemanse et al. 2007; see also chapter 2).

Modes of development and heritability

A further issue that has yet to receive sufficient attention is whether a "personality profile" at the genetic level consists of genetic variation in separate traits only with some overlap in genes due to pleiotropy or linkage disequilibrium, or whether there is genetic variation for the behavioral organization of the personality profile itself (Bell 2007; Belsky et al. 2007; Sinn et al. 2008a). Since genes code for proteins, the expression of a personality trait involves necessarily several levels of biological organization (i.e., protein interactions, neuroendocrine dynamics, behavioral expression), all of which are developmentally sensitive to genetic background and external environmental influences (Gottlieb 2007). The question therefore arises as to whether an individual's expression of behavior is the critical aspect of genetic evolutionary change or whether insight would be gained by a focus on the gene-frequency changes in the structural make-up of personality traits (i.e., neural systems and hormones underlaid by structural genes) that enables appropriate behavior to be expressed. Many discrete behavioral polymorphisms in nature (i.e., sneaker versus dominant males) are regulated by condition-dependent biochemical switches, influenced heavily by social and biotic state variables (Gross 1996; West-Eberhard 2003). In these threshold trait instances, the loci affecting a behavioral trait each have a relatively small effect on some structural trait that influences the developmental "switch" of the trait in question, for example, a biochemical product or age-dependent metabolic efficiency. In theory, there is no reason not to expect that similar threshold traits regulate personality expression, and in cases where personality traits are heavily regulated by condition-dependent properties, the appropriate level of genetic analyses would have to focus on the genetic influences and constraints of the *structural components* of behavioral expression, rather than the behavior *per se* (Belsky et al. 2007; Sinn et al. 2008a). This may help explaining why some personality traits are characterized by more additive genetic variation than others (see meta-analysis results below). In this case, different traits within the same personality profile of a population of animals may be characterized by different modes of development, with some traits being characterized as largely genotype-dependent and others (within the same individual) largely regulated by condition-dependent processes (Sinn et al. 2010).

GxE and what is a personality trait?

Van Oers and coauthors (2005a) had some significant insights for the study of animal personality from a quantitative genetic standpoint: they introduced reaction norms as a possible explanation for the variation reported on domain-generality and context-specificity of personality traits. It is well established that genetic and environmental effects can interact in such a way that the effect of any given environment on the phenotype is dependent on an individual's genotype (GxE interaction: Barton and Turelli 1989). GxE means that the phenotypic response to changing environments will differ among individuals such that phenotypic plasticity may itself be thought of as a heritable trait (Kruuk et al. 2008). For a given functional form of reaction norm, the genetic covariance matrix of associated parameters (the intercept and slope for a linear reaction norm) necessarily defines a corresponding G matrix of environment-specific traits. As a result, conclusions regarding reaction norms can equally be restated in terms of environment-specific traits: the presence of genetic variance for plasticity (reaction norm slope) necessarily means changing additive variance for environment-specific traits. This might also apply specifically in personality studies when one asks: *"how does one know whether one is studying two different personality traits or the same personality trait in different environments?"* In the former, one may or may not find phenotypic/genetic correlations, while in the latter, one should always find phenotypic/genetic correlations (unless reaction norms are crossed; Van Oers et al. 2005a; 2005b). Studies on genetic correlations, modes of development, and genotype-by-environment interactions of animal personality traits stand to contribute heavily to our understanding of "What is a personality trait?," which is still an outstanding issue in current studies of animal personality. Once again, we believe that answering basic questions in the quantitative genetics of animal personality stands to contribute heavily to our understanding of what constitutes a trait and what traits are the actual focus of selection.

Meta-analysis on the heritability of personality traits

Quantitative reviews of the primary literature are useful analytic tools to discover patterns across species and to generate useful hypotheses to inform and encourage further in-depth research. The heritability of animal personality traits has been qualitatively described by Van Oers et al. (2005). We undertook a formal meta-analysis of published studies reporting heritability estimates of animal personality traits. We did so by asking three basic questions: (1) Are personality traits heritable in nonhuman taxa? (or, in other words, are there large, moderate, small, or no measureable genetic effects on personality traits in nonhuman animals?) (2) Do heritability estimates derived from captive populations differ from those calculated when using individuals from natural, or wild, populations? (3) Are some personality traits more heritable than others?

DATA COLLECTION

To compile our data set, we first searched for published estimates of heritability of animal personality traits using the Web of Knowledge search engine on June 14, 2008, with the following search terms: animal personality and heritability, behavioral syndrome and heritability, temperament and heritability, coping style and heritability. After amassing unique studies from each of these database searches, we then examined the online animal personality bibliography compiled by Sam Gosling and colleagues (http://homepage.psy.utexas.edu/HomePage/Faculty/Gosling/bibliography.htm) and included any studies that were not found in the initial web searches. We then reviewed the reference list of each study identified in this initial primary search to identify those studies that had not previously been identified (secondary references). We did not attempt to identify further studies from reference lists of secondary references. Our initial collection of studies included both domesticated and wild species. While the influence of genetic variation on personality traits has most likely been altered by the domestication process itself, we chose to include these studies as there is a large and rich literature on the heritability of personality traits in domesticated species, and we attempted to account for this basic evolutionary difference in subsequent analyses (see below). Our data collection procedure resulted in 453 unique references.

We then removed studies from our data set if they (a) were conducted with humans, (b) were based on molecular studies of gene polymorphisms, (c) were not published in a peer-reviewed journal, (d) did not calculate novel heritability estimates, (e) were associated with working ability (e.g., racing in sport horses or thermal adaptation of beef cattle to tropical climes), (f) involved animal models of disease and/or psychiatric disorders, or (g) were not related to animal personality and heritability. The final list of 75 studies that met these criteria is given in Table 6.1.

We extracted the following data from each study for analyses and description: (1) year and author of publication, (2) taxonomic group being studied (to the level of phylogenetic class), (3) personality trait measured,

(4) method used to calculate heritability estimate (e.g., selection lines, variance components, or maximum likelihood animal models), (5) sample size, (6) age of subjects (subadult or sexually mature adults), (7) sex of subjects, and (8) whether a heritability estimate was derived from a population of a domesticated species, or whether it was derived from a wild population or offspring from wild-caught parents. We grouped the latter two categories since sample sizes of each group were small. Our assumption here is that estimates of heritability from offspring raised in the laboratory from wild-type parents approximate a lower bound on heritability in the wild population (Riska et al. 1989). For experiments that used selection lines to estimate genetic parameters, the study sample size was taken from the total number of individuals that were screened for behaviors during experiments. Four studies that reported estimates of heritability from captive populations of monkeys did not clearly fit our domestication/wild population distinction (Weiss et al. 2000; 2002; Fairbanks et al. 2004; Rogers et al. 2008). Therefore, we included these studies for omnibus tests of whether animal personality was heritable, but did not include them when testing for differences in heritability estimates between wild and domesticated species, or when testing for differences in heritability between personality traits within these same two categories.

COMPUTATION OF EFFECT SIZES

Instead of using raw heritability estimates for analyses, we converted heritability estimates into effect sizes in order to convert all estimates to a common, standardized measure. The effect size of each heritability estimate was calculated using Fisher's r to z transformation. Many studies reported more than one estimate of heritability, either of the same personality trait using a different method (i.e., a sire-only variance components model, a dam-only variance components model, and a combined model) or of multiple personality traits (i.e., boldness in two different contexts, or aggressiveness and boldness in separate contexts). An important statistical assumption of meta-analyses is the independence of observations, since lack of independence can increase type I error rates. In order to address our first question of interest ("Is personality heritable, and if so, what is the effect size?"), we calculated a single mean effect size for each study. In this case, each heritability estimate was converted to a Fisher's z value. Multiple Fisher's z values were then averaged within each study to give a single overall effect size for each study. These were then backtransformed to correlation coefficients, and reweighted effect sizes (using average sample sizes) were recalculated.

Two hundred fifty-two unique trait descriptors were used across studies and therefore, in order to group these data in a meaningful way to allow us to test our second question of interest ("Are some personality traits more heritable than others?"), we classified these unique trait descriptors into seven trait names, five of them based on the framework of Réale et al. 2007 (table 6.2). However, we were unable to identify an appropriate trait category for 39 estimates (table 6.1), and therefore these studies were removed for subsequent analyses on particular traits. We did not further investigate the trait "Disposition," since this omnibus trait name appeared to include almost all the other personality traits, and the context in which disposition was expressed was often unclear (e.g., generalized reactiveness/aggressiveness toward humans in domesticated species). In their review of current animal personality trait terminology, Réale et al. (2007) noted that their 5-trait framework is probably an oversimplification of the diversity of personality traits currently observed in animal taxa. Nevertheless, we found the 5-trait framework useful for simplifying the diversity of unique trait names found in our sampled studies. Table 6.1 lists both the original trait names given in each study and our reclassification of trait names used in analyses. To calculate effect sizes for our second question of interest, we again used Fisher's r to z transformation. When studies gave multiple estimates of heritability for the same personality trait, we first averaged the heritability estimate and the sample size across multiple estimates, and generated a single weighted effect size per trait per study.

ANALYSIS OF EFFECT SIZES

Each effect size calculated here was weighted as a function of sample size. To test whether the cumulative effect size was different from zero, 95% confidence intervals around the cumulative effect size were calculated. The homogeneity statistic (Q_t) was calculated to test whether the variation around the cumulative effect size was homogenous. Homogeneity can be considered an indication that the cumulative effect size was a true estimate of the population estimate. Since Q_t is chi-square distributed with degrees of freedom k − 1 (k = number of independent effect sizes from which the cumulative effect size is calculated), a Q_t that exceeds this critical value indicates that the variation is nonhomogeneous. For example, we expected significant heterogeneity for our first question of interest since we combined several different types of populations (domesticated versus wild populations) and several different personality traits.

Taking the mean estimate of the effect size of heritability from each single study allowed us to examine our first question of interest without violating statistical assumptions of independence. For our second question of interest,

Table 6.2. Definitions of trait names used in meta-analysis, starting with the Réale et al. 2007 framework.

Trait Name	Definition from Réale et al. 2007	Traits That We Included under This Rubric from Studies We Found Here
Shyness/ Boldness	An individual's reaction to any risky situation, but not new situations. Included reaction to potential dangers, such as predators and humans, and trait terms *docility*, *tameness*, and *fearfulness* in the specific context of reaction to humans.	Shyness/Boldness, Approach behavior, Docility, Freezing duration in fowl, Extraversion
Exploration-Avoidance	An individual's reaction to a new situation. This includes behavior toward a new habitat, new food, or novel objects.	Exploratory behavior, Open-field tests
Activity	The general level of activity of an individual.	Activity, Locomotion, Bleating and sniffing in sheep, Dustbathing activity in fowl
Sociability	An individual's reaction to the presence or absence of conspecifics (excluding aggressive behavior). Sociable individuals seek the presence of conspecifics, while unsociable individuals avoid conspecifics.	Sociability, Social Impulsivity
Aggression	An individual's agonistic reaction toward conspecifics.	Aggression
Dominance	Not included in Réale et al. 2007.	Dominance. We consider this to be more of a behavioral outcome rather than a personality trait per se. Personality traits contributing to Dominance outcomes could potentially include all 5 personality traits defined by Réale et al. 2007.
Disposition	Describes the overall behavioural disposition of an animal, usually in regard to an agricultural outcome. Not included in Réale et al. 2007.	Usually "disposition" was used without reference to context in which behavior was expressed, or was used in reference to behavior across a wide variety of contexts.

(*continued*)

Table 6.2. (*continued*)

Trait Name	Definition from Réale et al. 2007	Traits That We Included under This Rubric from Studies We Found Here
		We included the following trait names from domesticated studies under this trait rubric (usually to do with cattle and horses): Disposition, Temperament, Spirit (horses), Ease of handling, Flight speed, "Trouble-free workability," Nervousness, and "Response to stressful situation." We considered this trait to be an overall assessment of animal behavior, rather than a single personality trait per se. Personality traits contributing to Disposition could potentially include all 5 personality traits defined by Réale et al. 2007.

however, this approach was not possible owing to the fact that most studies reported multiple estimates for multiple personality traits, and the number of studies that reported only a single estimate did not allow for replication at the appropriate level for analyses (i.e., replicated estimates of traits from independent studies). For example, only 8 studies from wild populations estimated heritability for a single trait, and of these 8 studies 2 were of activity, 1 of aggression, 2 of boldness, 2 of exploration, and only 1 of sociability, making comparisons of independent observations impossible. Similarly, for estimates from domesticated studies that reported only 1 heritability estimate, 16 were of boldness, 3 of activity, 1 of aggression, 3 of dominance, and 1 of exploration. There is currently a lack of agreement about how to include multiple estimates taken from the same study in a meta-analysis. While there are statistical issues associated with pseudoreplication, to our knowledge no meta-analytic techniques have been developed that would allow us to account for our study design. In the case of animal personality studies, the default is that researchers often investigate and report on multiple traits. Therefore, we tentatively ignored the violation of independence for our second question of interest and interpreted overlapping 95% confidence intervals of effect sizes as conservative estimates of significant differences. We present these results with caution, and interpret them as an exploratory exercise to generate hypotheses about the patterns of genetic influence on different personality traits.

For both analyses, we assessed whether our data set was biased toward studies that reported significant heritability estimates (i.e., the "file drawer effect"), with Funnel Plots and Rosenthal's fail-safe numbers (Rosenthal 1991). The former can be used graphically to assess publication bias, while the latter values indicate the number of studies with effect sizes of zero that would be needed to reduce the observed effect size to a nonsignificant level (i.e., the overall omnibus average effect size is not different from zero). In addition to examining potential publication bias, for each analysis we investigated normality of estimates using standardized effect size vs. normal distribution quantile plots. All analyses were performed using MetaWin 2.1 (Rosenberg et al. 2000).

RESULTS

We obtained a total of 209 estimates of heritability from 14 different taxonomic groups, including insects, cephalopods, fish, several classes of birds, and mammals such as dogs, rodents, horses, and pigs (table 6.1). One hundred fifty-three estimates from 59 studies were from domesticated species, while 39 estimates from 12 studies were taken from wild populations or from lab-reared offspring from wild-caught parents. Sample sizes from these studies ranged from 29 to 232,064 individuals.

IS THERE A SIGNIFICANT GENETIC BASIS FOR ANIMAL PERSONALITY TRAITS?

The average heritability estimate for all personality traits taken across all studies was 0.26 (SE = 0.01), and the cumulative effect size, while small, was significantly different from zero (E = 0.18, L95%C.I. = 0.176, U95%C.I. = 0.187). Graphical analyses gave little support for a strong effect of statistical outliers, and removal of two studies whose standardized effect size fell outside the 95% C.I. of a normal distribution (Gervai and Csányi 1985; Visscher and Goddard 1995) did not change the results (E = 0.17, L95%C.I. = 0.169, U95%C.I. = 0.180). After re-including these two studies, there was, not unexpectedly, more heterogeneity in effect sizes than expected by sampling error alone ($Q_{t(70)}$ = 2363.2, P < 0.0001), indicating that other explanatory variables should be inspected (Cooper 1998). Rosenthal's fail-safe number was high, indicating the number of published studies of null results that would be needed to overturn this result was very high (Rosenthal's Method: 128198.2).

The average heritability estimate for personality traits in wild populations was 0.36 (SE = 0.05), and the average for domestic populations was 0.24 (SE = 0.01). While there was still significant heterogeneity within domesticated ($Q_{w(54)}$ = 1783.2, P < 0.0001) and wild ($Q_{w(12)}$ = 259.2, P < 0.0001) populations, variation in effect sizes between the two population groups exceeded this within group variation ($Q_{b(1)}$ = 262.4, P < 0.0001), indicating that wild populations and domesticated species had different patterns of genetic influence on personality traits. Wild populations had significantly higher heritability estimates than domesticated ones, with moderate to strong effect sizes in wild populations (E = 0.57, L95%C.I. = 0.51, U95%C.I. = 0.62) and small effect sizes in domesticated ones (E = 0.17, L95%C.I. = 0.17, U95%C.I. = 0.18). These results also held when we removed the two outliers identified above (E_{wild} = 0.37, L95%C.I. = 0.31, U95%C.I. = 0.43; $E_{domesticated}$ = 0.17, L95%C.I. = 0.16, U95%C.I. = 0.18).

ARE SOME PERSONALITY TRAITS MORE HERITABLE THAN OTHERS?
Average unweighted heritability estimates were 0.37 (SE = 0.09) for activity, 0.14 (SE = 0.05) for aggression, 0.20 (SE = 0.03) for boldness, 0.14 (SE = 0.05) for exploration/avoidance, and 0.21 (SE = 0.06) for dominance in domesticated populations. Examination of weighted effect sizes indicated a significant difference between the heritability of these traits (Q_b = 146.4, P < 0.0001). Aggression (E = 0.1644, L95%C.I. = 0.1304, U95%C.I. = 0.1983), dominance (E = 0.1898, 0.1197 to 0.2598), and activity (E = 0.1925, 0.1698 to 0.2153) had overlapping confidence intervals, while exploration (E = 0.2660, 0.2059 to 0.3261) had greater effect sizes than aggression but not dominance or activity. Boldness (E = 0.2678, 0.2612 to 0.2744) had significantly higher effect sizes than all other traits, except for exploration. Rosenthal's fail-safe number was very high, indicating that these results were not influenced by a publication bias toward positive results (Rosenthal's number = 54008.6).

Average unweighted heritability estimates were 0.39 (SE = 0.11) for activity, 0.28 (SE = 0.13) for aggression, 0.31 (SE = 0.15) for boldness, and 0.58 (SE = 0.22) for exploration/avoidance in wild populations. Examination of weighted effect sizes indicated a significant difference between the heritability of these traits (Q_b = 204.5, P < 0.0001). This was mainly due to the fact that exploration (E = 1.1052, L95%C.I. = 0.9543, U95%C.I. = 1.2562) and boldness (E = 0.9539, 0.8445 to 1.0634) had greater effect sizes than aggression (E = 0.3736, 0.1749 to 0.5724) or activity (E = 0.5950, 0.4882 to 0.7017). Rosenthal's fail-safe number was very high (7176.2).

CONCLUSIONS

Based on a range of studies on several phylogenetically distinct taxa with different evolutionary histories, our analysis suggests that there is sufficient evidence to conclude that there is genetic inheritance of animal personality traits. In wild populations of animals this genetic influence may be stronger than in domesticated populations, with moderate effect sizes of personality traits in wild populations (0.31 to 0.62) and small effect sizes in domesticated ones (0.16 to 0.18). Wild populations of animals can be characterized as having significant quantities of heritable personality variation available for evolution, while for domesticated species, genetic variation may have been eliminated through directional selection imposed by breeding programs.

Interestingly, our results also suggest that different animal personality traits may be characterized by different degrees of genetic influence. In both wild and domesticated populations, effect sizes for heritability of exploratory behavior and boldness were significantly greater than for activity, aggression, or dominance. In a recent meta-analysis on the fitness consequences of personality traits, Smith and Blumstein (2008) found that, in general, bolder individuals had increased reproductive success, particularly in males, but also incurred a survival cost. Exploration had a positive effect only on survival, whereas aggression had a positive effect on reproductive success but not on survival. Taken together, we can infer from these two syntheses that (a) the evolutionary history of genic selection on different personality traits within individuals may have differed/continues to differ fundamentally in wild populations, (b) some personality traits may conform to basic predictions of evolutionary trajectories better than others, and (c) animal personality traits have a genetic basis available for selection in the wild.

POTENTIAL LIMITATIONS

Our meta-analysis involved analyzing a mix of narrow- and broad-sense heritability estimates, with the latter estimates potentially being inflated by several potential confounding factors (Lynch and Walsh 1998). Thus, while it is clear that there is resemblance amongst relatives, and that this resemblance is due at least in part to genetic influences, our analysis was unable to shed light on the nature of this relationship. It is worth noting that some animal studies have begun to report narrow-sense heritabilities (e.g., Weiss et al. 2000; Drent et al. 2003; Quinn et al. 2009) or have even differentiated between various sources of variation, such as dominance and maternal genetic (e.g., Van Oers et al. 2004b), common (e.g., Weiss et al. 2000), and permanent environmental (e.g., Quinn et al. 2009) effects. Future meta-analyses using reports of narrow-sense estimates only may provide more accurate parameter estimates for understanding how well evolutionary model predictions fit real-life personality phenotypic patterns.

For both sets of analyses on differences in heritability of particular personality traits there was still significant within-group heterogeneity (Q_w), indicating that further explanatory factors remain to be explored. Indeed, our data collection methods identified several potentially important factors that may influence effect size estimates in our analyses. Significant within-group heterogeneity could also be due to sex-specific genetic effects, methodological effects (e.g., statistical models), age effects, or measurement effects (observational versus questionnaire measurement). Furthermore, in many studies, incomplete descriptions of environmental/contextual elements of behavioral assays limited our ability to clearly identify the functional context in which traits were expressed, which may also have been important. Clearly, any future primary reports of estimates of animal personality traits would enable subsequent tests for probable genetic interactions between many of the factors identified here, but for which we were unable to test owing to limitations in sample sizes.

Functional genomics and personality

Linking quantitative genetic variation in personality traits with polymorphisms in genes that code for this variance is essential for our understanding of the causes and consequences of personality trait diversity. For example, genomic approaches can be used to determine the molecular genetic basis of genetic correlations, whether resulting from linkage disequilibrium or pleiotropy, since linking genes or other DNA sequences to fitness ultimately requires genome sequence information. Molecular techniques now enable powerful tools for the genetic study of animal personality, since the identification of specific genetic loci influencing personality phenotypes should enable a much more precise understanding of what constitutes a personality trait and which traits may be the targets of selection. Here, we highlight some of the recent advances in candidate genes and QTL approaches and how high-throughput methodology and whole genome sequences can help us understand the evolutionary dynamics of personality gene expression.

GENOME INFORMATION: SEQUENCING

Until recently, obtaining large-scale genomic data has been a limiting step for progress on nonmodel species. New high-throughput technologies of genomics with the second-generation sequencing technologies such as 454, Solexa, or Solid have hugely decreased the unit time and cost of obtaining sequence data (Mardis 2008). Importantly, these new technologies promise to open up immense opportunities for large-scale sequencing initiatives in ecologically important species (Ellegren 2008a; 2008b; Van Bers et al. 2010). However, nonmodel studies can also use an increasing number of species that have been sequenced as reference genomes, which facilitates functional genomic studies in a growing number of ecologically relevant species from an animal personality standpoint. Most striking examples of current advances are the currently sequenced genomes of the zebra finch and the stickleback fish, both important species that are commonly used in personality research.

DETECTING THE GENES—THE QUEST FOR THE HOLY GRAIL

Two general strategies can be identified to pinpoint the regions of the genome that are of interest for the quantitative trait(s) of interest. First of all, there are bottom-up approaches, such as QTL mapping, in which associations are tested between variation in neutral markers in the genome and variation in the traits of interest. Depending on marker density, regions can be identified that cover a few to several hundred genes, which are potentially important genes for explaining variation in the traits of interest. A somewhat different but not mutually exclusive approach is the top-down approach. Here, information from other species or behaviors is used to specifically test the association between one gene polymorphism and the trait of interest. These candidate gene approaches can be very useful, but are always biased toward the genes with higher effect sizes, since genes of small effects, or genes that are important in interaction with other loci, will be detected only in more random scans of the whole genome.

A major advantage of studies on the genetic basis of personality differences in natural populations is that candidates for trait loci can be nominated on the basis of knowledge of similar phenotypes in model species, circumventing the tedious process of unprejudiced genome-wide approaches. Most likely candidates include the alleles of the dopamine 4 receptor gene (DRD4) and the serotonin transporter gene (SERT) (Savitz and Ramesar 2004). The DRD4 gene has been found to account for about 10% of the variation in novelty seeking in humans (Reif and Lesch 2003) and is expressed in parts of the cortex and in the hippocampus. For several domestic animal species, Lipp and coworkers found a relation between genetic variation in the infrapyramidal mossy fiber projection in the hippocampus and several behaviors (Lipp and Wolfer 1999). Several animal studies have looked at the relation between novelty seeking and DRD4 and in general found similar results (Ito et al. 2004; Fidler et al. 2007; Boehmler et al. 2007). Fidler and coworkers (2007) found that great tits selected for divergent levels of a combination of exploratory behavior and boldness differed from each other in the allele frequency of a polymorphism in the DRD4 gene. This association was also confirmed in a natural population, where the levels of exploratory behavior differed for birds with different genotypes (Fidler et al. 2007). In a QTL study, Gutierrez-Gil et al. (2008) identified several QTL regions. The most notable candidate gene found in a QTL region was the DRD4 gene, and it was located at the distal end of bovine chromosome 29, suggesting that DRD4 is probably one of the most important genes involved in variation in personality-related behavior. Studies looking at the serotonin transporter gene (SERT) have found a relation between a functional polymorphism in a regulatory sequence for this gene and anxiety (Eley and Plomin 1997; Reif and Lesch 2003; Gordon and Hen 2004). Other less-studied genes with possible effects on variation in personality include CCK (Ballaz et al. 2008), MAOA (Manuck et al. 2000), DRD2 (Noble 2003), the 5-HT2c receptor (Ebstein et al. 1997), the 5-HT2a receptor (Golimbet et al. 2002), and tyrosine hydroxylase (Persson et al. 2000). Unfortunately, results from these six genes are ambiguous (Savitz and Ramesar 2004). Such candidate genes potentially affect many traits, and therefore directly lead to genetic correlations between the traits involved. Interestingly, on physiological grounds one would expect the effects of a candidate gene on the different traits to be physiologically linked, and therefore lead to "structural pleiotropy," rather than to the formal pleiotropy usually referred to in quantitative genetics. Structural pleiotropy removes the potential for sign change in the genetic correlations between phenotypically plastic traits over environments (De Jong 1990).

Another potentially efficient strategy for the identification of personality genes is to analyze genes underlying orthologous QTL in model organisms (Willis-Owen and Flint 2007). With the use of genetic markers, quantitative trait loci (QTL) can be found that contribute to the variation in a quantitative trait like personality (Lynch and Walsh 1998). QTL analysis is used in a number of personality studies of both humans (Reif and Lesch 2003) and domestic animals (Flint and Corley 1996). The standard method to make genetic maps based upon molecular markers uses a structured F2 pedigree derived from inbred lines. Traditionally only one female and one male are used to produce many F1 and F2 crosses. For most nondomestic species, however, it is not possible to produce inbred lines, and the number of offspring that one pair can produce is fairly low compared with the traditional laboratory species used for this type of analysis. Therefore, further studies are needed to identify gene polymorphisms in wild animals that show associations with personality traits. Fortunately, genetic maps already can be developed for virtually any genome (Parsons and Shaw 2002). Thanks to recent advances in molecular and statistical methods (Erickson et al. 2004), the use of QTL studies on personality in natural populations has come a step closer (Van Bers et al. 2012).

COMPARATIVE GENOMICS

It is increasingly recognized that comparative genomics, in which sequences from two or more species are aligned and compared, is a powerful tool for detecting regions that evolve under negative or positive selection, thus indicating functionality. By examining genome sequences from multiple species, comparative genomics offers new insight into genome evolution and the way natural selection molds DNA sequence evolution (Ellegren 2008b). Adaptive evolution can be inferred, for example, from gene sequences showing sites of repeated nonsynonymous substitutions in multiple species alignments, an increased rate of nonsynonymous substitutions in divergence compared with diversity data, or a high frequency of derived alleles (Mitchell-Olds et al. 2007). Apart from comparing candidate genes between species (see above), a next step could be to use cross-species QTL concordance as a tool for QTL dissection. This technique is now mainly used in studies of mouse emotionality and human neuroticism (Willis-Owen and Flint 2007; Fullerton et al. 2008), but could also be used for comparing QTL results from model species with nonmodel species. A possible challenge of this method is that it seems improbable that the genetic determinants of traits were flawlessly preserved throughout evolution (Willis-Owen and Flint 2007), and it is therefore to be expected that the number of loci to be found will be highly dependent on the genetic distance of the two compared species.

Other, more advanced techniques at the transcript level include genome-wide transcriptional profiling with microarrays and molecular epigenetic studies. In the first technique, one quantifies the difference in amount of transcripts between two phenotypes. Microarrays have now been established as an essential tool for gene expression profiling in relation to physiology and development (Gibson 2002). Possible applications of microarray studies include the utilization for comparative methods to tease apart the functional relevance of correlations between gene expression and the development of a phenotype. But most interesting for personality research would be to use it to identify candidate genes as modifiers of traits. Although quantitative trait locus (QTL) mapping approaches are still in their infancy for animal personality research, and certainly for wild animal populations, it is now well recognized that this methodology is quite limited in its power to identify the actual genes that are responsible for quantitative variation.

Moreover, quantitative changes in mRNA can also help us understand how differences among populations might have evolved owing to selection, since differences in levels of transcription can mediate adaptive evolution (Roff 2007).

The quest to conceptually and methodologically incorporate genetics into animal personality research is now likely to become even more complex, as recent research suggests that epigenetic processes, too, could play a significant role in molecular evolution (Bossdorf et al. 2008). Epigenetics is defined as the study of heritable changes in gene expression and function that cannot be explained by changes in DNA sequence (Richards 2006; Bird 2007). These changes are based on a set of molecular processes that can activate, reduce, or disable the activity of particular genes: (i) methylation of cytosine residues in the DNA, (ii) remodeling of chromatin structure through chemical modification, and (iii) regulatory processes mediated by microRNAs, which are a class of small RNAs that are able to regulate gene expression (Bossdorf et al. 2008; see chapter 10).

Conclusions

(1) Although even the most basic reports of heritability estimates for personality traits in wild populations are still needed, we strongly recommend that, where possible, researchers should attempt to move on from just establishing a genetic basis for a trait through a "snapshot" in a single population at a single life-history stage. Understanding age- and sex-related influences (and perhaps their interaction) on the expression of genetic variation for different personality traits, and how this corresponds to life-stage specific selection and therefore evolutionary trajectories, would be a welcome progression for the field. Combined with this, understanding the relationships between age, sex, and social and genetic effects on behavior, and how these relationships may change under different environmental conditions, would continue to place the field at the leading edge of quantitative genetic and evolutionary biology.

(2) There is currently a great need to understand what influence nonadditive genetic effects, social indirect genetic effects, and recurrent environmental conditions have on the genetic architecture of animal personality traits, and therefore their inheritance. All of these factors will contribute to answering the question "what is a personality trait?" A deeper understanding of GxE interactions, what properties of the behavioral organization of an organism are the targets of selection,

and the role of heritable phenotypic plasticity are all fields of animal personality genetics that remain unexplored.

(3) Studies described in summary points 1 and 2 are both currently hampered by the fact that quantitative genetic analyses often require very large sample sizes, something that is often not possible for many animal personality researchers working with small populations. For example, some studies on personality differences within populations involve capture of most, if not all, individuals in a population, but still fall well below the requirements needed for a comprehensive genetic assessment of the traits in question (e.g., Sinn et al. 2008b; While et al. 2009). One of the major advances in quantitative genetics has been the use of mixed models for the estimation of variance components—in particular, a form of the mixed model known as the "animal model" (Lynch and Walsh 1998). In contrast to typical techniques used to estimate heritabilities, animal models incorporate information from sets of relatives other than parents and offspring (i.e., multigenerational pedigrees), and are not bound by assumptions of nonassortative mating, inbreeding, or selection, while allowing for unbalanced data sets (Kruuk 2004). Recent issues of leading journals have covered animal models in detail (see introduction, this book), so we have restricted our discussion of this technique here. Problems such as small sample sizes may be overcome by collecting long-term pedigree and behavioral data from wild populations; the use of animal models also allows for an explicit partitioning of the phenotypic and genotypic response to selection in wild populations through time, an essential component of our understanding of the evolution of animal personality traits. We expect that animal models used on long-term pedigreed data from wild populations will be one of the most exciting aspects of future animal personality genetics studies.

(4) While the basic assumption of many animal personality studies is that the mode of regulation for personality trait expression is genotype specific, past studies on discrete qualitative morphs in nature have shown that, more often than not, many behavioral morphs are condition dependent. Furthermore, most, if not all, phenotypic qualities of organisms are responsive to changing environmental conditions. However, currently almost nothing is known concerning the regulatory mechanisms of personality trait expression throughout organismal development, and the influences of environmental and social conditions on these regulatory mechanisms. This idea also extends to GxE interactions, context-specific versus domain-general personal-

ity expression, and the heritability of phenotypic plasticity. Further knowledge is sorely needed on the targets of selection for personality traits, and the heritability of developmental systems that enable appropriate lifestage-specific personality responses. In short, animal personality studies are well poised to address the question of "what is a trait?"

(5) Studies on animal personality traits within the field of molecular genetics and genomics now need to include natural populations. Molecular tools should be used not only to find variation on the genome that associates with variation in personality traits, but also to study GxE interactions and to identify regions with patterns of directional or stabilizing selection. Comparative genomics might therefore be of help in identifying general rules among species without having the problems that behavioral measurements have.

(6) Genetic approaches have proven to be crucial in answering questions about the evolutionary responses of life-history and morphological traits. Nevertheless, genetic studies on animal personality in natural populations are relatively scarce, and the field is, without doubt, in its infancy. Despite its nascent status, we have attempted to emphasize throughout our review how basic questions currently confronting the field of animal personality genetics also begin to address outstanding issues in evolutionary biology. We have also established a quantitative basis for the genetic influence on several widely identified animal personality traits: boldness, dominance, aggression, activity, and exploratory behavior. It is clear that much further work is needed to understand the relative influences of genes on animal personality traits, and the ways in which genetic, environmental (including cell-cell and protein-protein interactions), and social (indirect genetic effects) influences may alter the genetic architecture, and thus evolution of animal personality traits.

References

Adalsteinsson, S. 1977. Inheritance of temperament score in Icelandic riding ponies. *Journal of Agricultural Research in Iceland*, 9, 69–72.

Agyemang, K., Clapp, E., and Van Vleck, L. D. 1982. Components of variance of dairymen's workability traits among Holstein cows. *Journal of Dairy Science*, 65, 1334–1338.

Aitchison, T. E., Freeman, A. E., and Thomson, G. M. 1972. Evaluation of a type appraisal program in Holsteins. *Journal of Dairy Science*, 55, 840–844.

Alvarez, D., and Bell, A. M. 2007. Sticklebacks from streams are more bold than sticklebacks from ponds. *Behavioural Processes*, 76, 215–217.

Arnason, T. 1984. Genetic studies on conformation and performance of Icelandic toelter horses, 1: estimation of nongenetic effects and genetic parameters. *Acta Agriculturae Scandinavica*, 34, 409–427.

Arnason, T. H. 1979. Studies on traits in the Icelandic toelter horses, I: estimation of some environmental effects and genetic parameters. *Journal of Agricultural Research in Iceland*, 11, 81–93.

Astles, P. A., Moore, A. J., and Preziosi, R. F. 2006. A comparison of methods to estimate cross-environment genetic correlations. *Journal of Evolutionary Biology*, 19, 114–122.

Badyaev, A. V. 2005. Stress-induced variation in evolution: from behavioural plasticity to genetic assimilation. *Proceedings of the Royal Society of London B*, 272, 877–886.

Badyaev, A. V., and Martin, T. E. 2000. Individual variation in growth trajectories: phenotypic and genetic correlations in ontogeny of the house finch (*Carpodacus mexicanus*). *Journal of Evolutionary Biology*, 13, 290–301.

Bakker, T. C. M. 1994. Genetic correlations and the control of behavior, exemplified by aggressiveness in sticklebacks. *Advances in the Study of Behavior*, 23, 135–171.

Ballaz, S. J., Akil, H., and Watson, S. J. 2008. The CCK-system underpins novelty-seeking behavior in the rat: gene expression and pharmacological analyses. *Neuropeptides*, 42, 245–253.

Barton, N. H., and Keightley, P. D. 2002. Understanding quantitative genetic variation. *Nature Reviews Genetics*, 3, 11–21.

Barton, N. H., and Turelli, M. 1989. Evolutionary quantitative genetics—how little do we know? *Annual Review of Genetics*, 23, 337–370.

Beckman, D. W., Enns, R. M., Speidel, S. E., Brigham, B. W., and Garrick, D. J. 2007. Maternal effects on docility in Limousin cattle. *Journal of Animal Science*, 85, 650–657.

Beilharz, R. G., Butcher, D. F., and Freeman, A. E. 1966. Social dominance and milk production in Holsteins. *Journal of Dairy Science*, 49, 887–892.

Beilharz, R. G., and Cox, D. F. 1967. Genetic analysis of open field behavior in swine. *Journal of Animal Science*, 26, 988–990.

Bell, A. M. 2005. Behavioural differences between individuals and two populations of stickleback (*Gasterosteus aculeatus*). *Journal of Evolutionary Biology*, 18, 464–473.

———. 2007. Future directions in behavioural syndromes research. *Proceedings of the Royal Society of London B*, 274, 755–761.

Bell, A. M., and Sih, A. 2007. Exposure to predation generates personality in threespined sticklebacks (*Gasterosteus aculeatus*). *Ecology Letters*, 10, 828–834.

Belsky, J., Bakermans-Kranenburg, M. J., and Van IJzendoorn, M. H. 2007. For better and for worse: differential susceptibility to environmental influences. *Current Directions in Psychological Science*, 16, 300–304.

Benus, R. F., Bohus, B., Koolhaas, J. M., and Van Oortmerssen, G. A. 1990. Behavioural strategies of aggressive and nonaggressive male mice in response to inescapable shock. *Behavioural Processes*, 21, 127–141.

Bird, A. 2007. Perceptions of epigenetics. *Nature*, 447, 396–398.

Biro, P. A., and Post, J. R. 2008. Rapid depletion of genotypes with fast growth and bold personality traits from harvested fish populations. *Proceedings of the National Academy of Sciences U S A*, 105, 2919–2922.

Biro, P. A., Post, J. R., and Parkinson, E. A. 2003. From individuals to populations: prey fish risk-taking mediates mortality in whole-system experiments. *Ecology*, 84, 2419–2431.

Biro, P. A., and Stamps, J. A. 2008. Are animal personality traits linked to life-history productivity? *Trends in Ecology and Evolution*, 23, 361–368.

Blanchet, S., Paez, D. J., Bernatchez, L., and Dodson, J. J. 2008. An integrated comparison of captive-bred and wild Atlantic salmon (*Salmo salar*): implications for supportive breeding programs. *Biological Conservation*, 141, 1989–1999.

Blows, M. W., and Hoffmann, A. A. 2005. A reassessment of genetic limits to evolutionary change. *Ecology*, 86, 1371–1384.

Boake, C. R. B. (ed.). 1994. *Quantitative Genetic Studies of Behavioral Evolution*. Chicago: University of Chicago Press.

Boake, C. R. B., Arnold, S. J., Breden, F., Meffert, L. M., Ritchie, M. G., Taylor, B. J., Wolf, J. B., and Moore, A. J. 2002. Genetic tools for studying adaptation and the evolution of behavior. *American Naturalist*, 160, S143-S159.

Boehmler, W., Carr, T., Thisse, C., Thisse, B., Canfield, V. A., and Levenson, R. 2007. D4 dopamine receptor genes of zebrafish and effects of the antipsychotic clozapine on larval swimming behaviour. *Genes, Brain, and Behavior*, 6, 155–166.

Boenigk, K., Hamann, H., and Distl, O. 2005. Genetic analysis of the outcome of behavioural tests in puppies of the Hovawart dog. *Deutsche Tierarztliche Wochenschrift*, 112, 265–271.

———. 2006. Genetic influences on the outcome of the progeny tests for behaviour traits in Hovawart dogs. *Deutsche Tierarztliche Wochenschrift*, 113, 182–188.

Boissy, A., Bouix, J., Orgeur, P., Poindron, P., Bibe, B., and Le Neindre, P. 2005. Genetic analysis of emotional reactivity in sheep: effects of the genotypes of the lambs and of their dams. *Genetics Selection Evolution*, 37, 381–401.

Bossdorf, O., Richards, C. L., and Pigliucci, M. 2008. Epigenetics for ecologists. *Ecology Letters*, 11, 106–115.

Bouchard, T. J. 1994. Genes, environment, and personality. *Science*, 264, 1700–1701.

———. 2004. Genetic influence on human psychological traits: a survey. *Current Directions in Psychological Science*, 13, 148–151.

Bouchard, T. J., and Loehlin, J. C. 2001. Genes, evolution, and personality. *Behavior Genetics*, 31, 243–273.

Boyd, R., and Richerson, P. J. 1985. *Culture and the Evolutionary Process*. Chicago: University of Chicago Press.

Brydges, N. M., Colegrave, N., Heathcote, R. J. P., and Braithwaite, V. A. 2008. Habitat stability and predation pressure affect temperament behaviours in populations of three-spined sticklebacks. *Journal of Animal Ecology*, 77, 229–235.

Bulmer, M. G. 1989. Maintenance of genetic variability by mutation selection balance: a child's guide through the jungle. *Genome*, 31, 761–767.

Burrow, H. M. 1997. Measurements of temperament and their relationships with performance traits of beef cattle. *Animal Breeding Abstracts*, 65, 477–495.

———. 2001. Variances and covariances between productive and adaptive traits and temperament in a composite breed of tropical beef cattle. *Livestock Production Science*, 70, 213–233.

Burrow, H. M., and Corbet, N. J. 2000. Genetic and environmental factors affecting temperament of zebu and zebu-derived beef cattle grazed at pasture in the tropics. *Australian Journal of Agricultural Research*, 51, 155–162.

Burrow, H. M., Seifert, C. W., and Corbet N. J. 1988. A new technique for measuring temperament in cattle. *Proceedings of the Australian Society of Animal Production*, 17, 154–157.

Buss, D. M., and Greiling, H. 1999. Adaptive individual differences. *Journal of Personality*, 67, 209–243.

Charlesworth, D., and Charlesworth, B. 1987. Inbreeding depression and its evolutionary consequences. *Annual Review of Ecology and Systematics*, 18, 237–268.

Charmantier, A., and Garant, D. 2005. Environmental quality and evolutionary potential: lessons from wild populations. *Proceedings of the Royal Society of London B*, 272, 1415–1425.

Coltman, D. W., O'Donoghue, P., Hogg, J. T., and Festa-Bianchet, M. 2005. Selection and genetic (co)variance in bighorn sheep. *Evolution*, 59, 1372–1382.

Colvin, J., and Gatehouse, A. G. 1993. The reproduction-flight syndrome and the inheritance of tethered-flight activity in the cotton-bollworm moth, *Heliothis armigera*. *Physiological Entomology*, 18, 16–22.

Craig, J. V., Ortman, L. L., and Guhl, A. M. 1965. Genetic selection for social dominance ability in chickens. *Animal Behaviour*, 13, 114–131.

Crnokrak, P., and Roff, D. A. 1995. Dominance variance: associations with selection and fitness. *Heredity*, 75, 530–540.

De Jong, G. 1990. Quantitative genetics of reaction norms. *Journal of Evolutionary Biology*, 3, 447–468.

Detelleux, J., Volckaert, D., and Leroy, P. 1999. Evaluation génétique des caractères de conformation chez les bovins laitiers. *Annales De Medecine Veterinaire*, 143, 341–348.

Dickson, D. P., Barr, G. R., Johnson, L. P., and Wieckert, D. A. 1970. Social dominance and temperament of Holstein cows. *Journal of Dairy Science*, 53, 904–907.

Dingemanse, N. J., Both, C., Drent P. J., and Tinbergen, J. M. 2004. Fitness consequences of avian personalities in a fluctuating environment. *Proceedings of the Royal Society of London B*, 271, 847–852.

Dingemanse, N. J., Both, C., Drent, P. J., Van Oers, K., and Van Noordwijk, A. J. 2002. Repeatability and heritability of exploratory behaviour in wild great tits. *Animal Behaviour*, 64, 929–937.

Dingemanse, N. J., and Réale, D. 2005. Natural selection and animal personality. *Behaviour*, 142, 1159–1184.

Dingemanse, N. J., Van der Plas, F., Wright, J., Réale, D., Schrama, M., Roff, D. A., Van der Zee, E., and Barber, I. 2009. Individual experience and evolutionary history of predation affect expression of heritable variation in fish personality and morphology. *Proceedings of the Royal Society of London B*, 276, 1285–1293.

Dingemanse, N. J., Wright, J., Kazem, A. J. N., Thomas, D. K., Hickling, R., and Dawnay, N. 2007. Behavioural syndromes differ predictably between 12 populations of three-spined stickleback. *Journal of Animal Ecology*, 76, 1128–1138.

Dobzhansky, T., Ayala, F. J., Stebbins, G. L., and Valentine, J. W. 1977. *Evolution*. San Francisco: W. H. Freeman.

Drent, P. J., Van Oers, K., and Van Noordwijk, A. J. 2003. Realized heritability of personalities in the great tit (*Parus major*). *Proceedings of the Royal Society of London B*, 270, 45–51.

Duckworth, R. A. 2006a. Aggressive behaviour affects selection on morphology by influencing settlement patterns in a passerine bird. *Proceedings of the Royal Society of London B*, 273, 1789–1795.

———. 2006b. Behavioral correlations across breeding contexts provide a mechanism for a cost of aggression. *Behavioral Ecology*, 17, 1011–1019.

Duckworth, R. A., and Badyaev, A. V. 2007. Coupling of dispersal and aggression facilitates the rapid range expansion of a passerine bird. *Proceedings of the National Academy of Sciences U S A*, 104, 15017–15022.

Dusek, J. 1980. Estimates of coefficients of heritability of performance and exterior of horses in smaller populations. *Ziv Vyr*, 25, 71–80.

Ebstein, R. P., Segman, R., Benjamin, J., Osher, Y., Nemanov, L., and Belmaker, R. H. 1997. 5-HT2C (HTR2C) serotonin receptor gene polymorphism associated with the human personality trait of reward dependence: interaction with dopamine D4 receptor (D4DR) and dopamine D3 receptor (D3DR) polymorphisms. *American Journal of Medical Genetics*, 74, 65–72.

Eley, T. C., and Plomin, R. 1997. Genetic analyses of emotionality. *Current Opinions in Neurobiology*, 7, 279–284.

Ellegren, H. 2008a. Sequencing goes 454 and takes large-scale genomics into the wild. *Molecular Ecology*, 17, 1629–1631.

———. 2008b. Comparative genomics and the study of evolution by natural selection. *Molecular Ecology*, 17, 4586–4596.

Erf, D. F., Hanson, L. B., and Lawsteun, D. A. 1992. Inheritance and relationships of workability traits and yield for Holsteins. *Journal of Dairy Science*, 75, 1999–2007.

Erickson, D. L., Fenster, C. B., Stenoien, H. K., and Price, D. 2004. Quantitative trait locus analyses and the study of evolutionary process. *Molecular Ecology*, 13, 2505–2522.

Fairbanks, L. A., Newman, T. K., Bailey, J. N., Jorgensen, M. J., Breidenthal, S. E., Ophoff, R. A., Comuzzie, A. G., Martin, L. J., and Rogers, J. 2004. Genetic contributions to social impulsivity and aggressiveness in vervet monkeys. *Biological Psychiatry*, 55, 642–647.

Falconer, D. S., and Mackay, T. F. C. 1996. *Introduction to Quantitative Genetics*. New York: Longman.

Fidler, A. E., Van Oers, K., Drent, P. J., Kuhn, S., Mueller, J. C., and Kempenaers, B. 2007. DRD4 gene polymorphisms are associated with personality variation in a passerine bird. *Proceedings of the Royal Society of London B*, 274, 1685–1691.

Figueiredo, L. G. G., Eler, J. P., Mourao, G. B., Ferraz, J. B. S., Balieiro, J. C. d. C., and Mattos, E. C. d. 2005. Genetic analyses of temperament in a population of the Nelore breed. *Livestock Research for Rural Development*, 17, 1–7.

Fisher, R. A. 1930. *The Genetical Theory of Natural Selection*. Oxford: Oxford University Press.

Flint, J., and Corley, R. 1996. Do animal models have a place in the genetic analysis of quantitative human behavioural traits? *Journal of Molecular Medicine*, 74, 515–521.

Fordyce, G., and Goddard, M. E. 1984. Maternal influence on the temperament of *Bos indicus* cross cows. *Proceedings of the Australian Society of Animal Production*, 15, 345–348.

Fordyce, G., Goddard, M. E., and Seifert, G. W. 1982. The measurement of temperament in cattle and the effect of experience and genotype. *Proceedings of the Australian Society of Animal Production*, 14, 329–332.

Foster, W. W., Freeman, A. E., and Kuck, A. 1988. Linear type trait analysis with genetic parameter estimation. *Journal of Dairy Science*, 71, 223–231.

Francis, R. C. 1984. The effects of bidirectional selection for social dominance on agonistic behaviour and sex ratios in the paradise fish (*Macropodus opercularus*). *Behaviour*, 90, 25–45.

Fullerton, J. M., Willis-Owen, S. A. G., Yalcin, B., Shifman, S., Copley, R. R., Miller, S. R., Bhomra, A., Davidson, S., Oliver, P. L., Mott, R., and Flint, J. 2008. Human-mouse quantitative trait locus concordance and the dissection of a human neuroticism locus. *Biological Psychiatry*, 63, 874–883.

Gauly, M., Mathiak, H., and Erhardt, G. 2002. Genetic background of behavioural and plasma cortisol response to repeated short-term separation and tethering of beef calves. *Journal of Animal Breeding and Genetics*, 119, 379–384.

Gauly, M., Mathiak, H., Hoffmann, K., Kraus, M., and Erhardt, G. 2001. Estimating genetic variability in temperamental traits in German Angus and Simmental cattle. *Applied Animal Behaviour Science*, 74, 109–119.

Gerken, M., and Petersen, J. 1992. Heritabilities for behavioral and production traits in Japanese quail (*Coturnix coturnix japonica*) bidirectionally selected for dustbathing activity. *Poultry Science*, 71, 779–788.

Gervai, J., and Csányi, V. 1985. Behavior-genetic analysis of the paradise fish, *Macropodus opercularis*, I: characterization of the behavioral responses of inbred strains in novel environments: a factor analysis. *Behavior Genetics*, 15, 503–519.

Gibbons, M. E., Ferguson, A. M., and Lee, D. R. 2005. Both learning and heritability affect foraging behaviour of red-backed salamanders, *Plethodon cinereus*. *Animal Behaviour*, 69, 721–732.

Gibson, G. 2002. Microarrays in ecology and evolution: a preview. *Molecular Ecology*, 11, 17–24.

Goddard, M. E., and Beilharz, R. G. 1983. Genetics of traits which determine the suitability of dogs as guide-dogs for the blind. *Applied Animal Ethology*, 9, 299–315.

Golimbet, V. E., Affimova, M. V., Manandyan, K. K., Mitushina, N. G., Abramova, L. I., Kaleda, V. G., Oleichik, I. V., Yurov, Y. B., and Trubnikov, V. I. 2002. 5HTR2A gene polymorphism and personality traits in patients with major psychoses. *European Psychiatry*, 17, 24–28.

Gordon, J. A., and Hen, R. 2004. Genetic approaches to the study of anxiety. *Annual Review of Neuroscience*, 27, 193–222.

Gosling, S. D. 2001. From mice to men: what can we learn about personality from animal research? *Psychological Bulletin*, 127, 45–86.

———. 2008. Personality in nonhuman animals. *Social and Personality Psychology Compass*, 2, 985–1001.

Gottlieb, G. 2007. Probabilistic epigenesis. *Developmental Science*, 10, 1–11.

Gromko, M. H. 1995. Unpredictability of correlated response to selection: pleiotropy and sampling interact. *Evolution*, 49, 685–693.

Gross, M. R. 1996. Alternative reproductive strategies and tactics: diversity within sexes. *Trends in Ecology and Evolution*, 11, 263.

Guhl, A. M., Craig, J. V., and Mueller, C. D. 1960. Selective breeding for aggressiveness in chickens. *Poultry Science*, 38, 970–980.

Gutierrez-Gil, B., Ball, N., Burton, D., Haskell, M., Williams, J. L., and Wiener, P. 2008.

Identification of quantitative trait loci affecting cattle temperament. *Journal of Heredity*, 99, 629–638.

Han, E. N., and Gatehouse, A. G. 1993. Flight capacity: genetic determination and physiological constraints in a migratory moth *Mythimna separata*. *Physiological Entomology*, 18, 183–188.

Hayes, J. F. 1999. Heritability of temperament in Canadian Holsteins. *Stocarstvo*, 53, 175–179.

Hayman, B. I. 1960. The separation of epistatic from additive and dominance variation in generation means II. *Genetika*, 31, 133–146.

Hearnshaw, H., and Morris, C. A. 1984. Genetic and environmental effects on a temperament score in beef cattle. *Australian Journal of Agricultural Research*, 35, 723–733.

Hemsworth, P. H., Barnett, J. L., Treacy, D., and Madgwick, P. 1990. The heritability of the trait fear of humans and the association between this trait and subsequent reproductive performance of gilts. *Applied Animal Behaviour Science*, 25, 85–95.

Higgins, L. A., Jones, K. M., and Wayne, M. L. 2005. Quantitative genetics of natural variation of behavior in *Drosophila melanogaster*: the possible role of the social environment on creating persistent patterns of group activity. *Evolution*, 59, 1529–1539.

Hoffmann, A. A., Hallas, R., Sinclair, C., and Partridge, L. 2001. Rapid loss of stress resistance in *Drosophila melanogaster* under adaptation to laboratory culture. *Evolution*, 55, 436–438.

Hoffmann, A. A., and Merilä, J. 1999. Heritable variation and evolution under favourable and unfavourable conditions. *Trends in Ecology and Evolution*, 14, 96–101.

Hoffmann, A. A., and Parsons, P. A. 1991. *Evolutionary Genetics and Environmental Stress*. New York: Oxford University Press.

Houle, D. 1992. Comparing evolvability and variability of quantitative traits. *Genetics*, 130, 195–204.

Humphries, M. A., Fondrk, M. K., and Page, R. E. 2005. Locomotion and the pollen-hoarding behavioral syndrome of the honeybee (*Apis mellifera L.*). *Journal of Comparative Physiology A*, 191, 669–674.

Isles, A. R., Humby, T., Walters, E., and Wilkinson, L. S. 2004. Common genetic effects on variation in impulsivity and activity in mice. *Journal of Neuroscience*, 24, 6733–6740.

Ito, H., Nara, H., Inoue-Murayama, M., Shimada, M. K., Koshimura, A., Ueda, Y., Kitagawa, H., Takeuchi, Y., Mori, Y., Murayama, Y., Morita, M., Iwasaki, T., Ota, K., Tanabe, Y., and Ito, S. 2004. Allele frequency distribution of the canine dopamine receptor D4 gene exon III and I in 23 breeds. *Journal of Veterinary Medical Science*, 66, 815–820.

Keller, M. C., Coventry, W. L., Heath, A. C., and Martin, N. G. 2005. Widespread evidence for nonadditive genetic variation in Cloninger's and Eysenck's personality dimensions using a twin plus sibling design. *Behavior Genetics*, 35, 707–721.

Kenttamies, H., Nikkila, M., Miettinen, M., and Asikainen, J. 2006. Phenotypic and genetic parameters and responses in temperament of silver fox cubs in a selection experiment for confident behaviour. *Agricultural and Food Science*, 15, 340–349.

Kimura, M. 1958. On the change of population fitness by natural selection. *Heredity*, 12, 145–167.

Kruuk, L. E. B. 2004. Estimating genetic parameters in natural populations using the "animal model." *Philosophical Transactions of the Royal Society of London B*, 359, 873–890.

Kruuk, L. E. B., Slate, J., and Wilson, A. J. 2008. New answers for old questions: the evolutionary quantitative genetics of wild animal populations. *Annual Review of Ecology Evolution and Systematics*, 39, 525–548.

Lande, R. 1982. A quantitative genetic theory of life-history evolution. *Ecology*, 63, 607–615.

Lawstuen, D. A., Hansen, L. B., Steuernagel, G. R., and Johnson, L. P. 1988. Management traits scored linearly by dairy producers. *Journal of Dairy Science*, 71, 788–799.

Le Neindre, P., Trillat, G., Sapa, J., Menissier, F., Bonnet, J. N., and Chupin, J. M. 1995. Individual differences in docility in Limousin cattle. *Journal of Animal Science*, 73, 2249–2253.

Lewontin, R. C. 1974. *The Genetic Basis of Evolutionary Change*. New York: Columbia University Press.

Lindberg, S., Strandberg, E., and Swenson, L. 2004. Genetic analysis of hunting behaviour in Swedish flatcoated retrievers. *Applied Animal Behaviour Science*, 88, 289–298.

Lipp, H., and Wolfer, D. P. 1999 Natural genetic variation in hippocampal structures and behavior. In: *Neurobehavioral Genetics: Methods and Applications* (Jones, B. C., and Mormède, P., eds.), pp. 217–233. Boca Raton, FL: CRC Press.

Lynch, M., and Walsh, B. 1998. *Genetics and Analysis of Quantitative Traits*. Sunderland, MA: Sinauer Associates.

Mackenzie, S. A., Oltenacu, E. A. B., and Leighton, E. 1985. Heritability estimate for temperament scores in German shepherd dogs and its genetic correlation with hip dysplasia. *Behavior Genetics*, 15, 475–482.

Manuck, S. B., Flory, J. D., Ferrell, R. E., Mann, J. J., and Muldoon, M. F. 2000. A regulatory polymorphism of the monoamine oxidase-A gene may be associated with variability in aggression, impulsivity, and central nervous system serotonergic responsivity. *Psychiatry Research*, 95, 9–23.

Mardis, E. R. 2008. Next-generation DNA sequencing methods. *Annual Review of Genomics and Human Genetics*, 9, 387–402.

Mazer, S. J., and Damuth, J. 2001 Evolutionary significance of variation. In *Evolutionary Ecology: Concepts and Case Studies* (Fox, C. W., Roff, D. A., and Fairbairn, D. J., eds.), pp. 16–28. Oxford: Oxford University Press.

McCrae, R. R. 2007. Do we know enough to infer the evolutionary origins of individual differences? *European Journal of Personality*, 21, 616–618.

Meffert, L. M., Hicks, S. K., and Regan, J. L. 2002. Nonadditive genetic effects in animal behavior. *American Naturalist*, 160, S198-S213.

Merilä, J., and Sheldon, B. C. 1999. Genetic architecture of fitness and nonfitness traits: empirical patterns and development of ideas. *Heredity*, 83, 103–109.

———. 2000. Lifetime reproductive success and heritability in nature. *American Naturalist*, 155, 301–310.

———. 2001. Avian quantitative genetics. In: *Current Ornithology* (Nolan, V., Jr., and Thompson, C. F., eds.), pp. 179–255. New York: Plenum Press.

Mitchell-Olds, T., Willis, J. H., and Goldstein, D. B. 2007. Which evolutionary processes influence natural genetic variation for phenotypic traits? *Nature Reviews Genetics*, 8, 845–856.

Moretz, J. A., Martins, E. P., and Robison, B. D. 2007. Behavioral syndromes and the evolution of correlated behavior in zebrafish. *Behavioral Ecology*, 18, 556–562.

Morris, C. A., Cullen, N. G., Kilgour, R., and Bremner, K. J. 1994. Some genetic factors affecting temperament in *Bos taurus* cattle. *New Zealand Journal of Agricultural Research*, 37, 167–175.

Mourao, G. B., Bergmann, J. A. G., and Ferreira, M. B. D. 1998. Genetic differences and heritability estimates of temperament in Zebus and F-1 Holstein x Zebu females. *Brazilian Journal of Animal Science*, 27, 722–729.

Mousseau, T. A., and Roff, D. A. 1987. Natural selection and the heritability of fitness components. *Heredity*, 59, 181–197.

Naciri-Graven, Y., and Goudet, J. 2003. The additive genetic variance after bottlenecks is affected by the number of loci involved in epistatic interactions. *Evolution*, 57, 706–716.

Nettle, D. 2005. An evolutionary approach to the extraversion continuum. *Evolution and Human Behavior*, 26, 363–373.

———. 2006. The evolution of personality variation in humans and other animals. *American Psychologist*, 61, 622–631.

Noble, E. P. 2003. D2 dopamine receptor gene in psychiatric and neurologic disorders and its phenotypes. *American Journal of Medical Genetics Part B-Neuropsychiatric Genetics*, 116B, 103–125.

O'Bleness, G. V., Van Vleck, L. D., and Henderson, C. R. 1960. Heritabilities of some type appraisal traits and their genetic and phenotypic correlations with production. *Journal of Dairy Science*, 43, 1490–1498.

Ohta, T. 1992. The nearly neutral theory of molecular evolution. *Annual Review of Ecology and Systematics*, 23, 263–286.

———. 2002. Near-neutrality in evolution of genes and gene regulation. *Proceedings of the National Academy of Sciences U S A*, 99, 16134–16137.

Oikawa, T., Fudo, T., and Kaneji, K. 1989. Estimate of genetic parameters for temperament and body measurements of beef cattle. *Japanese Journal of Zootechnical Science*, 60, 894–896.

Oki, H., Kusunose, R., Nakaoka, H., Nishiura, A., Miyake, T., and Sasaki, Y. 2007. Estimation of heritability and genetic correlation for behavioural responses by Gibbs sampling in the Thoroughbred racehorse. *Journal of Animal Breeding and Genetics*, 124, 185–191.

Owens, I. P. F. 2006. Where is behavioural ecology going? *Trends in Ecology and Evolution*, 21, 356–361.

Parsons, Y. M., and Shaw, K. L. 2002. Mapping unexplored genomes: a genetic linkage map of the Hawaiian cricket *Laupala*. *Genetics*, 162, 1275–1282.

Penke, L., Denissen, J. J. A., and Miller, G. F. 2007. The evolutionary genetics of personality. *European Journal of Personality*, 21, 549–587.

Persson, M. L., Wasserman, D., Jonsson, E. G., Bergman, H., Terenius, L., Gyllander, A., Neiman, J., and Geijer, T. 2000. Search for the influence of the tyrosine hydroxylase (TCAT)(n) repeat polymorphism on personality traits. *Psychiatry Research*, 95, 1–8.

Phocas, F., Boivin, X., Sapa, J., Trillat, G., Boissy, A., and Le Neindre, P. 2006. Genetic correlations between temperament and breeding traits in Limousin heifers. *Animal Science*, 82, 805–811.

Pigliucci, M. 2007. Do we need an extended evolutionary synthesis? *Evolution*, 61, 2743–2749.

Price, T., and Langen, T. A. 1992. Evolution of correlated characters. *Trends in Ecology and Evolution*, 7, 307–310.

Quinn, J. L., Patrick, S. C., Bouwhuis, S., Wilkin, T. A., and Sheldon, B. C. 2009. Heterogeneous selection on a heritable temperament trait in a variable environment. *Journal of Animal Ecology*, 78, 1203–1215.

Réale, D., Festa-Bianchet, M., and Jorgenson, J. T. 1999. Heritability of body mass varies with age and season in wild bighorn sheep. *Heredity*, 83, 526–532.

Réale, D., Gallant, B. Y., LeBlanc, M., and Festa-Bianchet, M. 2000. Consistency of temperament in bighorn ewes and correlates with behaviour and life history. *Animal Behaviour*, 60, 589–597.

Réale, D., Reader, S. M., Sol, D., McDougall, P. T., and Dingemanse, N. J. 2007. Integrating temperament in ecology and evolutionary biology. *Biological Reviews*, 82, 291–318.

Reif, A., and Lesch, K. P. 2003. Toward a molecular architecture of personality. *Behavioural Brain Research*, 139, 1–20.

Rettew, D. C., Rebollo-Mesa, I., Hudziak, J. J., Willemsen, G., and Boomsma, D. I. 2008. Nonadditive and additive genetic effects on extraversion in 3314 Dutch adolescent twins and their parents. *Behavior Genetics*, 38, 223–233.

Richards, E. J. 2006. Opinion: inherited epigenetic variation: revisiting soft inheritance. *Nature Reviews Genetics*, 7, 395–402.

Riska, B., Prout, T., and Turelli, M. 1989. Laboratory estimates of heritabilities and genetic correlations in nature. *Genetics*, 123, 865–871.

Rodgers, R. J., Cao, B. J., Dalvi, A., and Holmes, A. 1997. Animal models of anxiety: an ethological perspective. *Brazilian Journal of Medical and Biological Research*, 30, 289–304.

Roff, D. A. 1996. The evolution of genetic correlations: an analysis of patterns. *Evolution*, 50, 1392–1403.

———. 1997. *Evolutionary Quantitative Genetics*. New York: Chapmann and Hall.

———. 2007. Contributions of genomics to life-history theory. *Nature Reviews Genetics*, 8, 116–125.

Roff, D. A., and Fairbairn, D. J. 2007. The evolution of trade-offs: where are we? *Journal of Evolutionary Biology*, 20, 433–447.

Rogers, J., Shelton, S. E., Shelledy, W., Garcia, R., and Kalin, N. H. 2008. Genetic influences on behavioral inhibition and anxiety in juvenile rhesus macaques. *Genes, Brain, and Behavior*, 7, 463–469.

Ruefenacht, S., Gebhardt-Henrich, S., Miyake, T., and Gaillard, C. 2002. A behaviour test on German Shepherd dogs: heritability of seven different traits. *Applied Animal Behaviour Science*, 79, 113–132.

Rushton, J. P., Bons, T. A., and Hur, Y. M. 2008. The genetics and evolution of the general factor of personality. *Journal of Research in Personality*, 42, 1173–1185.

Saetre, P., Strandberg, E., Sundgren, P. E., Pettersson, U., Jazin, E., and Bergstrom, T. F. 2006. The genetic contribution to canine personality. *Genes, Brain, and Behavior*, 5, 240–248.

Samore, A. B., Pagnacco, G., and Miglior, F. 1997. Genetic parameters and breeding values for linear type traits in the Haflinger horse. *Livestock Production Science*, 52, 105–111.

Sato, S. 1981. Factors associated with temperament of beef cattle. *Japanese Journal of Zootechnical Science*, 52, 595–605.

Savitz, J. B., and Ramesar, R. S. 2004. Genetic variants implicated in personality: a review of the more promising candidates. *American Journal of Medical Genetics Part B-Neuropsychiatric Genetics*, 131B, 20–32.

Sgro, C. M., and Hoffmann, A. A. 2004. Genetic correlations, trade-offs, and environmental variation. *Heredity*, 93, 241–248.

Sharma, J. S., and Khanna, A. S. 1980. Note on genetic group and parity differences in dairy temperament score of crossbred cattle. *Indian Journal of Animal Research*, 14, 127–128.

Sih, A., Bell, A. M., Johnson, J. C., and Ziemba, R. E. 2004. Behavioral syndromes: an integrative overview. *Quarterly Review of Biology*, 79, 241–277.

Silva, J. A. d. V. I., Matsunaga, M. E., Eler, J. P., and Ferraz, J. B. S. 2003. Genetic analysis of flight distance in a Nellore (*Bos taurus indicus*) herd. *ITEA Produccion Animal*, 99, 167–176.

Sinervo, B., and Svensson, E. 2002. Correlational selection and the evolution of genomic architecture. *Heredity*, 89, 329–338.

Sinn, D. L., Apiolaza, L. A., and Moltschaniwskyj, N. A. 2006. Heritability and fitness-related consequences of squid personality traits. *Journal of Evolutionary Biology*, 19, 1437–1447.

Sinn, D. L., Gosling, S. D., and Moltschaniwskyj, N. A. 2008a. Development of shy/bold behaviour in squid: context-specific phenotypes associated with developmental plasticity. *Animal Behaviour*, 75, 433–442.

Sinn, D. L., Moltschaniwskyj, N. A., Wapstra, E., and Dall, S. R. X. 2010. Are behavioural syndromes invariant? Spatiotemporal variation in shy/bold behavior in squid. *Behavioral Ecology and Sociobiology*, 64, 693–702.

Sinn, D. L., While, G. M., and Wapstra, E. 2008b. Maternal care in a social lizard: links between female aggression and offspring fitness. *Animal Behaviour*, 76, 1249–1257.

Sluyter, F., Van Oortmerssen, G. A., and Koolhaas, J. M. 1996. Genetic influences on coping behaviour in house mouse lines selected for aggression: effects of the Y chromosome. *Behaviour*, 133, 117–128.

Smith, B. R., and Blumstein, D. T. 2008. Fitness consequences of personality: a meta-analysis. *Behavioral Ecology*, 19, 448–455.

Sokolowski, M. B. 2001. *Drosophila*: genetics meets behaviour. *Nature Reviews Genetics*, 2, 879–890.

Stirling, D. G., Réale, D., and Roff, D. A. 2002. Selection, structure, and the heritability of behaviour. *Journal of Evolutionary Biology*, 15, 277–289.

Strandberg, E., Jacobsson, J., and Saetre, P. 2005. Direct genetic, maternal, and litter effects on behaviour in German Shepherd dogs in Sweden. *Livestock Production Science*, 93, 33–42.

Thompson, J. R., Freeman, A. E., Wilson, D. J., Chapin, C. A., Berger, P. J., and Kuck, A. 1981. Evaluation of a linear type program in Holsteins. *Journal of Dairy Science*, 64, 1610–1617.

Turkheimer, E. 1998. Heritability and biological explanation. *Psychological Review*, 105, 782–791.

Uller, T., Olsson, M., and Stahlberg, F. 2002. Variation in heritability of tadpole growth: an experimental analysis. *Heredity*, 88, 480–484.

Van Kampen, D. 1999. Genetic and environmental influences on preschizophrenic personality: MAXCOV-HITMAX and LISREL analyses. *European Journal of Personality*, 13, 63–80.

Van Bers, N. E. M., Van Oers, K., Kerstens, H. H. D., Dibbits, B. W., Crooijmans, R. P. M. A., Visser, M. E., and Groenen, M. A. M. 2010. Genome-wide SNP detection in the great tit *Parus major* using high throughput sequencing. *Molecular Ecology*, 19 (Suppl. 1), 89–99.

Van Bers, N. E. M., Santure, A. W., Van Oers, K., De Cauwer, I., Dibbits, B. W., Mateman, A. C., Crooijmans, R. P. M. A., Sheldon, B. C., Visser, M. E., Groenen, M. A. M., and Slate, J. 2012. The design and corss-population application of a geneome -wide SNP chip for the great tit *Parus major*. *Molecular Ecology Resources*, 12, 753–770.

Van Oers, K., De Jong, G., Drent, P. J., and Van Noordwijk, A. J. 2004a. Genetic correlations of avian personality traits: correlated response to artificial selection. *Behavior Genetics*, 34, 611–619.

Van Oers, K., De Jong, G., Van Noordwijk, A. J., Kempenaers, B., and Drent, P. J. 2005a. Contribution of genetics to the study of animal personalities: a review of case studies. *Behaviour*, 142, 1185–1206.

Van Oers, K., Drent, P. J., De Goede, P., and Van Noordwijk, A. J. 2004. Realized heritability and repeatability of risk-taking behaviour in relation to avian personalities. *Proceedings of the Royal Society of London B*, 271, 65–73.

Van Oers, K., Drent, P. J., De Jong, G., and Van Noordwijk, A. J. 2004b. Additive and nonadditive genetic variation in avian personality traits. *Heredity*, 93, 496–503.

Van Oers, K., Klunder, M., and Drent, P. J. 2005b. Context dependence of avian personalities: risk-taking behavior in a social and a nonsocial situation. *Behavioral Ecology*, 16, 716–723.

Van Oers, K., and Mueller, J. C. 2010. Evolutionary genomics of animal personality. *Philosophical Transactions of the Royal Society of London B*, 365, 3991–4000.

Van Oers, K., and Sinn, D. 2011. Towards a basis for the phenotypic gambit: advances in the evolutionary genetics of animal personality. In: *From Genes to Animal Behaviour, Primatology Monographs* (Weiss, A., Inoue-Murayama, M., Kawamura, S., and Inoue, E., eds.), pp. 165–183. Tokyo: Springer.

Van Vleck, L. D. 1964. Variation in type appraisal scores due to sire and herd effects. *Journal of Dairy Science*, 47, 1249–1256.

Varo, M. 1965. Some coefficients of heritability in horses. *Annals Agriculture Fenniae*, 4, 223–237.

Via, S., Gomulkiewicz, R., Dejong, G., Scheiner, S. M., Schlichting, C. D., and Van Tienderen, P. H. 1995. Adaptive phenotypic plasticity: consensus and controversy. *Trends in Ecology and Evolution*, 10, 212–217.

Via, S., and Lande, R. 1985. Genotype-environment interaction and the evolution of phenotypic plasticity. *Evolution*, 39, 505–522.

Visscher, P. M., and Goddard, M. E. 1995. Genetic parameters for milk yield, survival, workability, and type traits for Australian dairy cattle. *Journal of Dairy Science*, 78, 205–220.

Wehner, J. M., Radcliffe, R. A., and Bowers, B. J. 2001. Quantitative genetics and mouse behavior. *Annual Review of Neuroscience*, 24, 845–867.

Weigensberg, I., and Roff, D. A. 1996. Natural heritabilities: can they be reliably estimated in the laboratory? *Evolution*, 50, 2149–2157.

Weiss, A., King, J. E., and Enns, R. M. 2002. Subjective well-being is heritable and genetically correlated with dominance in chimpanzees (*Pan troglodytes*). *Journal of Personality and Social Psychology*, 83, 1141–1149.

Weiss, A., King, J. E., and Figueredo, A. J. 2000. The heritability of personality factors in chimpanzees (*Pan troglodytes*). *Behavior Genetics*, 30, 213–221.

West-Eberhard, M. J. 2003. *Developmental Plasticity and Evolution*. New York: Oxford University Press.

While, G. M., Sinn, D. L., and Wapstra, E. 2009. Female aggression predicts mode of paternity in a social lizard. *Proceedings of the Royal Society of London B*, 276, 2021–2029.

Wickham, B. W. 1979. Genetic parameters and economic values of traits other than production for dairy cattle. *Proceedings of the New Zealand Society of Animal Production*, 39, 180–193.

Willis, J. H., and Orr, H. A. 1993. Increased heritable variation following population bottlenecks: the role of dominance. *Evolution*, 47, 949–957.

Willis-Owen, S. A. G., and Flint, J. 2007. Identifying the genetic determinants of emotionality in humans: insights from rodents. *Neuroscience and Biobehavioral Reviews*, 31, 115–124.

Wilson, A. J., Gelin, U., Perron, M. C., and Réale, D. 2009. Indirect genetic effects and the evolution of aggression in a vertebrate system. *Proceedings of the Royal Society of London B*, 276, 533–541.

Wilson, A. J., Pemberton, J. M., Pilkington, J. G., Coltman, D. W., Mifsud, D. V., Clutton-Brock, T. H., and Kruuk, L. E. B. 2006. Environmental coupling of selection and heritability limits evolution. *PLoS Biology*, 4, 1270–1275.

Wilson, D. S. 1998. Adaptive individual differences within single populations. *Philosophical Transactions of the Royal Society of London B*, 353, 199–205.

Wilsson, E., and Sundgren, P. E. 1997. The use of a behaviour test for selection of dogs for service and breeding, 2: heritability for tested parameters and effect of selection based on service dog characteristics. *Applied Animal Behaviour Science*, 54, 235–241.

———. 1998. Behaviour test for eight-week old puppies: heritabilities of tested behaviour traits and its correspondence to later behaviour. *Applied Animal Behaviour Science*, 58, 151–162.

Wingfield, J. C., Visser, M. E., and Williams, T. D. 2008. Introduction: integration of ecology and endocrinology in avian reproduction: a new synthesis. *Philosophical Transactions of the Royal Society of London B*, 363, 1581–1588.

Wolf, J. B. 2001. Integrating biotechnology and the behavioral sciences. *Trends in Ecology and Evolution*, 16, 117–119.

Wolf, J. B., Brodie, E. D., and Moore, A. J. 1999. Interacting phenotypes and the evolutionary process, II: selection resulting from social interactions. *American Naturalist*, 153, 254–266.

Wolf, M., Van Doorn, G. S., Leimar, O., and Weissing, F. J. 2007. Life-history trade-offs favour the evolution of animal personalities. *Nature*, 447, 581–584.

Wright, D., Rimmer, L. B., Pritchard, V. L., Krause, J., and Butlin, R. K. 2003. Inter- and intrapopulation variation in shoaling and boldness in the zebrafish (*Danio rerio*). *Naturwissenschaften*, 90, 374–377.

7

What Is the Evidence that Natural Selection

Maintains Variation in Animal Personalities?

NIELS J. DINGEMANSE AND DENIS RÉALE

Introduction

Over the past few years, the study of animal personality has increasingly intrigued behavioral ecologists (Bell 2007; Réale et al. 2007; Biro and Stamps 2008; Sih and Bell 2008; Dingemanse et al. 2010b). Why is it that individuals from the same population often differ consistently in their behavior? And why is it that individuals also tend to differ in whole suites of behavioral traits?

Growing interest in the study of animal personality has been accompanied by a shift from an "individual optimum" approach to a "population optimum" approach in behavioral ecology. From an "individual optimum" perspective, natural and sexual selection are expected to favor a single optimal response to environmental variation within the same population (Wilson et al. 1994; Wilson 1998; Dall et al. 2004; Sih et al. 2004; Schuett et al. 2010). Accordingly, behavioral variation within populations is often interpreted as the product of individual phenotypic plasticity (i.e., each individual adjusts its behavior in response to changes in environmental conditions; Nussey et al. 2007) or measurement error (i.e., the failure to measure behavioral phenotypes accurately; Dall et al. 2004). In contrast, from the "population optimum" perspective, it is assumed that, under some circumstances, natural and sexual selection may favor the evolution of multiple responses to environmental challenges, thus resulting in within-population variation in the same behavioral trait, and in whole suites of behavioral traits.

In this chapter, we discuss the role of a number of evolutionary mechanisms maintaining behavioral variation and review the evidence that natural selection acts on animal personality (see also Dingemanse and Wolf 2010; Dingemanse et al. 2010b). We acknowledge that there is strong evidence for selection on personality traits in captivity (McDougall et al. 2006) and that most evidence for the link between personality and fitness comes from captive studies (Smith and Blumstein 2008). However, we choose to focus on studies conducted in the wild because our interest is in selection acting on wild animal populations. Finally, we discuss caveats in our knowledge on specific mechanisms and give future directions aimed at increasing our understanding of why variation in animal personality persists in the wild.

Defining the problem

We begin by defining the pattern of variation implied in the notion of animal personality. Most traits show phenotypic variation within the same population (Hallgrímsson and Hall 2005) but this is potentially due to different processes (Nussey et al. 2007; Dingemanse et al. 2010b) such as:

(i) individuals do not always express the same phenotype (resulting in within-individual variation; individual plasticity "E"),

(ii) individuals differ consistently from each other in their phenotype over time or across a range of contexts (resulting in between-individual variation "I"),

(iii) individuals differ in their plasticity (individual by environment interaction "I×E"; potentially resulting in both within- and between-individual variation), and/or

(iv) researchers fail to measure phenotypes accurately (measurement error "e," resulting in within-individual variation).

The total phenotypic variance is simply the sum of all variance components $(I + E + I×E + e)$. "I" and "I×E" can both be decomposed in permanent environmental (PE) and genetic (G) components, with $I = G + PE$ and $I×E = G×E + PE×E$ (each of these individual variance components can be partly heritable though not necessarily so; for further discussion, see Nussey et al. 2007; Dingemanse et al. 2010b). Animal personality refers specifically to one of these sources of phenotypic variation ("I"): the difference between individuals in their average level of behavior, that is, the portion of the variation that is consistently maintained over time or over different contexts (Réale et al. 2007; Dingemanse et al. 2010b).

The concept of animal personality has been used to refer to individual differences in the same behavior maintained over time (e.g., days, years), individual differences in the same behavior maintained in different contexts (e.g., low vs. high perceived predation risk), and/or interindividual associations between different behaviors (e.g., aggressiveness and exploratory behavior are correlated across individuals). In this chapter, we refer to the former two categories as variation in animal personality (Dall et al. 2004;

Dingemanse et al. 2010b). The latter category essentially refers to correlations between personality traits (Boon et al. 2007; Biro and Stamps 2008), which we call here behavioral syndromes (Clark and Ehlinger 1987; Sih et al. 2004) following Biro and Stamps (2008). Interestingly, these three types of individual (co)variation often go together, as exemplified by the existence of an "aggressiveness-boldness" syndrome in a wide variety of taxa (Réale et al. 2007; Stamps 2007; Wolf et al. 2007).

Now that we have introduced the notion of associations, or phenotypic correlations, between behavioral traits, it is important to note that a phenotypic correlation between two different behaviors does not necessarily represent evidence for a behavioral syndrome: phenotypic correlations reflect the joint effects of both between-individual correlations (what we call behavioral syndromes) and within-individual correlations. The latter type of correlation occurs either because a shift in one behavior within the same individual is associated with a shift in another behavior ("correlational plasticity," caused by environmental correlations) or because measurement errors are correlated across behaviors (Dingemanse et al. 2010b). Phenotypic correlations can exist between two behaviors in a population because within-individual variation in state ("condition") affects multiple aspects of the behavioral phenotype—for example, none of the behaviors might be repeatable (Bell and Stamps 2004). For instance, individual birds are known to increase both aggressiveness and boldness when they are in poor condition and under risk of starvation (Lima and Dill 1990). Because selection is measured as the covariance between individual phenotypic measures of a trait and fitness (Lande and Arnold 1983; see below), the measurement of natural, or sexual, selection acting on animal personality or behavioral syndromes requires one to separate the phenotypic behavioral (co)variation in between- and within-individual (co)variance components. In other words, whether we consider the same behavior over time (or contexts) or associations between behaviors, it is specifically the maintenance of the between-individual part of the phenotypic (co)variation in behavior that we seek to understand here.

How to measure selection on personality

How should we study personality variation within an adaptive framework? We think that the functional significance of personality variation can be understood only if we consider how natural, or sexual, selection acts on this variation (Dingemanse and Réale 2005; Schuett et al. 2010; Réale and Dingemanse 2010). In other words, we should try to understand whether selection favors situations where both phenotypic variation *and* behavioral consistency (i.e., a situation where each individual does not express the full range of behavioral traits values present in its population; Réale and Dingemanse 2010) are maintained at the same time (Schuett et al. 2010; Dingemanse et al. 2010b).

Evolutionary biologists have developed a framework for studying the evolution of complex phenotypes such as personality traits and behavioral syndromes: the phenotypic selection approach (Lande and Arnold 1983). This approach consists of measuring the covariance between standardized measures (proxies) of fitness and suites of phenotypic traits, enabling the detection of various forms of selection. These include both *linear* relationships between traits and fitness ("directional selection") as well as *nonlinear* ones. Nonlinear selection gradients can occur because selection favors the extremes of the trait distribution ("disruptive selection") or because selection goes against extremes ("stabilizing selection"); nonlinear selection is assessed by estimating a quadratic relationship between a trait and fitness. Another form of nonlinear selection is selection acting directly on the correlation between traits ("correlational selection"). In this case, the effect of a trait A on fitness varies along with another trait B so that particular combinations of the two traits give different fitness outcomes. This form of selection is measured by regressing fitness against the linear *and* quadratic terms of two traits in the same regression model while including also the interaction between the linear terms of the focal traits (thereby estimating the "correlational selection gradient"; Brodie et al. 1995). Note that correlational selection is different from the correlated response to selection, when one trait changes across generations as a result of selection acting on another, genetically associated trait (Falconer and Mackay 1996).

A growing number of studies have attempted to evaluate how selection acts on animal personality (for reviews see Dingemanse and Réale 2005; Réale et al. 2007; Smith and Blumstein 2008). In the current literature on this subject, however, there appears to be some confusion about what type of data are needed to assess how selection acts on personality variation *per se*. The majority of studies dealing with selection and personality report relationships between some repeatable phenotypic behavior (e.g., exploration behavior) and fitness, without distinguishing explicitly between how selection acts on the interindividual versus the intra-individual component (i.e., personality versus individual plasticity; Dingemanse and Réale 2005; Schuett et al. 2010). Arguably, such reports of covariance between raw behavioral phenotypes and fitness are important because they might help to clarify how selection acts on one component of personality variation: the phenotypic variance within the population (Dall 2004; Dingemanse and

Réale 2005; Schuett et al. 2010). Nevertheless, personality variation implies the existence of phenotypic variation *in combination with* limited behavioral plasticity (see above), and therefore the real challenge is to estimate how selection acts on both (Dingemanse et al. 2010b). It is worth noting that the question of selection on the plastic component of a personality trait makes sense only in the presence of individual variation in plasticity in the population; when all individuals react in the same way to changes in the environment there will be no material for selection on plasticity to act on (Nussey et al. 2007).

There are two different approaches researchers can apply to study selection on personality (Dingemanse et al. 2010b). In the first approach, we view the relationship describing the behavioral response of an individual over an environmental gradient (which could include "time" in terms of date or age) as the trait of interest (figure 7.1): this is the reaction norm approach (Fuller et al. 2005; Smiseth et al. 2008; Dingemanse et al. 2010b). Each individual is then characterized by its behavioral elevation (its behavior in the average, mean-centered environment) and its behavioral slope (its plasticity). The behavioral elevation of an individual can be viewed as its personality as defined in the behavioral ecology literature (for further discussion see Dingemanse et al. 2010b). Notably, it would also be possible to replace the behavioral slope for an index of within-individual variation (e.g., coefficient of variation) in cases where environmental contextual information is missing or where consistency (rather than slope) is of interest (Schuett et al. 2010; Dingemanse et al. 2010b). Adaptive personality variation implies that selection will help to maintain variation in behavioral elevation within the population while simultaneously inducing directional selection against behavioral plasticity or stabilizing selection in favor of limited plasticity.

The second approach views behavior as a collection of response (y) variables (character states), one for each environmental condition (or time period), and has therefore been termed the character state approach (Via et al. 1995) (figure 7.1). Application of this approach requires that repeated measurements of both the behavioral phenotype and its associated fitness consequences be collected for a set of individuals, either across time (e.g., early versus late in the season) or contexts (e.g., in the presence versus absence of predators). We then speak of personality variation when behavior in one context is correlated with behavior in another context (at the between-individual level). Here, adaptive personality variation implies that correlational selection will favor the association between the level of behavior at different points in time (or expressed in different contexts), thereby favoring individuals showing limited plasticity. Similarly, adaptive behavioral

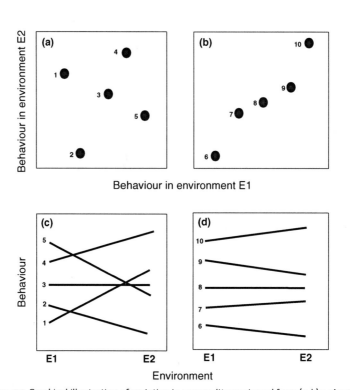

Figure 7.1. Graphical illustration of variation in personality as viewed from (a-b) a **character state perspective** and (c-d) a **reaction norm perspective**. In statistical terms, the two approaches are two sides of the same coin (Via et al. 1995): (a) and (c) depict the same fictional data as do (b) and (d). Personality does not exist in (a) and (c) because the correlation across contexts (E1, E2) is close to zero in (a), and rank order differences in behavior are not maintained across contexts in (c). Personality does exist in (b) and (d) because there is a tight cross-context correlation in (b), and individuals show limited plasticity in (d). Figure is reprinted and figure legends are modified after Figure 1 in Dingemanse (2007).

syndromes imply that selection favored the correlation between two or more behavioral traits by favoring particular combinations of phenotypes between the traits (e.g., Bell and Sih 2007). More generally, correlational selection is responsible for the evolution of genetic correlations—both by reinforcing or destroying correlations—between two or more traits and for the evolution of co-adaptation (Cheverud 1984; Sinervo and Svensson 2002).

Evolutionary mechanisms maintaining variation

One common way of measuring and visualizing selection is the phenotypic selection approach, which consists of looking at the covariance between a set of phenotypic traits and fitness (Lande and Arnold 1983; Brodie et al. 1995). With this method it is possible to study any type of selection pressure that might be responsible for the maintenance of personality traits in a population. Here we briefly summarize several possible selective mechanisms of variance conservation (Frank and Slatkin 1990). Temporal or spatial heterogeneity in environmental conditions (i.e., resource abundance or population density) can create antagonistic selection pressures on a trait within a generation, between successive generations, or at different locations. As a result, the total action of selection should average out and a flat fitness surface is expected, implying no net selection on personality traits. Frequency-dependent selection is a particular case of heterogeneous forms of selection, and happens when the fitness outcome of a phenotype (or strategy) depends on its relative abundance (i.e., frequency) in the population relative to other phenotypes (Maynard Smith 1982). In the presence of negative frequency-dependent selection, we would expect to observe a negative relationship between selection on a trait and the proportion of individuals with a specific phenotype of the trait in the population. Life-history trade-offs are also known to allow the maintenance of genetic variation (Roff 2002), and have been suggested to help maintain variation in personality (Wolf et al. 2007; Biro and Stamps 2008; Réale et al. 2009). In this case, we should expect that some particular personality traits have coevolved within particular life-history strategies and represent alternative adaptive strategies coexisting within the same population. For example, high aggressiveness and boldness might be associated with fast growth, early maturation and reproduction, and short lifespan, whereas low aggressiveness and shyness might be associated with extended development, delayed reproduction, and long lifespan. Other traits such as metabolic rate are then assumed to have coevolved with personality and life-history strategies (Biro and Stamps 2008; Careau et al. 2008). In the case of an association between personality traits and life-history strategies, we should expect antagonistic selection pressures on personality traits at different life stages. For example, selection favoring high value for a trait early in life should favor low value of that trait later in life. Antagonistic selection could also be observed between the sexes, and could help maintain variation in traits when the same genes affect the trait of males and females, but selection acts differently across sexes (Chippindale et al. 2001). Finally, correlational selection occurs when individuals with particular combinations of trait values show the highest fitness. Correlational selection can lead to a fitness "ridge" where a particular set of multivariate phenotypes shares the same fitness (figure 7.2). For instance, aggressive-and-bold or nonaggressive-and-shy individuals might live longer

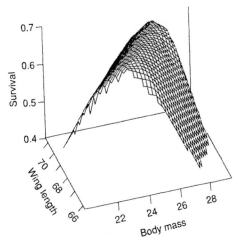

Figure 7.2. An example of correlational selection. Survival selection on body mass and wing length takes the form of a ridge of high fitness in sparrows (Schluter and Nychka 1994). Figure is reprinted from Figure 1 in Sinervo and Svensson (2002).

and/or reproduce more than aggressive-and-shy or nonaggressive-and-bold ones, with the former two categories performing equally well. Such a scenario would produce a "flat" ridge of high fitness, which can allow genetic variation in both traits to persist.

All these selective scenarios can potentially apply to personality traits and explain why those traits show high genetic and phenotypic variance (Réale et al. 2007). As we will see below, empirical tests of these hypotheses are not yet very common. Furthermore, these different selective scenarios do not automatically explain why individuals should vary in their plasticity. For example, frequency-dependent selection can explain the coexistence of strategies varying in their level of plasticity in a population (Luc-Alain Giraldeau, personal communication). However, limited plasticity may exist in a population as a result of either directional selection against plasticity or stabilizing selection for limited plasticity (Dingemanse et al. 2010b).

Empirical studies

In this section, we focus on empirical studies conducted in natural populations that evaluated the various evolutionary mechanisms for the maintenance of behavioral variation. Such studies can provide us with insights into why phenotypic variation in behavior is maintained within a population, though we realize that they cannot fully explain personality variation. This would require empirical studies documenting selection acting on behavioral consistency, which we briefly touch upon at the end of this section.

We found 11 studies estimating heterogeneous forms of selection acting on behavior within the same population (table 7.1). Ten of these (91%) reported some form of heterogeneous selection. These studies comprised a large diversity of taxa (birds, insects, mammals, reptiles), implying that heterogeneous selection is widespread in nature and appears to be the norm rather than the exception in the context of behavioral traits. Various forms of heterogeneous selection have been described.

First, temporal variation (i.e., from year to year) in selection has been detected to act on exploration behavior in two wild populations of great tits *Parus major* (Dingemanse et al. 2004; Quinn et al. 2009). In contrast, no temporal differences in selection could be found in a twelve-year study of Ural owls *Strix arulensis*, where selection instead favored aggressive females in most years (Kontiainen et al. 2009). Annual variation in selection has also been shown in mammals (bighorn sheep *Ovis canadensis* and North American red squirrels *Tamiasciurus hudsonicus*). In the red squirrels, aggressive females were favored in some (but not in other) years depending on the growth and survival of their offspring (Boon et al. 2007). Here, selection also acted on other correlated behaviors (e.g., activity), again fluctuating between years. In bighorn sheep, bold females were more likely to survive in years of high predation compared with shy ones, but selection did not act on this variation in years of low predation (Réale and Festa-Bianchet 2003).

Second, spatial variation in selection has been documented in British great tits, where fast-exploring individuals reproduced better than slow-exploring birds in low-density habitats of the same population in one of four years of study (Quinn et al. 2009). Experimental field studies in western bluebirds *Siala Mexicana* also revealed spatial variation in selection acting on aggressiveness: aggressive birds showed increased survival in newly settled populations but not in populations of older age (Duckworth and Badyaev 2007; Duckworth 2008).

Third, variation in the social environment appears to induce heterogeneous selection on exploration behavior in great tits (Dingemanse et al. 2004; Both et al. 2005; Van Oers et al. 2008). Here, frequency-dependent selection seems to act on the composition of breeding pairs in terms of exploration behavior. Assortatively paired individuals produced offspring in

Table 7.1. Studies estimating heterogeneous selection on behavioral traits in the wild.

Species	No. of Years	Stage/Sex[a]	Behavioral Trait	Fitness Measure[b]	Type of Heterogeneous Selection Considered[c]	Quadratic Effects of Behavior Considered[d]	Environmental Gradient Assessed	Description of Heterogeneous Selection If Detected[e]	Source
Insects									
Texas field cricket (*Gryllus texensis*)	1	A/M	Calling and mate searching	Mating frequency	Experimental	No	Number of same-sex competitors (low vs. high density)	Direct selection on calling and indirect selection on searching in low-density treatment only	Cade and Cade 1992
Birds									
Great tit (*Parus major*)	3	A/F&M	Exploration	Annual survival	Temporal	Yes	...	Year × sex × personality-dependent directional selection	Dingemanse et al. 2004
	4	A&J/F&M	Exploration	Annual survival	Temporal	Yes	...	Not detected	Quinn et al. 2009
	3	A/F	Exploration	Annual offspring recruitment	Temporal	Yes	...	Year × personality-dependent nonlinear selection	Dingemanse et al. 2004
	1	A/F&M	Exploration	Annual offspring recruitment	Social	No	Partner's personality	Own × Partner's personality-dependent selection	Dingemanse et al. 2004
	4	A/F&M	Exploration	Lifetime fecundity	Temporal	Yes	...	Not detected	Quinn et al. 2009
	4	A/F&M	Exploration	Lifetime fecundity	Spatial and temporal	No	Breeding density	Year × density × personality-dependent selection	Quinn et al. 2009
	4	A/F&M	Exploration	Lifetime fecundity	Spatial and temporal	No	Territory quality	Year × territory quality × personality-dependent selection in males only	Quinn et al. 2009

4	A/F&M	Exploration	Lifetime offspring recruitment	Spatial and temporal	No	Breeding density and Territory quality	Not detected	Quinn et al. 2009
4	A/F&M	Exploration	Lifetime offspring recruitment	Temporal	Yes	...	Year × sex × personality-dependent nonlinear selection	Quinn et al. 2009
4	A/F	Exploration	Offspring condition at independence	Social	Yes	Partner's personality	Own × Partner's personality-dependent selection	Both et al. 2005
2	A/M	Exploration	Occurrence of extra-pair young in own nest	Social	Yes	Partner's personality	Own × Partner's personality-dependent selection	Van Oers et al. 2008
Western bluebird (*Sialia mexicana*)								
4	A/M	Aggression towards predator	Annual reproductive success	Spatial	No	Population age	(Aggressive phenotypes had higher fitness in patches of younger age)	Duckworth 2008
Ural owl (*Strix arulensis*)								
12	A/F	Aggression towards predators	Annual survival and offspring recruitment	Temporal	Yes	...	Not detected	Kontiainen et al. 2009
Lizards								
Common lizard (*Lacerta vivipara*)								
1	J/MandF	Boldness and sociability	Annual survival	Experimental	No	Population density	Population density × sociability-dependent selection	Cote et al. 2008
Mammals								
Red squirrel (*Tamiasciurus hudsonicus*)								
4	A/F	Aggression and Exploration	Offspring growth rate	Temporal	No	...	Year × personality-dependent selection on exploration only	Boon et al. 2007
4	A/F	Aggression and Exploration	Offspring survival till independence	Temporal	No	...	Year × personality-dependent selection on aggression only	Boon et al. 2007

(continued)

Table 7.1. (continued)

Species	No. of Years	Stage/Sex[a]	Behavioral Trait[b]	Fitness Measure[b]	Type of Heterogeneous Selection Considered[c]	Quadratic Effects of Behavior Considered[d]	Environmental Gradient Assessed	Description of Heterogeneous Selection If Detected[e]	Source
	3	A/F	Aggression and Exploration	Offspring overwinter survival	Temporal	No		Year × personality-dependent selection on aggression only	Boon et al. 2007
Bighorn sheep (Ovis canadensis)	4	A/F	Boldness	Annual survival	Temporal	No	.....	(Direct selection in certain years only)	Réale and Festa-Bianchet 2003
	3	A/F	Docility	Annual survival	Temporal	No	.....	(Direct selection in certain years only)	Réale and Festa-Bianchet 2003
	18	A/M	Docility and boldness	Annual relative reproductive success	Life-history stage (age)	Yes	Age	Interaction between age and personality traits on reproductive success	Réale et al. 2009

[a] Stage: life-history stage for which fitness was determined (A = adult, J = juvenile), Sex: sexes included in the analyses (F = females, M = males, F&M = both).

[b] Fitness measures that researchers choose are largely determined by the specific life-history of their species (Annual offspring recruitment: the number of offspring produced in a single year that manage to recruit into the breeding population; Annual survival = survival from one year to the next; Lifetime fecundity: lifetime number of offspring reared till independence; Lifetime offspring recruitment: the lifetime number of offspring produced that manage to recruit into the breeding population; Mating frequency: number of females mated with within the same season; Offspring condition at independence: offspring body weight controlling for structural size few days before leaving the nest).

[c] We consider here temporal (usually annual), spatial, and social variation in environmental conditions. The last category consists primarily of examples of how one's partner's personality affects fitness. This category might be regarded as a nonphysical component of the spatial environment, since animals with certain personalities are more likely to breed in certain types of habitat (Both et al. 2005; Duckworth 2006).

[d] We refer here to cases where both linear and quadratic effects of personality were considered in the initial statistical model.

[e] Text in parentheses indicates studies where selection was detected in certain data sets (e.g., years) but not in others, but where heterogeneous selection was not tested statistically.

best condition (Both et al. 2005) as well as most recruits (Dingemanse et al. 2004). Moreover, male mating success was also affected by pair composition: assortatively paired males lost most paternity in their nest (Van Oers et al. 2008).

Experimentally induced variation in environmental conditions (e.g., density) will enable researchers to investigate the causal link between components of the environment and selection on behavior. Work on Texas field crickets *Gryllus texensis* showed that male mate-advertisement ("calling") behavior, an individual-bound heritable (i.e., personality) trait, is selected for in low-density (but not in high-density) treatments (Cade and Cade 1992). Similarly, sociable common lizards *Lacerta vivipara* survived better than less sociable conspecifics under high densities, but the pattern was reversed under low densities (Cote et al. 2008).

LIFE-HISTORY TRADEOFFS

A recent theoretical model by Wolf et al. (2007) has suggested that animals with different personalities should opt for different life-history strategies, with life-history tradeoffs helping to maintain this variation. When personality traits such as aggressiveness or exploration affect the chance of finding and monopolizing resources but also increase risk of mortality, personality traits can coevolve with general life-history strategies. The expectation is that exploratory-aggressive individuals will show early sexual maturity, fast reproduction, and short lifetime, whereas nonexploratory, nonaggressive individuals may delay and extend their reproductive life. The role of personality in life-history tradeoffs appears to have been studied empirically only in bighorn sheep, where some support for this idea was found (Réale et al. 2009). Aggressive bighorn rams reproduced earlier than docile rams but did not survived as long as docile rams, which enjoyed greater reproductive success early in life (a similar pattern was found for boldness). Furthermore, docility and boldness were negatively genetically correlated (i.e., bold individuals were also less docile), which may therefore result in a relatively flat fitness landscape. Biro and Stamps (2008) have provided the most complete review to date on the association between personality and life-history traits. Though there is good evidence for growth to be positively linked with activity, aggressiveness, and boldness, the evidence for the link with other life-history traits is less clear. Furthermore, the direction of the relationship may not be systematic and generalizable, but may depend on the past selection pressures associated with the ecology of a given species/population.

This form of selection occurs when the relationship between phenotype, or genotype, and fitness is negatively correlated across the sexes. In other words, we would expect to see an interaction between sex and the trait on fitness, so that values of the trait that are related to high fitness in the male are also associated with low fitness in females and *vice versa* (Chippindale et al. 2001). To our knowledge, very few studies have focused on this evolutionary mechanism for the maintenance of variation. A study on the Dutch great tits has revealed patterns of selection that are consistent with sexual antagonistic selection: in each of three years, survival selection acting on exploratory behavior was linear but differed qualitatively between the sexes (Dingemanse et al. 2004). In certain years, fast-exploring males survived best as did slow-exploring females, and in other years the pattern was reversed. Quantitative genetics analyses using artificial selection lines have not revealed sex-specific expression of exploratory behavior in this population (Van Oers et al. 2004), and therefore this example might represent a case of sexual antagonistic selection. A recent laboratory study on comb-footed spiders (*Anelosimus studiosus*) also suggests a role for sexual antagonistic selection in maintaining animal personality variation (Pruitt and Riechert 2009). In this species, sociable males outcompeted less sociable same-sex conspecifics in terms of access to females, thereby gaining higher mating success. In females, however, the pattern was reversed: sociable females were outcompeted by less sociable same-sex conspecifics in terms of access to males. Selection pressures thus favored high levels of sociability in males but low levels in females.

CORRELATIONAL SELECTION

Despite growing interest in the adaptive nature of correlations between personality traits (behavioral syndromes), relatively few studies have addressed the key question of whether selection favors behavioral correlations. In fact, we know of only one field study that both measured and detected correlational selection acting on a behavioral component of the phenotype. In his landmark study, Brodie (1992) showed that selection acted on the correlation between color and antipredator behavior in the garter snake, *Thamnophis ordinoides*. As expected, snakes that performed uninterrupted flight survived best when they had a striped pattern on their back, while snakes that fled evasively did best when they were either spotted or unmarked. To our knowledge, only one field study has attempted to estimate (but failed to find) correlational selection acting on behavioral syndromes. Boon et al.

(2007) studied the effects of activity and aggressiveness on various fitness components in red squirrels but could not detect a significant correlational selection gradient. Similarly, no correlational selection was detected to act on docility and boldness in either male (Réale et al. 2009) or female (Réale and Festa-Bianchet 2003) bighorn sheep. In the same way, Bell and Sih (2007) tested whether predation induced correlational selection favoring the aggressiveness-boldness syndrome in the laboratory, because population comparison had revealed that the syndrome existed only in predator-sympatric populations of stickleback *Gasterosteus aculeatus* (Bell 2005; Dingemanse et al. 2007; but see Dingemanse et al. 2010a). Again, correlational selection was not detected. However, it should be noted that these three studies were all based on sample sizes smaller than 150 individuals, which might not have been sufficient to document significant correlational selection (Kingsolver et al. 2001). In humans, correlational selection has been documented to act on the association between neuroticism and extraversion in a sample of Australian women (Eaves et al. 1990).

SELECTION FOR CONSISTENCY

The studies reviewed earlier in this section have all considered selection acting on behavior. In some but not all cases, the individual average phenotypic value of a trait was used. Unfortunately, very few studies have yet asked the question of whether selection acts on behavioral consistency. We have found only two field studies that have done so. Both studies measured the extent of behavioral plasticity of individuals using an index of within-individual variation as opposed to formally quantifying reaction norm slopes. The first study showed that chestnut-sided warbler *Dendroica pensylvanica* males that sing in a consistent manner have higher extra-pair success than males that sing more inconsistently (Byers 2007). The second study on tropical mockingbird *Mimus gilvus* also showed selection favoring consistency: males that sang consistently both were more dominant and had increased reproductive success compared with males that sang less consistently (Botero et al. 2009). Therefore, it appears as if selection might indeed deplete variation in behavioral plasticity, favoring individuals that are relatively stable in their behavior. However, this does not mean that directional selection favoring consistency is the general rule for other behavior traits. These two studies both represent a particular case (male bird song), where females could use consistency as an important criterion for mate choice. This interpretation should thus be considered premature because recent evidence suggests that personality and behavioral plasticity might often be linked (Dingemanse et al. 2010b), implying that selection on consistency might in fact reflect selection on personality and *vice versa*, and multivariate selection analysis is now needed to assess how selection acts on both components separately.

Conclusions

In this chapter, we have reviewed field studies that have investigated evolutionary mechanisms for the maintenance of personality variation. Our review revealed strong support for heterogeneous selection acting on behavioral traits, and we propose that this mechanism might be key to understanding the maintenance of personality variation in the wild. Interestingly, very little attention has been given to alternative mechanisms known to maintain variation: life-history trade-offs, sexual antagonistic selection, and correlational selection. The few studies that have addressed these mechanisms have, interestingly, found some support for them, but we argue here that unequivocal evidence would require more data from a wide variety of taxa.

Importantly, the majority of animal personality studies focused solely on selection acting on behavioral variation without differentiating between-versus within-individual variance components. We argue that such studies cannot explain personality variation *per se*. We therefore recommend that future studies move away from measuring selection on behavioral phenotypes and instead start collecting repeated measures of behavioral profiles of individuals over contexts or over their lifetime. Such data would enable one to decompose the behavioral variation in between-individual ("personality") and within-individual ("individual plasticity") components, enabling the analysis of how selection acts on both. The integration of personality and plasticity within the same framework would enable us to address more challenging questions concerning links between personality and plasticity and the adaptive nature of such associations (Sih and Bell 2008; Wolf et al. 2008; Dingemanse et al. 2010b).

References

Bell, A. M. 2005. Behavioral differences between individuals and two populations of stickleback (*Gasterosteus aculeatus*). *Journal of Evolutionary Biology*, 18, 464–473.
———. 2007. Future directions in behavioural syndromes research. *Proceedings of the Royal Society of London B*, 274, 755–761.
Bell, A. M., and Sih, A. 2007. Exposure to predation generates personality in threespined sticklebacks (*Gasterosteus aculeatus*). *Ecology Letters*, 10, 828–834.
Bell, A. M., and Stamps, J. A. 2004. Development of behavioural differences between

individuals and populations of sticklebacks, *Gasterosteus aculeatus. Animal Behaviour*, 68, 1339–1348.

Biro, P. A., and Stamps, J. A. 2008. Are animal personality traits linked to life-history productivity? *Trends in Ecology and Evolution*, 23, 361–368.

Boon, A. K., Réale, D., and Boutin, S. 2007. The interaction between personality, offspring fitness, and food abundance in North American red squirrels. *Ecology Letters*, 10, 1094–1104.

Botero, C. A., Rossman, R. J., Caro, L. M., Stenzler, L. M., Lovette, I. J., de Kort, S. R., and Vehrencamp, S. L. 2009. Syllable type consistency is related to age, social status, and reproductive success in the tropical mockingbird. *Animal Behaviour*, 77, 701–706.

Both, C., Dingemanse, N. J., Drent, P. J., and Tinbergen, J. M. 2005. Pairs of extreme avian personalities have highest reproductive success. *Journal of Animal Ecology*, 74, 667- 674.

Brodie, E. D. 1992. Correlational selection for color pattern and antipredator behavior in the garter snake *Thamnophis ordinoides. Evolution*, 46, 1284–1298.

Brodie, E. D., Moore, A. J., and Janzen, F. J. 1995. Visualizing and quantifying natural selection. *Trends in Ecology and Evolution*, 10, 313–318.

Byers, B. E. 2007. Extra-pair paternity in chestnut-sided warblers is correlated with consistent vocal performance. *Behavioral Ecology*, 18, 130–136.

Cade, W. H., and Cade, E. S. 1992. Male mating success, calling, and searching behavior at high and low densities in the field cricket, *Gryllus integer. Animal Behaviour*, 43, 49–56.

Careau, V., Thomas, D., Humphries, M. M., and Réale, D. 2008. Energy metabolism and animal personality. *Oikos*, 117, 641–653.

Cheverud, J. M. 1984. Quantitative genetics and developmental constraints on evolution by selection. *Journal of Theoretical Biology*, 110, 155–171.

Chippindale, A. K., Gibson, J. R., and Rice, W. R. 2001. Negative genetic correlation for adult fitness between sexes reveals ontogenetic conflict in *Drosophila. Proceedings of the National Academy of Science U S A*, 13, 1671–1675.

Clark, A. B., and Ehlinger, T. J. 1987. Pattern and adaptation in individual behavioral differences. In: *Perspectives in Ethology* (Bateson, P. P. G., and Klopfer, P. H., eds.), pp. 1–7. New York: Plenum.

Cote, J., Dreiss, A., and Clobert, J. 2008. Social personality trait and fitness. *Proceedings of the Royal Society of London B*, 275, 2851–2858.

Dall, S. R. X. 2004. Behavioural biology: fortune favours bold and shy personalities. *Current Biology*, 14, R470-R472.

Dall, S. R. X., Houston, A. I., and McNamara, J. M. 2004. The behavioural ecology of personality: consistent individual differences from an adaptive perspective. *Ecology Letters*, 7, 734–739.

Dingemanse, N. J. 2007. An evolutionary ecologist's view of how to study the persistence of genetic variation in personality. *European Journal of Personality*, 21, 593–596.

Dingemanse, N. J., Both, C., Drent, P. J., and Tinbergen, J. M. 2004. Fitness consequences of avian personalities in a fluctuating environment. *Proceedings of the Royal Society of London B*, 271, 847–852.

Dingemanse, N. J., Dochtermann, N. A., and Wright, J. 2010a. A method for exploring the structure of behavioural syndromes to allow formal comparison within and between data sets. *Animal Behaviour*, 79, 439–450.

Dingemanse, N. J., Kazem, A. J. N., Réale, D., and Wright, J. 2010b. Behavioural reaction norms: where animal personality meets individual plasticity. *Trends in Ecology and Evolution*, 25, 81–89.

Dingemanse, N. J., and Réale, D. 2005. Natural selection and animal personality. *Behaviour*, 142, 1165–1190.

Dingemanse, N. J., and Wolf, M. 2010. Recent models for adaptive personality differences: a review. *Philosophical Transactions of the Royal Society of London B*, 365, 3947–3958.

Dingemanse, N. J., Wright, J., Kazem, A. J. N., Thomas, D. K., Hickling, R., and Dawnay, N. 2007. Behavioural syndromes differ predictably between 12 populations of stickleback. *Journal of Animal Ecology*, 76, 1128–1138.

Duckworth, R. A. 2006. Aggressive behaviour affects selection on morphology by influencing settlement patterns in a passerine bird. *Proceedings of the Royal Society of London B*, 273, 1789–1795.

———. 2008. Adaptive dispersal strategies and the dynamics of a range expansion. *American Naturalist*, 172, S4-S17.

Duckworth, R. A., and Badyaev, A. V. 2007. Coupling of dispersal and aggression facilitates the rapid range expansion of a passerine bird. *Proceedings of the National Academy of Sciences U S A*, 104, 15017–15022.

Eaves, L. J., Martin, N. G., Heath, A. C., and Hewitt, J. K. 1990. Personality and reproductive fitness. *Behavioral Genetics*, 20, 563–568.

Falconer, D. S., and Mackay, T. F. C. 1996. *Introduction to Quantitative Genetics*. New York: Longman.

Frank, S. A., and Slatkin, M. 1990. Evolution in a variable environment. *American Naturalist*, 136, 244–260.

Fuller, T., Sarkar, S., and Crews, D. 2005. The use of norms of reaction to analyze genotypic and environmental influences on behavior in mice and rats. *Neuroscience and Biobehavioral Reviews*, 29, 445–456.

Hallgrímsson, B., and Hall, B. K. 2005. *Variation*. Burlington: Elsevier Academic Press.

Kingsolver, J. G., Hoekstra, H. E., Hoekstra, J. M., Berrigan, D., Vignieri, S. N., Hill, C. E., Hoang, A., Gilbert, P., and Beerli, P. 2001. The strength of phenotypic selection in natural populations. *American Naturalist*, 157, 245–261.

Kontiainen, P., Pietiäinen, H., Huttunen, K., Karell, P., Kolunen, H., and Brommer, J. E. 2009. Aggressive Ural owl mothers recruit more offspring. *Behavioral Ecology*, 20, 789–796.

Lande, R., and Arnold, S. J. 1983. The measurement of selection on correlated characters. *Evolution*, 37, 1210–1226.

Lima, S. L., and Dill, L. M. 1990. Behavioural decisions made under the risk of predation: a review and prospectus. *Canadian Journal of Zoology*, 68, 619–640.

Maynard Smith, J. 1982. *Evolution and the Theory of Games*. Cambridge: Cambridge University Press.

McDougall, P. T., Réale, D., Sol, D., and Reader, S. M. 2006. Wildlife conservation and animal temperament: causes and consequences of evolutionary change for captive, reintroduced, and wild populations. *Animal Conservation*, 9, 39–48.

Nussey, D. H., Wilson, A. J., and Brommer, J. E. 2007. The evolutionary ecology of individual phenotypic plasticity in wild populations. *Journal of Evolutionary Biology*, 20, 831–844.

Pruitt, J. N., and Riechert, S. E. 2009. Sex matters: sexually dimorphic fitness consequences of a behavioural syndrome. *Animal Behaviour*, 78, 175–181.

Quinn, J. L., Patrick, S. C., Bouwhuis, S., Wilkin, T. A., and Sheldon, B. C. 2009. Heterogeneous selection on a heritable temperament trait in a variable environment. *Journal of Animal Ecology*, 78, 1203–1215.

Réale, D., and Dingemanse, N. J. 2010. Personality and individual social specialisation. In: *Social Behaviour: Genes, Ecology, and Evolution* (Szekely, T., Moore, A., and Komdeur, J., eds.), pp. 417–441. Cambridge: Cambridge University Press.

Réale, D., and Festa-Bianchet, M. 2003. Predator-induced natural selection on temperament in bighorn ewes. *Animal Behaviour*, 65, 463–470.

Réale, D., Martin, J., Coltman, D. W., Poissant, J., and Festa-Bianchet, M. 2009. Male personality, life-history strategies, and reproductive success in a promiscuous mammal. *Journal of Evolutionary Biology*, 22, 1599–1607.

Réale, D., Reader, S. M., Sol, D., McDougall, P., and Dingemanse, N. J. 2007. Integrating temperament in ecology and evolutionary biology. *Biological Reviews*, 82, 291–318.

Roff, D. A. 2002. *Life History Evolution*. Sunderland: Sinauer Associates.

Schluter, D., and Nychka, D. 1994. Exploring fitness surfaces. *American Naturalist*, 143, 597–616.

Schuett, W., Tregenza, T., and Dall, S. R. X. 2010. Sexual selection and animal personality. *Biological Reviews*, 85, 217–246.

Sih, A., and Bell, A. M. 2008. Insights for behavioral ecology from behavioral syndromes. *Advances in the Study of Behavior*, 38, 227–281.

Sih, A., Bell, A. M., Johnson, J. C., and Ziemba, R. E. 2004. Behavioural syndromes: an integrative overview. *Quarterly Review of Biology*, 79, 241–277.

Sinervo, B., and Svensson, E. 2002. Correlational selection and the evolution of genomic architecture. *Heredity*, 89, 329–338.

Smiseth, P. T., Wright, J., and Kölliker, M. 2008. Parent-offspring conflict and co-adaptation: behavioural ecology meets quantitative genetics. *Proceedings of the Royal Society of London B*, 275, 1823–1830.

Smith, B. R., and Blumstein, D. T. 2008. Fitness consequences of personality: a meta-analysis. *Behavioral Ecology*, 19, 448–455.

Stamps, J. A. 2007. Growth-mortality tradeoffs and "personality traits" in animals. *Ecology Letters*, 10, 355–363.

Van Oers, K., Drent, P. J., de Jong, G., and Van Noordwijk, A. J. 2004. Additive and nonadditive genetic variation in avian personality traits. *Heredity*, 93, 496–503.

Van Oers, K., Drent, P. J., Dingemanse, N. J., and Kempenaers, B. 2008. Personality is associated with extra-pair paternity in great tits, *Parus major*. *Animal Behaviour*, 76, 555–563.

Via, S., Gomulkiewicz, R., de Jong, G., Scheiner, S. M., Schlichting, C. D., and Van Tienderen, P. H. 1995. Adaptive phenotypic plasticity: consensus and controversy. *Trends in Ecology and Evolution*, 10, 212–217.

Wilson, D. S. 1998. Adaptive individual differences within single populations. *Philosophical Transactions of the Royal Society of London B*, 353, 199–205.

Wilson, D. S., Clark, A. B., Coleman, K., and Dearstyne, T. 1994. Shyness and boldness in humans and other animals. *Trends in Ecology and Evolution*, 9, 442–446.

Wolf, M., Van Doorn, G. S., Leimar, O., and Weissing, F. J. 2007. Life-history trade-offs favour the evolution of animal personalities. *Nature*, 447, 581–585.

Wolf, M., Van Doorn, G. S., and Weissing, F. J. 2008. Evolutionary emergence of responsive and unresponsive personalities. *Proceedings of the National Academy of Sciences U S A*, 105, 15825–15830.

8

Frontiers on the Interface between Behavioral

Syndromes and Social Behavioral Ecology

ANDREW SIH

Introduction

The hallmark of a behavioral syndrome is behavioral consistency, both within and between individuals (Sih et al. 2004a, b). Individuals exhibit within-individual consistency when they behave in a consistent way through time or across situations; individuals then have a behavioral type (BT). Between-individual consistency occurs when individuals differ in their BT (e.g., some are consistently more aggressive, while others are consistently less aggressive than others), which would be reflected statistically as a behavioral correlation among individuals. That is, a behavioral syndrome can be quantified by a behavioral correlation across situations or contexts. Sih and Bell (2008) clarified that behavioral syndromes and BTs do NOT have to (1) be stable over a lifetime, or even over a large proportion of a lifetime; (2) have a genetic basis; (3) involve both multiple contexts and multiple situations; (4) be independent of social status or condition; (5) involve a dichotomy of behavioral types (e.g., bold vs. shy) as opposed to a continuous range of BTs, or (6) be associated with suboptimal behavior. The concept of a behavioral syndrome does not require animals to show little or no behavioral plasticity. While a behavioral syndrome might be more interesting or more important (for fitness or ecological or evolutionary outcomes) if it has a strong genetic basis, is stable over a lifetime, carries over across multiple contexts, and results in suboptimal behavior, these are not part of the definition of the concept.

To date, much of the ecologically based work on behavioral syndromes has focused on variation in boldness, aggressiveness, or activity per se. These three are often correlated (Huntingford 1976; Riechert and Hedrick 1993; Bell 2005; Johnson and Sih 2005). Boldness is often positively associated with an individual's exploratory tendency, and an individual's proactive vs. reactive coping style also relates to these other BT axes. Proactive individuals tend to be more exploratory, bold, active, and aggressive than reactive ones (Koolhaas et al. 1999; 2006). As noted by Stamps (2007), an ecologically important connection between these BT axes is that they all often result in both higher resource intake and higher mortality risk. That is, they can be viewed as alternative ways of taking risks to gain rewards. When two or more BT axes are correlated, then the behavioral syndrome is multidimensional. However, even if the different BT axes are uncorrelated, if individuals exhibit within- and between-individual consistency along any one BT axis (e.g., if some individuals are consistently more aggressive than others), this fits the definition of a behavioral syndrome: a behavioral syndrome can be one-dimensional.

While the above-mentioned BT axes relate closely to major fields of study in behavioral ecology (in particular, foraging and antipredator behavior, and competition), other widely studied areas of social behavioral ecology (e.g., mating behavior, cooperation, parental care, sociality, and group living) have not yet received as much attention from a behavioral syndrome view. The interface between behavioral syndromes and social behavioral ecology is one of the main, exciting frontiers in animal behavior (see Bergmüller and Taborsky 2007; 2010; Réale and Dingemanse 2010). The first goal of this chapter will thus be to examine the surprisingly understudied issue of how behavioral syndromes relate to mating and cooperation. Key issues include (1) getting more data on individual variation in mating or cooperative behavior and how these behaviors fit into an overall behavioral syndrome, and (2) partner choice based on the partner's BT (e.g., female choice based on the male's personality). Discussing these issues brings up the importance of social plasticity (behavioral responses to varying social conditions, including different social partners; see Dingemanse et al. 2010) and the related concepts of social sensitivity and social skill. Acknowledging social plasticity then leads to a general discussion of how variation in social situation (e.g., in the mix of BTs in a social group) might influence individual behavior, individual fitness, and group dynamics. How do social dynamics influence BTs and vice versa? My second goal is thus to organize and present recent and new ideas on what can be termed the *social ecology of behavioral syndromes.*

Behavioral syndromes and mating behavior

A fundamental and yet relatively unstudied issue for behavioral syndromes and mating behavior involves simply quantifying consistent, individual variation in mating tactics. For males, beyond the obvious cases where male mating tactics are closely tied to alternative morphologies (e.g., larger territorial males vs. smaller sneaky males; Shuster 1989; Sinervo and Lively 1996; Emlen 1997; Watters 2005), relatively few studies, to date, have analyzed whether males exhibit consistent BTs regarding their tendencies to be a territorial vs. a satellite male, to use courtship vs. coercion, or to focus on within-pair vs. extra-pair mating. Picture, for example, male damselflies that are physically capable of either defending a stationary mating territory or cruising around a pond searching for females. Do some males specialize in being territorial while others are specialized cruisers, or do all males switch between tactics depending, for example, on their current energy state? Even if all or most males utilize both strategies on a regular basis, do they differ consistently in percentage time cruising vs. being territorial? If ecological or social conditions change in ways that increase the benefit and thus prevalence of cruising, are rank order differences in percentage time cruising maintained? That is, even if all or most males shift to spend less time being territorial, do males that had previously been highly territorial maintain a tendency to spend more time defending a territory than other males?

For females, a growing literature documents at least short-term individual differences in mate preferences (Morris et al. 2003; Forstmeier and Birkhead 2004; Morris et al. 2006; Cummings and Mollaghan 2006; Holveck and Riebel 2010). Females might also vary, for example, in their tendency to resist male mating attempts (Lauer et al. 1996), in their mate choosiness *per se* (Forstmeier 2004), and in their propensity to engage in extra-pair copulations (Forstmeier 2007). While it can be laborious to test each individual female repeatedly, in the behavioral syndrome context, there is a great need for further study of within- and between-individual consistency among females in their mating tactics or preferences.

Going beyond mating behavior *per se*, do individual differences in mating tactics reflect differences in overall BT? Are mating tactics part of a behavioral syndrome? For example, male mating tactics can involve aggression toward other males in male-male competition, or toward females in sexual coercion. Females can also be aggressive toward other females or toward courting males, with sexual cannibalism being arguably the most extreme form of female aggression toward courting males. Is individual variation in the tendency to use these aggressive mating tactics, or in the degree of aggressiveness while engaging in these tactics, correlated to aggressiveness in other contexts (e.g., aggressiveness in attacking prey; Johnson and Sih 2005; Arnqvist and Henrikkson 1997)? Or, is aggressiveness associated with mating correlated to boldness with predators (Johnson and Sih 2005; 2007) or with poor parental care (Wingfield et al. 1990; Duckworth 2006)? If so, then selection favoring high aggressiveness in other contexts can spill over to cause inappropriately high aggressiveness in the mating context (e.g., males that win in male-male competition might engage in too much sexual coercion [Ophir et al. 2005], or females that are highly voracious foragers might be too willing to engage in sexual cannibalism [Johnson and Sih 2005]). Conversely, selection favoring aggressiveness in the mating context can result in inappropriate behavior outside of mating (e.g., poor parental care). In essence, behavioral carryovers associated with behavioral syndromes then represent another arena for possible conflicts between sexual selection and natural selection. Sexual selection favoring particular BTs might spill over to reduce viability outside of the mating context and vice versa.

Interestingly, if a male's mating tactics are part of his overall BT, and if his BT has fitness consequences (Dingemanse and Réale 2005; chapter 7), then females can potentially base adaptive mate choice on the male's BT (i.e., on his personality) using his mating tactics (e.g., courtship displays) as an indicator of his BT. The benefits of choosing males on the basis of their BT could come via either direct benefits or good genes.

With regard to direct benefits, in some systems, males provide females with direct benefits as part of the courtship display (e.g., nuptial gifts). In the current context, the more interesting situation is where direct benefits are deferred—where they do not come until after courtship, particularly if they do not come until after mating. If the benefits come later, females need a reliable indicator during courtship (or in any case, before mating) of the male's ability or tendency to provide benefits at a future time. In that case, a key issue is—what aspect of a male's display is a useful indicator (a credible promise; cf. Dall et al. 2004) of a future direct benefit? The fact that the male's personality or BT is, by definition, stable over time (or at least reasonably so) makes his BT, as revealed during courtship, a good candidate target for mate choice for deferred direct benefits.

An example of where a behavioral carryover may play a role in signaling direct benefits involves systems where males sometimes engage in sexual coercion (aggression during mating) that can be highly costly to females. Some females then actually prefer less aggressive males (that lose in male-male competition) presumably because their low aggressiveness in male-male contests carries over to mean less sexual coercion with the female (Ophir et al. 2005). Human females also often prefer nicer males over more aggressive ones, perhaps, in part, because of lower costs of sexual coercion (Miller 2007). Given that sexual coercion is relatively common in nature (Clutton-Brock and Parker 1995; Sih et al. 2002; Arnqvist and Rowe 2005), we might often see female preference for less aggressive, more affiliative males—where females use the male's behavior before mating as an indicator of his coercive tendencies.

A probably more widespread case of deferred direct benefits involves male parental care or other future cooperation with the female. The general question is: when a male "promises" to provide future benefits, why should a female believe that promise? The usual answer is that honesty is enforced by costs of signaling. Here, the alternative, behavioral syndromes–based hypothesis is that a male's courtship displays or other pre-mating behavioral interactions might reveal his BT that is a reasonably reliable indicator of his future parental care tendencies. For example, if there is a negative correlation between a male's aggressiveness and paternal care (e.g., Wingfield et al. 1990; Duckworth 2006), then a male's aggressiveness during male-male competition or courtship displays might be a useful indicator of his future lack of cooperation in parental care. Alternatively, his affiliative behavior during an extended courtship process might reliably indicate his affiliative commitment to future parental care, and his sensitivity to female signals during courtship might correlate with his sensitivity to offspring or female needs during parental care (Miller 2007).

Another possibility is female choice for good genes based on the male's BT as the indicator of good genes. A key fact here is that BT is typically moderately heritable (Penke et al. 2007; see chapter 6), so a male's BT should influence his offspring BTs that can affect offspring fitness. If offspring environments are at least somewhat predictable (perhaps because females choose or manipulate their offspring's environments; Davis and Stamps 2004), so females can predict what BTs will be adaptive for their offspring, then, in theory, females should choose males accordingly by BT. For example, Godin and Dugatkin (1996) showed that female guppies prefer bolder males (that engage in more predator inspection) perhaps because bolder males both have higher feeding rates in the absence of predators and are more skilled predator inspectors. The idea is that these abilities will reliably be favored in offspring. If, on the other hand, offspring environments are not predictable, then females can choose males that are behaviorally flexible in order to produce offspring that can cope with a range of environments. Or, if success in social interactions inherently requires social sensitivity (the ability to understand social signals and respond appropriately), then females should prefer males whose displays indicate high social sensitivity (e.g., Patricelli et al. 2002; 2006). This could, in part, explain the human female preference for males who are funny (Bressler and Balshine 2006).

Of course, the female's own BT should also play a role in her mate choice. The BT of her offspring should depend not just on the male's BT, but also on the female's BT. Furthermore, the success of bi-parental care can depend on the interaction between male and female BTs—on behavioral compatibility or complementarity (e.g., Dingemanse et al. 2004; Both et al. 2005; Spoon et al. 2006). Thus, in theory, female choice should involve a female BT x male BT interaction. To date, outside of humans, few studies have investigated this possibility (e.g., Groothuis and Carere 2005). Intriguingly, a recent paper showed that extra-pair mating rates (i.e., choice of actual mating partners as opposed to social partners) in great tits depended on the similarity of male and female BTs (Van Oers et al. 2008).

A key aspect of female choice based on male compatibility or complementarity is the high likelihood that different females will prefer different male BTs. Along similar lines, if females choose male BTs on the basis of anticipated future ecological or social conditions for offspring, then given spatial or temporal variability in these conditions, again, different females will likely prefer different male BTs. Variation among females in mate preferences is crucial because it is one mechanism that can help explain the maintenance of genetic variation in male traits favored by sexual selection.

To date, humans are the main system where investigators have quantified effects of personality on mate choice. Miller (2007) summarized an extensive literature on what he called *sexual selection for "moral virtues,"* many of which are aspects of personality that plausibly relate to fidelity, cooperation, parental care ability, or general social skill. In my view, a shortcoming of Miller's (2007) article is its lack of consideration of mechanisms of sexual selection that favor traits, including personality traits, that are not morally virtuous. Mating systems, indeed social systems in general, often feature an interplay between cooperation vs. conflict, and between honest indicators vs. deception. The BT corollary of the sexy son hypothesis suggests that nonvirtuous BT traits (e.g., involving successful deception or coercion, not just of the female, but in social interactions, in general) should, under the right circumstances, be preferred by females. Again, a male BT x female BT interaction would not be surprising. Overall, further study of the interface between behavioral syndromes and mating behavioral ecology and sexual selection is a very exciting area that should yield important insights.

Behavioral syndromes and cooperation

Cooperation has been the subject of a great deal of study in both behavioral ecology and human psychology (Nowak 2006; Wilson and Wilson 2007); however, to date, only a few studies have related behavioral syndromes and cooperation. Simple theory on cooperation models the evolution of interactions among individuals that are either pure cooperators, pure defectors, or conditional cooperators (e.g., following a tit-for-tat, TFT, strategy). Few studies on animals, however, have actually quantified individual variation in cooperativeness (Bergmüller and Taborsky 2007; Rutte and Taborsky 2007; Bergmüller et al. 2010).

In the syndrome context, a key question is: do individual differences in tendency to cooperate carry over across multiple contexts? Are the same individuals cooperative or even altruistic in social foraging, in group vigilance, in resource sharing, and in cooperative breeding or shared parental care? If males show a general cooperative tendency, then a female can use a male's earlier help in group foraging or vigilance as an indicator of his future help as a bi-parental care partner. In some situations, individuals probably cooperate owing to kin selection, whereas in others, the notion is that cooperation reflects reciprocal altruism, either via direct reciprocity (if you scratch my back, I'll scratch yours) or indirect reciprocity, where altruistic acts build a reputation for cooperation that results in others doing favors for the altruist. Is a tendency to cooperate with kin correlated to a tendency to be altruistic toward non-kin? Emphasizing possible suboptimal consequences of behavioral syndromes, are individuals with a highly cooperative BT overly cooperative and exploited by others? Or for a more detailed syndromes hypothesis—might frequent cooperation with kin spill over to result in excess, nonoptimal cooperation with non-kin, or vice versa? To my knowledge, these issues have rarely, if ever, been quantified.

Going beyond cooperative behavior *per se*, the behavioral syndromes approach suggests the need to understand correlations between cooperative behavior and other aspects of BT. One might imagine that social individuals might be more cooperative, while asocial or aggressive individuals might be less cooperative. An example of a study that quantified correlations between cooperative behavior and other BT axes is the work by Bergmüller and Taborsky (2007; 2010) on a cooperatively breeding Lake Tanganyika cichlid fish where alternative cooperative helping behaviors fit in as part of an overall behavioral and life-history syndrome (see also Bergmüller et al. 2010). A "disperser" type that was more aggressive and exploratory provided help with territorial defense (an aggressive form of helping), but not territory maintenance, whereas individuals that stayed in the group nest were less aggressive and exploratory, and helped with territory maintenance (a nonaggressive form of helping) but not territory defense.

A second way of looking at the interplay between cooperation and other

BT axes acknowledges that behavior in cooperative scenarios might often reflect the intersection of several behavioral tendencies. Consider, for example, predator inspection in schooling fish (e.g., guppies, sticklebacks), where individuals leave the school and approach predators apparently to gain information about the risk posed by the predator. When individuals inspect in pairs, they might engage in reciprocal altruism where individuals take turns being the one that is closer to the predator (taking greater risks while generating information for both members of the pair; Milinski 1987; Dugatkin and Alfieri 1991). Investigators have been interested in whether the dynamics of predator inspection fit a simple tit-for-tat (TFT) model, where individuals cooperate (take the lead) as long as their partner cooperates, but defect (stop inspecting) when the partner defects. A given individual's behavior during predator inspection could depend on its (1) cooperative tendency, (2) schooling tendency, (3) social sensitivity, and also (4) boldness *per se*. Social sensitivity should include "punishment" (reducing inspection if the partner has recently defected), but also "forgiveness" (resuming inspection if the partner resumes being cooperative). Cooperativeness should be assessed after accounting statistically for these other behavioral tendencies. Of course, the notion that expressed behavior might be influenced by multiple BT axes applies not just to predator inspection, but to behavior, in general (Sih and Bell 2008; Uher 2011).

The integrative view just described notes that if individuals differ in cooperative tendency, then individuals should adjust their behavior—both their tendency to interact at all with a given potential partner, and their tendency to cooperate with that partner—on the basis of the potential partner's cooperative tendencies. Obviously, individuals should not be unconditionally cooperative. If they are unconditionally cooperative, then others can exploit that tendency by engaging in overt or subtle cheating. This, in turn, favors the evolution of social sensitivity (to evaluate the trustworthiness of potential social partners; McNamara et al. 2009). Intriguingly, the model of McNamara et al. (2009) found that the evolution of cooperation and social sensitivity were linked—the existence and maintenance of variation in cooperativeness favored the evolution of social sensitivity and vice versa. One issue that McNamara et al. (2009) did not address is the possibility of correlations between cooperative tendency and social sensitivity. Perhaps cooperative individuals should be more sensitive since they are otherwise open to be cheated. Alternatively, it could be that deceptive individuals should be more sensitive since they live and die by their ability to be successfully deceptive.

The social ecology of behavioral syndromes

An auxiliary theme of the above discussion on social behaviors such as mating and cooperation was the notion that the behavior and/or fitness of a focal individual might often depend on the BT of potential social partners (mating partners or partners in cooperation). The BT of a social partner is one aspect of any individual's social situation. Other aspects include the group's density or sex ratio, the mix of BTs in the group, and the group's social structure (the pattern of interaction among individuals). I next discuss how the broader social situation influences behavior and fitness, and vice versa. I consider the following issues: (1) key aspects of social situation and new tools for quantifying social situation, (2) how social situation affects the fitness of different BTs, (3) patterns of social sensitivity or plasticity in response to varying social situations and their implications for the concept of behavioral syndromes, (4) adaptive social plasticity or social skill, and (5) effects of the mix of BTs on group dynamics and group fitness.

QUANTIFYING VARIATION IN SOCIAL SITUATION

As noted above, one component of an individual's social situation involves the traits of the individual's social partner. A broader view of social situation, the one that I will focus on for most of the rest of the chapter, involves the characteristics of the overall social group. For example, a key aspect of social situation is sex ratio and associated components: male and female density. In a classic paper, Emlen and Oring (1977) suggested that sex ratio plays a key role in governing mating systems and sexual selection. Since then, numerous papers have considered how variation in sex ratio influences mating tactics and mating patterns (e.g., Sih and Krupa 1995; Head and Brooks 2006; Kokko and Jennions 2008; Gosden and Svensson 2009). Another major aspect of social situation is the group's relative frequency of different BTs— for example, the frequency of more vs. less aggressive individuals (hawks vs. doves), or of cooperators vs. defectors, or producers vs. scroungers. A fundamental assumption of game theory is that the mix of BTs in a group should have important impacts on individual fitness and group dynamics.

Beyond sex ratio and the frequency of different BTs, much of our thinking about social biology implicitly assumes that social structure (e.g., the pattern of who interacts with whom) is important for both individual and group outcomes. Animal behaviorists, however, often do not quantify social structure in any detail. Most game theory assumes very simple social structure; for example, simple game models assume that on average BTs (e.g., hawks/doves) experience the average frequency of all BTs in the population. In reality, individuals often interact with only a limited subset of the population. Individuals generally differ in how many social partners they have, and perhaps in the relative frequency of BTs experienced. Although many behavioral studies have collected data on social structure, often we have not analyzed or presented these data in much detail. Recently, however, animal behaviorists have increasingly used a well-developed framework from the social sciences for characterizing and analyzing social structure—social network (SN) theory (Scott 2000). Several conceptual overview papers (Wey et al. 2008; Sih et al. 2009) and a recent book (Croft et al. 2008) have highlighted the value of using SN metrics to characterize the group's social structure and each individual's position in the SN, including the number of interacting partners (degree), number of partners of your partners (reach), the tendency for your partners to also interact with each other (clustering coefficient), and whether short paths between other individuals go through a focal individual (centrality or betweenness). For sexual selection, SN metrics can be used to characterize each male's interactions with both other males and with females, and a male's SN traits can help explain his subsequent mating success (McDonald 2007). To date, except for humans (e.g., Burt et al. 1998), few studies have looked at how different BTs might differ in their average SN position (Pike et al. 2008; Croft et al. 2009). We might, for example, expect more active, proactive, or social individuals to have more social partners, and more bold or exploratory individuals to have perhaps a greater tendency to bridge across different social groups. Differences between BTs in average SN metrics might then affect their success and influence in the social group.

EFFECTS OF SOCIAL SITUATION ON FITNESS OF DIFFERENT BTS

As noted above, game theory assumes that in social groups, the fitness of behavioral types (e.g., of hawks vs. doves, or territorials vs. satellites) is frequency dependent—that is, it depends on the mix of behavioral types in the group (Maynard Smith 1982; Dugatkin and Reeve 1998; Costantini 2005; Sinervo and Calsbeek 2006). This basic scenario holds for many theoretical behavioral dichotomies—for example, hawk/dove, producer/scrounger, co-operator/defector, as well as for more complex games like rock/paper/scissors. Although this is a fundamental tenet of game theory that has guided our thinking on social behavior for the past 35 years, surprisingly few studies have experimentally manipulated the frequency of behavioral types to examine actual effects on fitness or on behavioral dynamics (Beauchamp et al. 1997; Sih and Watters 2005; Flack et al. 2006). The exception involves morphologically based alternative mating types (AMTs), for example, larger territorial individuals vs. smaller satellites. Here, because behavioral types are easy to identify, studies have indeed examined frequency-dependent fitnesses associated with the different types. But even here, few experimental studies have manipulated the relative frequency of these AMTs (Warner and Hoffman 1980; Sinervo and Calsbeek 2006). Now that behavioral syndromes are receiving more attention, a key suggestion is to experimentally manipulate the frequency of BTs to better understand how their behavior and fitness depends on the group's social composition (mix of BTs in the group, Sih and Watters 2005).

To test game theory assumptions on how fitness depends on the mix of BTs in the social group, we need a methodology for quantifying these fitness effects. Social selection theory (Wolf et al. 1999) provides a quantitative framework for relating both individual traits (e.g., BT) and the group's social composition to individual fitness. The basic idea extends the regression approach for quantifying natural selection and sexual selection on traits (e.g., Arnold and Wade 1984a, b) by adding the social group's mean trait value (e.g., mean aggressiveness) as an independent variable in the regression equation of traits (individual and group) on fitness. The method partitions natural and sexual selection gradients (relationships between the individual's traits and fitness) from social selection gradients (the relationship between the interacting group's mean trait value and individual fitness). Selection on a focal trait then also depends on social selection—the product of the social selection gradient and the covariance between the individual's trait and the social group's trait. For example, if more aggressive individuals ("hawks") tend to be in social groups that have lots of other aggressive individuals (a positive covariance between the individual and the social group's trait values), and if being in aggressive groups tends to reduce fitness (a negative social selection gradient), then this group composition effect can have an important influence on the overall fitness of hawks. Other social group traits can be similarly incorporated statistically. In frequency-dependent games, individual fitness should depend on the interaction between the individual's BT and the social group's mean BT. This is handled by adding an interaction term into the regression equation. Social selection and related indirect genetic effects have generated considerable interest in evolutionary biology; however, they are only just starting to be used by behavioral ecologists (but see Sinervo and Calsbeek 2006; Bleakley et al. 2007).

Given that fitness in various social environments (e.g., different sex ratios or different mixes of BTs) depends on behavior in those environments, and that individuals typically experience multiple environments in their lifetime, a key underappreciated issue is the importance of behavioral consistency versus behavioral plasticity across environments in determining overall fitness. For example, in some systems, different BTs (e.g., more vs. less aggressive) are favored by sexual selection in different social situations. If individuals have consistent BTs, then different individuals should have high mating success in the different situations. We might then have strong sexual selection within each situation, and yet, across situations pooled, we might see little variation in mating success. On the other hand, if the same general behaviors are favored in different social situations, then the persistence of consistent BTs across situations increases variation in mating success and sexual selection; "winners are winners" and "losers are losers" in all social situations. While many studies probably have data tracking individuals across different social situations, few studies to date have looked at how behavioral consistency vs. plasticity influences overall fitness.

EFFECTS OF SOCIAL SITUATION ON BEHAVIOR: SOCIAL PLASTICITY AND SENSITIVITY

A classic, perhaps caricatured view of standard behavioral ecology assumes that individuals exhibit optimal behavioral responses to environmental variation, including variation in social situations. Alternatively, simple game theory models often assume either that individuals have pure BTs (i.e., no plasticity, pure hawks or pure doves), or no BTs (i.e., all individuals follow the same optimal probabilistic or condition-dependent evolutionary stable strategy, ESS) (Dugatkin and Reeve 1998; Maynard Smith 1982). In contrast, the behavioral syndromes approach emphasizes individual differences in mean BT, often with individuals exhibiting limited (less than optimal) plasticity. These two views can be contrasted and visualized using a reaction norm presentation (figure 8.1; Sih et al. 2004a, b called these *plasticity plots*). Standard behavioral ecology, in its simplest form, assumes one optimal individual reaction norm often with a steep slope (figure 8.1a; assuming that optima are quite different in the different environments). In contrast, behavioral syndromes, in their simplest form, posit that individuals show parallel reaction norms that differ in mean BT, but not in plasticity (slope; figure 8.1b). In reality, individuals often differ in both mean BT and in behavioral plasticity (Briffa et al. 2008; figure 8.1c). I next consider various issues regarding social plasticity (i.e., social sensitivity)—behavioral responses to variations in social situations—including (1) social sensitivity as a BT, (2) the related concept of choosiness as a BT, (3) implications of social sensitivity for the concept of behavioral syndromes, and (4) effects of social

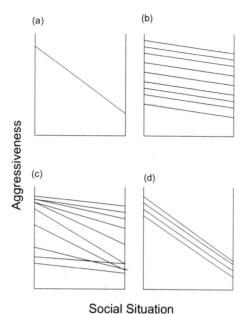

Figure 8.1. Four hypothetical scenarios on how social situation might influence aggressiveness. Social situations might, for example, be different social partners that differ in rank, different sex ratios, or different proportions of aggressive individuals in the social group. The key is that optimal aggressiveness decreases as we go from left to right.
(a) Simple, optimality-based behavioral ecology predicts that all individuals show the optimal pattern of behavioral plasticity. (b) A simple, behavioral-syndromes view assumes that individuals differ in behavioral type (BT) where all exhibit the same amount of limited (less than optimal) plasticity, but individuals differ in average aggressiveness. (c) A scenario where individuals differ in both average BT and plasticity where more extreme BTs are less plastic. (d) A scenario where individuals differ in average BT, but the social situation plays a large role in governing expressed behaviors.

sensitivity on fitness, including the possibility of adaptive social plasticity that I refer to as social skill.

The idea that plasticity or sensitivity might be a key part of a BT, perhaps a BT axis in itself, is not new. The long-standing literature on proactive/reactive coping styles emphasizes individual variation in environmental sensitivity (Koolhaas et al. 1999; 2006). Reactive individuals are highly sensitive to changes in their environment. In contrast, proactive individuals follow set behavioral routines and are relatively insensitive to environmental changes. These differences in coping style are often associated with genetically based differences in neuroendocrine profiles (Koolhaas et al. 1999;

2006; chapter 12). The differences in sensitivity are also related to other BT axes. Sensitive, reactive individuals tend to be more fearful, whereas proactive individuals tend to be bolder and aggressive. Coping styles have been studied in some detail in laboratory rodents (Koolhaas et al. 1999; 2006), farm animals (Hessing et al. 1993), great tits (chapter 12), rainbow trout, and primates (Suomi 1987; Capitanio et al. 1998). A parallel literature in humans (Aron 1996; Boyce and Ellis 2005; Jawer 2005) notes that variation in sensitivity might be associated with variation in habitat and job choice (highly sensitive people avoid highly stimulating situations), in fine-scale behavior (e.g., preferred volume level while listening to music), in other aspects of personality (e.g., creativity), and in mental and physical well-being (e.g., extreme sensitivity might be associated with depression, migraine headaches, and suppressed immune systems).

The recent growth of interest in animal personality has brought the literature on coping styles to the attention of behavioral ecologists, which has, in turn, resulted in new insights on the evolution of behavioral sensitivity (Wolf et al. 2008). Nonetheless, to date, with the exception of the work on great tits, few studies in behavioral ecology have quantified either individual variation in environmental or social sensitivity, or its effect on behavior or performance. A few studies have examined effects of the group's social composition on behavioral plasticity within groups. Some found that individuals do not retain their BTs when they are placed in a social group (e.g., Mottley and Giraldeau 2000). Other studies show that BTs are largely maintained (e.g., aggressive individuals stay relatively aggressive, or AMTs do not modify their behavior) regardless of the group's social composition (Van Erp-Van der Kooij et al. 2003; Sih and Watters 2005). However, even in the studies where most individuals maintained their BT, some individuals showed substantial behavioral plasticity. For example, Sih and Watters (2005) created groups of water striders that differed in average male aggressiveness. They found that although, in general, hyper-aggressiveness was only seen in groups made up primarily of highly aggressive males, one hyper-aggressive male emerged in a group that was created by putting together very unaggressive males. Apparently, one male that was unaggressive in a mixed social background became much more aggressive when it was surrounded by males that were all relatively passive. Clearly, more study is needed to better understand variation among BTs in their social plasticity in response to the group's social composition.

In the syndrome context, additional interesting issues revolve around whether social sensitivity carries over across different tasks and contexts. For example: (1) within a mating context, is sensitivity in mate choice (in the ability to evaluate different mates appropriately) correlated to sensitivity in choosing the time or place to search for mates? Or, (2) is sensitivity in mate choice correlated to sensitivity to subtle signals in the male-female interplay that results in successful mating? (3) Going beyond mating, is sensitivity in the mating context correlated to social sensitivity in other contexts—for example, in partner choice and adjustments to social situations or partners in the context of cooperation or competition? And, (4) going beyond social situations, is social sensitivity in one or more social situations correlated to sensitivity relative to other fitness-related options—for example, habitat choice, or diet choice? Finally, is sensitivity correlated to other aspects of personality? The coping style literature suggests that sensitivity is negatively related to boldness and aggressiveness. An alternative possibility is that extreme BTs (very bold or very shy individuals, very aggressive or very unaggressive ones) are generally less plastic, while at least some individuals with intermediate BTs might be most likely to be highly plastic (figure 8.1c).

Interestingly, logic suggests that correlations between aspects of sensitivity could be either positive or negative. If individuals vary in general sensitivity, then sensitivity should be positively correlated across different tasks or situations. Alternatively, if sensitivity is costly (sensitivity is sometimes invoked as a common cost of plasticity, DeWitt et al. 1998), for example, if it draws on a finite pool of attention (Dukas 1998), then we might expect negative correlations between sensitivity in different tasks. For example, sensitivity toward potential mates might draw attention away from, and thus reduce sensitivity toward, food or predators. Clearly, empirical studies are needed to evaluate actual correlations between sensitivity in different tasks in natural systems.

A related concept is choosiness. Although choice, for example, mate choice, is a major topic in behavioral ecology, few studies have reported on not just the female's preference (e.g., whether she prefers larger males) but also her choosiness (e.g., the strength of her preference for either male). And, few have retested the same females and looked for consistent variation among individual females in choosiness. In the syndrome context, interesting questions are whether females are repeatable in their choosiness across trials in the same basic situation, and whether they are consistent in their degree of mate choosiness across different situations (e.g., different types of focal male traits, or different male abundances). Going beyond mating *per se* to other ecological contexts, the behavioral syndromes approach asks whether the same individuals that are choosier than others about mates are also relatively choosy about other social partners, about their food choices, or about aspects of habitat use. In humans, we feel that we know that people differ in choosiness about their diets, music, movies or TV shows, brands of clothing or electronic equipment, or their mating partners. The question is—is choosiness correlated across these different situations? As with sensitivity, if there is a general choosiness syndrome then choosiness might be positively correlated across situations; however, given that choice takes time and energy (to gather information and evaluate choices), it is also conceivable that choosiness might be negatively correlated across contexts. In either case, as with other personality-related correlations, a choosiness correlation could result in suboptimal behavior. An individual that is generally very choosy across many situations will likely be too choosy in some situations. Or, an individual that has enough energy to be choosy in only one or a few situations might be inappropriately unselective in another situation.

Social interactions, plasticity, and the existence and concept of behavioral syndromes

Recent studies have suggested that social interactions might play a role in explaining the existence of behavioral syndromes, that is, in explaining behavioral consistency per se. The issue derives from the observation that in a changing world, in which optimal behaviors can be very different under different circumstances, behavioral consistency can sometimes be associated with suboptimal behavior. Animals that have a bold or aggressive BT are sometimes too bold or too aggressive (e.g., Sih et al. 2003; Johnson and Sih 2005; Duckworth 2006) and conversely, animals that have a shy, unaggressive BT sometimes miss out on opportunities that they could otherwise have. Similarly, the positive correlation between boldness and aggressiveness can result in spillovers across contexts that can be associated with suboptimal behavior (Riechert and Hedrick 1993; Johnson and Sih 2007). If that is the case, then why do animals exhibit consistent BTs and behavioral correlations? Why not be optimally plastic in all circumstances? Early on, most hypotheses to explain behavioral syndromes focused on proximate constraints associated with a stable, common underlying proximate mechanism (Sih et al. 2004a, b). More recently, several adaptive hypotheses have been suggested (summarized in Sih and Bell 2008; Sih 2011; chapter 9), often involving the benefits of specializing on a BT coupled with costs of switching BTs (Dall et al. 2004), or positive feedback loops between behavior and an underlying state variable (Sih and Bell 2008; Wolf et al. 2008; Luttbeg and Sih 2010; chapter 9). Some hypotheses focus, in particular, on social implications of behavioral consistency and behavioral correlations—on the benefits of social predictability or social specialization. Next, I briefly summarize these ideas.

One explanation for behavioral consistency revolves around the social benefits of being predictable in systems with repeated social interactions (cooperative or competitive). Being predictable has costs—predictable individuals can be exploited (cheated); however, in the right social circumstances, where credible threats or promises yield important benefits, behavioral consistency can be favored (Dall et al. 2004; McNamara et al. 2008). For example, in aggressive situations, individuals clearly benefit if they can win contests by simply using threats without engaging in escalated fights. Game theory suggests that threats should cause an opponent to back off without a fight if the threat is backed by superior fighting ability (Maynard Smith 1982). In principle, however, a weaker but irrationally aggressive individual could cause a rational, stronger competitor to back down (to avoid the costs of a fight) by threatening to escalate fights despite the costs. Such a threat, however, should work only if it is credible, if the overaggressive individual is indeed predictably overaggressive. Similarly, in situations involving reciprocal altruism, individuals should be more likely to do favors for others who are dependably cooperative (who make credible promises to reciprocate). Social predictability can be a strong force in systems with repeated social interactions with individual recognition and reputational effects (knowledge of who is aggressive or cooperative). If social predictability is indeed a major factor in explaining the existence of behavioral syndromes, then species that better fit the required social scenario should exhibit "more personality" than less social species—more behavioral consistency, particularly for BT axes that relate to social interactions.

An adaptive social explanation not just for behavioral consistency, but also for an overall behavioral syndrome involving a suite of correlated behaviors, revolves around the notion of specialized social niches (Bergmüller and Taborsky 2007; 2010). The idea is that within a population, there are alternative ways to succeed (multiple ecological or social niches). Specializing on one aspect of a social niche carries along with it a suite of correlated social and ecological behavioral and life-history tactics that together form an adaptive integrated package. Bergmüller and Taborsky's (2007) example discussed earlier involved a correlation between dispersal tendency, exploratory tendency, boldness, aggressiveness, and cooperative helping tactics with different packages (different behavioral syndromes) for helpers that stay versus those that disperse. This basic idea parallels Stamps' (2007) idea that boldness and aggressiveness might often be positively correlated because these two BT axes are functionally related—they both represent ways of gaining more resources while taking greater risks. Thus individuals that have chosen a high risk/high gain life history will be both bold and aggressive, while those that have chosen a low risk/low gain lifestyle will be both cautious and unaggressive. The striking thing about the bold-aggressive correlation is that while it often occurs, it does not always occur. The correlation between boldness and aggressiveness appears to depend on predation risk (Dingemanse et al. 2007; Bell and Sih 2007) and perhaps on resource levels or general habitat favorability (Carere et al. 2005; Luttbeg and Sih 2010). For suites of correlated behaviors favored by social interactions, the question remains—what kinds of social conditions, in particular, favor social behavioral syndromes?

If an individual's social niche determines its BT and a suite of associated behaviors, then it may be argued that this calls into question the entire concept of a behavioral syndrome. Some studies have found, for example, that social situation (and associated social plasticity) might be more important than "inherent BTs" in determining behavioral consistency and behavioral correlations (figure 8.1d). For example, in some systems, dominance rank might be the main determinant of a suite of social behaviors—not just contest behavior, but also foraging, mating, antipredator, and/or parental behavior (i.e., a behavioral syndrome). As long as the dominance hierarchy is stable, individuals exhibit apparently consistent BTs. However, if individuals change their rank, or if the social system overall is disrupted, individuals can change their behavior accordingly. This raises the question of whether these individuals truly exhibit a BT. Individuals do show behavioral consistency, but it is associated with their social rank. Nelson et al. (2009) suggest that either this is not a behavioral syndrome or it is a syndrome in only a "weak sense." It is perhaps just a semantic point, but my view is that the fact that the BT has an underlying mechanism, in this case social rank, does not make it not a BT. In other cases, the mechanism underlying the BT might be the individual's physiology or its morphology, or energy or life-history state. Indeed, an individual's BT always has some underlying mechanism. If that mechanism changes, the BT might change as well. If we accept that hormonally based BTs are real, why not accept that socially governed BTs are real, too?

In any case, instead of debating whether or not socially governed BTs are real, a more fruitful line of study might focus, as noted above, on better understanding why in some situations, individuals exhibit strong BTs independent of the social situation (figure 8.1b), whereas in other cases, social situation plays a larger role in governing behavior (figure 8.1d). Recent studies on humans suggest that two main outcomes might emerge from a positive feedback between the social system and the BTs expressed by individuals in that system. Heine and Buchtel (2009) reviewed the extensive literature contrasting social systems and personality in Western (primarily, American) and East Asian (primarily Japanese) cultures. In brief, in Western societies, people are more individualistic and less ruled by social constraints. Social position is more fluid. Personality assays show strong signals of what I call a BT (figure 8.1b). These parts fit together. Weak social constraints allow individuals to express a stronger, individual BT, and the fact that people place high value on their individualism (including their BT) keeps social constraints weak. In contrast, in East Asian societies, social constraints are stronger. People often do not view themselves as having a consistent BT. Instead, they recognize that their behavior is largely situation dependent (figure 8.1d). Am I an aggressive person? It depends entirely on whether I am with my peers, my underlings, my boss, or my parents. Breaking the social rules is taboo. Again, the parts fit together. Because social constraints are strong, individuals exhibit substantial socially-governed social plasticity, which allows strong social constraints to be stable. Of course, I have oversimplified a much richer, nuanced contrast between these cultures; nonetheless, the literature on humans suggests a potentially general framework where social structure determines the costs and benefits of behavioral consistency vs. social plasticity that shapes the structure of personality, which in turn allows the maintenance of stable social structures.

Adaptive social plasticity or social skill

The previous sections included some examples of situations in which animals appear to show adaptive social plasticity. I next explore the concept of adaptive social plasticity, which I call "social skill," in more detail. To organize my presentation, I use the time-honored tradition of following a sequence of steps that lead to a successful outcome. For example, to gain mating success, in step 1, a male must choose a place and time to search for females. This behavioral decision determines his encounter rate with different females. In step 2, males choose which potential mates to court or otherwise attempt to mate with (i.e., male mate choice). Finally, having chosen a potential partner, in step 3, males often display their size, ornaments, and/ or courtship vigor (Andersson 1994; Andersson and Simmons 2006) and attempt to mate. Most studies of sexual selection focus on the relatively static morphological traits (e.g., male size, color, ornaments) or on condition-dependent behaviors (e.g., courtship vigor) expressed in step 3. While these traits clearly often play a role in sexual selection, the percentage of variance in mating success explained is often quite low. Here, I discuss the idea that unexplained variation in mating success might be due, in part, to individual variation in social skill expressed in one or more stages of the overall mating sequence.

A notable example of social skill in step 3 is provided by the work by Patricelli et al. (2002; 2006) on bowerbirds. Male bowerbirds display for females in front of elaborate bowers. Patricelli et al. (2002; 2006) used a robot female that they could control to evaluate the relative ability of different males to adjust their courtship intensity to signals from the female. They found that males that displayed very intensely regardless of signals of interest (or not) from females tended to scare females away. Most notably, their quantitative analysis revealed that a large proportion of the variance in male mating success could be explained by the male's sensitivity (and adjustment) to female signals (Patricelli et al. 2002; 2006), suggesting that skillful social sensitivity plays a large role in sexual selection. Anecdotally, it seems obvious that this is also true in humans—both male and female choice depends not just on the potential partner's static traits, but also on their social skill. Anecdotes about social skill in other mating systems abound in seminars and natural history accounts. The suggestion here is that attempts to study this quantitatively should prove rewarding.

An example of the importance of social skill in stage 2 involves hyper-aggressive males in water striders (Sih and Watters 2005). Males show individual variation in their response to females, to males, and to male-female pairs. Obviously, males should only attempt to mate with females. They should not attempt to mate with males, and have almost no success at separating pairs in order to take over a female. Most males are sensitive to the gender of other water striders—that is, they attempt to mate with females, but not with males or pairs; however, some males are hyperaggressive—they expend a great deal of effort toward trying to mate forcibly with not just females, but also males or pairs. Quantitative analyses showed that hyperaggressiveness has important negative effects on mating success in water striders. Does this happen in other species? Over the years, I have heard anecdotes and often seen photographs of males attempting to mate (and in some cases, mating) with inappropriate partners (e.g., with males, or with females of other species, or even with dead females, or inanimate objects that resemble conspecific females). In some unisexual systems (with only females), females are sexual parasites that must mate with males of other, closely related species to reproduce, but do not use the male's sperm to fertilize their eggs (e.g., Gabor and Aspbury 2008). Males gain no benefit from mating with these females; i.e., females rely on males making mistakes in step 2 of the mating sequence. Even if most males in most systems rarely mate with the wrong species, or rarely attempt to mate with males, they might still vary substantially in their ability to efficiently choose females that are both of high quality and willing to mate with them. My suggestion again is that it would be useful to quantify individual variation in this aspect of social skill in more animal taxa.

In step 1, if social situation has important effects on mating success (as evaluated using social selection methods, see above), then sexual selection should favor males that exhibit adaptive social situation choice, that is, that prefer social situations that yield better mating success. For example, males that inhabit areas with a more favorable sex ratio may enjoy high mating success largely because of their favorable social situation. A more complex scenario arises if the mating success of different BTs depends, as assumed by game theory, on the frequency of those different BTs. The preferred mix of BTs should then depend on the male's BT. My unpublished work on water striders suggests an intriguing possibility that has received little or no attention in other systems—that individual males differ in their ability to choose favorable sites for accessing females. Whereas some males disperse readily from unfavorable social situations (male-biased pools), others stay for long periods in these pools despite having little or no mating success. Variation among males in ability to find good social situations is a form of social skill that could play a major role in explaining variation in mating success. Although this idea seems reasonable and potentially very important, it has rarely been studied quantitatively.

In general, adaptive situation choice can generate diversity both by driving the evolution of specialization (and ultimately, speciation) and by allowing the maintenance of variation (Wcislo 1989; Wilson and Yoshimura 1994). For example, if bold individuals do well in habitat X, but not Y, and vice versa for shy individuals, then both can do well and persist if they each primarily use their optimal habitats. Social situation choice, however, is made more complicated by the fact that the sum of individual social situation choices determines the observed social compositions in different groups. In a world where each individual is choosing its social group, it is possible that no one gets its preferred social situation. Hawks that prefer to associate with doves might not be able to do so if doves avoid associating with hawks. And, doves that prefer to avoid hawks might not be able to do so if hawks attempt to join groups of doves. Further study of adaptive BT-dependent social situation choice and how the aggregate choices of members of a population together determine the distribution of social situations available, as well as social situations experienced by different BTs, should prove insightful.

Finally, note that from the syndrome perspective, an interesting issue involves behavioral correlations for aspects of social skill. Social skill might be related to social sensitivity. As discussed earlier, depending on the underlying mechanism, social sensitivity might be either positively or negatively correlated across the stages of the mating sequence. Similarly, social skill might be positively correlated across stages if some individuals are generally more socially sensitive than others. Those males might choose social situations well, choose mating partners efficiently, and display their favorable traits or respond to females skillfully. Alternatively, social skill in one stage can compensate for poor success in another stage; males that are not highly preferred and thus perform in a mediocre way in stage 3 can perhaps compensate by being particularly good at social situation choice. In addition, social skill might be related to other aspects of BT. More aggressive or proactive males might be less discriminating, and thus might waste more time attempting to mate with inappropriate partners (Sih and Watters 2005). These sorts of possibilities seem likely but, as with other issues about social skill, have rarely been studied quantitatively.

Effects of the mix of BTs on group dynamics and population fitness

Without using behavioral syndromes or animal personality terminology, ecologists have long recognized that differences among species in their average BT (e.g., in the species' boldness, aggressiveness, or sociability) can have important effects on species interactions and population success in different environments. Above and beyond effects of a species' *average* BT, emphasis has been placed on effects of the population's *variance* in BTs, or on the BT of a particular subset of individuals on group outcomes. Population variance among individuals in BT can be important via mechanisms that require relatively little social interaction. For example, individual differences in BT represent within-species niche variation that can significantly reduce intraspecific competition (Bolnick et al. 2003). For the same range of total species' niche breadth, intraspecific competition should be weaker if individuals have different, specialized BTs as opposed to being all generalists with the same BT. Second, variation among individuals in BT can be important for a species' response to environmental change. Even if a species' average BT is not well matched to the new environment, if the species has sufficient variation in BT, enough individuals might have the right BT to cope. Having the right BT can allow an individual to show an immediate adaptive response to environmental change. In addition, genetically based BTs are potentially important both because they can evolve, and because they can shape the evolution of other traits (e.g., morphology; West-Eberhard 2003).

A final idea that has a larger social component, and in a sense combines the previous two ideas, involves the importance of social/ecological complementarity in enhancing a group's response to environmental change. In socially complex systems, different BTs might play different complementary roles that are critical for the overall population response to environmental change, as well for population performance, in general. For example, it is important to actively enhance variation in BTs for species conservation, animal reintroductions into the wild, and the management of zoo populations (Watters and Meachem 2007; chapter 13).

Other effects of BTs on group dynamics or fitness revolve around the importance, in many social systems, of the BT of particular individuals that have disproportionately large effects on the rest of the group. Sih and Watters (2005) called these *keystone individuals*. Examples include alpha males or females (assuming that they are not only dominant, but also interactive and influential), conflict mediators in pigtailed macaques (Flack et al. 2006), facilitators of social cohesion in dolphins (Lusseau and Newman 2004), hyperaggressive males in water striders (Sih and Watters 2005), bridge individuals that facilitate information flow between social groups that are otherwise largely unconnected (Burt et al. 1998; Burt 2004), and super disease-spreaders that, by having an unusual amount of social contacts, facilitate spread of disease (Newman 2003). Social scientists have developed (Scott 2000), and animal behaviorists have recently begun to use, social network metrics and theory to quantify the relative importance of different individuals in group dynamics (Wey et al. 2007; Croft et al. 2008; Sih et al. 2009). In the behavioral syndrome context, a key issue is the effect of the personality of keystone individuals in shaping overall dynamics. In baboons, a long-term field study found that a group's social dynamics were strongly influenced by the personality of the dominant males (Sapolsky 2006). When dominant males that were highly aggressive died (because of eating tainted meat) and were survived by more affiliative males, this changed the dynamics of the group for more than two decades, long after the original surviving males had passed on. In humans, studies have looked for examples of what kinds of personalities end up being bridges across cliques (Burt et al. 1998), and how the personality of these bridges might shape information flow. Overall, however, as with many other topics in the social ecology of BTs, there is a great need for more work on the importance of the BTs of keystone individuals in shaping group dynamics and success.

Conclusions

Although there have now been hundreds of studies on animal personalities or behavioral syndromes, relatively little of this work has focused on the interface between behavioral syndromes and social behavioral ecology. Here, I discussed a selected set of topics on this interface.

(1) First, I focused on how the behavioral syndromes approach suggests new issues to study for two major areas in behavioral ecology—mating behavior and cooperation. I noted the need for more study on how mating behavior or cooperation in particular situations might be part of a larger behavioral syndrome. If mating or cooperative behaviors exhibit behavioral consistency, and in particular, if they are part of a broader behavioral type (BT), then we might expect individuals to exhibit partner choice based on the potential partner's personality. While this idea seems obvious (humans obviously do it), surprisingly little work on nonhumans has examined how, for example, female mate choice might be determined by the mate's BT.

(2) I then explored various issues revolving around variation in social situations. Differences among potential social partners are a simple aspect of variation in a social situation. Other aspects of a social situation that are generally thought to be important include the social group's sex ratio, the frequency or mix of BTs in the social group, and the group's social structure (the pattern of interactions among members of the social group). Behavioral ecologists are finally beginning to quantify social structure more rigorously using social network metrics.

(3) Variation in social situation should have important effects on the relative fitness of different BTs. Although this is a long-standing assumption of game theory, surprisingly few experiments have identified BTs, and then experimentally manipulated their relative frequency in social groups to quantify how the mix of BTs in a social group influences relative fitness. A statistical methodology that is rarely used in animal behavior, but that should be useful for quantifying effects of social situation on fitness, is social selection theory.

(4) Variation in social situations probably influences not only fitness, but also the behaviors themselves, that is, individuals often exhibit social plasticity or social sensitivity. I discussed the notion that real animals

might often fall somewhere in between the assumptions of classic behavioral ecology and classic behavioral syndromes; real individuals might vary both in average BT and in behavioral plasticity. The idea that individuals vary in plasticity or sensitivity is a core part of the coping styles literature. Still, there are numerous issues about sensitivity as a BT that remain largely unstudied. In essence, we currently know little about whether sensitivity in one context is correlated to sensitivity (or BT) in other contexts. Similarly, we know little about individual variation in choosiness as a behavioral syndrome.

(5) Social interactions have implications for the existence and concept of behavioral syndromes. First, social interactions can be important in providing an adaptive explanation for why individuals exhibit behavioral consistency at all, and why particular social behaviors might be correlated. However, if social interactions have major effects on behavior (e.g., if social rank largely governs a stable BT), then this raises questions, which I attempted to answer, about the entire concept of animal personality. More interestingly, I suggested a hypothesis for how social structure, social constraint, and BTs might be linked by a positive feedback mechanism.

(6) I next discussed adaptive social plasticity, which I termed *social skill*. I outlined a 3-step framework for organizing ideas on social skill, presented examples of where individual variation in social skill appears to be quantifiably important, and discussed ideas on how variation among individuals in social skill might play an important, understudied role in determining overall success—that is, mating success.

(7) Finally, I discussed how variance among individuals in BT, and the BT of particular keystone individuals, might have important effects on the group's overall dynamics and group fitness.

References

Andersson, M. 1994. *Sexual Selection*. Princeton, NJ: Princeton University Press.

Andersson, M., and Simmons, L. W. 2006. Sexual selection and mate choice. *Trends in Ecology and Evolution*, 21, 296–302.

Arnold, S. J., and Wade, M. J. 1984a. On the measurement of natural and sexual selection: theory. *Evolution*, 38, 709–719.

———. 1984b. On the measurement of natural and sexual selection: applications. *Evolution*, 38, 720–734.

Arnqvist, G., and Henriksson, S. 1997. Sexual cannibalism in the fishing spider and a model for the evolution of sexual cannibalism based on genetic constraints. *Evolutionary Ecology*, 11, 255–273.

Arnqvist, G., and Rowe, L. 2005. *The Natural History of Sexual Conflict*. Princeton, NJ: Princeton University Press.

Aron, E. N. 1996. *The Highly Sensitive Person: How to Thrive When the World Overwhelms You.* New York: Carol Publishing Group.

Beauchamp, G., Giraldeau, L. A., and Ennis, N. 1997. Experimental evidence for the maintenance of foraging specializations by frequency-dependent choice in flocks of spice finches. *Ethology Ecology and Evolution*, 9, 105–117.

Bell, A. M. 2005. Differences between individuals and populations of threespined stickleback. *Journal of Evolutionary Biology*, 18, 464–473.

Bell, A. M., and Sih, A. 2007. Exposure to predation generates personality in threespined sticklebacks. *Ecology Letters*, 10, 828–834.

Bergmüller, R., Schürch, R., and Hamilton, I. M. 2010. Evolutionary causes and consequences of consistent individual variation in cooperative behaviour. *Philosophical Transactions of the Royal Society of London B*, 365, 2751–2764.

Bergmüller, R., and Taborsky, M. 2007. Adaptive behavioural syndromes due to strategic niche specialization. *BMC Ecology*, 7, 12. doi:10.1186/1472–6785/7/12.

———. 2010. Animal personality due to social niche specialisation. *Trends in Ecology and Evolution*, 25, 504–511.

Bleakley, B. H., Parker, D. J., and Brodie, E. D. 2007. Nonadditive effects of group membership can lead to additive group phenotypes for anti-predator behaviour of guppies, *Poecilia reticulata*. *Journal of Evolutionary Biology*, 20, 1375–1384.

Bolnick, D. I., Svanback, R., Fordyce, J. A., Yang, L. H., Davis, J. M., Hulsey, C. D., and Forister, M. L. 2003. The ecology of individuals: incidence and implications of individual specialization. *American Naturalist*, 161, 1–28.

Both, C., Dingemanse, N. J., Drent, P. J., and Tinbergen, J.M. 2005. Pairs of extreme avian personalities have highest reproductive success. *Journal of Animal Ecology*, 74, 667–674.

Boyce, W. T., and Ellis, B. J. 2005. Biological sensitivity to context, I: an evolutionary-developmental theory of the origins and functions of stress reactivity. *Development and Psychopathology*, 17, 271–301.

Bressler, E. R., and Balshine, S. 2006. The influence of humor on desirability. *Evolution and Human Behavior*, 27, 29–39.

Briffa, M., Rundle, S. D., and Fryer, A. 2008. Comparing the strength of behavioural plasticity and consistency across situations: animal personalities in the hermit crab *Pagurus bernhardus*. *Proceedings of the Royal Society of London B*, 275, 1305–1311.

Burt, R. S. 2004. Structural holes and good ideas. *American Journal of Sociology*, 110, 349–399.

Burt, R. S., Jannotta, J. E., and Mahoney, J. T. 1998. Personality correlates of structural holes. *Social Networks*, 20, 63–87.

Capitanio, J. P., Mendoza, S. P., and Lerche, N. W. 1998. Individual differences in peripheral blood immunological and hormonal measures in adult male rhesus macaques (*Macaca mulatta*): evidence for temporal and situational consistency. *American Journal of Primatology*, 44, 29–41.

Carere, C., Drent, P. J., Koolhaas, J. M., and Groothuis, T. G. G. 2005. Epigenetic effects on personality traits: early food provisioning and sibling competition. *Behaviour*, 142, 1329–1355.

Clutton-Brock, T. H., and Parker, G. A. 1995. Sexual coercion in animal societies. *Animal Behaviour*, 49, 1345–1365.

Costantini, D. 2005. Animal personalities in a competitive game theory context. *Ethology, Ecology, and Evolution*, 17, 279–281.

Croft, D. P., James, R., and Krause, J. 2008. *Exploring Animal Social Networks*. Princeton, NJ: Princeton University Press.

Croft, D. P., Krause, J., Darden, S. K., Ramnarine, W., Faria, J. J., and James, R. 2009. Behavioural trait assortment in a social network: patterns and implications. *Behavioral Ecology and Sociobiology*, 63, 1495–1503.

Cummings, M., and Mollaghan, D. 2006. Repeatability and consistency of female preference behaviours in a northern swordtail, *Xiphophorus nigrensis*. *Animal Behaviour*, 72, 217–224.

Dall, S. R. X., Houston, A. I., and McNamara, J. M. 2004. The behavioural ecology of personality: consistent individual differences from an adaptive perspective. *Ecology Letters*, 7, 734–739.

Davis, J. M., and Stamps, J. A. 2004. The effect of natal experience on habitat preferences. *Trends in Ecology and Evolution*, 19, 411–416.

DeWitt, T. J., Sih, A., and Wilson, D. S. 1998. Costs and limits of phenotypic plasticity. *Trends in Ecology and Evolution*, 13, 77–81.

Dingemanse, N. J., Both, C., Drent, P. J., and Tinbergen, J. M. 2004. Fitness consequences of avian personalities in a fluctuating environment. *Proceedings of the Royal Society of London B*, 271, 847–852.

Dingemanse, N. J., Kazem, A. J. N., Réale, D., and Wright, J. 2010. Behavioural reaction norms: animal personality meets individual plasticity. *Trends in Ecology and Evolution*, 25, 81–89.

Dingemanse, N. J., and Réale, D. 2005. Natural selection and animal personality. *Behaviour*, 142, 1159–1184.

Dingemanse, N. J., Thomas, D. K., Wright, J., Kazem, A. J. N., Koese, B., Hickling, R., and Dawnay, N. 2007. Behavioural syndromes differ predictably between twelve populations of three-spined stickleback. *Journal of Animal Ecology*, 76, 1128–1138.

Duckworth, R. A. 2006. Behavioral correlations across breeding contexts provide a mechanism for a cost of aggression. *Behavioral Ecology*, 17, 1011–1019.

Dugatkin, L. A., and Alfieri, M. 1991. Tit-for-tat in guppies: the relative nature of cooperation and defection during predator inspection. *Evolutionary Ecology*, 5, 300–309.

Dugatkin, L. A., and Reeve, H. K. 1998. *Game Theory and Animal Behavior*. Oxford: Oxford University Press.

Dukas, R. 1998. Constraints on neural processing and their effects on behavior. In: *Cognitive Ecology: The Evolutionary Ecology of Information Processing and Decision Making* (Dukas, R., ed.), pp. 89–127. Chicago: University of Chicago Press.

Emlen, D. J. 1997. Alternative reproductive tactics and male dimorphism in the horned beetle (Coleoptera: Scarabaeidae). *Behavioral Ecology and Sociobiology*, 41, 335–341.

Emlen, S. T., and Oring, L. W. 1977. Ecology, sexual selection, and the evolution of mating systems. *Science*, 197, 215–223.

Flack, J. C., Girvan, M., de Waal, F. B. M., and Krakauer, D. C. 2006. Policing stabilizes construction of social niches in primates. *Nature*, 439, 426–429.

Forstmeier, W. 2004. Female resistance to male seduction in zebra finches. *Animal Behaviour*, 68, 1005–1015.

———. 2007. Do individual females differ intrinsically in their propensity to engage in extra-pair copulation? *PLoS One*, 2, e952. doi:10.1371/journal.pone.0000952.

Forstmeier, W., and Birkhead, T. R. 2004. Repeatability of mate choice in the zebra finch: consistency within and between females. *Animal Behaviour*, 68, 1017–1028.

Gabor, C. R., and Aspbury, A. S. 2008. Nonrepeatable mate choice by male sailfin mollies, *Poecilia latipinna*, in a unisexual-bisexual mating complex. *Behavioral Ecology*, 19, 871–878.

Godin, J.-G. J., and Dugatkin, L.A. 1996. Female mating preference for bold males in the guppy, *Poecilia reticulata*. *Proceedings of the National Academy of Sciences U S A*, 93, 10262–10267.

Gosden, T. P., and Svensson, E. I. 2009. Density-dependent male mating harassment, female resistance, and male mimicry. *American Naturalist*, 173, 709–721.

Groothuis, T. G. G., and Carere, C. 2005. Avian personalities: characterization and epigenesis. *Neuroscience and Biobehavioral Reviews*, 29, 137–150.

Head, M. L., and Brooks, R. 2006. Sexual coercion and the opportunity for sexual selection in guppies. *Animal Behaviour*, 71, 515–522.

Heine, S. J., and Buchtel, E. E. 2009. Personality: the universal and the culturally specific. *Annual Review of Psychology*, 60, 369–394.

Hessing, M. J. C., Hagelso, A. M., Van Beek, J. A. M., Wiepkeme, P. R., Schouten, W. G. P., and Krukow, R. 1993. Individual behavioural characteristics in pigs. *Applied Animal Behaviour Science*, 37, 285–295.

Holveck M.-J., and Riebel K. 2010. Low-quality females prefer low-quality males when choosing a mate. *Proceedings of the Royal Society of London B*, 277, 153–160.

Huntingford, F. A. 1976. The relationship between antipredator behaviour and aggression among conspecifics in the three-spined stickleback. *Animal Behaviour*, 24, 245–260.

Jawer, M. 2005. Environmental sensitivity: a neurobiological phenomenon? *Seminars in Integrative Medicine*, 3, 104–109.

Johnson, J. C., and Sih, A. 2005. Precopulatory sexual cannibalism in fishing spiders (*Dolomedes triton*): a role for behavioral syndromes. *Behavioral Ecology Sociobiology*, 58, 390–396.

———. 2007. Fear, food, sex and parental care: a syndrome of boldness in the fishing spider, *Dolomedes triton*. *Animal Behaviour*, 74, 1131–1138.

Kokko, H., and Jennions, M. D. 2008. Parental investment, sexual selection, and sex ratios. *Journal of Evolutionary Biology*, 21, 919–948.

Koolhaas, J. M., Korte, S. M., De Boer, S. F., Van Der Vegt, B. J., Van Reenen, C. G., Hopster, H., De Jong, I. C., Ruis, M. A. W., and Blokhuis, H. J. 1999. Coping styles in animals: current status in behavior and stress-physiology. *Neuroscience and Biobehavioral Reviews*, 23, 925–935.

Koolhaas, J. M., De Boer, S. F., and Buwalda, B. 2006. Stress and adaptation. *Current Directions in Psychological Science*, 15, 109–112.

Lauer, M. J., Sih, A., and Krupa, J. J. 1996. Male density, female density, and intersexual conflict in a stream-dwelling insect. *Animal Behaviour*, 52, 929–939.

Lusseau, D., and Newman, M. E. J. 2004. Identifying the role that animals play in their social networks. *Proceedings of the Royal Society of London B*, 271, S477-S481.

Luttbeg, B., and Sih, A. 2010. Risk, resources and state-dependent adaptive behavioural syndromes. *Philosophical Transactions of the Royal Society of London B*, 365, 3977–3990.

Maynard Smith, J. 1982. *Evolution and the Theory of Games*. Cambridge: Cambridge University Press.

McDonald, D. B. 2007. Predicting fate from early connectivity in a social network. *Proceedings of the National Academy of Sciences U S A*, 104, 10910–10914.

McNamara, J. M., Stephens, P. A., Dall, S. R. X., and Houston, A. I. 2009. Evolution of trust and trustworthiness: social awareness favours personality differences. *Proceedings of the Royal Society of London B*, 276, 605–613.

Milinski, M. 1987. Tit-for-tat in sticklebacks and the evolution of cooperation. *Nature*, 325, 433–435.

Miller, G. F. 2007. Sexual selection for moral virtues. *Quarterly Review of Biology*, 82, 97–125.

Morris, M. R., Nicoletto, P. F., and Hesselman, E. 2003. A polymorphism in female preference for a polymorphic male trait in the swordtail fish *Xiphophorus cortezi*. *Animal Behaviour*, 65, 45–52.

Morris, M. R., Rios-Cardenas, O., and Tudor, M.S. 2006. Larger swordtail females prefer asymmetrical males. *Biology Letters*, 2, 8–11.

Mottley, K., and Giraldeau, L. A. 2000. Experimental evidence that group foragers can converge on predicted producer-scrounger equilibria. *Animal Behaviour*, 60, 341–350.

Nelson, X. J., Wilson, D. R., and Evans, C. S. 2009. Behavioral syndromes in stable social groups: an artifact of external constraints? *Ethology*, 114, 1154–1165.

Newman, M. E. J. 2003. The structure and function of complex networks. *SIAM Review*, 45, 167–256.

Nowak, M. A. 2006. Five rules for the evolution of cooperation. *Science*, 314, 1560–1563.

Ophir, A. G., Persaud, K. N., and Galef, B. G. 2005. Avoidance of relatively aggressive male Japanese quail (*Coturnix japonica*) by sexually experienced conspecific females. *Journal of Comparative Psychology*, 119, 3–7.

Øverli, O., Sorensen, C., Pulman, K. G. T., Pottinger, T. G., Korzan, W., Summers, C. H., and Nilsson, G. E. 2007. Evolutionary background for stress-coping styles: relationships between physiological, behavioral, and cognitive traits in nonmammalian vertebrates. *Neuroscience and Biobehavioral Reviews*, 31, 396–412.

Patricelli, G. L., Uy, J. A. C., Walsh, G., and Borgia, G. 2002. Male displays adjusted to female's response—macho courtship by the satin bowerbird is tempered to avoid frightening the female. *Nature*, 415, 279–280.

Patricelli, G. L., Coleman, S. W., and Borgia, G. 2006. Male satin bowerbirds, *Ptilonorhynchus violaceus*, adjust their display intensity in response to female startling: an experiment with robotic females. *Animal Behaviour*, 71, 49–59.

Penke, L., Denissen, J. J. A., and Miller, G. F. 2007. The evolutionary genetics of personality. *European Journal of Personality*, 21, 549–587.

Pike, T.W., Samanta, M., Lindstrom, J., and Royle, N.J. 2008. Behavioural phenotype affects social interactions in an animal network. *Proceedings of the Royal Society of London B*, 275, 2515–2520.

Réale, D., and Dingemanse, N. J. 2010. Personality and individual social specialisation. In: *Social Behaviour: Genes, Ecology and Evolution* (Szekely, T., Moore, A., and Komdeur, J., eds.), pp. 417–441. Cambridge: Cambridge University Press.

Riechert, S. E., and A. V. Hedrick. 1993. A test for correlations among fitness-linked behavioural traits in the spider *Agelenopsis aperta*. *Animal Behaviour*, 46, 669–675.

Rutte, C., and Taborsky, M. 2007. General reciprocity in rats. *PLoS Biology*, 7, 1421–1425.

Sapolsky, R. M. 2006. Social cultures among nonhuman primates. *Current Anthropology*, 47, 641–656.

Scott, J. 2000. *Social Network Analysis: A Handbook*. New York: Sage Publishing.

Shuster, S. M. 1989. Male alternative reproductive strategies in a marine isopod crustacean—the use of genetic markers to measure differences in fertilization success among alpha-males, beta-males, and gamma-males. *Evolution*, 43, 1683–1698.

Sih, A. 2011. Behavioral syndromes: a behavioral ecologist's view on the evolutionary and ecological implications of animal personalities. In: *Personality and Temperament in Nonhuman Primates* (Weiss, A., King, J. E., and Murray, L., eds.), pp. 313–336. New York: Springer.

Sih, A., and Bell, A. M. 2008. Insights for behavioral ecology from behavioral syndromes. *Advances in the Study of Behavior*, 38, 227–281.

Sih, A., Bell, A. M., and Johnson, J. C. 2004a. Behavioral syndromes: an ecological and evolutionary overview. *Trends in Ecology and Evolution*, 19, 372–378.

Sih, A., Bell, A. M., Johnson, J. C., and Ziemba, R. 2004b. Behavioral syndromes: an integrative overview. *Quarterly Review of Biology*, 79, 241–277.

Sih, A., Hanser, S. F., McHugh, K. A. 2009. Social network theory: new insights and issues for behavioral ecologists. *Behavioral Ecology and Sociobiology*, 63, 975–988.

Sih, A., Kats, L. B., and Maurer, E. F. 2003. Behavioral correlations across situations and the evolution of antipredator behaviour in a sunfish-salamander system. *Animal Behaviour*, 65, 29–44.

Sih, A., and Krupa, J. J. 1995. Interacting effects of predation risk, sex ratio, and density on male/female conflicts and mating dynamics of stream water striders. *Behavioral Ecology*, 6, 316–325.

Sih, A., Lauer, M., and Krupa, J. J. 2002. Path analysis and relative importance of male-female conflict, female choice, and male-male competition in water striders. *Animal Behaviour*, 63, 1079–1089.

Sih, A., and Watters, J. V. 2005. The mix matters: behavioural types and group dynamics in water striders. *Behaviour*, 142, 1417–1431.

Sinervo, B., and Calsbeek, R. 2006. The developmental, physiological, neural, and genetical causes and consequences of frequency-dependent selection in the wild. *Annual Review of Ecology and Systematics*, 37, 581–610.

Sinervo, B., and Lively, C. M. 1996. The rock-paper-scissors game and the evolution of alternative male strategies. *Nature*, 380, 240–243.

Spoon, T. R., Millam, J. R., and Owings, D. H. 2006. The importance of mate behavioural compatibility in parenting and reproductive success by cockatiels, *Nymphicus hollandicus*. *Animal Behaviour*, 71, 315–326.

Stamps, J. A. 2007. Growth-mortality tradeoffs and "personality traits" in animals. *Ecology Letters*, 10, 355–363.

Suomi, S. J. 1987. Genetic and maternal contributions to individual differences in rhesus monkey biobehavioral development. In: *Perinatal Development: A Psychobiological Perspective* (Krasnegor, N., Blass, E., Hofer, M., and Smotherman, W., eds.), pp. 397–420. New York: Academic Press.

Uher, J. 2011. Personality in nonhuman primates: what can we learn from human personality psychology? In: *Personality and Temperament in Nonhuman Primates* (Weiss, A., King, J. E., and Murray, L., eds.), pp. 41–76. New York: Springer.

Van Erp-Van der Kooij, E., Kuijpers, A. H., Van Eerdenburg, F. J. C. M., Dieleman, S. J., Blankenstein, D. M., and Tielen, M. J. M. 2003. Individual behavioural characteristics in pigs—influences of group composition but no differences in cortisol responses. *Physiology and Behavior*, 78, 479–488.

Van Oers, K., Drent, P. J., Dingemanse, N. J., and Kempenaers, B. 2008. Personality is associated with extra-pair paternity in great tits, *Parus major*. *Animal Behaviour*, 76, 555–563.

Warner, R. R., and Hoffman, S. G. 1980. Population density and the economics of territorial defense in a coral reef fish. *Ecology*, 61, 772–780.

Watters, J. V. 2005. Can the alternative male tactics "fighter" and "sneaker" be considered "coercer" and "cooperator" in Coho salmon? *Animal Behaviour*, 70, 1055–1062.

Watters, J. V., and Meehan, C. L. 2007. Different strokes: can managing behavioral types increase postrelease success? *Applied Animal Behaviour Science*, 102, 364–379.

Wcislo, W. T. 1989. Behavioral environments and evolutionary change. *Annual Review of Ecology and Systematics*, 20, 137–169.

West-Eberhard, M. J. 2003. *Developmental Plasticity and Evolution*. Oxford: Oxford University Press.

Wey, T., Blumstein, D. T., Shen, W., and Jordan, W. 2008. Social network analysis of animal behaviour: a promising tool for the study of sociality. *Animal Behaviour*, 75, 333–344.

Wilson, D. S., and Wilson, E.O. 2007. Rethinking the theoretical foundation of sociobiology. *Quarterly Review of Biology*, 82, 327–348.

Wilson, D. S., and Yoshimura, J. 1994. On the coexistence of specialists and generalists. *American Naturalist*, 144, 692–707.

Wingfield, J. C., Hegner, R. E., Dufty, A. M., and Ball, G. F. 1990. The challenge hypothesis—theoretical implications for patterns of testosterone secretion, mating systems, and breeding strategies. *American Naturalist*, 136, 829–846.

Wolf, J. B., Brodie, E.D., and Moore, A. J. 1999. Interacting phenotypes and the evolutionary process, II: selection resulting from social interactions. *American Naturalist*, 153, 254–266.

Wolf, M., Van Doorn, G. S., and Weissing, F. J. 2008. Evolutionary emergence of responsive and unresponsive personalities. *Proceedings of the National Academy of Sciences U S A*, 105, 15825–15830.

9

The Evolution of Animal Personalities

MAX WOLF, G. SANDER VAN DOORN,

OLOF LEIMAR, AND FRANZ J. WEISSING

Introduction

In many animal species, individuals of the same sex, age, and size differ consistently in whole suites of correlated behavioral tendencies, comparable with human personalities (Clark and Ehlinger 1987; Digman 1990; Gosling 2001; Sih et al. 2004a). Individual birds, for example, often differ consistently in the way they explore their environment, and these differences are associated with differences in boldness and aggressiveness (Groothuis and Carere 2005). In rodents such as mice and rats, individuals differ consistently in the way they deal with environmental challenges, and such differences encompass exploration, attack, avoidance, and nest-building behavior (Koolhaas et al. 1999). Interestingly, personality differences are often associated with morphological (Ehlinger and Wilson 1988), physiological (Korte et al. 2005), and cognitive (Reddon and Hurd 2009) differences among individuals (see also chapter 12). In this chapter we focus on the evolutionary causes of animal personalities (Wilson 1998; Buss and Greiling 1999; Dall et al. 2004). What are the factors promoting the evolution of personalities? And how do these factors shape the structure (what type of traits are associated with each other?) and the ontogenetic stability of personalities?

Understanding the evolution of animal personalities (henceforth personalities) requires a shift in our thinking about animal behavior (Wilson 1998). While behavioral ecologists have traditionally "atomized" the organism into single behavioral traits that are studied in isolation (Gould and Lewontin 1979), the study of personalities requires a more holistic approach for at least two reasons. First, personalities refer to suites of correlated traits (Sih et al. 2004a, b) that are stable across part of the ontogeny of individuals (different authors use different criteria for such stability, ranging from weeks to years, Sih and Bell 2008). Consequently, interdependencies between multiple different traits (e.g., the relationship between the boldness of an individual and its aggressiveness and exploration behavior) and the same trait expressed at different points during ontogeny (e.g., the relationship between juvenile and adult aggressiveness) have to be taken into account. Second, on a proximate level, trait correlations are often caused by genetic (Mackay 2004), hormonal (Ketterson and Nolan 1999), or cognitive (Rolls 2000) mechanisms affecting multiple traits at the same time. In order to understand the evolution of such trait correlations, we need an approach that integrates mechanisms and adaptation (Tinbergen 1963).

Personalities refer to differences in suites of correlated behavioral traits that are stable over part of the ontogeny of individuals. In some cases, personality differences are associated with differences in state (McNamara and Houston 1996), that is, with differences in the morphological (Ehlinger and Wilson 1988), physiological (Koolhaas et al. 1999), or cognitive (Howard et al. 1992) characteristics or with environmental conditions (Wilson 1998) that individuals face. Some state differences are readily observable (e.g., differences in size, sex, or position in dominance hierarchy) while others are less conspicuous (e.g., differences in nutrition, stress responsiveness, or experience).

The state of an individual may affect the cost and benefits of its actions, and thus its optimal behavior (McNamara and Houston 1996; Houston and McNamara 1999; Clark and Mangel 2000). In such cases, individuals benefit from adjusting their behavior to their current state by expressing state-dependent behavior (condition-dependent behavior, phenotypic plasticity). Importantly for personalities, single states often affect the costs and benefits of multiple behavioral traits at the same time (McNamara and Houston 1996). Differences in states thus provide a potentially powerful explanation for differences in suites of correlated behavioral traits.

Explaining personalities in terms of differences in state, however, requires us to provide answers to two basic questions. First, why do individuals differ in states in the first place? In many cases, the maintenance of such differences seems puzzling. Why, for example, should individuals differ in physiological characteristics such as stress responsiveness (Aron and Aron 1997; Koolhaas et al. 1999) or basal metabolic rate (Careau et al. 2008)? Second, why are such differences stable over time? Many states (e.g., energy reserves, experience, parental investment received, future fitness expectation) are affected by many different factors, including an individual's own behavior. Such states are potentially highly variable over time. Why, then, should differences in states be stable over time and what are the mechanisms that give rise to such stability?

Differences in state provide a plausible explanation for personality differences (Dall et al. 2004; Sih and Bell 2008), but only a partial one. Personality differences can also be observed for individuals that do not seem to differ in states relevant to the observed behavioral differences. In fact, many empirical studies that report personality differences control for state differences among individuals (Verbeek et al. 1994; Dingemanse et al. 2002; 2007; Bell 2007). The observation of personalities among individuals that do not differ in states is particularly puzzling. First, why should individuals differ in their response to the same problem (e.g., how to explore the environment) when facing identical costs and benefits associated with behavioral actions? Should we not rather expect that any given problem has a unique optimal behavioral solution? Second, why are the responses to different problems sometimes correlated with each other (Clark and Ehlinger 1987; Digman 1990; Gosling 2001; Sih et al. 2004a)? Why should a bold individual, for example, be more aggressive than a shy conspecific, and why should bold individuals tend to remain bold throughout ontogeny? Such limited plasticity seems especially surprising since behavior, in contrast to many morphological features, is often thought to be highly plastic (but see DeWitt et al. 1998), and such plasticity would indeed seem to be advantageous (Wilson 1998; Dall et al. 2004).

These are the questions we address in this chapter. We will organize our discussion under the two main themes of variation and correlation, reflecting two main aspects of personalities. We first focus on the causes of variation within populations. In particular, we discuss how random causes, frequency-dependent selection, spatiotemporal variation in the environment, and non-equilibrium dynamics can give rise to variation in behavior and states underlying behavior. We then focus on the two types of behavioral correlations that define personalities, that is, correlations over time and across contexts. In particular, we will discuss the role of the architecture of behavior, stable state differences, and social conventions in causing stable behavioral correlations.

Causes of variation

Individuals can differ substantially in their behavioral responses when confronted with the same problem (e.g., how to explore the environment, how to respond to a predator), and such behavioral variation is a key feature of personalities (Clark and Ehlinger 1987; Digman 1990; Gosling 2001; Sih et al. 2004a). Behavioral variation can take various forms. In some cases, variation has a broad unimodal distribution (e.g., as in the case of variation in many human personality traits, Nettle 2006), while in other situations a small number of discrete variants coexist (e.g., variation in mating strategies,

Gross 1996). As discussed above, behavioral variation may or may not be associated with variation in states among individuals. Moreover, behavioral variation may or may not be associated with genetic variation (Wilson 1994; Bouchard and Loehlin 2001).

In this section we focus on the ultimate causes of the emergence and maintenance of behavioral variation. Since our focus is on adaptive behavioral differences, we will not discuss processes that give rise to nonadaptive behavioral variation such as, for example, mutation. As we have emphasized above, adaptive behavioral variation is often caused by variation in states and state-dependent behavior, and we briefly discuss the two main routes to state differences among individuals, evolved vs. random state differences. We then discuss two basic and well-studied mechanisms that can give rise to adaptive variation in behavior, frequency-dependent selection and spatiotemporal variation in the environment. We conclude this section by discussing how non-equilibrium dynamics can give rise to variation among individuals.

DIFFERENCES IN STATES

State differences among individuals are ubiquitous: pick any two individuals within a population and typically you find that these individuals differ in some aspects of their morphological, physiological, cognitive, or environmental condition. These state differences are an important source of adaptive behavioral differences. In many situations, aspects of the state of an individual are under the direct control of that individual. Individuals typically have, for example, the option to choose among different environmental conditions (e.g., habitats, social environments) or the opportunity to fine-tune certain aspects of their physiology (e.g., stress responsive system, basal metabolic rate). Why should individuals that do not initially differ in states make different decisions? At first sight, one would perhaps expect that there is one optimal option among different states. State differences among individuals, however, need not always reflect adaptation. In many situations, aspects of the state of an individual are affected by factors that are not under the individual's control. Differences in states arise whenever these factors differ between individuals. One individual, for example, grows up in a rich environment while another grows up in a poor environment. One individual finds a high-quality food source and thus increases its nutritional condition while another one does not find such a food source. One individual gets infected by a parasite while another one does not. Examples of state differences caused by such random factors abound.

In groups of foraging animals, individuals typically have the choice between two different behavioral roles (Giraldeau and Beauchamp 1999): either search for food sources on their own ("producers") or exploit food sources discovered by others ("scroungers"). In this case, the benefits associated with a phenotype depend on the frequency of that phenotype in the population (Barnard and Sibly 1981): the higher the frequency of scroungers in a group, the less beneficial this role becomes, since more scroungers compete for fewer resources. Such situations give rise to so-called negative frequency-dependent selection (Maynard Smith 1982), a form of selection that is known to be an important source of variation within populations (Heino et al. 1998; Dugatkin and Reeve 2000; Sinervo and Calsbeek 2006).

In situations with negative frequency-dependence selection, selection acts to increase the frequency of rare phenotypes within populations. In its simplest form, as in the producer-scrounger example above (Barnard and Sibly 1981), this rarity advantage gives rise to two phenotypes that coexist in stable frequencies within a population. Moreover, whenever individuals have the choice between more than two phenotypes, negative frequency dependence can give rise to situations in which any number of phenotypes can coexist in stable frequencies. When negative frequency dependence interacts with positive frequency dependence, as for example in the so-called rock-paper-scissors games (Maynard Smith 1982), selection can give rise to a dynamic equilibrium in which multiple phenotypes coexist at continuously changing frequencies (Sinervo and Lively 1996; Sinervo and Calsbeek 2006).

Negative frequency-dependent selection can, as in the producer-scrounger situation, give rise to adaptive behavioral variation among individuals that initially do not differ in states. The same process can also give rise to adaptive state differences among individuals. For example, the benefits of a particular physiological or cognitive architecture (e.g., a particular level of stress responsiveness, a particular learning rule) might depend on how common this architecture is in the population, thus promoting the coexistence of different architectures (Wolf et al. 2008). Similarly, the benefits of being in a particular environment (e.g., territory, habitat, or social position) might depend on the frequency with which other individuals choose this environment, thus promoting the coexistence of individuals in different environmental states (Ens et al. 1995; Wilson 1998).

Negative frequency-dependent selection is a common phenomenon in social interactions (Maynard Smith 1982, Svensson and Sheldon 1998; Dugatkin and Reeve 2000). It occurs, for example, in interactions between individuals in which adopting a different phenotype confers an advantage over the interacting partner. Such situations occur in agonistic interactions, as for example hawk-dove–like encounters (Maynard Smith 1982), in which the aggressive hawk strategy is beneficial whenever the opponent plays dove, whereas the nonaggressive dove strategy is beneficial whenever the opponent plays hawk. They also occur in more cooperative interactions, when social partners benefit from diversifying into different behavioral roles that complement each other (Clark and Ehlinger 1987). The benefits of such behavioral complementarity can be caused by various mechanisms. Choosing different behavioral roles may help, for example, to avoid competition between partners, to reap the benefits of behavioral specialization, or to reduce the risk associated with a certain strategy.

Negative frequency dependence can also be caused more indirectly (Kokko and Lopez-Sepulcre 2007) via competition for different types of resources that have density-dependent benefits (i.e., the benefits of a resource decrease with the number of individuals that compete for that type of resource). Such density dependence gives rise to negative frequency dependence: the more individuals compete for a particular type of resource (e.g., a territory, habitat, mate), the less beneficial it becomes. Density-dependent resource competition can thus promote the coexistence of individuals that exploit different resources (Wilson 1998).

Negative frequency-dependent selection can thus give rise to adaptive variation in states and/or behavior among individuals. In principle, this variation might or might not be associated with genetic variation (Maynard Smith 1982; Wilson 1994). Consider, for example, a situation in which two phenotypes coexist with frequencies of 30% and 70% (e.g., producer and scroungers, individuals with a low and a high basal metabolic rate, or slow and fast learners). This phenotypic variation can arise in a population of genetically identical individuals that adopt each of the phenotypes randomly but with the same probability (choose one phenotype in 30%, the other in 70% of the cases), as for example in the case of environmental sex determination, where mixed broods arise despite the fact that individuals do not differ genetically with respect to sex determination. However, variation can also arise in a genetically polymorphic population in which a fixed proportion of individuals adopts each of the phenotypes (30% of the individuals choose one phenotype, 70% the other). Individual differences in foraging behavior in the larvae of the fruitfly (*Drosophila melanogaster*) provide a good example of the latter situation (Fitzpatrick et al. 2007). In natural populations, a dimorphism in foraging strategies can be observed ("rover" vs. "sitter" individuals). This dimorphism is based on a single major gene polymorphism that is maintained by negative frequency-dependent selection; both the rover and the sitter allele attain their highest relative fitness when rare in the population.

Natural selection shapes the phenotype of individuals to match their environment, and in many natural situations, the environment and thus the optimal phenotype varies in space or in time. What is the expected evolutionary outcome in such a situation? In particular, should we expect that, as it is often thought to be the case (Nettle 2006; Koolhaas et al. 2007; Penke et al. 2007), environmental variation promotes phenotypic variation within populations? And if so, should we expect that such phenotypic variation is associated with genetic variation? The answers to these questions depend on the situation (Hedrick 1976; 1986; Seger and Brockmann 1987; Moran 1992; Leimar 2005) and, in particular, on whether the population faces spatial or temporal variation in the environment (for an alternative classification, see Frank and Slatkin 1990 and Donaldson-Matasci et al. 2008) and on how well individuals can match their phenotype to their environment.

To understand the importance of phenotype-environment matching, consider first a situation in which individuals can match their phenotype to their environment in an error- and cost-free manner, be it via habitat choice, habitat tracking and limited migration (i.e., the environment is chosen to match the phenotype), phenotypic plasticity (i.e., the phenotype is chosen to match the environment), or a combination of these processes. In such a situation evolution is expected to result in perfect phenotype-environment matching. No variation is maintained within each environment. This example is certainly extreme and unrealistic (DeWitt et al. 1998); it illustrates, however, that environmental variation can give rise to phenotypic variation only within environments in situations with limited phenotype-environment matching.

Consider now the most basic scenario of spatial variation. A population inhabits an environment with two types of habitats, so that different phenotypes are favored within each habitat. As we have just seen, if perfect phenotype-environment matching is possible, no variation within environments can be maintained. This is different in situations in which there is an intermediate degree of phenotype-environment matching, that is, in situations where habitat choice (or habitat tracking) or phenotypic plasticity is possible but not perfect. Individuals might, for example, make errors when choosing habitats. In such situations, phenotypic variation can be maintained both at a population level and within each habitat (Seger and Brockmann 1987). The reason for this is as follows. Since some degree of phenotype-environment matching is possible, coexisting genotypes (or phenotypes of a plastic genotype) experience different environments, and each genotype will, on average, experience more often the environment in which it is favored. Variation within environments arises since phenotype-environment matching is not perfect. The resulting phenotypic variation can in principle be due to plasticity, genetic polymorphism, or a combination of both factors.

A good example of adaptive variation caused by spatial variation in the environment is provided by the bluegill sunfish (*Lepomis macrochirus*) that inhabit North American freshwater lakes (Ehlinger and Wilson 1988). In these populations, consistent individual differences in foraging tactics (e.g., hover duration, pattern of movement) have been described. It turns out that these differences can be associated with differences in habitat use: the most efficient foraging tactic depends on whether an individual is in littoral or open water zones, and individuals that employ different tactics are preferentially (but not exclusively, thus phenotype-environment matching is not perfect) found in the habitat that fits their foraging tactic best. Interestingly, differences in foraging tactics are associated with rather subtle morphological differences (e.g., fin size, fin placement) between individuals, which again tend to favor one habitat over the other. We will return to this point below when discussing the causes of consistency. Spatial variation need not correspond to differences in the abiotic environment, as above, but can be induced by variation in the biotic environment of individuals. It has been suggested, for example, that variation in boldness within animal species can be maintained by the fact that different habitats vary in their degree of risk (Wilson 1998).

In addition to spatial variation, temporal variation in environmental conditions has also been suggested as contributing to the maintenance of personality differences. Contrary to the intuition of many biologists, however, it is not always the case that temporal variation will result in the coexistence of different strategies. Exactly as with spatial variation, the evolutionary effects of temporal fluctuations depend on population regulation, the degree of phenotype-environment matching, the costs of plasticity, and many other details of the biological system under consideration. To grasp this, consider a simple scenario of temporal fluctuations. Within a generation, all individuals within a population face the same environment, but the environment varies across generations, and different environments favor different phenotypes. As we have seen above, whenever individuals can adjust their phenotype to their current environment in an error- and cost-free manner, no variation is maintained within environments. However, unlike for spatial variation, genetic variation cannot be maintained in situations with limited phenotype-environment matching, at least as long as generations are non-overlapping (see the discussion of bet-hedging below for how purely phenotypic variation can be maintained in such a scenario). This is because all genotypes face exactly the same environment, there is no frequency dependence, and among any number of potential genotypes there will always be one that achieves the highest (geometric mean) fitness (Seger and Brockmann 1987). Many species, however, are iteroparous and have overlapping generations; in this situation then, temporal fluctuations can maintain genetic polymorphisms (Ellner and Hairston 1994). As we model species where the lifetime of individuals (and thus the generation overlap) progressively increase, temporal fluctuations tend to average out within the lifetime of a single individual and the model comes closer and closer to a temporal analogue of a spatial model without habitat choice, which can maintain genetic polymorphisms through local density dependence (Levene 1953).

A well-known feature of temporally fluctuating environments, with or without overlapping generations, is that so-called bet-hedging genotypes are selectively favored (Seger and Brockmann 1987); these are genotypes that switch during development stochastically between two or more phenotypes. A single bet-hedging genotype thus gives rise to a mixture of phenotypes (e.g., aggressive and nonaggressive individuals, individuals with a low and a high stress responsiveness; see Bergmüller and Taborsky 2010). This can be seen as a risk-spreading strategy, since no matter how the environment turns out, some of the bet-hedging phenotypes are well adapted. A diversifying bet-hedger can reduce its variance in fitness in an optimal way and thereby increase its geometric mean fitness. Bet-hedging can thus explain the coexistence of different personalities; however, since the variation caused by bet-hedging is only phenotypic (i.e., all phenotypes have the same genotype), bet-hedging alone cannot account for the observation that personalities are moderately heritable (Bouchard and Loehlin 2001).

In summary, in species with non-overlapping generations, temporal fluctuations can maintain only phenotypic variation. In species with overlapping generations, genetic variation can be maintained as well, but bet-hedging strategies are selectively favored (Leimar 2005). As should have been clear from the above discussion, however, the mere presence of temporal variation is by no means sufficient for explaining variation. To give an example, Dingemanse and colleagues (2004) studied a population of great tits (*Parus major*) for which environmental conditions (masting of beeches) varied across years. They found that different behavioral types were favored depending on the environmental condition, which in turn explained the maintenance of variation in this population (for other examples in which temporal variation in environmental conditions may explain personality variation, see Réale and Festa-Bianchet 2003 and chapter 7).

NON-EQUILIBRIUM DYNAMICS

Until now our analysis has been based on the premise that natural selection gives rise to an equilibrium in which strategies coexist in stable frequencies. It is not at all clear, however, that the dynamics of selection will lead to such an equilibrium. There are plenty of examples where the dynamics of frequency-dependent selection (e.g., Weissing 1991), competition (e.g., Huisman and Weissing 1999), and sexual selection (e.g., Van Doorn and Weissing 2006) lead to oscillations and often even chaotic dynamics. This has important implications since non-equilibrium conditions generally have a much higher potential for maintaining variation than the long-term constancy induced by equilibrium conditions (e.g., Huisman et al. 2001; Van Doorn and Weissing 2006).

A good example for non-equilibrium coexistence is the covariation of dispersal and colonizing ability observed in many species (e.g., Chitty 1967; Duckworth and Badyaev 2007). In such species, some individuals disperse while others are philopatric, potentially reflecting a bet-hedging strategy. Dispersers typically have a phenotype that allows them to colonize unoccupied space, but this same phenotype is selectively disadvantageous under crowded conditions (e.g., Duckworth and Kruuk 2009). Such a "colonizer" phenotype could probably not persist under constant and stable equilibrium conditions. In a perturbed environment, however, where empty spaces are created once in a while, the colonizers can flourish because they can exploit these opportunities. Once the empty spaces are filled, however, the settlers succumb to their own success, since they create an environment that can be more efficiently exploited by alternative phenotypes that do better under crowded conditions. At each point in space, there is ongoing directional selection, but in the population as a whole both types of strategies can stably coexist (Duckworth and Badyaev 2007).

Causes of correlations

Up to now we have focused on the causes of behavioral differences in response to environmental problems (e.g., how to explore an environment, how to respond to a predator). Personalities, however, entail much more than simple behavioral differences among individuals (Clark and Ehlinger 1987; Digman 1990; Gosling 2001; Sih et al. 2004a). First, personalities refer to behavioral differences that are stable through part of the ontogeny of individuals (time-consistency of behavior), that is, individuals that score relatively high (low) in a given behavioral situation often tend to score relatively high (low) in the same situation at later points in time. Second, personalities refer to differences that extend to whole suites of correlated behaviors, that is, correlated variation in functionally different contexts (e.g., antipredator behavior is correlated with agonistic behavior). Both types of correlations indicate behavioral inflexibilities (Wilson 1998; Dall et al. 2004) in the sense that the behavior that an individual exhibits at one point in time and in one particular context is predictive of the same individual's behavior at later points in time and in different contexts. Why did evolution give rise to such behavioral inflexibilities when a flexible structure of behavior would seem to be more advantageous? To answer this question, we first need to explore why, in some cases, evolution gives rise to architectures of behavior that result in apparently maladaptive behavioral correlations. We then focus on state variables as a cause of behavioral correlations in general and discuss two main sources for the stability of state differences over time, that is, stable state differences and positive feedback mechanisms. We conclude this section by discussing how social conventions can give rise to adaptive behavioral correlations.

ARCHITECTURE OF BEHAVIOR

On a proximate level, the behavioral phenotype of an individual is affected by its architecture of behavior, that is, the genetic, physiological, neurobiological, and cognitive systems underlying behavior. This architecture, in turn, gives rise to behavioral correlations whenever multiple traits are affected by a common underlying mechanism. Such common mechanisms are ubiquitous. Examples include pleiotropic genes (Mackay 2004), hormones (Ketterson and Nolan 1999), and neurotransmitters (Bond 2001) that affect multiple traits at the same time; emotions (Rolls 2000); and simple behavioral rules that are used for a variety of different but related problems (Todd and Gigerenzer 2000).

It has been shown, for example, that the consistency of aggressiveness through ontogeny in the three-spined stickleback (*Gasterosteus aculeatus*) is caused by pleiotropic genes (Bakker 1986). Pleiotropic genes are also thought to be responsible (Riechert and Hedrick 1993; Maupin and Riechert 2001) for the positive correlation between agonistic behavior, antipredator behavior, and superfluous killing in an American desert spider (*Agelenopsis aperta*). The negative correlation between mating effort and parental effort in several bird species is mediated by the hormone testosterone (McGlothlin et al. 2007). And finally, the fearfulness of an individual affects its reaction to a range of potentially threatening situations, including persistent dangers in its habitat, novelty, and interactions with conspecifics (Boissy 1995).

Behavioral correlations can thus be the result of a relatively rigid architecture of behavior. The resulting behavioral associations appear adaptive in some cases (see below). In others, however, they give rise to apparently maladaptive behaviors (Sih et. al 2004b). It might, for example, be advantageous for a female spider to show high levels of aggression toward territorial intruders, but she might also kill and consume all potential mates during courtship and as a consequence be left unmated at the time of egg laying (Arnqvist and Henriksson 1997). Similarly, it might be advantageous for salamander larvae to be active in the absence of predatory cues but not advantageous if the larvae are also active in the presence of such cues (Sih et al. 2003). In other words, rigid behavioral architectures can explain behavioral correlations at the proximate level, but from an ultimate perspective, one is tempted to ask why such rigid behavioral architectures persist over evolutionary time. Especially in cases where a rigid architecture gives rise to apparently maladaptive behavior, one would expect evolution to uncouple unfavorable behavioral associations.

The evolution of a more flexible behavioral architecture might in principle be prevented for two types of reasons. First, a more flexible architecture might be advantageous (i.e., an individual with such an architecture would achieve a higher fitness than an individual with a more rigid architecture) but not attainable by evolution. Such a situation can occur because the evolutionary transition from one complex phenotype (here: rigid architecture) to another complex phenotype (more flexible architecture) is typically not possible in one step but requires several intermediate steps. A more flexible architecture of behavior might, for example, require a novel hormonal system that cannot directly (i.e., with a small number of mutations) emerge from the present hormonal system. The architectures associated with these intermediate hormonal systems, however, might be disadvantageous to the individual. In other words, the evolution of a more flexible behavioral architecture might be prevented by the crossing of an adaptive valley of the fitness landscape. Such a situation occurs if the current behavioral correlations correspond to a local peak in the fitness landscape, reflecting the fact that the involved traits are to some extent well adapted to each other.

Second, a more flexible behavioral architecture might not be advantageous. In such situations, evolution is expected to give rise to adaptive behavioral canalization (e.g., West-Eberhard 2003; Edgell et al. 2009). Such a situation can occur if a more flexible architecture is associated with costs (costs of plasticity such as the costs associated with the acquisition of information, DeWitt et al. 1998) that are not outweighed by the corresponding benefits. In both cases, correlations can persist even though they give rise to behavioral traits that, when viewed in isolation from their architectural basis, might appear maladaptive.

STABLE STATE VARIABLES

Many aspects of an individual's state affect the cost and benefits of multiple behavioral traits at the same time (McNamara and Houston 1996). State differences in combination with state-dependent behavior (condition-dependent behavior, phenotypic plasticity) thus provide an explanation for adaptive behavioral correlations of apparently unrelated behavioral traits. However, state differences do not immediately explain why individuals should be consistent over time. Put differently, why should initial state differences among individuals be relatively stable over time? In this section we discuss two main determinants of consistency of states, inherently stable state variables and positive feedback mechanisms between state and behavior.

Inherently stable state variables. As discussed above, random causes, frequency-dependent selection, and spatiotemporal variation can give rise to populations with variation in states among individuals. Whenever a change of state among these variants is associated with substantial costs, such situations can result in consistent differences in state and, consequently, consistent differences in whole suites of (state-dependent) traits that are affected by this state. In some situations, differences in states are associated with differences in morphological and physiological characteristics that are costly to change. Consider, for example, sex differences. In many animal species, frequency-dependent selection maintains two sexes at constant proportions within populations. These equilibrium proportions, however, can in principle be maintained in populations in which individuals change their sex over time. Such a sex change, however, is often associated with substantial costs to the individual (caused by the necessary morphological and physiological changes). We thus typically observe stable (life-long) sex differences among individuals, which are, in turn, associated with whole suites of correlated behavioral traits. In humans (Costa et al. 2001), for example, women typically score higher than males on traits related to the agreeableness axis (e.g., cooperativeness, empathy, trust), while in many other animal species, sex differences exist in parental care and courtship behavior (Kelley 1988).

In some situations, evolution gives rise to populations in which individuals are distributed into a small number of discrete size classes (Brockmann 2001). A change among these variants is typically associated with substantial costs, which in turn favors consistency in size and thus consistency in behavioral traits that are affected by body size. A common phenomenon, for example, is the use of fighting and sneaking as alternative mating tactics depending on body size, as observed in dung beetles, bees, and many other species (Gross 1996).

Morphological and physiological differences, which can be changed only with substantial costs, need not be as conspicuous as in the case of sex or size differences. As discussed above (Ehlinger and Wilson 1988), within populations of bluegill sunfish (*Lepomis macrochirus*) individuals differ in morphological characteristics that are functional either in the littoral or in open water zone (e.g., fin size, fin placement). Such (stable) differences are associated with consistent differences in behavioral traits such as foraging tactics. Interestingly, the underlying morphological differences are not obvious to an observer; in fact, sunfish have been studied for many years without any recognition of the adaptive nature of these differences (Wilson 1998).

Inherently stable state differences need not be associated with morphological and physiological characteristics that are costly to change. In some cases stability is caused, at least in part, by factors external to the individual. Human societies, for example, encompass a large diversity of professions (e.g., teachers, managers, bureaucrats). Although it is in principle possible for an individual to change its profession, such changes are typically very costly to the individual (e.g., in terms of required training or education). As a result, individuals typically stick to their profession, once chosen. Differences in professions, in turn, are often associated with consistent differences in suites of correlated behavioral traits. Human leaders, for example, are more extrovert and more conscientious than nonleaders (Judge et al. 2002); entrepreneurs are more conscientious and open, but less neurotic and agreeable, than managers (Zhao and Seibert 2006).

Self-reinforcing feedback loops. Many aspects of an individual's state are much more labile than the ones discussed above. Consider, for example, the energy reserves of an individual, the experience that an individual has with a certain situation, or the future fitness expectations of an individual. These states are, like many others, labile since they can easily be changed by many different factors, including the individual's own behavior. Labile states can, like the inherently stable states, affect multiple behavioral traits at the same time, thus explaining suites of correlated behavioral traits. But why should labile states be stable over time?

In some situations the state and behavior of individuals are linked by a positive feedback (Sih et al. 2004a, b; Wolf et al. 2008): initial state differences give rise to differences in behavior, which act to stabilize or even increase the initial state differences. Such positive feedback mechanisms, in turn, can give rise to consistent individual differences in labile states and behavioral traits that are associated with these states. An important positive feedback is the feedback between behavior and an individual's experience. Individuals often get better at certain activities with increased experience (Rosenzweig and Bennett 1996); in other words, learning, training, and skill formation reduce the costs or increase the benefits of the same action when this action is repeated, which in turn favors consistency in this behavior (Wolf et al. 2008). Positive feedbacks via experience can give rise to consistent individual differences in single behavioral traits. Animals often learn how to recognize predators (Griffin 2004), which in turn makes it less costly to explore and forage a risky habitat for individuals that did this before. Under such conditions, whenever variation in risk-taking behavior is maintained within populations (Wilson 1998; Wolf et al. 2007), positive feedback acts to promote consistent individual differences in risk-taking behavior. Positive feedbacks can also give rise to consistent differences in suites of correlated behavioral traits. Experience gained in one context, for example, can affect the cost and benefits of behavioral actions in another context and thus give rise to a cross-context association of behavioral traits. Individuals that learn to assess the strength of conspecific competitors might, for example, at the same time get better at assessing the risk associated with predators.

Positive feedback does not need to act via behavior directly. The costs and benefits of behavioral traits that are related to resource acquisition (e.g., aggression, boldness), for example, often depend on an individual's characteristics, such as its resource-holding potential, and this interaction can give rise to a positive feedback loop (Sih and Bell 2008). An individual's physical strength, for example, enhances its fighting ability, which in turn may result in more resources being available to the individual (e.g., access to food and better nutrition) that further enhance its strength.

Positive feedback can also act via physiological characteristics of the individual. It has been suggested, for example, that in many animal species, deviations from a once-chosen growth rate are costly to the individual (Stamps 2007; Biro and Stamps 2008). Compensatory growth, for example, often comes at the cost of increased risk of disease, higher mortality rates, or decreased physiological capacity later in life (Mangel and Munch 2005). Similarly, deviations from a once-chosen basic metabolic rate or stress responsiveness might be costly to the individual. In such situations, a once-chosen physiological characteristic (growth rate, basic metabolic rate, stress responsiveness) affects the costs and benefits of future physiological characteristics such that maintaining a once-chosen set-up is advantageous. This, in turn, favors consistency in suites of traits that are associated with these characteristics. Differences in growth rates, for example, affect the cost and benefits of many traits that are related to food intake, such as aggression and boldness (Stamps 2007).

SOCIAL CONVENTIONS

An adaptive association of different behaviors does not necessarily reflect an underlying state variable that affects the costs and benefits of these behaviors. Rather, behavioral traits (be it the same trait expressed at different points during ontogeny or different traits) can get associated with each other because individuals in a population follow an adaptive behavioral rule (or convention) that favors the association of these traits. When an individual is confronted with another individual in a social interaction, for example, its behavior may be dependent on the behavioral history of the other individual. For example, when A and B interact in a hawk-dove–like encounter (Maynard Smith 1982), individuals might follow the rule "if the opponent played hawk before, choose dove; otherwise choose hawk." Such an eavesdropping strategy (Johnstone 2001) makes sense whenever there is some consistency in the behavior of individuals. Just as consistency favors eavesdropping, so eavesdropping favors consistency whenever it is beneficial for individuals to be predictable (but see Dall et al. 2004; McNamara et al. 2008). This interaction between consistency and eavesdropping can thus give rise to populations in which individuals follow a behavioral rule that favors consistency and, as a result, individuals show consistent behavior.

Do such conventions arise in natural situations? The so-called winner-loser effect (Chase et al. 1994) might be a good example of that. It is well known that winners of previous contests are more likely to win again (and losers are more likely to lose again), even against different opponents and in situations in which there are no asymmetries between the opponents. According to a survey across several taxa (Rutte et al. 2006), when there are no other asymmetries between opponents, the probability of winning a subsequent contest is almost doubled for previous winners, but is reduced more than five times for previous losers even against different opponents. Winner-loser effects are currently not well understood, but one possible explanation is that an individual's prior success acts as a "random historical asymmetry" that is used to settle the conflict (Parker 1974; Maynard Smith and Parker 1976; Hammerstein 1981; Van Doorn et al. 2003a, b). Social conventions that favor consistency are not limited to aggressive interactions.

An influential idea in explaining cooperation is that individuals should make their behavior dependent on an image score of the other individual (Nowak and Sigmund 1998; Leimar and Hammerstein 2001), a measure of how cooperative the other individuals has been in the past. Such image scoring can, in turn, favor consistency in cooperative behavior.

Conclusions

In this chapter, we have provided an overview of the mechanisms that promote the evolution of personalities. We focused on two basic questions: which factors promote adaptive behavioral variation within populations, and which factors promote adaptive correlations between behavioral traits across contexts or over time (table 9.1).

Factors promoting variation and correlations can interact in various ways (figure 9.1). Random causes, frequency-dependent selection, spatiotemporal variation, and non-equilibrium dynamics can give rise to variation in states among individuals. Whenever such states affect the cost and benefits of behavioral traits, state differences promote state-dependent behavior and thus adaptive behavioral variation; whenever the costs and benefits of multiple behavioral traits are affected at the same time, the result is adaptive behavioral variation that is correlated across contexts. In principle, behavioral variation should be stable as long as the underlying variation in states is stable. Whenever states are very costly to change and thus inherently stable (e.g., sex and size differences), variation is expected to be stable. In the case of labile states (e.g., energetic state, experience with a certain behavior), variation can be stabilized via positive feedback mechanisms between behavior and states: variation in states gives rise to variation in behavior that acts to stabilize or increase initial state differences.

Spatiotemporal variation, frequency-dependent selection, and non-equilibrium dynamics can give rise to behavioral variation that is not associated with underlying state differences and state-dependent behavior (e.g., producers and scroungers, hawks and doves). Such variation can be stabilized over time via positive feedback mechanisms: the initial variation in behavior gives rise to differences in states that act to stabilize the behavioral differences (e.g., producers gain experience that makes producing more beneficial). Such feedbacks can also extend to multiple behavioral traits, thus giving rise to adaptive behavioral correlations (e.g., experience gained in the producer-scrounger context can affect the cost and benefits of behavior in different contexts). Alternatively, behavioral variation in single

Table 9.1. Causes of adaptive behavioral variation, of adaptive behavioral correlations across contexts, and of adaptive behavioral consistency over time.

(a) Causes of adaptive behavioral variation
- State differences[1]
- Frequency-dependent selection
- Spatiotemporal variation
- Non-equilibrium dynamics

(b) Causes of adaptive behavioral correlations across contexts
- Architecture of behavior
- States[1] affecting behavior in multiple contexts
- Social conventions favoring the association of traits

(c) Causes of adaptive time-consistency in behavior
- Architecture of behavior
- Inherently stable states[1]
- Labile states[1] that are stabilized via positive feedbacks
- Social conventions favoring time-consistency

[1] In combination with state-dependent behavior.

or multiple behavioral traits can be stabilized via social conventions, which favor consistency in behavior.

The analysis presented here suggests that personalities, and animal behavior in general, require a holistic approach to be fully understood. Rather than "atomizing the organism" (Gould and Lewontin 1979) into single behavioral traits that are studied in isolation, studying multiple different traits in concert is necessary: behavior in one context (e.g., antipredator, mating, fighting, parental care) can often be understood only when taking the interdependencies with past and future behavior in the same and other contexts into account. The understanding of personalities also requires an integration of mechanism and adaptation (Tinbergen 1963). As we have emphasized above, behavioral correlations are often caused by the architecture of behavior, that is, by the genetic, physiological, neurobiological, and cognitive systems underlying behavior. In order to understand such correlations, we thus have to understand the coevolution of behavior and its underlying mechanisms. Finally, personalities in themselves may have consequences for the evolutionary process (Wilson 1998; Dall et al. 2004). For example, in social contexts, the existence of variation in one trait often selects for variation in another trait. Variation in cooperativeness, for example, can select

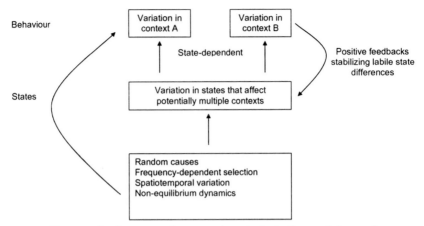

Figure 9.1. Random causes, frequency-dependent selection, spatiotemporal variation, and non-equilibrium dynamics can act on states or directly on behavior. Differences in states in combination with state-dependent behavior can give rise to adaptive behavioral differences. Behavioral differences, in turn, can be stabilized via positive feedback mechanisms between behavior and states.

for variation in choosiness, which in turn selects for cooperativeness (Mc-Namara et al. 2008). Variation in cooperativeness can thus trigger a coevolutionary process of cooperativeness and choosiness that gives rise to very different evolutionary outcomes (here: high levels of cooperation) when compared with situations without such initial variation.

References

Arnqvist, G., and Henriksson, S. 1997. Sexual cannibalism in the fishing spider and a model for the evolution of sexual cannibalism based on genetic constraints. *Evolutionary Ecology*, 11, 255–273.

Aron, E. N., and Aron, A. 1997. Sensory-processing sensitivity and its relation to introversion and emotionality. *Journal of Personality and Social Psychology*, 73, 345–368.

Bakker, T. C. M. 1986. Aggressiveness in sticklebacks (*Gasterosteus aculeatus*): a behavior-genetic study. *Behaviour*, 98, 1–144.

Barnard, C. J., and Sibly, R. M. 1981. Producers and scroungers: a general model and its application to captive flocks of house sparrows. *Animal Behaviour*, 29, 543–550.

Bell, A. M. 2007. Future directions in behavioural syndromes research. *Proceedings of the Royal Society of London B*, 274, 755–761.

Bergmüller, R., and Taborsky, M. 2010. Animal personality due to social niche specialisation. *Trends in Ecology and Evolution*, 25, 504–511.

Biro, P. A., and Stamps, J. A. 2008. Are animal personality traits linked to life-history productivity? *Trends in Ecology and Evolution*, 23, 361–368.

Boissy, A. 1995. Fear and fearfulness in animals. *Quarterly Review of Biology*, 70, 165–191.

Bond, A. J. 2001. Neurotransmitters, temperament, and social functioning. *European Neuropsychopharmacology*, 11, 261–274.

Bouchard, T. J., and Loehlin, J. C. 2001. Genes, evolution, and personality. *Behavior Genetics*, 31, 243–273.

Brockmann, H. J. 2001. The evolution of alternative strategies and tactics. *Advances in the Study of Behavior*, 30, 1–51.

Buss, D. M., and Greiling, H. 1999. Adaptive individual differences. *Journal of Personality*, 67, 209–243.

Careau, V., Thomas, D., Humphries, M. M., and Réale, D. 2008. Energy metabolism and animal personality. *Oikos*, 117, 641–653.

Chase, I. D., Bartolomeo, C., and Dugatkin L. A. 1994. Aggressive interactions and intercontest interval: how long do winners keep winning? *Animal Behaviour*, 48, 393–400.

Chitty, D. 1967. The natural selection of self-regulatory behaviour in animal populations. *Proceedings of the Ecological Society of Australia*, 2, 51–78.

Clark, A. B., and Ehlinger, T. J. 1987. Pattern and adaptation in individual behavioral differences. In: *Perspectives in Ethology* (Bateson, P. P. G., and Klopfer, P. H., eds.), pp. 1–47. New York: Plenum Press.

Clark, C.W., and Mangel, M. 2000. *Dynamic State Variable Models in Ecology*. Oxford: Oxford University Press.

Costa, P. T., Terraciano, A., and McCrae, R. 2001. Gender differences in personality traits across cultures: robust and surprising findings. *Journal of Personality and Social Psychology*, 81, 322–331.

Dall, S. R. X., Houston, A. I., and McNamara, J. M. 2004. The behavioural ecology of personality: consistent individual differences from an adaptive perspective. *Ecology Letters*, 7, 734–739.

DeWitt, T. J., Sih, A., and Wilson, D. S. 1998. Costs and limits of phenotypic plasticity. *Trends in Ecology and Evolution*, 13, 77–81.

Digman, J. M. 1990. Personality structure: emergence of the Five-Factor Model. *Annual Review of Psychology*, 41, 417–440.

Dingemanse, N. J., Both, C., Drent, P. J., Van Oers, K., and Van Noordwijk, A. J. 2002. Repeatability and heritability of exploratory behaviour in great tits from the wild. *Animal Behaviour*, 64, 929–938.

Dingemanse, N. J., Both, C., Drent, P. J., and Tinbergen, J. M. 2004. Fitness consequences of avian personalities in a fluctuating environment. *Proceedings of the Royal Society of London B*, 271, 847–852.

Dingemanse, N. J., Wright, J., Kazem, A. J. N., Thomas, D. K., Hickling, R., Dawnay, N. 2007. Behavioural syndromes differ predictably between 12 populations of three-spined stickleback. *Journal of Animal Ecology*, 76, 1128–1138.

Donaldson-Matasci, M. C., Lachmann, M., and Bergstrom C. T. 2008. Phenotypic diversity as an adaptation to environmental uncertainty. *Evolutionary Ecology Research*, 10, 493–515.

Duckworth, R. A., and Badyaev, A. V. 2007. Coupling of dispersal and aggression facilitates the rapid range expansion of a passerine bird. *Proceedings of the National Academy of Sciences U S A*, 104, 15017–15022.

Duckworth, R. A., and Kruuk, L. E. B. 2009. Evolution of genetic integration between dispersal and colonization ability in a bird. *Evolution*, 63, 968–977.

Dugatkin, L. A., and Reeve, H. K. 2000. *Game Theory and Animal Behavior*. Oxford: Oxford University Press.

Edgell, T. C., Lynch, B. R., Trussell, G. C., and Palmer, R. A. 2009. Canalization in natural populations. *American Naturalist*, 174, 434–440.

Ehlinger, T. J., and Wilson, D. S. 1988. Complex foraging polymorphism in bluegill sunfish. *Proceedings of the National Academy of Sciences U S A*, 85, 1878–1882.

Ellner, S., and Hairston, N. G. 1994. Role of overlapping generations in maintaining genetic variation in a fluctuating environment. *American Naturalist*, 143, 403–417.

Ens, B. J., Weissing, F. J., and Drent, R. H. 1995. The despotic distribution and deferred maturity: two sides of the same coin. *American Naturalist*, 146, 625–650.

Fitzpatrick, M. J., Feder, E., Rowe, L., and Sokolowski, M. B. 2007. Maintaining a behaviour polymorphism by frequency-dependent selection on a single gene. *Nature*, 447, 210–212.

Frank, S. A., and Slatkin, M. 1990. Evolution in a variable environment. *American Naturalist*, 136, 244–260.

Giraldeau, L. A., and Beauchamp, G. 1999. Food exploitation: searching for the optimal joining policy. *Trends in Ecology and Evolution*, 14, 102–106.

Gosling, S. D. 2001. From mice to men: what can we learn about personality from animal research? *Psychological Bulletin*, 127, 45–86.

Gould, S. J., and Lewontin, R. C. 1979. Spandrels of San Marco and the Panglossian paradigm—a critique of the adaptationist program. *Proceedings of the Royal Society of London B*, 205, 581–598.

Griffin, A. S. 2004. Social learning about predators: a review and prospectus. *Learning and Behavior*, 32, 131–140.

Groothuis, T. G. G., and Carere, C. 2005. Avian personalities: characterization and epigenesis. *Neuroscience and Biobehavioral Reviews*, 29, 137–150.

Gross, M. R. 1996. Alternative reproductive strategies: diversity within the sexes. *Trends in Ecology and Evolution*, 11, 92–98.

Hammerstein, P. 1981. The role of asymmetries in animal contests. *Animal Behaviour*, 29, 193–205.

Hedrick, P. W. 1976. Genetic variation in a heterogeneous environment, II: temporal heterogeneity and directional selection. *Genetics*, 84, 145–50.

———. 1986. Genetic polymorphism in heterogeneous environments: a decade later. *Annual Review of Ecology and Systematics*, 17, 535–66.

Heino, M., Metz, J. A. J., and Kaitala, V. 1998. The enigma of frequency-dependent selection. *Trends in Ecology and Evolution*, 13, 367–370.

Houston, A. I., and McNamara, J. M. 1999. *Models of Adaptive Behaviour*. Cambridge: Cambridge University Press.

Howard, R., Fenwick, P., Brown, D., and Norton, R. 1992. Relationship between CNV asymmetries and individual differences in cognitive performance, personality, and gender. *International Journal of Psychophysiology*, 13, 191–197.

Huisman, J., and Weissing, F. J. 1999. Biodiversity of plankton by species oscillations and chaos. *Nature*, 402, 407–410.

Huisman, J., Johansson, A. M., Folmer, E. O., and Weissing, F. J. 2001. Towards a solution of the plankton paradox: the importance of physiology and life history. *Ecology Letters*, 4, 408–411.

Johnstone, R. A. 2001. Eavesdropping and animal conflict. *Proceedings of the National Academy of Sciences U S A*, 98, 9177–9180.

Judge, T. A., Bono, J. E., Ilies, R., and Gerhardt, M. W. 2002. Personality and leadership: a qualitative and quantitative review. *Journal of Applied Psychology*, 87, 765–780.

Kelley, D. B. 1988. Sexually dimorphic behaviors. *Annual Review of Neuroscience*, 11, 225–251.

Ketterson, E. D., and Nolan, V. 1999. Adaptation, exaptation, and constraint: a hormonal perspective. *American Naturalist*, 154, S4–S25.

Kokko, H., and Lopez-Sepulcre, A. 2007. The ecogenetic link between demography and evolution: can we bridge the gap between theory and data? *Ecology Letters*, 10, 773–782.

Koolhaas, J. M., Korte, S. M., De Boer, S. F., Van Der Vegt, B. J., Van Reenen, C. G., Hopster, H., De Jong, I. C., Ruis, M. A. W., and Blokhuis, H. J. 1999. Coping styles in animals: current status in behavior and stress-physiology. *Neuroscience and Biobehavioral Reviews*, 23, 925–935.

Koolhaas, J. M., De Boer, S. F., Buwalda, B., and Van Reenen, K. 2007. Individual variation in coping with stress: a multidimensional approach of ultimate and proximate mechanisms. *Brain, Behavior, and Evolution*, 70, 218–226.

Korte, S. M., Koolhaas, J. M., Wingfield, J. C., and McEwen, B. S. 2005. The Darwinian concept of stress: benefits of allostasis and costs of allostatic load and the trade-offs in health and disease. *Neuroscience and Biobehavioral Reviews*, 29, 3–38.

Leimar, O. 2005. The evolution of phenotypic polymorphism: randomized strategies versus evolutionary branching. *American Naturalist*, 165, 669–681.

Leimar, O., and Hammerstein, P. 2001. Evolution of cooperation through indirect reciprocity. *Proceedings of the Royal Society of London B*, 268, 745–753.

Levene, H. 1953. Genetic equilibrium when more than one ecological niche is available. *American Naturalist*, 87, 331–33.

Mackay, T. F. C. 2004. The genetic architecture of quantitative traits: lessons from *Drosophila*. *Current Opinions in Genetics and Development*, 14, 253–257.

Mangel, M., and Munch, S. B. 2005. A life-history perspective on short- and long-term consequences of compensatory growth. *American Naturalist*, 166, E155-E176.

Maupin, J. L., and Riechert, S. E. 2001. Superfluous killing in spiders: a consequence of adaptation to food-limited environments? *Behavioral Ecology*, 12, 569–576.

Maynard Smith, J. 1982. *Evolution and the Theory of Games*. Cambridge: Cambridge University Press.

Maynard Smith, J., and Parker, G. A. 1976. The logic of asymmetric contests. *Animal Behaviour*, 24, 159–175.

McGlothlin, J. W., Jawor, J. M., and Ketterson, E. D. 2007. Natural variation in a testosterone-mediated trade-off between mating effort and parental effort. *American Naturalist*, 170, 864–875.

McNamara, J. M., and Houston, A. I. 1996. State-dependent life-histories. *Nature*, 380, 215–211.

McNamara, J. M., Barta, Z., Fromhage, L., and Houston, A. I. 2008. The coevolution of choosiness and cooperation. *Nature*, 451, 189–192.

Moran, N. A. 1992. The evolutionary maintenance of alternative phenotypes. *American Naturalist*, 139, 971–989.

Nettle, D. 2006. The evolution of personality variation in humans and other animals. *American Psychologist*, 61, 622–631.

Nowak, M. A., and Sigmund, K. 1998. Evolution of indirect reciprocity by image scoring. *Nature*, 393, 573–577.

Parker, G. A. 1974. Assessment strategy and the evolution of fighting behaviour. *Journal of Theoretical Biology*, 47, 223–243.

Penke, L., Dennissen, J. J. A., and Miller, G. F. 2007. The evolutionary genetics of personality. *European Journal of Personality*, 21, 549–587.

Réale, D., and Festa-Bianchet, M. 2003. Predator-induced natural selection on temperament in bighorn ewes. *Animal Behaviour*, 65, 463–470.

Reddon, A. P., and Hurd, P. L. 2009. Individual differences in cerebral lateralization are associated with shy-bold variation in the convict cichlid. *Animal Behaviour*, 77, 189–193.

Riechert, S. E., and Hedrick, A. V. 1993. A test of correlations among fitness-related behavioral traits in the spider, *Agelenopsis aperta* (Araneae, Agelinadae). *Animal Behaviour*, 46, 669–675.

Rolls, E. T. 2000. Précis of the brain and emotion. *Behavioural and Brain Sciences*, 23, 177–233.

Rosenzweig, M.R., and Bennett, E. L. 1996. Psychobiology of plasticity: effects of training and experience on brain and behavior. *Behavioural Brain Research*, 78, 57–65.

Rutte, C., Taborsky, M., and Brinkhof, M. W. G. 2006. What sets the odds of winning and losing? *Trends in Ecology and Evolution*, 21, 16–21.

Seger, J., and Brockmann, H. J. 1987. What is bet-hedging? In: *Oxford Surveys in Evolutionary Biology* (vol. 4, Harvey, P., and Partridge, L., eds.), pp. 182–211. Oxford: Oxford University Press.

Sih, A., Kats, L. B., and Maurer, E. F. 2003. Behavioural correlations across situations and the evolution of antipredator behaviour in a sunfish-salamander system. *Animal Behaviour*, 65, 29–44.

Sih, A., Bell, A. M., Johnson, J. C., and Ziemba, R. 2004a. Behavioral syndromes: an integrative overview. *Quarterly Review of Biology*, 79, 241–277.

Sih, A., Bell, A. M., and Johnson, J. C. 2004b. Behavioral syndromes: an ecological and evolutionary overview. *Trends in Ecology and Evolution*, 19, 372–378.

Sih, A., and Bell, A. M. 2008. Insights for behavioral ecology from behavioral syndromes. *Advances in the Study of Behavior*, 38, 227–281.

Sinervo, B., and Calsbeek, R. 2006. The developmental, physiological, neural, and genetical causes and consequences of frequency-dependent selection in the wild. *Annual Review of Ecology and Systematics*, 37, 581–610.

Sinervo, B., and Lively, C. 1996. The rock-paper-scissors game and the evolution of alternative male strategies. *Nature*, 380, 240–43.

Stamps, J. A. 2007. Growth-mortality tradeoffs and "personality" traits in animals. *Ecology Letters*, 10, 355–363.

Svensson, E., and Sheldon, B. C. 1998. The social context of life-history evolution. *Oikos*, 83, 466–477.

Tinbergen, N. 1963. On aims and methods of ethology. *Zeitschrift fuer Tierpsychologie*, 20, 410–433.

Todd, P. M., and Gigerenzer, G. 2000. Precis of simple heuristics that make us smart. *Behavioral and Brain Sciences*, 23, 727–780.

Van Doorn, G. S., Hengeveld, G. M., and Weissing, F. J. 2003a. The evolution of social dominance, I: two-player models. *Behaviour*, 140, 1305–1332.

Van Doorn, G. S., Hengeveld, G. M., and Weissing, F. J. 2003b. The evolution of social dominance, II: multiplayer models. *Behaviour*, 140, 1333–1358.

Van Doorn, G. S., and Weissing, F. J. 2006. Sexual conflict and the evolution of female preferences for indicators of male quality. *American Naturalist*, 168, 743–757.

Verbeek, M. E. M., Drent, P. J., and Wiepkema, P. R. 1994. Consistent individual differences in early exploratory behavior of male great tits. *Animal Behaviour*, 48, 1113–1121.

Weissing, F. J. 1991. Evolutionary stability and dynamic stability in a class of evolutionary normal form games. In: *Evolution and Game Dynamics*. Vol. 1 of *Game Equilibrium Models*. (Selten, R., ed.), pp. 29–97. Berlin: Springer-Verlag.

West-Eberhard, M. J. 2003. *Developmental Plasticity and Evolution*. Oxford: Oxford University Press.

Wilson, D. S. 1994. Adaptive genetic variation and human evolutionary psychology. *Ethology and Sociobiology*, 15, 219–235.

———. 1998. Adaptive individual differences within single populations. *Philosophical Transactions of the Royal Society of London B*, 353, 199–205.

Wolf M., Van Doorn G. S., Leimar O., and Weissing F. J. 2007. Life-history trade-offs favour the evolution of animal personalities. *Nature*, 447, 581–584.

Wolf M., Van Doorn G. S., and Weissing F. J. 2008. Evolutionary emergence of responsive and unresponsive personalities. *Proceedings of the National Academy of Sciences U S A*, 105, 15825–15830.

Zhao, H., and Seibert, S. E. 2006. The Big Five personality dimensions and entrepreneurial status: a meta-analytical review. *Journal of Applied Psychology*, 91, 259–271.

III

DEVELOPMENT OF PERSONALITIES AND

THEIR UNDERLYING MECHANISMS

10

Ontogeny of Stable Individual Differences

Gene, Environment, and Epigenetic Mechanisms

JAMES P. CURLEY AND IGOR BRANCHI

Introduction

Stable individual differences in behavioral responses occur within species and can be observed even when comparing same-sex and same-age individuals within the same population. A behavioral response is considered as stable when, though its absolute value might change across different conditions, the rank order of individuals displaying that behavioral response does not. For instance, some individuals are more bold, aggressive, or active than others independently from the situation to which they are exposed. In addition, behavioral responses are often associated with each other. For instance, boldness and aggression have been reported to be positively correlated in selected animal populations (Bell and Stamps 2004), meaning that those individuals that are bold are usually also aggressive (Sih et al. 2004a, b). The associations among behaviors, which can be considered as "packages" reflecting how an individual deals with environmental challenges, have been referred to as behavioral syndromes, temperaments, or personalities (for reviews see Gosling, 2001; 2008; Sih et al. 2004a, b; Groothuis and Carere 2005; Bell 2007; Koolhaas et al. 2007; Réale et al. 2007).

There have been extensive studies of stable individual differences in behavior across a wide variety of species, including invertebrates (Maupin and Riechert 2001; Brodin and Johansson 2004; chapter 1), reptiles (Stapley and Keogh 2005), birds (Drent et al. 2003; Carere et al. 2005; chapter 3), and mammals (Benus et al. 1991; Koolhaas et al. 1999; chapter 4). Among these studies, those performed on laboratory rodents provide an important contribution to our understanding of the origins of these differences (see also chapter 15). The extensive knowledge of the genetics, physiology, and behavior of rats and mice permits a more mechanistic approach to the study of variations in behavior. Indeed, studies of laboratory rodents have provided a different perspective from ecological studies and, within the rigorously controlled conditions offered by the laboratory setting, a number of issues concerning stable individual differences can be elucidated. These range from the identification of underlying neurobiological mechanisms to the characterization of the role played by genetic and environmental factors during ontogeny.

In this chapter we will discuss the mechanisms through which stable individual differences in neurobiology and behavior emerge during development. In particular, we will focus on evidence from studies of laboratory rodents, which illustrate the role of the early environment in shaping stable variation in gene expression and behavior. This discussion includes maternal effects (see also chapter 11), the influence of peer interactions during infancy, and the juvenile period of development. In addition, we will discuss the role of gene-environment interactions and epigenetic mechanisms in mediating the variation in behavior that emerges across the life span and accounts for stable characteristics within an individual. Finally, we will briefly describe the ontogeny of behavioral responses in an evolutionary framework to better understand the adaptive value of plasticity in the development of these individual differences.

Influence of early life experiences

MOTHER-OFFSPRING INTERACTIONS

In mammals, from the time of conception until weaning, the primary physical, nutritional, and social environment for the majority of young is centered upon the mother. For the entire duration of this intimate relationship the development of the offspring's brain and behavior is intrinsically linked to the cues that they receive about their external environment from this primary caregiver. In this section, we shall discuss how variations in the quality of care received throughout this period have long-term consequences for the emergence of stable individual differences in behavior in offspring.

Events occurring at birth, and even before fertilization, are able to exert long-term effects on offspring phenotype. The generation of oocytes occurs over several years, and levels of gene transcription and translation are very high prior to oocyte maturation in order to "stockpile" proteins and mRNA that are essential for early zygotic development (Wassarman and Kinloch 1992). Until the 2-cell embryo stage, zygote genes are silenced through chromatin-mediated suppression and all protein synthesis is regulated solely via maternal mRNAs (Nothias et al. 1995; De Sousa et al. 1998). Significantly, this oocyte gene expression is altered according to changes in maternal phenotype such as maternal nutrition and age, and may underlie subsequent developmental changes in the offspring associated with these environmental conditions (Janny and Menezo 1996; Eichenlaub-Ritter 1998; Symonds et al. 2001).

Following implantation of the zygote into the uterine wall and the production of the placenta, the embryo places increasing metabolic demands on the mother, who provides oxygen, nutrients, waste removal, and protection from toxins and other harmful chemicals such as high levels of stress hormone (Weissgerber and Wolfe 2006). As this is also a period of extensive early embryonic brain development, stress experienced by the mother during gestation has long-term consequences for behavioral development. One experimental paradigm that has been used extensively to study the mechanisms of such prenatal maternal effects is the exposure of pregnant rodent dam to psychosocial stressors (e.g., overcrowding or changing the social composition of cages) and physical stressors (e.g., exposing the dams to loud noises or restraining them in a tube from which they are unable to move) (Maccari and Morley-Fletcher 2007). Following such an exposure, the maternal hypothalamic-pituitary-adrenal (HPA) axis is activated and glucocorticoids are released into the maternal bloodstream from the adrenal cortex. Although the placenta does contain enzymes such as 11-β-hydroxysteroid dehydrogenase-2 (11-βHSD-2), which are able to inactivate glucocorticoids, particularly high levels of stress hormones may overwhelm the capacity of the enzymatic conversion and prevent the buffering of the developing embryo (Edwards et al. 1996; Seckl 2004). Offspring born to dams who have undergone these prenatal stressors exhibit long-term stable changes in phenotype, including being more hyperactive, neophobic, and impaired on cognitive tasks; showing enhanced locomotor activity following administration of amphetamine, morphine, and cocaine; as well as exhibiting deficits in social behavior (Weinstock et al. 1988; Patin et al. 2005). These changes in behavior are dependent upon the type, timing, and intensity of the stressor as well as the offspring's gender, and are accompanied by permanent changes in various neuroendocrine systems of offspring exposed prenatally to stress including increased plasma corticosterone (Stohr et al. 1998), elevated CRH (corticotrophin-releasing hormone) mRNA in the amygdala, lower monoamine and catecholamine turnover, altered estrogen and androgen receptor distributions, and impaired dopaminergic and glutamergic functioning (Kaiser and Sachser 2005; Darnaudery and Maccari 2008; Weinstock 2008). It must be recognized, however, that while fetal exposure to prenatal maternal glucocorticoid secretion is certainly a mediator of these effects, alterations in maternal behavior postnatally as a result of the gestational maternal stress may also result in similar changes to offspring development (Moore and Power 1986; Champagne and Meaney 2006).

Variation in the ways that postparturient rodent mothers interact and care for their young has long been recognized as a critical mediator of stable individual differences in offspring behavior. One of the simplest models used to study the influence of postpartum mother-infant contact is the maternal separation paradigm in which dams are separated from pups for varying durations (Cirulli et al. 2003; 2009; Pryce and Feldon 2003). At the extreme, in artificial rearing studies, pups are entirely removed from their mother on day 3 postpartum and then reared in complete social isolation (Hall 1975). Under this complete postnatal maternal deprivation, adult offspring are more fearful (e.g., they make fewer open-arm entries in an elevated plus maze), hyperactive, and cognitively impaired, and exhibit deficits in social behavior (Gonzalez et al. 2001; Gonzalez and Fleming 2002; Lovic and Fleming 2004). Artificially reared female offspring also exhibit impairments in maternal care when they are adults, displaying reduced maternal licking/ grooming and other forms of contact with their own pups (Fleming et al. 2002).

The paradigm most commonly used, especially to investigate the developmental origins of stress reactivity, usually termed *maternal deprivation*, involves separation of dams from their pups daily for at least three hours per day starting after birth and lasting until about day 14 postpartum. In addition to decreasing the overall level of maternal care (Macrì and Würbel 2006; Boccia et al. 2007), this procedure also induces short-term distress in the pups, indicated by their increased corticosterone release and ultrasonic vocalizations (Hofer et al. 1993; Lehmann et al. 2002). Generally, offspring reared in this manner exhibit increased neophobia, decreased exploration, elevated CRH mRNA in the paraventricular nucleus (PVN), an increased corticosterone response after stress, and less hippocampal glucocorticoid receptor mRNA (Plotsky and Meaney 1993; Meaney et al. 1996; Lehmann et al. 1999). Offspring also show other long-term behavioral and neuroendocrine changes such as impairments in cognitive ability (e.g., they show increased latencies on the Morris Water Maze) and decreased hippocampal synaptophysin levels (Lehmann et al. 2002). Maternally separated offspring have also been shown to exhibit increased locomotor activity and self-administration of addictive drugs such as cocaine, ethanol, and amphetamine (Meaney et al. 2002; Brake et al. 2004; Kikusui et al. 2005; Moffett et al. 2007).

In an alternative separation paradigm, pups are removed daily from the dam for 10–20 minutes before being returned to the home-cage. In this instance, the reunion of pups and dam is associated with increased levels of maternal care such as licking/grooming and nursing (Levine 1957; Lee and Williams 1974; Liu et al. 1997) This procedure is commonly referred to as *neonatal handling* or *brief maternal separation* and has typically been found to lead to the opposite effects of the extended maternal separation paradigm. For instance, handled offspring typically exhibit an attenuated response to stress, elevated levels of hippocampal glucocorticoid receptor (GR) mRNA, improved cognitive ability, increased levels of social behavior, and reduced behavioral responses to drugs of addiction (Meaney and Aitken 1985; Meaney et al. 1989; 1991; Lehmann et al. 2002). In addition to handling and maternal separation, various other procedures using a variety of separation durations and schedules have shown long-term stable changes in offspring behavior and neurobiology. However, a major criticism of these separation paradigms, and especially of maternal deprivation, is that findings are often inconsistent and appear to be dependent upon many factors, such as the duration and timing of the separation as well the condition and environment of the pups and dam during the separations (Pryce and Feldon 2003; Macrì and Würbel 2006; Cirulli et al. 2009). Moreover, it is not entirely clear exactly which aspect of maternal separation (altered maternal care, elevated pup stress, or the reduction in warmth or nutrition experienced by pups) is responsible for the induced alterations in offspring phenotype (Macrì and Würbel 2007).

Recently, more ethologically relevant studies of individual differences in postpartum maternal care have been used to advance our understanding of how more subtle variations in early life experiences can permanently alter pup brain development and behavior. During late pregnancy, females are exposed to increasing levels of hormones such as estrogen, oxytocin, and prolactin that prime the maternal brain to allow for postpartum maternal care. These maternal behaviors include nest building, retrieval, nursing, licking and grooming of pups, and maternal defense of the nest (Rosenblatt 1967; Fleming and Rosenblatt 1974; Rosenblatt 1975; Gammie 2005; Shoji and Kato 2006). The retrieval of pups, the grouping of them into a nest, and the crouching over the litter by the female is critical in allowing for the provision of heat to these altricial pups (Croskerry et al. 1978; Leon et al. 1978) and permits pups to access the dam's nipples to initiate suckling (Stern and Johnson 1990). The licking and grooming of pups, particularly the anogenital region, stimulates urination and defecation (Rosenblatt and Lehrman 1963), increases the motor activity of pups enhancing their ability to attach to nipples, and also serves to regulate brain and body temperature (Sullivan and Hall 1988; Sullivan et al. 1988). For the mother, licking the pups' urine provides a mechanism to reclaim salt and water that have been lost through lactation (Gubernick and Alberts 1983; 1985). Lactating females will also show other behaviors, such as an increase in aggressive behavior toward nonfamiliar intruders to the nest site (Lonstein and Gammie 2002; Gammie 2005).

Although all dams must show these behaviors for their offspring to survive and develop, not all dams show them to the same degree and frequency. In particular, it has been shown that lactating dams, both within and between rodent strains, differ during the first week postpartum in the levels of nursing and licking/grooming of pups and that these differences shape offspring phenotype (Champagne and Meaney 2007). For instance, the differences in adult blood pressure between offspring of spontaneously hypertensive (SHR) and Wistar Kyoto (WKY) rats are due not just to genetic differences but in part to differences in maternal licking/grooming, retrieval of pups, and nursing posture exhibited by these two strains (Myers et al. 1989). Moreover, the phenotype of offspring of different mouse strains can be shifted via cross-fostering when the strains differ in their levels of maternal care (Francis et al. 2003; Priebe et al. 2005). However, the role of individual differences in one aspect of maternal care, maternal licking/grooming, in modulating offspring gene expression, neurobiology, and behavior has been explored most thoroughly in the studies of Michael Meaney and colleagues in Long Evans rats (Fish et al. 2004).

Female Long Evans rats can be characterized as being high or low in licking/grooming (LG) behavior on the basis of the mean frequency of this behavior during the first week postpartum. This characterization is very stable, as dams will continue to exhibit high or low LG to their subsequent litters (Champagne et al. 2003). By cross-fostering pups between high and low-LG dams, it is possible to investigate how individual differences in this postpartum care shape offspring behavioral traits such as exploration, fear, anxiety, cognition, reward-seeking, and social behavior. For instance, compared with male pups reared by mothers who exhibit low levels of LG, male pups reared by mothers who are high-LG are more exploratory in a novel environment and have a number of associated neural changes. These high-LG offspring have reduced plasma adrenocorticotropin and corticosterone in response to stress, increased hippocampal glucocorticoid receptor mRNA, lower levels of hypothalamic CRH mRNA, an increased density of benzodiazepine receptors in the amygdala, and an altered pattern of GABA subunit receptor distribution (Liu et al. 1997; 2000; Francis et al. 1999; Caldji et al. 2000). Adult male offspring of high-LG dams also show improved abilities on cognitive tests of spatial learning and memory compared with male offspring of low-LG dams. Consistent with this, high-LG male offspring have higher hippocampal brain-derived neurotrophic factor (BDNF) mRNA, as well as elevated levels of choline acetyltransferase and synaptophysin, both of which regulate efficient synaptic neurotransmission (Liu et al. 2000). There are also changes in sensitivity in the dopaminergic system, with sons of low-LG dams being more seeking of rewards such as sucrose and ethanol (Zhang et al. 2005). Furthermore, these natural variations in maternal care have been shown to induce changes in the social, reproductive, and sexual behavior of female offspring. Daughters of high-LG dams are more maternal to their own offspring, exhibiting higher levels of nursing and LG, but are less sexual, being slower to enter lordosis and having fewer and less frequent matings with a male (Cameron et al. 2008). These behavioral differences are related to an altered patterning of estrogen and oxytocin receptors in the hypothalamus and amygdala. These studies have provided some elegant examples of how small variations in a critical early life experience can have long-term consequences on shaping stable individual differences in offspring behavior and neurobiology.

Although individual differences in maternal care are stable across time (Champagne et al. 2003; Champagne and Meaney 2007), they are also plastic in that a dam may change her levels of nursing and LG in response to variations in the external physical and social environment. Through such alterations in maternal care, a dam is able to signal the quality of the external environment that offspring are likely to enter at weaning and thereby inducing them to proceed down alternative pathways of development (see last section of this chapter, titled "Epigenetic routes of environmental influence on development," for a more thorough discussion). For example, female rats that are exposed to predator odor for just one hour on the day of parturition exhibit higher levels of LG than those dams that are not exposed. This higher LG then leads to female offspring themselves exhibiting higher levels of maternal care to their own offspring as well as having higher levels of estrogen receptor alpha mRNA in the medial preoptic area (MPOA) (McLeod et al. 2007). Similarly, in lactating mice who are exposed to a rat predator odor, active maternal care is increased and male offspring display reduced levels of fear in adulthood (Coutellier et al. 2008). In addition to exposure to predators, another variation in the physical environment of rodents that can alter patterns of maternal care is the accessibility to food. In laboratory rats this has been investigated by requiring dams to forage in a separate cage from their nesting cage and by varying the timing and availability of food supplies. It has been found that those rats that are forced to forage away from the nest show a more effective and efficient nursing style and female offspring are consequently less fearful and have a lower HPA activity than offspring of dams that are not required to leave the home-cage (Macrì and Würbel 2007). Congruent with studies in primates (Coplan et al. 2005), rat offspring born to dams that are required, in an unpredictable manner, sometimes to forage and sometimes not, are more fearful and have a higher HPA activity as adults. Thus, when dams are faced with an unpredictable challenging environment, their maternal care decreases and their offspring develop increased fear responses. Similarly, female rat dams that are housed after weaning in isolation in impoverished conditions exhibit considerably lower levels of LG toward their offspring compared with dams who are housed in social groups in cages with high levels of enrichment (Champagne and Meaney 2007). As expected, the exploratory behavior of male offspring is shifted as a consequence of this environmentally induced change in maternal care (Champagne and Meaney 2007). It is worth noting that in many of these reported studies the plastic changes in offspring behavior are sex-specific. The reasons why one sex may show an induced change in one paradigm but not another are not yet fully known, though it has been proposed that whichever sex is most likely to remain in their natal area and therefore going to inhabit the environment in which their mothers are (and thus the environment which is being predicted to them through their maternal behavior) would be the one most susceptible to such early life experiences (Coutellier and Würbel 2009).

PEER INTERACTION AND COMMUNAL NESTING

Though the relevance of mother-offspring interactions for the development of the individual has been widely studied, less attention has been paid to peer-peer interactions and their long-term consequences. Social complexity and peer interactions are essential features of the ecological niche of the developing mammal and have a major impact on shaping stable individual differences (Crowcroft and Rowe 1963; D'Udine and Alleva 1980; Berman et al. 1997; Suomi 2005; Branchi 2009).

Research on nonhuman primates has provided important contributions to the characterization of the role of peer interaction during early ontogeny. In the 1960s and 1970s, Harry Harlow carried out a number of studies in rhesus macaques (*Macaca mulatta*), showing that peer social stimulation during plastic postnatal phases leads to the development of complex social behavior and emotional responses (Harlow 1969; Harlow and Suomi 1971; Suomi 1979). Monkeys reared by their mothers but totally deprived of peer contact failed to develop essential social skills, such as the ability to play with peers or to cope with aggression, and traversed abnormal developmental trajectories (Harlow 1969). More recent studies performed in the field have confirmed these results, showing that infant rhesus monkeys living in groups in which they have few opportunities to interact with non-kin age-mates show play behavior that is less frequent, briefer in duration, and less positive in affective tone and, at adulthood, their relationships with non-kin age-mates are relatively infrequent and often hostile (Berman et al. 1997; Suomi 2005).

Peer interaction represents a prominent feature of the early environment for mice and rats as well (Laviola and Terranova 1998; Branchi 2009). Among the features of the early rodent social environment that have been investigated, litter gender composition has been reported as a factor deeply affecting pup development (Laviola and Terranova 1998). Male mice reared only with female siblings display high levels of aggressive behavior at adulthood compared with males reared with males (Namikas and Wehmer 1978; Mendl and Paul 1991a). These effects have been ascribed to the reported gender differences in the development of social behavior (Tahakashi and Lore 1983), which include the differential onset of approach response and play behavior (Laviola and Alleva 1995; Laviola and Terranova 1998). It has also been suggested that the litter gender composition effects may be mediated, at least in part, by different levels of sibling competition for maternal milk (Mendl and Paul 1991b; Branchi et al. 2009). Another factor determining the early social environment in studies using genetically modified mice is the genotype ratio. In these studies, when newborns are produced through mating of two heterozygous mice, litters are composed of a varying number of wild-type, heterozygous, and homozygous offspring. Since animals having different genotypes display different behavior, the relative number of individuals of each genotype determines the social milieu in which each pup of a given litter develops (Holmes et al. 2005). Studies on mice bearing the estrogen receptor alpha null mutation clearly showed that not only the sex ratio but also the genotype ratio markedly affects the behavior at adulthood (Crews et al. 2004; Crews 2008).

An experimental manipulation that has been used to investigate the role of the early social environment on adult behavior is the communal nest (CN) (Branchi 2009). This manipulation allows the study of both peer and mother-offspring interactions, though disentangling the effects of these two components is a complex issue that warrants further studies. The CN consists of females that combine their pups in a single nest, sharing caregiving behavior. Rearing pups in a CN occurs very frequently in feral mouse (*Mus musculus*) populations: semi-naturalistic and field studies demonstrated that up to 90% of females may rear their offspring in communal nests (Crowcroft and Rowe 1963; Manning et al. 1995). Relatedness and familiarity among females are key factors in determining the creation and success of the CN (König 1994). Sharing of parental responsibilities by multiple individuals occurs in a number of social species (Gittleman 1985; König 1997). Indeed, communal nesting and nursing leads to a number of benefits, including allogrooming, cooperative foraging and feeding, improved defense, and enhanced thermoregulation. However, it also involves a number of costs, such as higher risk for infanticide, increased competition for food, and increased parasite transmission and visibility to predators (Hayes 2000; Ebensperger 2001). In a CN, developing pups show a faster rate of growth (Sayler and Salmon 1969) and the overall levels of maternal behavior are higher compared with standard laboratory-rearing conditions (Sayler and Salmon 1971; Branchi et al. 2006a).

The social stimulation provided by the CN has a major impact on the development of a number of adult behaviors, ranging from social behavior to the emotional response. Adult CN mice, compared with those reared in the standard nest (SN), display a higher propensity to interact socially and more promptly achieve a well-defined social role (i.e., dominant or subordinate) when tested in a social interaction test. In this test, which exploits the tendency of an adult male mouse to defend its own territory and establish a hierarchy, mice lacking communal nesting experience need several encounters to fully display a behavioral profile typical of a dominant or subordinate male, while mice reared in a communal nest display a well-defined social role already on the first encounter (Branchi et al. 2006a). Furthermore, home-cage observations indicated that CN mice display social responses in a way that is appropriate in an eco-ethological perspective: CN mice are more aggressive than SN mice, but only when social hierarchy needs to be established (D'Andrea et al. 2007). The social competencies of CN mice, and in particular the relevance of social environment, have also been investigated with regard to their emotional responses in different social contexts (Branchi and Alleva 2006). In one paradigm, experimental subjects are exposed to an anxiogenic environment (i.e., an elevated plus maze), in two different social conditions: alone or paired with a familiar conspecific. While SN mice do not behave differently when exposed to the apparatus in the two social conditions, CN mice show significantly less anxiety-like behavior in the paired than in the alone condition. Thus, the social context appears to influence more markedly the emotional response of CN than SN mice (Branchi and Alleva 2006).

Being reared in a CN also affects the emotional response in adulthood. However, such modifications are context dependent. Adult CN mice displayed increased anxiety-like behavior when exposed to a physical challenge, such as the elevated plus maze or the open field (Branchi et al. 2006b). In contrast, in a social interaction test, CN mice are less fearful than SN mice when interacting with an unfamiliar conspecific, showing reduced levels of social anxiety–like behavior (Branchi et al. 2006a). Finally, when exposed to the forced swim test, CN mice show higher levels of immobility, considered as an index of vulnerability to depression, though the interpretation of this test is highly debated (Petit-Demouliere et al. 2005; Branchi et al. 2006b).

There have been a number of studies investigating the mechanisms underlying the effects of being reared in a CN. In line with the behavioral data, these studies suggest that in adulthood, CN animals have more pronounced brain plasticity. Indeed, the early social stimulation provided by the CN condition markedly affects the levels of neurotrophins (Branchi et al. 2006a, b). NGF (nerve growth factor) and BDNF (brain-derived neurotrophic factor) levels are increased respectively up to 5- and 8-fold in different brain areas, including the hippocampus and hypothalamus, in CN mice compared with SN mice. An increase of neurotrophin levels in the hippocampus is suggestive of changes in the neurogenesis rate in the dentate gyrus of the hippocampus (Castren 2005). Indeed, CN mice show a marked increase in the number of newly generated brain cells (Branchi et al. 2006b). In particular, when compared with SN mice, CN mice show no difference in brain cell proliferation but an increase in survival, in line with the hypothesis that BDNF signaling is required mainly for the long-term survival and less for proliferation of newborn brain cells (Sairanen et al. 2005).

Thus the CN peer/maternal environment results in long-term changes in brain and behavior. However, this model has produced conflicting results, as some studies reported an increased emotional response in CN mice (Sayler and Salmon 1971; Curley et al. 2009). Discrepancies in the CN protocols, such as the time at which multiple females are placed in a single communal nest, may account for these differences. In particular, it appears that adult offspring are more emotional in an open-field or an elevated plus maze (but not in a social interaction test) when the mothers are put together some days before parturition (Branchi and Alleva 2006; Branchi et al. 2006b), with opposite effects observed when mothers are put together immediately after parturition (Curley et al. 2009).

REROUTING THE INFLUENCE OF THE EARLY LIFE EXPERIENCES
It is worth noting that the influence of early experiences can be, at least partially, reversed or compensated. For instance, animal studies reveal that postnatal manipulations, such as fostering or handling, lead to a recovery from the effects of prenatal stress on HPA axis development and the adult coping

response to stress (Maccari et al. 1995; Vallee et al. 1997), and postweaning housing conditions such as enrichment partially reverse the effects of treatments imposed on behavior either prenatally (Laviola et al. 2004) or over the first days of life (Francis et al. 2002). Meaney and colleagues investigated whether such reversibility at the level of the behavioral response corresponds to a reversibility of those cellular and molecular effects triggered by early life experiences. Alternatively, it may reflect a "compensatory" effect acting through mechanisms different from those that affect perinatal brain development, which might then offset the effects of early trauma. Though these processes are not mutually exclusive, the results showed limited reversibility at the level of cellular mechanism, suggesting that compensation processes may underlie the recovery as a result of environmental stimulation (Francis et al. 2002).

Rerouting developmental trajectories deviated by early experiences by exposing individuals to environmental stimulation has been demonstrated also in studies on nonhuman primates. Harlow and Suomi (1971) showed that rhesus monkeys displaying severe deficits in virtually every aspect of social behavior induced by isolation during the first 6 months of life can recover from such trauma. In particular, when 6-month-old socially isolated monkeys are exposed to 3-month-old normally reared monkeys, they achieve essentially complete social recovery. The 3-month-old monkeys are very active in social interaction, thus providing an important stimulation to isolated subjects that contributes to the development of normal social behavior. A great challenge for future studies on the ontogeny of stable individual differences in behavior is to further elucidate the epigenetic and neural mechanisms through which these types of reversibility or compensation of early experiences occur.

Interaction between genes and environment in shaping behavioral responses

It is clear that experiences occurring prenatally, early postnatally, and through juvenile social interactions can have sustained long-term effects on behavior. Though much of this evidence comes from rodent models in which there is no genetic variation between individuals, both genetic and environmental variation and their bidirectional interaction are responsible for producing individual variations in phenotype (figure 10.1). Indeed, the appreciation that all phenotypes, including behavior, arise not in a deterministic fashion but through the interaction of genetic and environmental variables throughout development has gained much ground in the past

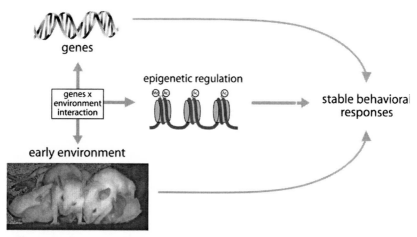

Figure 10.1. Gene-environment interaction, that is, the process by which the environment and the genes moderate each other's effect in shaping the phenotype. The epigenetic regulations play an important role in mediating the influence of gene-environment interactions on brain development and on the ontogeny of stable behavioral differences.

20 years. This is in part due to Gilbert Gottlieb, whose probabilistic epigenesis theory has laid much of the theoretical framework for this way of thinking (Gottlieb 1998; 2007). This theory emphasizes that the development of phenotype is a series of continuous and probabilistic interactions between organismic (internal) factors, such as genetic and neural activity, and contextual factors, such as environmental experiences (Gottlieb 2007). Related to this theory is the "norm of reaction" concept, according to which the phenotypic outcome of any particular genotype cannot be predicted from simply observing the phenotype of that genotype in two different environments and inferring the likely phenotype that will develop in a third environment. Thus, a given genotype does not simply set upper and lower limits on possible phenotypic outcomes, but through unique interactive developmental processes with the environment may give rise to a whole range of potential phenotypes (Gupta and Lewontin 1982; Platt and Sanislow 1988; Fuller et al. 2005; Gottlieb 2007).

Among the first studies illustrating the effects of such a gene-environment interaction on behavior is the one by Cooper and Zubek (1958). These authors used two rat lines selectively bred for high and low performance in a maze. Since the two rat lines were reared in identical environments, genetic differences accounted for the clear difference in their performance (Tryon 1942). However, when the rats were reared in a physically and socially impoverished vs. an enriched environment, the difference in performance between the two lines disappeared, indicating that the environment profoundly modulated the effects of genetic differences on this behavioral response (Cooper and Zubek 1958).

Both laboratory and field studies of rhesus monkeys have shown that 5%–10% of the individuals in a particular population exhibit impulsive behavior and poor social competence (Suomi 1997). Such variation in behavior was associated with cerebrospinal fluid (CSF) levels of 5-hydroxyindoleacetic acid (5-HIAA)—a metabolite of the chemical messenger serotonin—and thus to variation in serotonin function. Monkeys showing high levels of aggression and a lack of positive affiliative behaviors had significantly reduced CSF levels of 5-HIAA (Higley et al. 1996). These behavioral and neurochemical differences are, in turn, associated with the polymorphism of the serotonin transporter gene (5-HTT) (Bennett et al. 2002); individuals carrying one short allele of the 5-HTT gene-linked polymorphic region, which confers low transcription efficiency, have lower levels of 5-HIAA and exhibit more aggressive behaviors than those homozygous for the long allele. However, this association was found only in young rhesus monkeys that were maternally deprived during early postnatal life, whereas individuals reared by their mothers showed similar behavioral and neurochemical profiles independently from the 5-HTT gene polymorphism. These results show a gene by environment interaction, wherein a given gene of an individual gives rise to a phenotype according to the early experiences of that individual (Barr et al. 2003; Caspi and Moffitt 2006; Suomi 2006).

In mouse studies, the importance of the gene-environment interactions has been illustrated by investigations of behavioral differences between different mouse strains reared and tested following identical procedures in different laboratories across the globe (Crabbe et al. 1999; Bohannon 2002; Wolfer et al. 2004; Wahlsten et al. 2006). Though these studies were aimed at evaluating the reproducibility of behavioral genetic studies performed in different laboratories, they found that the pattern of behavioral differences on standard tests of anxiety, activity, and memory among strains varied substantially according to the location where the test was performed. Although the genetic factors mediating these effects remain unknown, these studies illustrate the important point that the genetic background interacts with the environment in shaping the adult individual's behavior.

Several other studies have investigated the impact of gene-environment interactions occurring during very early postnatal phases on the development of stable individual differences. For example, a study has shown that the behavior of control and knockout mice lacking the 5-HT1A receptor or the 5-HT1B receptor changed according to the genotype, and thus the phenotype, of the mother (Weller et al. 2003). Another example is provided by a series of studies by Lipp and colleagues working with a mouse model of Alzheimer's disease bearing a targeted mutation that results in low expression of a shortened betaAPP protein (betaAPP[delta/delta]) (Muller et al. 1994; Tremml et al. 1998; 2002). In early studies of these mice, the authors reported that such a genetic modification significantly reduced learning abilities in adulthood (Muller et al. 1994). However, in a subsequent study, these mice were subjected to neonatal handling, which eliminated the difference in learning abilities between mutant and control mice. In a third experiment (Tremml et al. 2002), the researchers demonstrated that this early postnatal stimulation clearly modulated the effects associated with the genetic mutation, illustrating that learning deficits were determined by a gene-environment interaction.

The interaction between genes and the environment occurs not only during development but also into adulthood, though the effects are usually less pervasive (Rosenzweig 2003; Bateson et al. 2004). For instance, knockout mice that lack a subunit of the N-methyl-D-aspartate (NMDA) receptor in selected brain areas show clear alterations of neural plasticity and are profoundly impaired in learning and memory tasks (Tsien et al. 1996; Rampon et al. 2000). However, such alterations can be reverted by a 15-day (from 1.5 to 2 months of life) daily training in an enriched environment. Thus, the memory deficits observed in mutant mice may be reversible according to the environmental circumstances (Rampon et al. 2000). Moreover, genetically modified mouse models of Huntington's disease, in which the *huntingtin* gene expression has been modified, develop motor and cognitive impairments as well as progressive neurodegeneration in selected brain areas. These abnormalities can be effectively counteracted by the exposure to an enriched environment for few months from weaning to adulthood (Van Dellen et al. 2000; Hockly et al. 2002; Spires et al. 2004).

These studies illustrate that variations in genotype do not lead directly to variations in phenotype, but through interactions with variations in the physical and social environment different developmental pathways are selected. As we have discussed, though the environment is able to produce changes throughout ontogeny, the earlier such influences occur the greater their effect. This view of behavioral development is essentially akin to that of Waddington, who used the metaphor of an "epigenetic landscape" whereby in response to environmental cues ontogeny proceeds down alternative pathways preselected by genetic factors, though movement between pathways becomes increasingly more difficult with time (Jablonka and Lamb 2002; 2005). In the following section, we shall discuss a more modern use of the term epigenetics and discuss some mechanisms underlying the effects of the interplay between environmental and genetic factors regulating development.

Epigenetic routes of environmental influence on development

As described in the previous section, the environment and genome interact during development to lead to stable individual differences in behavior. One mechanism through which these effects can occur is epigenetic modifications to the genome, with variations in gene expression occurring because of changes in the way in which the transcriptional machinery can bind to chromatin. It is increasingly clear that long-term changes in chromatin structure and gene expression can occur as a result of variations in early life experiences leading to altered brain development and behavioral traits. However, to understand the pathways through which this occurs it is necessary to understand something of the molecular basis of these epigenetic modifications (for references and further information see Fuks 2005; Mersfelder and Parthun 2006; Berger 2007; Kouzarides 2007).

Within cell nuclei, DNA is contained within individual nucleosomes that are comprised of 147 base pairs of nucleotides wrapped around eight histone proteins (two copies each of H2A, H2B, H3, and H4 proteins) (Kouzarides 2007). Nucleosomes are separated by approximately 80 base pairs and are linked together via a fifth histone protein H1. Nucleosomes, along with other scaffolding proteins and enzymes, make up chromatin, which exists in two forms. Heterochromatin is condensed chromatin and prevents RNA polymerase II from directly accessing DNA and promoting gene transcription, whereas euchromatin is looser and allows access of the transcriptional machinery. Different epigenetic modifications to the DNA, to the histone proteins, and to the amino acid tails of these proteins, as well as interaction between histones, DNA, and other transcription factors, coactivators, and repressor molecules, dictate which state chromatin is in and the overall level of gene transcription within each cell. These processes are critical to the regulation of cellular differentiation, as the epigenetic status of one cell is inherited by its daughter cells following mitosis.

Numerous covalent modifications to histone proteins (particularly H3 and H4) that alter chromatin structure have been described (figure 10.2) (Mersfelder and Parthun 2006; Berger 2007; Kouzarides 2007; Keverne and Curley 2008). For instance, histone acetylation is the addition of an acetyl group to lysine residues of histone protein tails and is catalyzed by histone acetyltransferase (HAT) enzymes. This process neutralizes the positive

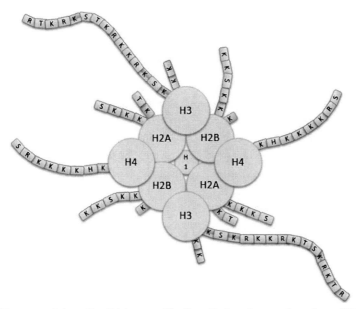

Figure 10.2. Schematic of histone modifications. Each nucleosome is made up of DNA wrapped around two sets of four histone proteins (H2A, H2B, H3, H4) and a core histone (H1). Each histone protein has tails that are comprised of up to several hundred amino acid residues (represented here by different letters, e.g., H is histidine, K is lysine, R is arginine, S is serine, and T is threonine). At several of these amino acids various different covalent chemical modifications may occur, such as acetylation, phosphorylation, methylation, and ubiquitination; see text for explanation of each of these. These histone modifications induce changes in the chemical and physical structure of both histone proteins and DNA methylation, which enables tight regulation of gene transcription. In reality there are many more amino acid residues per tail, with those shown here being the most common covalent modifications.

charge of the histone tail making it less attracted to the negatively charged DNA and thereby loosening chromatin and leading to increased gene expression. The opposite reaction is to remove acetyl groups by a process known as histone deacetylation, which is catalyzed by histone deacetylases (HDACs), leading to more condensed chromatin and reduced gene expression. The other covalent modifications may be transcription activating or repressing depending upon the particular amino acid residue. Histone phosphorylation, the addition of a phosphate, is catalyzed by protein kinase enzymes and occurs at serine and threonine amino acids. It acts to reduce the positive charge of histones, thereby relaxing the DNA within the chromatin and increasing gene expression, as well as to further increase the efficacy of HATs leading to phophoacetylation of histones. Different lysine and arginine amino acid residues on the histone proteins may also have between one and three methyl groups added through histone methylation being catalyzed by histone methyltransferases (HMTs) and reversed by histone demethylases (HDMs). Histone methylation acts to prevent further acetylation and phosphorylation of histones and is typically gene transcription repressing, but may also be gene activating at particular residues. Other histone modifications include ubiquitylation, SUMOlyation, ADP-ribosylation, biotinylation, citrullination, deamination, and proline isomerization, with several others probably yet to be discovered. The interaction between each of these post-translation histone modification processes as well as their ability to recruit further transcription repressors and activators leads to a tight control over gene expression.

Another modification to chromatin, and one that can lead to the establishment of long-term fixed changes in gene expression, is the methylation of DNA (Fuks 2005). Essentially this process is the transfer of a methyl group from S-adenosyl methionine to the 5' position of cytosine residues within the nucleotide sequence, particular to those at dinucleotide CpG sequences, and is catalyzed by DNA methyltransferases (DNMTs); DNMT 3a and 3b in the case of *de novo* methylation and DNMT 1 in the case of maintenance of DNA methylation. Increasing levels of methylation of the DNA generally act to suppress gene transcription and interact with histone modifications in two ways. First, the deacetylation and methylation of certain histone proteins (e.g., H3 lysine residue 9) is associated with increased DNA methylation through the recruitment of DNMTs. Second, increased DNA methylation leads to the addition of CpG binding domain proteins (MDBs) such as methyl CP-binding protein 2 (MeCP2) that in turn associate with HDACs and MHT (methyl halide transferase), resulting in further reduced gene transcription as a consequence of increased histone deacetylation and methylation.

It has long been established that chromatin modifications such as those described above, as well as changes to the three-dimensional structure of chromatin such as nucleosome sliding (Becker 2004), have critical roles in shaping the fate of cell lineages and the phenotype of individual cells and tissues during development. However, it is only in the past five years that it has become increasingly clear that these processes may also play a crucial role in mediating the effect of external environments upon the developing nervous system. Work by Meaney and colleagues has shown how variations in the quality of maternal care early in life can shape stable differences in fear and stress responses of offspring as a result of individual variations in the epigenetic modification of glucocorticoid receptor (GR) gene promoters

(Fish et al. 2004; Szyf et al. 2005). Offspring who receive low levels of LG from their mothers have been shown to develop an elevated stress response and to be more neophobic, which is related to decreased expression of GR in the hippocampus. In particular, low levels of the hippocampal GRs reduces the negative feedback exerted by corticosterone on the HPA axis. As a consequence, the exposure to a stressing condition leads to a higher activation of the HPA axis, a higher release of glucocorticoids, and slower clearance of them. Importantly, these effects are observable even when offspring are cross-fostered between high and low LG as the phenotype is dependent upon the foster mother's behavior rather than that of the biological mother. Moreover, DNA methylation within exon 1_7 of the GR promoter has also been shown to vary as a function of maternal care, with increased DNA methylation in the hippocampus of adult male offspring of low-LG dams compared with high-LG dams at this promoter (Weaver et al. 2004) (figure 10.3). Furthermore, it was discovered that these DNA methylation differences were not present in offspring immediately before birth but emerged only during the first week postpartum and were sustained thereafter. This evidence strongly indicates that the stable individual differences in behavior and gene expression are related to the variations in early life environments experienced by the two groups of offspring.

Further evidence that DNA methylation alterations are causally related to the observed long-term changes in offspring phenotype has been provided via pharmacological manipulations of the epigenome. The HDAC inhibitor trichostatin-A (TSA) is a compound that has the overall effect of reducing methylation, as it actively promotes DNA demethylation by preventing histones from losing their acetylation status. Intra-cerebroventricular (ICV) infusion of TSA into adult male rats led to the offspring of low-LG dams exhibiting increased exploration, an attenuated stress-induced corticosterone response, and increased expression of hippocampal GR mRNA compared with offspring of low-LG dams that received vehicle treatment. TSA-treated low offspring were also indistinguishable from offspring of high-LG dams that had received either TSA or vehicle (Weaver et al. 2004; 2006). In the reverse experiment, adult offspring of high-LG dams that received an ICV infusion of methionine (a methyl donor that promotes overall levels of DNA methylation) exhibited decreased exploration, increased corticosterone response to stress, and decreased levels of GR mRNA in the hippocampus (Weaver et al. 2005). Therefore, through pharmacological manipulation of the overall levels of DNA methylation including at the GR promoter, it is possible to shift these stable individual differences in behavior in adult animals.

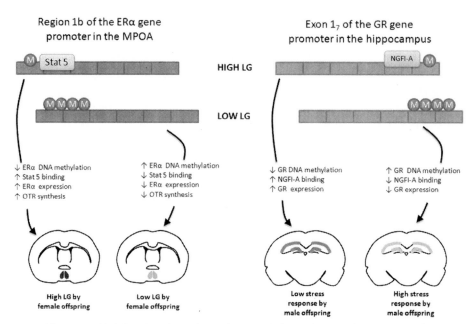

Figure 10.3. Variations in early rearing environment induce epigenetic changes at specific gene promoters. Lower levels of licking/grooming (LG) behavior exhibited by dams have been shown to lead in their offspring to increased levels of DNA methylation (M in circles) at specific regions of two gene promoters (estrogen receptor alpha [ERα] and glucocorticoid receptor [GR]) in specific brain regions. Offspring who received higher levels of LG have reduced methylation and increased binding of proteins that induce gene transcription such as signal transducer and activator of transcription 5 (Stat5) and nerve growth factor inducible protein A (NGFI-A). Alterations in levels of DNA methylation lead to long-term changes in gene expression and receptor expression and consequently behavioral phenotypes. MPOA indicates medial preoptic area; OTR, oxytocin receptor.

The question still remains, though, as to what exactly is the signal pathway from environmental experience (variation in the tactile stimulation provided by maternal LG) to the altered chromatin modifications in offspring. A site-specific analysis of the DNA methylation pattern of the GR promoter gene identified a differentially methylated NGFI-A (nerve growth factor inducible protein A) transcription factor binding site in the offspring of high-LG versus low-LG dams (Weaver et al. 2004; 2005) (figure 10.3). Following this discovery, it was found that the binding of NGFI-A to the hippocampal GR promoter in the hippocampus is actually lower in the offspring of low-LG dams. It was also found that ICV administration of methionine decreased the binding of NGFI-A to this promoter, further indicating that NGFI-A binding was reduced with increasing DNA methylation. It has also recently been demonstrated that NGFI-A induces activity of the GR promoter in a transient reporter assay, and in a transient transfection assay that NGFI-A overexpression results in histone acetylation and DNA demethylation of the same promoter (Weaver et al. 2007). While there is obviously a lot of work to be done to complete the intracellular pathways, it is a significant advancement in our understanding of how an environmental stimulus can interact with the genome during development.

The significant issue for behavioral biologists is whether this environment-epigenetic interaction is representative of behavioral development generally, or whether somehow the maternally mediated GR expression and promoter DNA methylation is unique. To resolve this question, Meaney and colleagues have provided evidence that natural variations in maternal care are able to epigenetically modify DNA methylation at other steroid receptors leading to stable individual differences in the social behavior of offspring (Champagne et al. 2006). They found that offspring raised by high-LG dams exhibited high levels of licking/grooming themselves to their own offspring when they became mothers, and that these elevated levels of maternal LG were correlated with increased levels of oxytocin receptor (OTR) density and estrogen receptor alpha (ERα) expression in the MPOA. In lactating dams, OTR levels are upregulated in response to ERα activity, and indeed low-LG dams have reduced sensitivity to estrogen-induced upregulation of oxytocin receptors. This was found to be related to elevated levels of ER alpha DNA methylation at several sites within the 1b promoter region of low-LG offspring, including a binding site for the Stat5 protein, which acts as a signal transducer and activator of transcription (figure 10.3). Chromatin immunoprecipitation assays have confirmed that increased levels of DNA methylation do indeed lead to reduced Stat5 immunoreactivity demonstrating that the epigenetic modification of this gene promoter has functional consequences for transcription factors that induce gene expression (Frasor et al. 2001; Champagne et al. 2006). Hence, reduced levels of maternal care received early in life induce changes in the epigenetic status of the ERα gene promoter, which leads to the long-term reduced expression of ERα and OTR in the MPOA, which then account for the lower LG exhibited by these females when they are adult.

These studies provide good evidence that social experiences induce epigenetic modifications at two gene promoters (GR and ERα) during early development, probably acting via different intracellular pathways. However, what about other neuropeptide and steroid receptors that underlie the neural regulation of stable individual differences in behaviors such as exploration, neophobia, social interactions, and reward-seeking? Do their receptors undergo similar modifications resulting in stable differences in adult behavior? Currently we do not know the answer to this question, as searching for modifications at individual gene promoters related to specific early environmental changes is timely and costly. However, we do know that neural mechanisms underlying several behavioral phenotypes that are plastically shifted by early experiences can also be epigenetically modified by specific environmental stimuli, suggesting that epigenetic mechanisms could potentially play a role in the ontogeny of these stable individual differences.

One example comes from studies of the dopaminergic mechanisms underlying addictive-type behaviors. Long after administration of addictive substances such as ethanol and cocaine, the drugs and other contextual cues can have powerful effects on behavior. Alterations in early life experiences have been found to alter dramatically these behaviors in adulthood. For instance, rat offspring that are maternally separated for upwards of 3 hours daily for the first two weeks of life exhibit elevated ethanol intake when they are adults (Ploj et al. 2003), whereas rats that are handled (i.e., separated for 15 minutes daily and experience a subsequent increase in LG from their mothers) tend to consume less ethanol (Moffett et al. 2007; Francis and Kuhar 2008). Maternal separation also alters cocaine-induced locomotor activity in rats and mice as well as behavioral sensitization to cocaine (Brake et al. 2004; Kikusui et al. 2005). Moreover, maternally separated rats have been found to show the highest levels of self-administration of cocaine as adults while handled rats show the lowest self-administration levels (Moffett et al. 2007). While the dopaminergic system is not the only one regulating these behaviors, these changes have been associated with higher levels of dopamine D_1 receptors and D_2 receptors in the hippocampus and ventral tegmental area respectively in handled offspring (Ploj et al. 2003; Moffett et al. 2007). Moreover, D_3 receptor mRNA is reduced in the nucleus accumbens of both separated and handled offspring (Brake et al. 2004) and levels of the dopamine transporter are significantly lower in the nucleus accumbens of separated offspring (Meaney et al. 2002). Finally, natural variations in maternal LG have been found to inversely correlate with offspring self-administration of ethanol and cocaine in adulthood (Francis and Kuhar 2008), and offspring of low-LG dams have been found to have a blunted medial prefrontal cortex release of dopamine in response to stress, lower levels of COMT (an enzyme that degrades catecholamines such as dopamine) immunoreactivity (Zhang et al. 2005), and reduced motivation to engage in socially rewarding behaviors such as LG of pups in response to dopamine release (Champagne et al. 2004) .

Recently, it has been demonstrated that exposure to drugs such as ethanol and cocaine induces chromatin modifications. In rats, the promoters for FosB, BDNF, and Cdk5 become acetylated at the H3 histone following chronic cocaine administration, with the acetylation at the BDNF promoter continuing long after exposure (Kumar et al. 2005). This same procedure also results in an increase in MeCP2 and MBD1 in the adult brain, but different changes occur following acute administration when there is a dynamic increase in H4 acetylation at the cfos and fosB genes and also phosphorylation on H4 proteins at the cfos promoter. The behavioral effects of cocaine administration, such as rats exhibiting a conditioned place preference, are inhibited when the rats are treated with HDACs and are elevated when treated with HDAC inhibitors, demonstrating the importance of acetylation in regulating these behaviors (Colvis et al. 2005; Kumar et al. 2005). The chronic administration of other drugs, such as amphetamine, also leads to chromatin modifications such as H4 hyperacetylation in the mouse striatum, and the treatment of mice with HDAC inhibitors enhances amphetamine-induced behavioral sensitization (Kalda et al. 2007). Therefore, there is an emerging picture that epigenetic modifications of chromatin play a significant role in regulating the addictive-like behaviors of rodents, and it seems plausible that the long-term changes in these behaviors induced by early experiences may occur through this type of mechanism.

In addition to the effects of cocaine on the epigenetic status of BDNF, other environmental variables have also been shown to have chromatin remodeling effects at this gene. For instance, when rats are induced to have mild seizures, they later show increased H4 acetylation at the P2 promoter as well as elevated levels of phosphoacetylation of H3 histones at the cfos gene in the hippocampus (Tsankova et al. 2004). Following chronic seizures, the H3 proteins of the BDNF gene show increased acetylation. In another test of chronic social defeat, where subjects are exposed to an aggressive opponent on successive days in a test arena and are defeated, subjects avoid contact with all conspecifics but this can be reversed following treatment with antidepressants such as imipramine. It has been shown that social defeat leads to a reduction in gene expression of BDNF in the hippocampus and a four-fold increase in H3 demethylation at the lysine 27 residue of the P3 and P4 promoters (Tsankova et al. 2006). However, after chronic imipramine antidepressant treatment, such socially defeated subjects have elevated expression of BDNF in the hippocampus, which is associated with a two-fold increase in H3 acetylation at these same promoters.

The above data would suggest that the large class of neurotrophins, including BDNF, which play a role in regulating neural and synapse growth and maturation are strong targets for epigenetically mediated effects of early social experiences on long-term changes in development (Cirulli et al. 2003; 2009). Indeed, long-term changes in BDNF levels have been found in adult animals exposed to different early life experiences. For instance, the expression of BDNF mRNA is downregulated in the prefrontal cortex and striatum of adult rats who were maternally separated for two weeks postpartum (Roceri et al. 2004). Furthermore, the offspring of high-LG dams have increased hippocampal BDNF mRNA as well as elevated hippocampal choline acetyltransferase and synaptophysin, which are thought to affect the improved cognitive abilities of these animals compared with the offspring of low-LG dams (Liu et al. 2000). Similarly, mice reared in a communal nest and receiving extra maternal and peer social enrichment exhibit higher levels of BDNF in adulthood and have longer survival of BRDU (5-bromo-2'-deoxyuridine) positive cells in the hippocampus (Branchi et al. 2006a, b). It remains to be discovered whether stable individual differences in behavior are mediated via epigenetic modifications to neurotrophin gene promoters, but accumulating evidence suggests that environmental experiences may exert their effects through this mechanism.

As we have discussed throughout this chapter, long-term stable individual differences in personality traits such as neophobia, exploration, aggression, reward-seeking, and fear are the product of numerous interactions between genes and environments during development. We have presented here some of the most recent data showing that these stable differences are in part regulated by differences in gene expression that are the result of long-term stable modifications to chromatin, specifically DNA methylation and covalent bond changes to histone proteins. Many important questions remain, however, of which three are especially significant. First, the exact cellular pathways that induce specificity in these epigenetic modifications following early experience need to be further understood, especially with regard to how the same environmental experience may lead to increased gene expression in one brain region but suppressed gene expression in another. Second, we have yet to discover whether similar epigenetic mechanisms underlie stable individual differences in other behaviors. For instance, it has long been established that variations in maternal care lead to individual differences in behavior that are mediated by alterations in serotonergic, GABAergic, glutamergic, opioid, oxytocinergic, vasopressinergic, and various other neural pathways. Are these altered by similar or different epigenetic mechanisms to GR, ER, and BDNF? Third, in terms of the development of stable individual differences in behavior, most research has focused on the immediate postnatal phase as the key mediator of these epigenetic changes. As we have previously discussed, stable differences in behavioral traits between individuals emerge throughout development, so that there are many different opportunities for adaptation and change. A critical issue for the future of this research is to understand how these environmentally induced supposedly stable epigenetic changes may be mechanistically altered during development—can DNA methylation changes at specific promoters be reversed, or are these behavioral changes mediated via epigenetic alterations at other gene promoters? It is an exciting future for behavioral biologists interested in the study of the developmental origins of individual variation, as we are finally beginning to understand how the environment can induce changes in ontogeny through acting directly upon the genome. Yet, given that within any mammalian cell there may be up to 500 million nucleosomes, the major problem for future research is unlikely to be discovering epigenetic modifications but rather to be understanding which ones are the most significant in shaping neural and behavioral development.

Adaptive significance of developmental plasticity

An important issue to consider regards the adaptive significance of stable individual differences in behavior. There are two broad evolutionary approaches toward investigating this phenotypic plasticity; the first takes a behavioral ecological approach to understanding the existence of variation in behavioral syndromes at a population level (see chapters 6, 7, 8, and 9). The second, which we shall briefly discuss here, has its origins in behavioral biology and human epidemiology and examines why individuals in different environments would develop different phenotypes including alternative behavioral profiles.

All organisms have the capacity to develop according to alternative trajectories determined by cues received early in life, although with time the degree of plasticity diminishes and development becomes more canalized. As mentioned above, this is what Waddington referred to as the epigenetic landscape, emphasizing that environmental triggers are able to drive development toward different phenotypes within the constraints of the existing genetic architecture (Waddington 1957; Jablonka and Lamb 2002). If these early environmental cues are predictive of the adult environment then development is adaptive, but if they are not then development may be totally disruptive or an organism may have a suboptimal phenotype—this has been coined the "match-mismatch" hypothesis (Gluckman et al. 2005). Adaptive developmental plasticity has been observed across diverse taxa and there are several well-known examples; for instance, the desert locust *Schistocerca gregaria* is shy, cryptic, sedentary, and nocturnal under low population density but bold, gregarious, active, and diurnal under high densities (see West-Eberhard 2003 for other examples).

Developmental plasticity has also been studied in human epidemiology where longitudinal studies have been carried out on the effect of birth weight (a proxy for maternal environmental quality during gestation) on later health, disease, and behavior. In particular, low birth weight has been associated with a multitude of long-term negative health outcomes, such as increased risk of diabetes, cardiovascular disease, osteoporosis, and obesity (Barker 2002). It appears that for a developing fetus the low energy supply experienced induces epigenetic changes that ready themselves for a nutritionally poor postnatal environment by storing energy intake as glucose and fat. This would be adaptive if the adult environment was also nutritionally poor, but many low-birth-weight children live in nutritionally rich adult environments and consequently the metabolic changes induced early in life lead to a maladaptive outcome. This is especially true in instances when children were conceived during famines but then grew up in a normal environment, or for children born to mothers from the developing world but who now live in the developed world. Conversely, children who have higher birth weights but develop in nutritionally poor adult environments appear to be more susceptible than low-birth-weight individuals to developing rickets (Chali et al. 1998); for example, smaller males were more likely than larger males to survive the severely nutritionally poor environments of concentration camps during World War II (Lucas 1991).

These studies from both behavioral biology and human epidemiology highlight the important principle that individual differences in phenotype may be adaptive when the adult environment matches the early environment, but may be maladaptive when it does not. In this chapter we have illustrated how mammals adjust their ontogeny according to cues received from both prenatal and postnatal environments and that this plasticity is facilitated through epigenetic changes. However, the question still remains as to whether these changes are indeed adaptive, as this is difficult to demonstrate conclusively in the laboratory, though several of the findings reported in this chapter certainly make sense intuitively. For example, offspring born to prenatally stressed dams that provide less licking/grooming may expect to emerge into a more adverse environment and therefore the decreased exploratory behavior and increased neophobia they exhibit as adults could be adaptive. Further work is needed in rodents to understand the adaptive significance of environmentally induced individual differences in behavioral types, especially to assess whether different behavioral phenotypes are adaptive or maladaptive when adult and early environments are matched or mismatched.

Conclusions

Understanding the mechanisms that mediate individual differences in behavioral responses provides a strategy for studying the biological basis of behavioral syndromes and personality in animals. Converging evidence from studies of laboratory rodents implicates the early environment in regulating long-term changes in neurobiology and behavior resulting in stable variation in phenotype. In particular, it is evident that mother-infant and peer-peer social interactions can have pervasive developmental effects. These environmental effects can also serve to augment the effects of genetically derived variation, providing evidence for the importance of considering gene-environment interactions when attempting to understand the origins of behavior. More recently, molecular approaches to the study of individual differences have been developed that highlight the role of epigenetic modification in sustaining the effects of environmental experience. Finally, a future challenge for those interested in studying individual differences in behavior will be to understand those behaviors and suites of correlated behaviors that are adaptive or maladaptive under matched versus mismatched adult and early life environments.

References

Barker, D. J. 2002. Fetal programming of coronary heart disease. *Trends in Endocrinology and Metabolism*, 13, 364–368.

Barr, C. S., Newman, T. K., Becker, M. L., Parker, C. C., Champoux, M., Lesch, K. P., Goldman, D., Suomi, S. J., and Higley, J. D. 2003. The utility of the nonhuman primate: model for studying gene by environment interactions in behavioral research. *Genes, Brain, and Behavior*, 2, 336–340.

Bateson, P., Barker, D., Clutton-Brock, T. H., Deb, D., D'Udine, B., Foley, R. A., Gluckman, P., Godfrey, K., Kirkwood, T., Lahr, M. M., McNamara, J., Metcalfe, N. B., Monaghan, P., Spencer, H. G., and Sultan, S. E. 2004. Developmental plasticity and human health. *Nature*, 430, 419–421.

Becker, P. B. 2004. The chromatin accessibility complex: chromatin dynamics through nucleosome sliding. *Cold Spring Harbor Symposia on Quantitative Biology*, 69, 281–287.

Bell, A. M. 2007. Future directions in behavioural syndromes research. *Proceedings of the Royal Society of London B*, 274, 755–761.

Bell, A. M., and Stamps, J. A. 2004. Development of behavioural differences between individuals and populations of sticklebacks, *Gasterosteus aculeatus. Animal Behaviour*, 68, 1339–1348.

Bennett, A. J., Lesch, K. P., Heils, A., Long, J. C., Lorenz, J. G., Shoaf, S. E.,

Champoux, M., Suomi, S. J., Linnoila, M. V., and Higley, J. D. 2002. Early experience and serotonin transporter gene variation interact to influence primate CNS function. *Molecular Psychiatry*, 7, 118–122.

Benus, R. F., Bohus, B., Koolhaas, J. M., and Van Oortmerssen, G. A. 1991. Heritable variation for aggression as a reflection of individual coping strategies. *Experientia*, 47, 1008–1019.

Berger, S. L. 2007. The complex language of chromatin regulation during transcription. *Nature*, 447, 407–412.

Berman, C. M., Rasmussen, K. L., and Suomi, S. J. 1997. Group size, infant development, and social networks in free-ranging rhesus monkeys. *Animal Behaviour*, 53, 405–421.

Boccia, M. L., Razzoli, M., Vadlamudi, S. P., Trumbull, W., Caleffie, C., and Pedersen, C. A. 2007. Repeated long separations from pups produce depression-like behavior in rat mothers. *Psychoneuroendocrinology*, 32, 65–71.

Bohannon, J. 2002. Animal models: can a mouse be standardized? *Science*, 298, 2320–2321.

Brake, W. G., Zhang, T. Y., Diorio, J., Meaney, M. J., and Gratton, A. 2004. Influence of early postnatal rearing conditions on mesocorticolimbic dopamine and behavioural responses to psychostimulants and stressors in adult rats. *European Journal of Neuroscience*, 19, 1863–1874.

Branchi, I. 2009. The mouse communal nest: investigating the epigenetic influences of the early social environment on brain and behavior development. *Neuroscience and Biobehavioral Reviews*, 33, 551–559.

Branchi, I., and Alleva, E. 2006. Communal nesting, an early social enrichment, increases the adult anxiety-like response and shapes the role of social context in modulating the emotional behavior. *Behavioural Brain Research*, 172, 299–306.

Branchi, I., D'Andrea, I., Gracci, F., Santucci, D., and Alleva, E. 2009. Birth spacing in the mouse communal nest shapes adult emotional and social behavior. *Physiology and Behavior*, 96, 532–539.

Branchi, I., D'Andrea, I., Fiore, M., Di Fausto, V., Aloe, L., and Alleva, E. 2006a. Early social enrichment shapes social behavior and nerve growth factor and brain-derived neurotrophic factor levels in the adult mouse brain. *Biological Psychiatry*, 60, 690–696.

Branchi, I., D'Andrea, I., Sietzema, J., Fiore, M., Di Fausto, V., Aloe, L., and Alleva, E. 2006b. Early social enrichment augments adult hippocampal BDNF levels and survival of BrdU-positive cells while increasing anxiety- and "depression"-like behavior. *Journal of Neuroscience Research*, 83, 965–973.

Brodin, T., and Johansson, F. 2004. Conflicting selection pressures on the growth/predation risk trade-off in a damselfly. *Ecology*, 85, 2927–2932.

Caldji, C., Francis, D., Sharma, S., Plotsky, P. M., and Meaney, M. J. 2000. The effects of early rearing environment on the development of GABA(A) and central benzodiazepine receptor levels and novelty-induced fearfulness in the rat. *Neuropsychopharmacology*, 22, 219–229.

Cameron, N. M., Fish, E. W., and Meaney, M. J. 2008. Maternal influences on the sexual behavior and reproductive success of the female rat. *Hormones and Behavior*, 54, 178–184.

Carere, C., Drent, P. J., Koolhaas, J. M., and Groothuis, T. G. G. 2005. Epigenetic effects on personality traits: early food provisioning and sibling competition. *Behaviour*, 142, 1335–1361.

Caspi, A., and Moffitt, T. E. 2006. Gene-environment interactions in psychiatry: joining forces with neuroscience. *Nature Reviews Neuroscience*, 7, 583–590.

Castren, E. 2005. Is mood chemistry? *Nature Reviews Neuroscience*, 6, 241–246.

Chali, D., Enquselassie, F., and Gesese, M. 1998. A case-control study on determinants of rickets. *Ethiopian Medicine Journal*, 36, 227–234.

Champagne, F. A., and Meaney, M. J. 2006. Stress during gestation alters postpartum maternal care and the development of the offspring in a rodent model. *Biological Psychiatry*, 59, 1227–1235.

———. 2007. Transgenerational effects of social environment on variations in maternal care and behavioral response to novelty. *Behavioral Neuroscience*, 121, 1353–1363.

Champagne, F. A., Francis, D. D., Mar, A., and Meaney, M. J. 2003. Variations in maternal care in the rat as a mediating influence for the effects of environment on development. *Physiology and Behavior*, 79, 359–371.

Champagne, F. A., Chretien, P., Stevenson, C. W., Zhang, T. Y., Gratton, A., and Meaney, M. J. 2004. Variations in nucleus accumbens dopamine associated with individual differences in maternal behavior in the rat. *Journal of Neuroscience*, 24, 4113–4123.

Champagne, F. A., Weaver, I. C. G., Diorio, J., Dymov, S., Szyf, M., and Meaney, M. J. 2006. Maternal care associated with methylation of the estrogen receptor-alpha-1b promoter and estrogen receptor-alpha expression in the medial preoptic area of female offspring. *Endocrinology*, 147, 2909–2915.

Cirulli, F., Berry, A., and Alleva, E. 2003. Early disruption of the mother-infant relationship: effects on brain plasticity and implications for psychopathology. *Neuroscience and Biobehavioral Reviews*, 27, 73–82.

Cirulli, F., Francia, N., Berry, A., Aloe, L., Alleva, E., and Suomi, S. J. 2009. Early-life stress as a risk factor for mental health: a role for neurotrophins from rodents to nonhuman primates. *Neuroscience and Biobehavioral Reviews*, 33, 573–585.

Colvis, C. M., Pollock, J. D., Goodman, R. H., Impey, S., Dunn, J., Mandel, G., Champagne, F. A., Mayford, M., Korzus, E., Kumar, A., Renthal, W., Theobald, D. E., and Nestler, E. J. 2005. Epigenetic mechanisms and gene networks in the nervous system. *Journal of Neuroscience*, 25, 10379–10389.

Cooper, R. M., and Zubek, J. P. 1958. Effects of enriched and restricted early environments on the learning ability of bright and dull rats. *Canadian Journal of Psychology*, 12, 159–164.

Coplan, J. D., Altemus, M., Mathew, S. J., Smith, E. L., Sharf, B., Coplan, P. M., Kral, J. G., Gorman, J. M., Owens, M. J., Nemeroff, C. B., and Rosenblum, L. A. 2005. Synchronized maternal-infant elevations of primate CSF CRF concentrations in response to variable foraging demand. *CNS Spectrums*, 10, 530–536.

Coutellier, L., Friedrich, A. C., Failing, K., Marashi, V., and Würbel, H. 2008. Effects of rat odour and shelter on maternal behaviour in C57BL/6 dams and on fear and stress responses in their adult offspring. *Physiology and Behavior*, 94, 393–404.

Coutellier, L., and Würbel, H. 2009. Early environment cues affect object recognition in adult female but not adult male C57BL/6 mice. *Behavioural Brain Research*, 203, 312–315.

Crabbe, J. C., Wahlsten, D., and Dudek, B. C. 1999. Genetics of mouse behavior: interactions with laboratory environment. *Science*, 284, 1670–1672.

Crews, D. 2008. Epigenetics and its implications for behavioral neuroendocrinology. *Frontiers in Neuroendocrinology*, 29, 344–357.

Crews, D., Fuller, T., Mirasol, E. G., Pfaff, D. W., and Ogawa, S. 2004. Postnatal environment affects behavior of adult transgenic mice. *Experimental Biology and Medicine*, 229, 935–939.

Croskerry, P. G., Smith, G. K., and Leon, M. 1978. Thermoregulation and the maternal behaviour of the rat. *Nature*, 273, 299–300.

Crowcroft, P., and Rowe, F. P. 1963. Social organization and territorial behavior in the wild house mice. *Proceedings of the Zoological Society of London*, 140, 517–531.

Curley, J. P., Davidson, S., Bateson, P., and Champagne, F. A. 2009. Social enrichment during postnatal development induces transgenerational effects on emotional and reproductive behavior in mice. *Frontiers in Behavioral Neuroscience*, 3, 1–14.

D'Andrea, I., Alleva, E., and Branchi, I. 2007. Communal nesting, an early social enrichment, affects social competences but not learning and memory abilities at adulthood. *Behavioural Brain Research*, 183, 60–66.

Darnaudery, M., and Maccari, S. 2008. Epigenetic programming of the stress response in male and female rats by prenatal restraint stress. *Brain Research Reviews*, 57, 571–585.

De Sousa, P. A., Caveney, A., Westhusin, M. E., and Watson, A. J. 1998. Temporal patterns of embryonic gene expression and their dependence on oogenetic factors. *Theriogenology*, 49, 115–128.

Drent, P. J., Van Oers, K., and Van Noordwijk, A. J. 2003. Realized heritability of personalities in the great tit (*Parus major*). *Proceedings of the Royal Society of London B*, 270, 45–51.

D'Udine, B., and Alleva, E. 1980. On the teleonomic study of maternal behaviour. In: *Dialectics of Biology and Society in the Production of Mind: The Dialectics of Biology Group* (Muir, A., and Rose, S., eds.), pp. 50–61. London: Allison and Busby.

Ebensperger, L. A. 2001. A review of the evolutionary causes of rodent group-living. *Acta Theriologica*, 46, 115–144.

Edwards, C. R. W., Benediktsson, R., Lindsay, R. S., and Seckl, J. R. 1996. 11 beta-hydroxysteroid dehydrogenases: key enzymes in determining tissue-specific glucocorticoid effects. *Steroids*, 61, 263–269.

Eichenlaub-Ritter, U. 1998. Genetics of oocyte ageing. *Maturitas*, 30, 143–169.

Fish, E. W., Shahrokh, D., Bagot, R., Caldji, C., Bredy, T., Szyf, M., and Meaney, M. J. 2004. Epigenetic programming of stress responses through variations in maternal care. *Annals of the New York Academy of Sciences*, 1036, 167–180.

Fleming, A. S., and Rosenblatt, J. S. 1974. Maternal behavior in the virgin and lactating rat. *Journal of Comparative and Physiological Psychology*, 86, 957–972.

Fleming, A. S., Kraemer, G. W., Gonzalez, A., Lovic, V., Rees, S., and Melo, A. 2002. Mothering begets mothering: the transmission of behavior and its neurobiology across generations. *Pharmacology Biochemistry and Behavior*, 73, 61–75.

Francis, D. D., Diorio, J., Liu, D., and Meaney, M. J. 1999. Nongenomic transmission across generations of maternal behavior and stress responses in the rat. *Science*, 286, 1155–1158.

Francis, D. D., Diorio, J., Plotsky, P. M., and Meaney, M. J. 2002. Environmental enrichment reverses the effects of maternal separation on stress reactivity. *Journal of Neuroscience*, 22, 7840–7843.

Francis, D. D., and Kuhar, M. J. 2008. Frequency of maternal licking and grooming correlates negatively with vulnerability to cocaine and alcohol use in rats. *Pharmacology Biochemistry and Behavior*, 90, 497–500.

Francis, D. D., Szegda, K., Campbell, G., Martin, W. D., and Insel, T. R. 2003. Epigenetic sources of behavioral differences in mice. *Nature Neuroscience*, 6, 445–446.

Frasor, J., Barkai, U., Zhong, L., Fazleabas, A. T., and Gibori, G. 2001. PRL-induced ERalpha gene expression is mediated by Janus kinase 2 (Jak2) while signal transducer and activator of transcription 5b (Stat5b) phosphorylation involves Jak2 and a second tyrosine kinase. *Molecular Endocrinology*, 15, 1941–1952.

Fuks, F. 2005. DNA methylation and histone modifications: teaming up to silence genes. *Current Opinion in Genetics and Development*, 15, 490–495.

Fuller, T., Sarkar, S., and Crews, D. 2005. The use of norms of reaction to analyze genotypic and environmental influences on behavior in mice and rats. *Neuroscience and Biobehavioral Reviews*, 29, 445–456.

Gammie, S. C. 2005. Current models and future directions for understanding the neural circuitries of maternal behaviors in rodents. *Behavioral and Cognitive Neuroscience Reviews*, 4, 119–135.

Gittleman, J. L. 1985. Functions of communal care in mammals. In: *Evolution: Essays in Honour of John Maynard Smith* (Greenwood, P. J., Harvey, P. H., Slatkin, M., eds.), pp. 187–205. Cambridge: Cambridge University Press.

Gluckman, P. D., Hanson, M. A., Spencer, H. G., and Bateson, P. 2005. Environmental influences during development and their later consequences for health and disease: implications for the interpretation of empirical studies. *Proceedings of the Royal Society of London B*, 272, 671–677.

Gonzalez, A., and Fleming, A. S. 2002. Artificial rearing causes changes in maternal behavior and c-fos expression in juvenile female rats. *Behavioral Neuroscience*, 116, 999–1013.

Gonzalez, A., Lovic, V., Ward, G. R., Wainwright, P. E., and Fleming, A. S. 2001. Intergenerational effects of complete maternal deprivation and replacement stimulation on maternal behavior and emotionality in female rats. *Developmental Psychobiology*, 38, 11–32.

Gosling, S. D. 2001. From mice to men: what can we learn about personality from animal research? *Psychological Bulletin*, 127, 45–86.

———. 2008. Personality in nonhuman animals. *Social and Personality Psychology Compass*, 2, 985–1001.

Gottlieb, G. 1998. Normally occurring environmental and behavioral influences on gene activity: from central dogma to probabilistic epigenesis. *Psychological Review*, 105, 792–802.

———. 2007. Probabilistic epigenesis. *Developmental Science*, 10, 1–11.

Groothuis, T. G. G., and Carere, C. 2005. Avian personalities: characterization and epigenesis. *Neuroscience and Biobehavioral Reviews*, 29, 137–150.

Gubernick, D. J., and Alberts, J. R. 1983. Maternal licking of young: resource exchange and proximate controls. *Physiology and Behavior*, 31, 593–601.

———. 1985. Maternal licking by virgin and lactating rats: water transfer from pups. *Physiology and Behavior*, 34, 501–506.

Gupta, A. P., and Lewontin, R. C. 1982. A study of reaction norms in natural populations of *Drosophila pseudoobscura*. *Evolution*, 36, 934–948.

Hall, W. G. 1975. Weaning and growth of artificially reared rats. *Science*, 190, 1313–1315.

Harlow, H. F. 1969. Agemate or peer affectional system. *Advances in the Study of Behavior*, 2, 333–383.

Harlow, H. F., and Suomi, S. J. 1971. Social recovery by isolation-reared monkeys. *Proceedings of the National Academy of Sciences U S A*, 68, 1534–1538.

Hayes, L. D. 2000. To nest communally or not to nest communally: a review of rodent communal nesting and nursing. *Animal Behaviour*, 59, 677–688.

Higley, J. D., King, S. T., Jr., Hasert, M. F., Champoux, M., Suomi, S. J., and Linnoila, M. 1996. Stability of interindividual differences in serotonin function and its relationship to severe aggression and competent social behavior in rhesus macaque females. *Neuropsychopharmacology*, 14, 67–76.

Hockly, E., Cordery, P. M., Woodman, B., Mahal, A., Van Dellen, A., Blakemore, C., Lewis, C. M., Hannan, A. J., and Bates, G. P. 2002. Environmental enrichment slows disease progression in R6/2 Huntington's disease mice. *Annals of Neurology*, 51, 235–242.

Hofer, M. A., Brunelli, S. A., and Shair, H. N. 1993. The effects of 24-hr maternal separation and of litter-size reduction on the isolation-distress response of 12-day-old rat pups. *Developmental Psychobiology*, 26, 483–497.

Holmes, A., le Guisquet, A. M., Vogel, E., Millstein, R. A., Leman, S., and Belzung, C. 2005. Early life genetic, epigenetic, and environmental factors shaping emotionality in rodents. *Neuroscience and Biobehavioral Reviews*, 29, 1335–1346.

Jablonka, E., and Lamb, M. J. 2002. The changing concept of epigenetics. *Annals of the New York Academy of Sciences*, 981, 82–96.

———. 2005. *Genetic, Epigenetic, Behavioral, and Symbolic Variation in the History of Life.* Cambridge: Bradford Books/MIT Press.

Janny, L., and Menezo, Y. J. 1996. Maternal age effect on early human embryonic development and blastocyst formation. *Molecular Reproduction and Development*, 45, 31–37.

Kaiser, S., and Sachser, N. 2005. The effects of prenatal social stress on behaviour: mechanisms and function. *Neuroscience and Biobehavioral Reviews*, 29, 283–294.

Kalda, A., Heidmets, L. T., Shen, H. Y., Zharkovsky, A., and Chen, J. F. 2007. Histone deacetylase inhibitors modulates the induction and expression of amphetamine-induced behavioral sensitization partially through an associated learning of the environment in mice. *Behavioural Brain Research*, 181, 76–84.

Keverne, E. B., and Curley, J. P. 2008. Epigenetics, brain evolution, and behaviour. *Frontiers in Neuroendocrinology*, 29, 398–412.

Kikusui, T., Isaka, Y., and Mori, Y. 2005. Early weaning deprives mouse pups of maternal care and decreases their maternal behavior in adulthood. *Behavioural Brain Research*, 162, 200–206.

König, B. 1994. Fitness effects of communal rearing in house mice: the role of relatedness versus familiarity. *Animal Behaviour*, 48, 1449–1457.

———. 1997. Cooperative care of young in mammals. *Naturwissenschaften*, 84, 95–104.

Koolhaas, J. M., de Boer, S. F., Buwalda, B., and Van Reenen, K. 2007. Individual variation in coping with stress: a multidimensional approach of ultimate and proximate mechanisms. *Brain, Behavior, and Evolution*, 70, 218–226.

Koolhaas, J. M., Korte, S. M., De Boer, S. F., Van Der Vegt, B. J., Van Reenen, C. G., Hopster, H., De Jong, I. C., Ruis, M. A., and Blokhuis, H. J. 1999. Coping styles in animals: current status in behavior and stress-physiology. *Neuroscience and Biobehavioral Reviews*, 23, 925–935.

Kouzarides, T. 2007. Chromatin modifications and their function. *Cell*, 128, 693–705.

Kumar, A., Choi, K. H., Renthal, W., Tsankova, N. M., Theobald, D. E., Truong, H. T., Russo, S. J., Laplant, Q., Sasaki, T. S., Whistler, K. N., Neve, R. L., Self, D. W., and Nestler, E. J. 2005. Chromatin remodeling is a key mechanism underlying cocaine-induced plasticity in striatum. *Neuron*, 48, 303–314.

Laviola, G., and Alleva, E. 1995. Sibling effects on the behavior of infant mouse litters (*Mus domesticus*). *Journal of Comparative Psychology*, 109, 68–75.

Laviola, G., and Terranova, M. L. 1998. The developmental psychobiology of behavioural plasticity in mice: the role of social experiences in the family unit. *Neuroscience and Biobehavioral Reviews*, 23, 197–213.

Laviola, G., Rea, M., Morley-Fletcher, S., Di Carlo, S., Bacosi, A., De Simone, R., Bertini, M., and Pacifici, R. 2004. Beneficial effects of enriched environment on adolescent rats from stressed pregnancies. *European Journal of Neuroscience*, 20, 1655–1664.

Lee, M. H. S., and Williams, D. I. 1974. Changes in licking behaviour of rat mother following handling of young. *Animal Behaviour*, 22, 679–681.

Lehmann, J., Pryce, C. R., Bettschen, D., and Feldon, J. 1999. The maternal separation paradigm and adult emotionality and cognition in male and female Wistar rats. *Pharmacology Biochemistry and Behavior*, 64, 705–715.

Lehmann, J., Pryce, C. R., Jongen-Relo, A. L., Stohr, T., Pothuizen, H. H., and Feldon, J. 2002. Comparison of maternal separation and early handling in terms of their neurobehavioral effects in aged rats. *Neurobiology of Aging*, 23, 457–466.

Leon, M., Croskerry, P. G., and Smith, G. K. 1978. Thermal control of mother-young contact in rats. *Physiology and Behavior*, 21, 790–811.

Levine, S. 1957. Infantile experience and resistance to physiological stress. *Science*, 126, 405.

Liu, D., Diorio, J., Day, J. C., Francis, D. D., and Meaney, M. J. 2000. Maternal care, hippocampal synaptogenesis, and cognitive development in rats. *Nature Neuroscience*, 3, 799–806.

Liu, D., Diorio, J., Tannenbaum, B., Caldji, C., Francis, D., Freedman, A., Sharma, S., Pearson, D., Plotsky, P. M., and Meaney, M. J. 1997. Maternal care, hippocampal glucocorticoid receptors, and hypothalamic-pituitary-adrenal responses to stress. *Science*, 277, 1659–1662.

Lonstein, J. S., and Gammie, S. C. 2002. Sensory, hormonal, and neural control of maternal aggression in laboratory rodents. *Neuroscience and Biobehavioral Reviews*, 26, 869–888.

Lovic, V., and Fleming, A. S. 2004. Artificially reared female rats show reduced prepulse inhibition and deficits in the attentional set shifting task: reversal of effects with maternal-like licking stimulation. *Behavioural Brain Research*, 148, 209–219.

Lucas, A. 1991. Programming by early nutrition in man. In: *The Childhood Environment and Adult Disease* (Bock, G. R., and Whelan, J., eds.), pp. 38–55. Chichester, England: Wiley.

Maccari, S., and Morley-Fletcher, S. 2007. Effects of prenatal restraint stress on the hypothalamus-pituitary-adrenal axis and related behavioural and neurobiological alterations. *Psychoneuroendocrinology*, 32, S10–S15.

Maccari, S., Piazza, P. V., Kabbaj, M., Barbazanges, A., Simon, H., and Le Moal, M. 1995. Adoption reverses the long-term impairment in glucocorticoid feedback induced by prenatal stress. *Journal of Neuroscience*, 15, 110–116.

Macrì, S., and Würbel, H. 2006. Developmental plasticity of HPA and fear responses in rats: a critical review of the maternal mediation hypothesis. *Hormones and Behavior*, 50, 667–680.

———. 2007. Effects of variation in postnatal maternal environment on maternal behaviour and fear and stress responses in rats. *Animal Behaviour*, 73, 171–184.

Manning, C. J., Dewsbury, D. A., Wakeland, E. K., and Potts, W. K. 1995. Communal nesting and communal nursing in house mice, *Mus musculus domesticus. Animal Behaviour*, 50, 741–751.

Maupin, J. L., and Riechert, S. E. 2001. Superfluous killing in spiders: a consequence of adaptation to food-limited environments? *Behavioral Ecology*, 12, 569–576.

McLeod, J., Sinal, C. J., and Perrot-Sinal, T. S. 2007. Evidence for non-genomic transmission of ecological information via maternal behavior in female rats. *Genes, Brain, and Behavior*, 6, 19–29.

Meaney, M. J., and Aitken, D. H. 1985. The effects of early postnatal handling on hippocampal glucocorticoid receptor concentrations: temporal parameters. *Brain Research*, 354, 301–304.

Meaney, M. J., Brake, W., and Gratton, A. 2002. Environmental regulation of the development of mesolimbic dopamine systems: a neurobiological mechanism for vulnerability to drug abuse? *Psychoneuroendocrinology*, 27, 127–138.

Meaney, M. J., Aitken, D. H., Bhatnagar, S., and Sapolsky, R. M. 1991. Postnatal handling attenuates certain neuroendocrine, anatomical, and cognitive dysfunctions associated with aging in female rats. *Neurobiology of Aging*, 12, 31–38.

Meaney, M. J., Aitken, D. H., Sharma, S., Viau, V., and Sarrieau, A. 1989. Postnatal handling increases hippocampal type II glucocorticoid receptors and enhances adrenocortical negative-feedback efficacy in the rat. *Neuroendocrinology*, 50, 597–604.

Meaney, M. J., Diorio, J., Francis, D. D., Widdowson, J., LaPlante, P., Caldji, C., Sharma, S., Seckl, J. R., and Plotsky, P. M. 1996. Early environmental regulation of forebrain glucocorticoid receptor gene expression: implications for adrenocortical responses to stress. *Developmental Neuroscience*, 18, 49–72.

Mendl, M., and Paul, E. S. 1991a. Litter composition affects parental care, offspring growth, and the development of aggressive behavior in wild house mice. *Behaviour*, 116, 90–108.

———. 1991b. Parental care, sibling relationships, and the development of aggressive behavior in 2 lines of wild house mice. *Behaviour*, 116, 11–41.

Mersfelder, E. L., and Parthun, M. R. 2006. The tale beyond the tail: histone core domain modifications and the regulation of chromatin structure. *Nucleic Acids Research*, 34, 2653–2662.

Moffett, M. C., Vicentic, A., Kozel, M., Plotsky, P., Francis, D. D., and Kuhar, M. J. 2007. Maternal separation alters drug intake patterns in adulthood in rats. *Biochemical Pharmacology*, 73, 321–330.

Moore, C. L., and Power, K. L. 1986. Prenatal stress affects mother-infant interaction in Norway rats. *Developmental Psychobiology*, 19, 235–245.

Muller, U., Cristina, N., Li, Z. W., Wolfer, D. P., Lipp, H. P., Rulicke, T., Brandner, S., Aguzzi, A., and Weissmann, C. 1994. Behavioral and anatomical deficits in mice homozygous for a modified beta-amyloid precursor protein gene. *Cell*, 79, 755–765.

Myers, M. M., Brunelli, S. A., Shair, H. N., Squire, J. M., and Hofer, M. A. 1989.

Relationships between maternal behavior of SHR and WKY dams and adult blood pressures of cross-fostered F1 pups. *Developmental Psychobiology*, 22, 55–67.

Namikas, J., and Wehmer, F. 1978. Gender composition of the litter affects behavior of male mice. *Behavioral Biology*, 23, 219–224.

Nothias, J. Y., Majumder, S., Kaneko, K. J., and DePamphilis, M. L. 1995. Regulation of gene expression at the beginning of mammalian development. *Journal of Biological Chemistry*, 270, 22077–22080.

Patin, V., Lordi, B., Vincent, A., and Caston, J. 2005. Effects of prenatal stress on anxiety and social interactions in adult rats. *Developmental Brain Research*, 160, 265–274.

Petit-Demouliere, B., Chenu, F., and Bourin, M. 2005. Forced swimming test in mice: a review of antidepressant activity. *Psychopharmacology*, 177, 245–255.

Platt, S. A., and Sanislow, C. A. 1988. Norm-of-reaction: definition and misinterpretation of animal research. *Journal of Comparative Psychology*, 102, 254–261.

Ploj, K., Roman, E., and Nylander, I. 2003. Long-term effects of maternal separation on ethanol intake and brain opioid and dopamine receptors in male Wistar rats. *Neuroscience*, 121, 787–799.

Plotsky, P. M., and Meaney, M. J. 1993. Early, postnatal experience alters hypothalamic corticotropin-releasing factor (CRF) mRNA, median eminence CRF content, and stress-induced release in adult rats. *Molecular Brain Research*, 18, 195–200.

Priebe, K., Romeo, R. D., Francis, D. D., Sisti, H. M., Mueller, A., McEwen, B. S., and Brake, W. G. 2005. Maternal influences on adult stress and anxiety-like behavior in C57BL/6J and BALB/cJ mice: a cross-fostering study. *Developmental Psychobiology*, 47, 398–407.

Pryce, C. R., and Feldon, J. 2003. Long-term neurobehavioural impact of the postnatal environment in rats: manipulations, effects, and mediating mechanisms. *Neuroscience and Biobehavioral Reviews*, 27, 57–71.

Rampon, C., Tang, Y. P., Goodhouse, J., Shimizu, E., Kyin, M., and Tsien, J. Z. 2000. Enrichment induces structural changes and recovery from nonspatial memory deficits in CA1 NMDAR1-knockout mice. *Nature Neuroscience*, 3, 238–244.

Réale, D., Reader, S. M., Sol, D., McDougall, P. T., and Dingemanse, N. J. 2007. Integrating animal temperament within ecology and evolution. *Biological Reviews*, 82, 291–318.

Roceri, M., Cirulli, F., Pessina, C., Peretto, P., Racagni, G., and Riva, M. A. 2004. Postnatal repeated maternal deprivation produces age-dependent changes of brain-derived neurotrophic factor expression in selected rat brain regions. *Biological Psychiatry*, 55, 708–714.

Rosenblatt, J. S. 1967. Nonhormonal basis of maternal behavior in the rat. *Science*, 156, 1512–1514.

———. 1975. Prepartum and postpartum regulation of maternal behaviour in the rat. *Ciba Foundation Symposia*, 5, 17–37.

Rosenblatt, J. S., and Lehrman, D. S. 1963. Maternal behavior of the laboratory rat. In: *Maternal Behavior in Mammals* (Rheingold, H., ed.), pp. 8–57. New York: Wiley.

Rosenzweig, M. R. 2003. Effects of differential experience on the brain and behavior. *Developmental Neuropsychology*, 24, 523–540.

Sairanen, M., Lucas, G., Ernfors, P., Castren, M., and Castren, E. 2005. Brain-derived neurotrophic factor and antidepressant drugs have different but coordinated effects on neuronal turnover, proliferation, and survival in the adult dentate gyrus. *Journal of Neuroscience*, 25, 1089–1094.

Sayler, A., and Salmon, M. 1969. Communal nursing in mice: influence of multiple mothers on the growth of the young. *Science*, 164, 1309–1310.

———. 1971. An ethological analysis of communal nursing by the house mouse (*Mus musculus*). *Behaviour*, 40, 60–85.

Seckl, J. R. 2004. Prenatal glucocorticoids and long-term programming. *European Journal of Endocrinology*, 27, 74–78.

Shoji, H., and Kato, K. 2006. Maternal behavior of primiparous females in inbred strains of mice: a detailed descriptive analysis. *Physiology and Behavior*, 89, 320–328.

Sih, A., Bell, A. M., and Johnson, J. C. 2004a. Behavioral syndromes: an ecological and evolutionary overview. *Trends in Ecology and Evolution*, 19, 372–378.

Sih, A., Bell, A. M., Johnson, J. C., and Ziemba, R. E. 2004b. Behavioral syndromes: an integrative overview. *Quarterly Review of Biology*, 79, 241–277.

Spires, T. L., Grote, H. E., Varshney, N. K., Cordery, P. M., Van Dellen, A., Blakemore, C., and Hannan, A. J. 2004. Environmental enrichment rescues protein deficits in a mouse model of Huntington's disease, indicating a possible disease mechanism. *Journal of Neuroscience*, 24, 2270–2276.

Stapley, J., and Keogh, J. S. 2005. Behavioral syndromes influence mating systems: floater pairs of a lizard have heavier offspring. *Behavioral Ecology*, 16, 514–520.

Stern, J. M., and Johnson, S. K. 1990. Ventral somatosensory determinants of nursing behavior in Norway rats, I: effects of variations in the quality and quantity of pup stimuli. *Physiology and Behavior*, 47, 993–1011.

Stohr, T., Wermeling, D. S., Szuran, T., Pliska, V., Domeney, A., Welzl, H., Weiner, I., and Feldon, J. 1998. Differential effects of prenatal stress in two inbred strains of rats. *Pharmacology Biochemistry and Behavior*, 59, 799–805.

Sullivan, R. M., and Hall, W. G. 1988. Reinforcers in infancy: classical conditioning using stroking or intra-oral infusions of milk as UCS. *Developmental Psychobiology*, 21, 215–223.

Sullivan, R. M., Wilson, D. A., and Leon, M. 1988. Physical stimulation reduces the brain temperature of infant rats. *Developmental Psychobiology*, 21, 237–250.

Suomi, S. J. 1979. Peers, play, and primary prevention in primates. In: *Primary Prevention of Psychopathology: Social Competence in Children* (Kent, M., and Rolf, J., eds.), pp. 127–149. Hanover, NH: Press of New England.

Suomi, S. J. 1997. Early determinants of behaviour: evidence from primate studies. *British Medical Bulletin*, 53, 170–184.

Suomi, S. J. 2005. Mother-infant attachment, peer relationships, and the development of social networks in rhesus monkeys. *Human Development*, 48, 67–79.

Suomi, S. J. 2006. Risk, resilience, and gene x environment interactions in rhesus monkeys. *Annals of the New York Academy of Sciences*, 1094, 52–62.

Symonds, M. E., Budge, H., Stephenson, T., and McMillen, I. C. 2001. Fetal endocrinology and development: manipulation and adaptation to long-term nutritional and environmental challenges. *Reproduction*, 121, 853–862.

Szyf, M., Weaver, I. C., Champagne, F. A., Diorio, J., and Meaney, M. J. 2005. Maternal programming of steroid receptor expression and phenotype through DNA methylation in the rat. *Frontiers in Neuroendocrinology*, 26, 139–162.

Tahakashi, L. K., and Lore, R. K. 1983. Play-fighting and the development of agonistic behaviour in male and female rats. *Aggressive Behavior*, 9, 217–227.

Tremml, P., Lipp, H. P., Muller, U., and Wolfer, D. P. 2002. Enriched early experiences of mice underexpressing the beta-amyloid precursor protein restore spatial learning capabilities but not normal openfield behavior of adult animals. *Genes, Brain, and Behavior*, 1, 230–241.

Tremml, P., Lipp, H. P., Muller, U., Ricceri, L., and Wolfer, D. P. 1998. Neurobehavioral development, adult openfield exploration, and swimming navigation learning in mice with a modified beta-amyloid precursor protein gene. *Behavioural Brain Research*, 95, 65–76.

Tryon, R. C. 1942. Individual differences. In: *Comparative Psychology* (2nd ed., Moss, F. A., ed.), pp. 330–365. New York: Prentice-Hall.

Tsankova, N. M., Kumar, A., and Nestler, E. J. 2004. Histone modifications at gene promoter regions in rat hippocampus after acute and chronic electroconvulsive seizures. *Journal of Neuroscience*, 24, 5603–5610.

Tsankova, N. M., Berton, O., Renthal, W., Kumar, A., Neve, R. L., and Nestler, E. J. 2006. Sustained hippocampal chromatin regulation in a mouse model of depression and antidepressant action. *Nature Neuroscience*, 9, 519–525.

Tsien, J. Z., Huerta, P. T., and Tonegawa, S. 1996. The essential role of hippocampal CA1 NMDA receptor-dependent synaptic plasticity in spatial memory. *Cell*, 87, 1327–1338.

Vallee, M., Mayo, W., Dellu, F., Le Moal, M., Simon, H., and Maccari, S. 1997. Prenatal stress induces high anxiety and postnatal handling induces low anxiety in adult offspring: correlation with stress-induced corticosterone secretion. *Journal of Neuroscience*, 17, 2626–2636.

Van Dellen, A., Blakemore, C., Deacon, R., York, D., and Hannan, A. J. 2000. Delaying the onset of Huntington's in mice. *Nature*, 404, 721–722.

Waddington, C. H. 1957. *The Strategy of the Genes*. London: Allen and Unwin.

Wahlsten, D., Bachmanov, A., Finn, D. A., and Crabbe, J. C. 2006. Stability of inbred mouse strain differences in behavior and brain size between laboratories and across decades. *Proceedings of the National Academy of Sciences U S A*, 103, 16364–16369.

Wassarman, P. M., and Kinloch, R. A. 1992. Gene expression during oogenesis in mice. *Mutation Research*, 296, 3–15.

Weaver, I. C., Meaney, M. J., and Szyf, M. 2006. Maternal care effects on the hippocampal transcriptome and anxiety-mediated behaviors in the offspring that are reversible in adulthood. *Proceedings of the National Academy of Sciences U S A*, 103, 3480–3485.

Weaver, I. C., D'Alessio, A. C., Brown, S. E., Hellstrom, I. C., Dymov, S., Sharma, S., Szyf, M., and Meaney, M. J. 2007. The transcription factor nerve growth factor-inducible protein a mediates epigenetic programming: altering epigenetic marks by immediate-early genes. *Journal of Neuroscience*, 27, 1756–1768.

Weaver, I. C., Cervoni, N., Champagne, F. A., D'Alessio, A. C., Sharma, S., Seckl, J. R., Dymov, S., Szyf, M., and Meaney, M. J. 2004. Epigenetic programming by maternal behavior. *Nature Neuroscience*, 7, 847–854.

Weaver, I. C. G., Champagne, F. A., Brown, S. E., Dymov, S., Sharma, S., Meaney, M. J., and Szyf, M. 2005. Reversal of maternal programming of stress responses in adult offspring through methyl supplementation: altering epigenetic marking later in life. *Journal of Neuroscience*, 25, 11045–11054.

Weinstock, M. 2008. The long-term behavioural consequences of prenatal stress. *Neuroscience and Biobehavioral Reviews*, 32, 1073–1086.

Weinstock, M., Fride, E., and Hertzberg, R. 1988. Prenatal stress effects on functional development of the offspring. *Progress in Brain Research*, 73, 319–331.

Weissgerber, T. L., and Wolfe, L. A. 2006. Physiological adaptation in early human pregnancy: adaptation to balance maternal-fetal demands. *Applied Physiology of Nutrition and Metabolism*, 31, 1–11.

Weller, A., Leguisamo, A. C., Towns, L., Ramboz, S., Bagiella, E., Hofer, M., Hen, R., and Brunner, D. 2003. Maternal effects in infant and adult phenotypes of 5HT1A and 5HT1B receptor knockout mice. *Developmental Psychobiology*, 42, 194–205.

West-Eberhard, M. J. 2003. *Developmental Plasticity and Evolution*. Oxford: Oxford University Press.

Wolfer, D. P., Litvin, O., Morf, S., Nitsch, R. M., Lipp, H. P., and Würbel, H. 2004. Laboratory animal welfare: cage enrichment and mouse behaviour. *Nature*, 432, 821–822.

Zhang, T. Y., Chretien, P., Meaney, M. J., and Gratton, A. 2005. Influence of naturally occurring variations in maternal care on prepulse inhibition of acoustic startle and the medial prefrontal cortical dopamine response to stress in adult rats. *Journal of Neuroscience*, 25, 1493–1502.

11

Parental Influences on Offspring Personality

Traits in Oviparous and Placental Vertebrates

TON G. G. GROOTHUIS AND DARIO MAESTRIPIERI

Introduction

Although the study of animal personality has become a flourishing field over the past decade, the ontogeny of personality has received relatively little attention (see Stamps and Groothuis 2010a, b). This may be due to several reasons: first, the main goal of early studies was to explore the very existence of animal personality and describe its nature rather than understand its development; second, trait stability over time is one defining characteristic of personality while, by definition, ontogeny deals with change over time; third, early studies often used selection lines, emphasizing the role of genes rather than environment in shaping animal personality. However, all developmental processes are the result of a continuous interaction between genes and other internal and external factors, and evidence for one does not exclude evidence for the others. Moreover, insight into the development of personality is important for several reasons: first, it may enhance our knowledge of the degree of heritability and flexibility of animal personalities, their developmental constraints, and their life-history–dependent changes, which is extremely useful for understanding the ecology and evolution of animal personality; second, it may reveal underlying and perhaps age-specific mechanisms useful for understanding the physiology and neurobiology of animal personality (for a more extensive discussion on this topic, see Stamps and Groothuis 2010a, b).

Chapter 10 in this book addresses many aspects of the ontogeny of animal personality, and particularly what studies of laboratory rodents have revealed about the molecular mechanisms through which early experience can result in long-term epigenetic modifications of behavior and neuroendocrine function. This chapter focuses on a particular aspect of ontogeny: maternal effects on personality development. As mentioned above, organisms develop under the interaction of genes and environment and parents not only transfer the genetic information to their offspring but can also affect the environment in which they develop. Parental or maternal effects refer to influences of the parental phenotype on the offspring phenotype, which occur independent of offspring genotype (Mousseau and Fox 1998; Maestripieri and Mateo 2009). Since it is the mother that provisions the egg, and in many species mothers provide more parental care than fathers, parental effects are often labeled maternal effects. When they were first discovered, such effects were seen as annoying noise in breeding programs or a disturbance of development. Given our greater understanding of adaptive developmental plasticity and our recent interest in variation in gene expression, many maternal effects are now viewed as adaptations (Mousseau and Fox 1998; Maestripieri and Mateo 2009; Schwabl and Groothuis 2010).

Maternal effects can be genetic or environmental. When the individual variation in the maternal phenotypic traits that affect the offspring's phenotype has a significant genetic basis, this variation can be subject to natural selection. In other words, natural selection can favor the evolution of maternal genes whose effects are expressed in the offspring's phenotype, that is, "maternal-effect genes." Genetic maternal effects can have profound effects on the rate of evolutionary change, since the parents transmit not only their genes to the offspring, but also the phenotypic consequences of these genes. This may also lead to evolution in the opposite direction to selection (e.g., Kirkpatrick and Lande 1989). Environmental maternal effects occur when the individual variation in the maternal phenotypic traits that affect the offspring's phenotype originates in the parent's environment. Environmental maternal effects allow parents to affect their offspring depending on the information received from the environment, thereby adjusting offspring development in relation to prevailing conditions and maximizing parental fitness. Maternal effects, both genetic and environmental, can lead to parent-offspring similarities and thereby inflate estimates for genetic heritability. Phenotypic similarities between parents and offspring can include similarities in personalities and their behavioral and physiological components. Although studies of rodents and primates have provided evidence for intergenerational transmission of social and maternal behavior from mothers to daughters (see below), little is known about the intergenerational transmission of personality dimensions via genetic or environmental maternal effects.

There are many different types of maternal effects. They include postnatal food provisioning and other forms of parental care, which affect offspring body condition, metabolism, and, later, behavior. Nutritional maternal effects can be relevant for personality development since it has been suggested that personality traits should be adjusted to an individual's morphology and physiology (Stamps 2007). Maternal effects can also occur in the social domain. For example, in cercopithecine monkeys, the mother's dominance rank can affect the growth rate, timing of first reproduction, and adult behavior of the offspring (Maestripieri 2009). Both in oviparous and placental vertebrates, maternal effects may involve social facilitation of learning and imprinting processes concerning the appearance of the parents, or the habitat in which the young are raised, or the food provided by the mother, affecting later social preferences, habitat preferences, and food preferences (e.g., Mateo 2009). Some maternal effects are more indirect: for example, parental nest site choice may determine the social environment of the offspring, affecting in turn hormone production, which may have long-term consequences for behavior (e.g., Ros et al. 2002). Other maternal effects are even subtler and easily overlooked. This is especially the case of prenatal maternal effects. For example, the claim made in some studies that differences between animals of different artificial selection lines must be genetic since cross-fostering after birth or hatching was performed does not take prenatal maternal effects into account. Prenatal effects can be strong, since the mother bestows her eggs with different amounts of nutrients and bioactive substances such as hormones, immune factors, and antioxidants known to affect offspring development. However, these effects do not need to be exclusively maternal. For example, fathers may affect hormone levels in the mothers that, in turn, affect egg hormone levels, or males in some fish species may provide eggs with their own hormones transferred via their urine (Groothuis et al. 2005). Prenatal hormonal maternal effects are especially interesting for personality development (Groothuis and Carere 2005) since hormones are known to have pleiotropic effects and coordinate the developmental programs of a wide range of traits. Moreover, early exposure to steroid hormones may have long-term "organizational" effects on the phenotype, which include brain structure and function as well as the activity of other physiological and neurochemical systems.

Unfortunately, only a handful of studies have investigated maternal effects on the behavior of offspring from the perspective of personality research. We define personality as behavioral differences among individuals of the same species that are consistent over time and across contexts (see introduction to this book). Using this definition, personality is not an easy concept to quantify. First, to account for consistency over time behavior should be measured repeatedly, which is often difficult to do in animal studies, particularly in the field. Second, personality does not refer to absolute rates or frequencies of behavior, but to an individual's rates relative to those of others. Absolute rates of behavior may vary dramatically in the course of ontogeny, while differences between individuals in their relative rates of behavior may remain constant. Third, behavior needs to be measured in different contexts. This may include both situations in which animals perform the same type of behavior in different contexts (for example, aggression toward conspecifics and toward predators) and situations in which animals perform different types of behavior in different contexts (for example, aggression toward conspecifics and latency to approach a novel object).

Despite the difficulty of studying personality development in animals, some studies have documented how correlations between different personality traits change across ontogeny (e.g., Dingemanse et al. 2002; Bell and Stamps 2004; Carere et al. 2005b; Weiss et al. 2007), or have experimentally tested whether a specific type of experience at a given age affects correlations between personality traits later in life (e.g., Sluyter et al. 1996; Benus 1999; Carere et al. 2005a; Bell and Sih 2007; Frost et al. 2007). Only a few of these studies, however, have considered maternal effects. Therefore, in this chapter, we will also discuss studies of maternal effects on the development of single personality traits: that is, behavioral traits that are commonly recognized as being an important component of personality in a particular species. In addition, we will consider maternal effects on physiological systems that are likely involved in personality, such as hormones and the serotonergic and dopaminergic systems (e.g., Van Hierden et al. 2002; Fidler et al. 2007) (see figure 11.1.).

Perhaps owing to the greater similarities between humans and other mammals compared with between humans and oviparous vertebrates, the

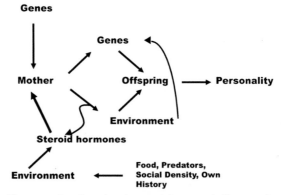

Figure 11.1. Genetic and environmental maternal effects on the development of offspring personality and their underlying hormonal mechanisms in oviparous and placental vertebrates.

literature on maternal effects on personality (a concept originally borrowed from the human literature) is much more extensive for mammals such as primates and rodents than for oviparous organisms such as fish, reptiles, or birds. Therefore, in each of the sections of this chapter, we will first discuss parental influences on offspring personality traits in rodents and primates and then in oviparous vertebrates. Another reason for discussing oviparous and placental vertebrates separately is the different focus of research on maternal effects in these taxa. A significant fraction of the work on hormone-mediated maternal programming of the fetus in humans and other mammals has been conducted in the context of health research focusing on factors predicting vulnerability to disease. This work has had a strong emphasis on stress and glucocorticoid hormones and has often employed laboratory rodents or nonhuman primates as animal models for humans. Although research with mammals, particularly in the field, has also considered maternal effects as evolutionary adaptations (Maestripieri and Mateo 2009), adaptation has been the exclusive focus of virtually all research on maternal effects in oviparous vertebrates. Since in birds the embryo develops outside the mother's body, and egg provisioning takes place in a well-defined period, descriptive and experimental research on hormone-mediated prenatal effects is strongly facilitated. Bird studies, in contrast to those on mammalian species, have focused mainly on gonadal hormones, especially androgens, which are present in substantial quantities in eggs, while evidence for substantial levels of glucocorticoids is lacking (Groothuis et al. 2005; Rettenbacher et al. 2009).

In the following sections, we will first briefly describe personality dimensions in placental and oviparous vertebrates, and then discuss prenatal and postnatal maternal effects on personality development. In the last section we present some general conclusions and discuss avenues for future research.

Personality dimensions in placental and oviparous vertebrates
RODENTS AND PRIMATES
In rodents, individual differences in behavior that may be attributed to personality have been investigated almost exclusively in the context of laboratory studies of rats and mice. As discussed in chapters 10 and 15, the measures of behavior for which there are stable differences between individuals as well as differences between genetic strains are as follows: anxiety (measured by thigmotaxis and defecation in an open field, or proportion of time spent in closed/dark areas of the elevated plus maze or light-dark box), exploration or locomotor activity (measured as distance traveled in an open

field apparatus), defensive burying of a probe that provides a mild electric shock, aggressive behavior (measured by the latency to attack an intruder in a resident-intruder test), and sociability (measured by the tendency to play in rats or the frequency of contact and proximity between socially housed individuals) (see chapter 15 for references). In a few studies using mouse genetic lines selected for different coping strategies, additional behaviors have been found to differ between the lines, such as adjustment of the biological clock to changing light-dark regimen, and performance in a maze after changing details in the maze. This had led to the idea that these selection lines, and perhaps the coping strategies of different personalities in general, differ in the extent to which behavior is guided by external cues (reactive animals) or internal cues (routine formation in proactive animals: see Benus et al. 1987; 1988; 1990; Koolhaas et al. 1999).

Similar to those in rodents, the main personality dimensions identified in nonhuman primates include anxiety, exploration, locomotor activity, aggressiveness, and sociability (Gosling et al. 2003). These aspects of primate personalities have been studied either with behavioral measures or with subjective ratings. Anxiety is sometimes referred to as Emotionality, Fearfulness, or simply Reactivity, and is generally conceptualized as emotional reactivity to the environment (Higley and Suomi 1989; Clarke and Boinski 1995). In some studies Anxiety has been measured by rates of displacement activities such as scratching, or by vigilance behavior (e.g., Maestripieri et al. 1992; Maestripieri 2000), while in others it has been assessed with subjective ratings (e.g., Bolig et al. 1992). High- and low-reactive monkeys have sometimes been referred to as "uptight" and "laid-back" (e.g., Bolig et al. 1992; Suomi 1987; 1991). Similar to Anxiety, Exploration is also conceptualized as reactivity to novelty, but instead of being measured with emotional behavior or with subjective ratings, it has been measured by the tendency to explore novel environments, or to approach and interact with novel objects. Individuals who score high in Exploration have been described as curious, inquisitive, enterprising, or risk-prone (e.g., Buirski et al. 1978; Simpson et al. 1989; Simpson and Datta 1990; Fairbanks and McGuire 1988; King and Figueredo 1997). Related to Exploration, but without the novelty component, is an animal's general activity level as a personality dimension. This has been measured by amount of time spent locomoting, or amount of space covered during locomotion, or general levels of body movements (Reite and Short 1980; Bard and Gardner 1996) and has been rated on subjective traits as active or inactive (McGuire et al. 1994). Aggressiveness and sociability have been identified as social dimensions of personality in primates. Aggressiveness has been assessed by such behavioral measures as attacking another individual (Capitanio 1984) and ratings of traits like irritable or aggressive (Chamove et al. 1972; Martau et al. 1985; Bolig et al. 1992). Sociability has been measured by such behaviors as the proximity to other group members (Byrne and Suomi 1995) and frequency of social encounters (Chamove et al. 1972) and rated on traits like sociable, playful, or solitary (Gold and Maple 1994; Clarke and Snipes 1998).

There is a general consensus that individual differences in personality dimensions among adult rodents and nonhuman primates are likely to be accompanied by differences in neurochemical and neuroendocrine substrates regulating emotional and social processes. Neuropeptides and hormones of the hypothalamic-pituitary-adrenal (HPA) axis such as corticotropin-releasing hormone (CRH), adrenocorticotropic hormone (ACTH), and cortisol, along with the brain monoamine neurotransmitters norepinephrine, serotonin, and dopamine would be likely candidates, as these substances play an important role in the regulation of emotional and behavioral processes and their concentrations can be relatively stable over long periods of time (Maestripieri 2010). Neuropeptides such as oxytocin and beta-endorphins may be responsible for individual differences in sociability (Maestripieri 2010).

OVIPAROUS VERTEBRATES

Personality in oviparous vertebrates is generally quantified on the basis of standard behavioral tests and not assessed with subjective ratings. These tests often involve measuring reactivity to novelty, such as a novel object or novel food; reactivity is measured in terms of latency to approach the object, or distance from which the object is inspected, or frequency of alarm calls given in response to the object. Other tests include exploration in a novel environment, in which patterns of locomotor activity or patterns or latencies of visits to different compartments of the experimental apparatus are scored. These scores are often used as measures of boldness or shyness (e.g., fish: Coleman and Wilson 1998; birds: Drent et al. 2003; Riedstra 2003). Fear is sometimes measured as the number of trials necessary to induce tonic immobility (a predatory response in birds) or the duration of the time interval in which the animal stays immobile (e.g., Van Hierden et al. 2002). Risk assessment has been measured by the time taken by a bird to return to a place where food is provided, after it has been startled (Van Oers et al. 2004; 2005a, b). In the social domain, personality-related behavioral traits are assessed by measuring the latency to attack, or the number of threat displays and attacks in a standard intruder test (e.g., birds: Carere et al. 2005b) as well as the reaction to the odor of a conspecific (lizards: Cote and Clobert 2007).

The physiological or molecular bases of individual differences in behavioral responsiveness to personality tests in oviparous vertebrates are only beginning to be investigated. Differences in corticosterone basal levels or in corticosterone responses to stress may be related to differences in fearfulness, while differences in early androgen exposure may underlie differences in boldness (Groothuis and Carere 2005). Recently, it has been suggested that genetic polymorphisms in dopamine receptors may be related to personality differences in a song bird species (Fidler et al. 2007).

Prenatal maternal stress and offspring personality development
RODENTS AND PRIMATES

There is a large body of research literature on the effects of maternal stress experienced during gestation on the behavioral and neuroendocrine development of the offspring in rodents (see also chapter 10). Experimental studies of laboratory rats and mice have used a wide range of stressors (e.g., unpredictable noise, restraint stress, cold water immersion, or electric shock to the tail), at different stages of gestation, and of varying levels of intensity and duration. The effects of gestational stress on offspring personality dimensions such as anxiety or exploration vary greatly in relation to the type and intensity of stress and the timing of its occurrence. In some cases, male and female offspring are differentially affected by the stress. Finally, in many cases, the developmental effects of the prenatal stress (i.e., the effects of maternal stress on the embryo or fetus) are observed at some ages (usually early in life) but not in others (see Welberg and Seckl 2001; Kofman 2002; Maccari et al. 2003 for reviews). General locomotor activity and exploration may or may not be reduced in prenatally stressed rodents, while sociability is generally reduced. The most consistently reported effect of prenatal stress is greater anxiety or fearfulness in response to novelty. Physiologically, this behavioral effect is accompanied by chronic hyperreactivity of the HPA axis, including elevated glucocorticoid hormones, as well as by alterations of brain neurotransmitter systems (Maccari et al. 2003; Kapoor et al. 2006). Some rodent studies, however, have shown that maternal gestational stress can result in hyporeactivity of the HPA axis and lowered, rather than higher, glucocorticoid hormone levels in the offspring. Increased anxiety and fearfulness, along with HPA axis hyperreactivity, may be adaptive for offspring that will develop in a harsh or unpredictable environment. Although the same has been argued also for reduced anxiety and HPA hyporeactivity, unequivocal evidence that this adaptive view is valid is still lacking. Whether adaptive or maladaptive, the effects of maternal gestational stress on offspring personality development and its physiological substrates are probably mediated by stress-induced increased secretion of maternal glucocorticoids during gestation, which in turn affect prenatal brain and neuroendocrine development in the offspring (Welberg and Seckl 2001).

The effects of maternal gestational stress on offspring behavioral and neuroendocrine development in nonhuman primates have been investigated in several studies by Schneider and collaborators (see Schneider et al. 2002, for a review). Most of these studies investigated the effects of unpredictable noise during pregnancy on infant development in rhesus monkeys. Prenatally stressed individuals showed reduced exploratory behavior, increased reactivity to novelty (e.g., more disturbance behavior, stereotypies, self-clasping, and freezing), and lower sociability in various testing situations. Some of these effects were observed not only in infancy, but up to 4 years later. As in rodents, these behavioral modifications were accompanied by enhanced HPA axis basal activity and reactivity to stress, as well as by alterations in brain monoamine neurotransmitters. Furthermore, as in rodents, these effects appeared to be mediated by maternal glucocorticoid responses to stress during pregnancy (see Schneider et al. 2002, for a review).

Taken together, the results of rodent and primate studies, along with those of human research, suggest that the development of mammalian personalities is very sensitive to prenatal programming from the environment, that mothers mediate the influence of the environment on offspring personality development, and that they do so through altered secretion of glucocorticoid hormones and the subsequent effects of maternal hormones of offspring neural, neuroendocrine, and neurochemical development.

OVIPAROUS VERTEBRATES

Studies on the effects of prenatal maternal stress on offspring phenotype have been carried out in fish, lizards, and birds in three different ways: exposure of egg-producing females to stressors; treating egg-producing or pregnant females with the steroid hormones corticosterone or cortisol under the assumption that maternal stress induces elevation of these hormones; and exposing freshly laid eggs to elevation of these hormones, under the assumption that elevation of stress hormones in the maternal circulation during episodes of maternal stress is reflected in the egg. These assumptions are reasonable since some evidence exists that exposure to stressors during egg production elevates glucocorticoids in the egg (in fish: heterospecific interactions or crowding: McCormick 1999; 2006; 2009; in birds: exposure to predator: Saino et al. 2005; prevention of habituation to humans: Bertin et al. 2008; see also Hayward and Wingfield 2004; Hayward et al. 2005). However, recent studies indicate that the corticosterone measured in the egg may in fact be another hormone (birds: Rettenbacher et al. 2009) or a conjugated and inactive form of corticosterone (turtles: Paitz and Bowden 2008). Nevertheless, mothers may change the composition of their eggs when exposed to stress through mechanisms other than stress hormones. This may occur, for example, by decreasing testosterone concentrations in the egg (Bertin et al. 2008).

Most studies have analyzed the effects of maternal prenatal stress on offspring body mass, growth, survival, or morphology, and not on personality traits. There are nevertheless a few studies of birds whose findings are partly consistent with the results of mammalian studies. In quails, mothers prevented from habituating to human disturbance produced offspring showing low emotional reactivity compared with those of habituated mothers (Bertin and Richard-Yris 2004; Richard-Yris et al. 2005). Low emotional reactivity was expressed in several different contexts, such as exposure to a novel object, reaction to humans, tonic immobility, and behavior in an open field. The hypothesis that this effect is mediated by maternal induction of elevated levels of corticosterone in the egg is consistent with the finding in the same species that egg injections with corticosterone lower the HPA axis activity in the chick (Hayward et al. 2006). However, two other reports contradict this result: in one of them, maternal stress induced higher HPA activity in quail offspring (Hayward and Wingfield 2004), while in the other egg injections with corticosterone increased fearfulness (as measured by tonic immobility) after handling in the domestic chick (Janczak et al. 2007). The inconsistent results of the egg injections may be due to different dose-response relationships in different species and/or to the administration of extremely high levels of hormones. The other inconsistencies between studies may suggest that maternal stress enhances levels of corticosterone in the egg in some cases but not in others, and that stressed mothers affect physiological variables in the egg other than just corticosterone levels. Consistent with these hypotheses, a study in quails has shown that selection for fearfulness, as measured by tonic immobility, gave rise to mothers producing eggs with lower androgen and progesterone concentrations, and these maternal hormones can have a wide array of effects on chick behavior (see below). Finally, in a viviparous lizard, treatment of the mother with corticosterone and the mother's own body condition jointly affected the offspring's dispersal tendency, a trait related to personality in birds (Meylan et al. 2002). Offspring were also more attracted to the odor of the mother (De Fraipont et al. 2000), which may be a personality trait in this species (Cote and Clobert 2007). So, as in rodents, maternal stress affects the embryo, and the timing and the way in which this occurs can affect subsequent offspring development.

Other prenatal maternal effects

In addition to prenatal exposure to stress and elevated levels of glucocorticoid hormones, prenatal exposure to varying levels of the gonadal hormones androgens and estrogen can also potentially affect personality development. In humans and monkeys, elevated levels of maternal androgens can masculinize daughters both morphologically and behaviorally, for example, by increasing rough and tumble play, aggressiveness, and other personality traits (for a review see Cohen-Bendahan et al. 2005).

It has been suggested that natural variation in maternal androgens affects the behavioral phenotype of the offspring also in wild animals. There are some wild mammalian species, such as spotted hyenas, in which the females exhibit masculinized genitalia and behavior, showing penis-like structures; are dominant over males; and show increased levels of aggression. There is some debate as to whether pregnant mothers of these species show elevated levels of androgens, which may be responsible for the masculinization of their daughters. Hormone manipulations of pregnant females indicate that the development of the female pseudopenis is only partly depending on maternal hormones, while correlative evidence suggests that maternal androgens do affect offspring behavior (Holekamp and Dloniak 2009).

Another example of within-sex differentiation due to prenatal exposure to gonadal steroids is the effect of the intra-uterine position of embryos in mice, gerbils, and pigs. In gerbils, female embryos positioned in between two males have been shown to be influenced by the hormones of their neighboring brothers, in terms of their morphology, endocrine system, and behavior such as aggression and mating (e.g., Clark and Galef 1995; Vandenbergh 2009). At first glance this might seem an effect of siblings and not a maternal effect. However, the sex ratio of the litter is determined by the mother (and interestingly by the intra-uterine position too, potentially inducing intergenerational effects), and this determines the chance that a fetus is positioned in between two brothers.

To test experimentally the effect of prenatal maternal environment on the development of personality, Sluyter et al. (1996) transplanted mouse embryos of a selection line for short attack latencies to females of a selection line for long attack latencies. These two selection lines differ in a wide array of behavioral traits and have been characterized as proactive and reactive animals (Koolhaas et al. 1999). There was no effect of transplantation on aggression, suggesting that prenatal maternal effects do not play a role in this case of personality differentiation. However, these mice have been selected for many generations, so perhaps some sensitivity to gene-environment interactions may have been lost (Stamps and Groothuis 2010a, 2010b).

The study of the influence of maternal androgens on offspring phenotype is a flourishing field in behavioral ecology using oviparous vertebrates as study organisms (Groothuis et al. 2005; Gil 2008; Schwabl and Groothuis 2010). Freshly laid eggs of fish, reptiles, and birds contain substantial concentrations of maternal androgens that vary systematically between species, between females, in relation to clutch number or position of the egg in the clutch, and in relation to environmental factors. Behavioral effects of these maternal androgens have been studied almost exclusively in birds. Effects on chick behavior include begging for food, alertness, sibling competition, and aggression (Muller et al. 2009), and tonic immobility and open field behavior (Daisley et al. 2005; Okuliarova et al. 2007), the latter three being part of personality traits in birds (Carere et al. 2005a). In line with the finding that early exposure to androgens can have long-lasting organizing effects, exposure to elevation of androgens injected in eggs before the start of incubation (mimicking elevation of maternal androgen concentrations) also has long-lasting influences on behavior (Eising et al. 2006; Partecke and Schwabl 2008).

Further evidence for maternal effects on personality development in birds comes from selection lines for behavior in which additional behaviors were accidentally co-selected for, giving rise to different personality profiles. Interestingly, these selection lines also show differences in concentrations of maternal egg hormones. In quails selected for different levels of social reinstatement behavior, androgen concentrations in the eggs differed between the lines (Gil and Faure 2007). Likewise, in the same species, selection lines for high or low fearfulness produced eggs that differed in both androgens and progesterone concentrations (Bertin et al. 2009). In the great tit, selection lines for boldness and shyness differed in their allocation of androgens over the laying sequence of the eggs in a clutch (Groothuis et al. 2008). In chickens selected for high and low feather pecking, in which the lines also differed in tonic immobility and in manual restraint tests, differences in yolk androgen concentrations were also found (Riedstra 2003). These data indicate that selection for certain behaviors may have taken place in conjunction with selection for maternal hormone deposition in the egg. Given the pleiotropic effects of maternal androgens on behavior (e.g., Groothuis and Schwabl 2008), this raises the possibility that these hormones play a causal role in the development of different behavioral profiles between these selection lines. This can be tested by selection on maternal androgen deposition and testing its consequences for behavior. Recently, evidence has been obtained that hormone deposition in the egg has a heritable component in wild populations of birds (Tschirren et al. 2009).

In birds, maternal androgens may induce a male-biased sex ratio and corticosterone a female-biased sex ratio (Goerlich et al. 2009). The sex ratio in the clutch may determine the level of sibling competition and this may affect the future phenotype of the offspring (see below). In organisms with environmental sex determination, such as some species of turtles, lizards, and geckos, environmental cues such as temperature affect androgen conversion to estrogen, and thereby the sex of the offspring. Eggs of such species contain substantial levels of gonadal hormones, and it has been suggested that by hormone allocation mothers may affect the sex ratio of their offspring (Janzen et al. 1998), although conclusive experimental evidence for this is still lacking. Apart from sex determination, mothers of these species may also affect within-sex differentiation. In species with temperature-dependent sex determination, the nest site choice of the mother determines the temperature at which the eggs will be incubated, and this affects not only the sex ratio, but also in some cases individual differentiation within the sex. Although the sex ratio varies with temperature, the same sex can be produced by different temperatures; in the leopard gecko, females produced at the temperature that induces mostly females are less male-like than those produced at a temperature that induces mostly males. These females differ in morphology, aggressive behavior, sexual behavior, and attractivity to males and may therefore represent different personalities (Crews and Groothuis 2005).

Finally, in an oviparous tree lizard, early exposure to testosterone and/or progesterone affects the male reproductive strategy, with high levels of this hormone leading to territorial males that monopolize females and low levels leading to nomadic and less aggressive sneaker males (for a review see Moore et al. 1998). These experiments have shown that there is a sensitive period for hormone-guided differentiation and that this period begins at least the first day after hatching; these studies have also suggested that reproductive strategies show heritable variation. It is therefore conceivable that this sensitive period is already present in the egg and that maternal egg hormones influence this within-sex differentiation with carryover effects to the next generation, but this has not yet been experimentally demonstrated.

Several environmental factors have been shown to influence the transfer of maternal androgens to the egg, such as food abundance, parasite prevalence, mate quality, social competition, and habitat characteristics (for reviews see Groothuis et al. 2005; Gil 2008). This potentially provides the mother with additional tools to adjust the personality of her offspring to environmental conditions.

Postnatal maternal stress
RODENTS AND PRIMATES

As discussed in chapter 10, a large number of studies have investigated the effects of temporary or permanent separation from the mother on offspring behavioral and physiological development in rodents and primates. In the early "neonatal handling" studies of rats conducted by Levine and others, pups were briefly (3–15 min) separated from their mothers and handled every day during the first 2 weeks of life (e.g., Levine 1957). Handled pups exhibited lower anxiety and reduced HPA responses to stress later in life (Liu et al. 1997; Macrì et al. 2004; Pryce et al. 2005). In contrast, longer (3–6 hours) daily maternal separations resulted in increased fearfulness and HPA stress responses later in life (Plotsky and Meaney 1993; Macrì et al. 2004; Pryce et al. 2005). The effects of neonatal handling on offspring development in rats were hypothesized to be the result of changes in maternal behavior induced by the brief daily separations (Liu et al. 1997; Meaney 2001; but see Macrì and Würbel 2006), and therefore neonatal handling falls into the category of maternal postpartum stress. In fact, these separations were associated with elevated levels of maternal crouching over the pups and licking and grooming after reunion (see below, for the effects of high licking and grooming on offspring development). The effects of longer separations on offspring behavioral and physiological development, however, do not appear to be mediated by the mother but to result from direct effects of stress on the offspring's phenotype (Macrì and Würbel 2006). Discussing such stress effects is beyond the scope of this chapter.

Another paradigm used to investigate the effects of postpartum maternal stress on offspring personality development in rodents has involved manipulation of food availability. For example, Macrì and Würbel (2007) manipulated spatial access to food (a moderate environmental challenge in which food was placed either in the home cage or in a novel cage) or temporal access to food (a more severe environmental challenge in which mothers were subjected to periods of food restriction) in lactating females. The moderate challenge resulted in reduced anxiety and lower HPA reactivity in the offspring, while the severe challenge resulted in increased behavioral and HPA reactivity in the offspring (Macrì and Würbel 2007; Macrì et al. 2008). Macrì and Würbel (2007) argued that different phenomena were at play in the moderate vs. the severe challenge, as in the case of short vs. long maternal separations: in the former case, the effects on the offspring were mediated by changes in maternal behavior, whereas in the latter case, environmental stress affected offspring phenotype directly, through adverse effects on pup homeostasis. They also recognized the possibility that offspring HPA reactivity is a function of a single underlying factor, maternal glucocorticoid hormones, with a U-shaped relation between offspring HPA reactivity and maternal glucocorticoids (see also Macrì and Würbel 2006; Macrì et al. 2008).

A large number of studies of primates have shown that permanent separation from the mother after birth and rearing in isolation or with peers has dramatic long-term consequences for offspring behavioral and physiological development. The effects of isolation- or peer-rearing, however, have often been inconsistent among different studies and it is not clear whether they could be attributed to sensory, motor, or social deprivation and other confounding environmental variables (see Maestripieri and Wallen 2003). Since the effects of these manipulations are not maternally mediated, they will not be discussed in this chapter.

Studies of macaques and other monkeys have investigated the developmental effects of a single or repeated maternal separation(s), lasting a few days or weeks. When rhesus macaque infants who experienced brief separations from their mothers were tested in a novel environment, they showed significantly greater disturbance of behavior, lower exploration, and lower manipulation of novel objects than controls (Spencer-Booth and Hinde 1969; Hinde et al. 1978). Stevenson-Hinde et al. (1980a, b) investigated personality ratings in previously separated infants but found no clear effects of previous separations on traits such as excitability or sociability at 1 or 2 years of age. Suomi et al. (1981) studied rhesus infants that were subjected to repeated 4-day maternal separations conducted between the third and ninth month of life: they found some differences in later infant behavior upon reunion with their mothers, but no clear effects of separations on personality dimensions. Studies by Reite, Caine, Capitanio, and colleagues reported that pigtail macaque infants who experienced a 10-day maternal separation at 4–7 months of age were rated as less sociable than controls at 3–4 years of age (Caine et al. 1983; Capitanio and Reite 1984). Separated infants also showed more disturbance behavior and longer latency to retrieve food in a novel environment later in life (Capitanio et al. 1986). Finally, separated infants showed physiological changes, such as reduced heart rate and body temperature (Reite et al. 1989), as well as alterations in immune function later in life (Laudenslager et al. 1985). Reite (1987) suggested that possible long-term alterations in maternal behavior following early separation and reunion may have been the mechanism underlying the effects of early separation on later infant behavior and physiology but no relevant data were available. Thus, it is possible that the effects of early separation could be a direct consequence of this event being an acute stressful experience. In marmosets, repeated daily separations similar to neonatal handling in rats resulted in reduced behavioral and cortisol responses to novelty and higher urinary levels of norepinephrine and dopamine (Pryce et al. 2005). However, no clear evidence in favor of the maternal mediation hypothesis was available.

A direct attempt to manipulate maternal behavior and examine the effects on offspring development was made by altering accessibility and amounts of food in bonnet macaques (Rosenblum and Paully 1984; Andrews and Rosenblum 1991). Several mother-infant pairs were exposed to different foraging treatments: Low Foraging Demand (LFD), High Foraging Demand (HFD), and Variable Foraging Demand (VFD). In the LFD condition animals had access to ad libitum food, and such food could be retrieved without effort. In the HFD condition, animals had access to 6 times less food than the LFD animals. Finally, in the VFD condition, animals were exposed to a 2-week alternation of HFD and LFD. The treatment period began when infant ages ranged from 4 to 17 weeks and lasted 14 weeks. Shortly after the foraging treatments, mothers and infants were tested for 1 hour in a novel room for 4 consecutive days. Mother-infant pairs in all groups spent more time in contact in the novel room but the increase was more marked for the VFD pairs. VFD infants also showed less object exploration and play. It was not clear, however, whether differences in mother-infant interactions were the result of differences in maternal behavior or infant activity, or both. At the age of 2.5–3.5 years, after the infants had been permanently separated from their mothers and housed in peer groups, they were tested in a novel environment and the VFD individuals were reported as being less sociable and less assertive/dominant than the LFD individuals (Andrews and Rosenblum 1991). In subsequent studies, VFD individuals exhibited a number of differences from LFD or HFD individuals in terms of physiological variables or responses to pharmacological manipulations, suggesting hyperreactivity of the HPA axis and blunted noradrenergic responsivity (Rosenblum et al. 1994; Coplan et al. 1996; 1998; however, see Maestripieri and Wallen 2003, for a critical reinterpretation of these findings).

The effects of the VFD treatment on infant behavior and neuroendocrine function were assumed to be mediated by alterations in maternal behavior. As a result of the environmental manipulation, VFD mothers presumably became more anxious, erratic, dismissive, less responsive to their infants' signals, and less likely to engage in the "intense compensatory patterns typical of normal mothers following periods of dyadic disturbance" (Coplan et al. 1995; 1996; 1998). The empirical evidence that maternal behavior was affected by the foraging treatment, however, was weak. For example, there were no significant differences in foraging between the HD and LD conditions for the VFD animals (Andrews and Rosenblum 1991; Rosenblum and Paully 1984), and foraging occurred only less than 10% of the time in both conditions (Andrews and Rosenblum 1991). There were also few or no differences in mother-infant interactions between VFD and LFD animals. VFD and LFD pairs did not differ in the percentage of contacts broken by mothers, contacts made by mothers, and maternal rejections (Andrews and Rosenblum 1991). Thus, the causes of long-term behavioral and physiological alterations in VFD animals remain unclear.

Another example of environmental manipulations of maternal stress and maternal behavior in nonhuman primates involved the introduction of new adult males into well-established captive groups of vervet monkeys. Fairbanks and McGuire (1987) showed that following the introduction of new males, vervet monkey mothers became more protective of their infants, presumably because of the increased risk of infanticide, or male aggression, or simply the social instability resulting from this manipulation. Increased maternal protectiveness affected the personality development of the offspring, as these offspring showed increased anxiety/fearfulness and lower tendency to explore when tested for responsiveness to novelty months or years later (Fairbanks and McGuire 1987; 1988). Differences in maternal behavior have also been documented in monkeys in social environments that differ in terms of risk of infant kidnapping or infant harassment from conspecifics (rhesus macaques: Maestripieri 2001), or following periods of time with frequent capture procedures (rhesus macaques: Berman 1989), or in relation to changes in predation risk or food availability (vervet monkeys: Hauser and Fairbanks 1988). Although the effects of these stress-induced alterations in maternal behavior on offspring personality development were not investigated in these studies, our knowledge of the relation between interindividual variation in maternal behavior and offspring personality development (see below) suggests that these effects are likely to occur.

In rodents, handling-induced alterations in maternal behavior and their consequences for offspring development have been interpreted as adaptive maternal effects. In this view, stress-induced variation in maternal behavior serves as an environmental cue in response to which pups adjust their personality to their future environment (e.g., Macrì and Würbel 2007). In primates, even more than in rodents, maternal behavior exhibits a high degree of plasticity in response to environmental conditions. In most cases, changes in parenting behavior induced by maternal stress affect some aspects of personality development in the offspring, notably in the direction of enhanced anxiety, behavioral inhibition, and physiological responsiveness to stress. These behavioral and physiological changes in the offspring may be adaptive and prepare the offspring to deal with the stressful environments in which they will presumably find themselves later in life. Evidence for a relation between stress-induced alterations in maternal behavior and offspring personality development in both rodents and nonhuman primates is consistent with a large body of work on naturally occurring individual differences in maternal behavior and offspring development in these mammals (see below).

OVIPAROUS VERTEBRATES

In contrast to studies of rodents and primates, studies of oviparous vertebrates have not systematically investigated the effects of postnatal maternal stress on offspring behavioral or physiological development. Nutritional stress in the mother may induce nutritional stress in her offspring, and the latter topic is discussed below. Brood size (determined by the mother) affects brood competition and chick hormone levels, which in turn may influence hormone levels in the eggs of these offspring (Gil and Faure 2007), resulting in intergenerational effects. It is likely that, in species with parental care such as birds, stress-resistant mothers may enhance their offspring's capacity to handle stress, and therefore affect personality-related behavioral traits such as neophobia, but to our knowledge this has not been studied systematically. Some evidence exists, however, suggesting that in birds, as in mammals, early postnatal exposure to glucocorticoids affects the stress response later in life (Spencer and Verhulst 2007).

Maternal behavior
RODENTS AND PRIMATES

Maternal behavior in laboratory rats and mice includes nest building, retrieval, nursing, licking and grooming of pups, and maternal defense of the nest. In Long Evans rat females, the frequency of licking/grooming (LG) behavior observed during the first week postpartum is normally distributed. Females can be characterized as being High or Low in licking/grooming (LG), and this characterization is highly stable within each female (Meaney 2001; Champagne and Curley 2009). Several studies have demonstrated an association between levels of LG and the exploratory behavior of the offspring, with the offspring of High LG females being more exploratory in a novel environment. These differences in behavior emerge early in life and persist into adulthood. Cross-fostering studies have demonstrated that the differences in the offspring behavior result from maternal behavior received rather than reflect genetic similarities between mothers and offspring (Francis et al. 1999; Champagne et al. 2003). The effects of maternal behavior on offspring behavior are mediated by long-term alterations in stress-sensitive physiological systems and neural areas, including reduced plasma adrenocorticotropin and corticosterone in response to stress, elevated hippocampal glucocorticoid receptor mRNA, elevated hypothalamic CRH mRNA, and increased density of benzodiazepine receptors in the amygdala of offspring of High LG mothers compared with the offspring of Low LG mothers (Champagne and Curley 2009). As extensively discussed in chapter 10, these physiological alterations, in turn, result from epigenetic regulation of gene expression in the offspring (i.e., changes in chromatin and DNA structure that alter gene expression, without involving changes to the sequence of DNA). Such epigenetic regulation can result in the long-term silencing of gene expression achieved through differential DNA methylation.

In adulthood, the female offspring of High LG rat mothers exhibit high levels of maternal LG toward their own offspring whereas the daughters of Low LG mothers are themselves low in LG (Francis et al. 1999; Champagne et al. 2003). Cross-fostering female offspring between High and Low LG mothers has confirmed the role of experience in mediating this intergenerational transmission of maternal behavior (Francis et al. 1999; Champagne et al. 2003). Although it cannot be ruled out that the shaping of offspring personality in infancy through exposure to maternal behavior may contribute to the expression of similar maternal behavior in adulthood, studies have shown that epigenetic regulation of genes for estrogen and oxytocin receptors in the female brain, which are known to regulate maternal behavior, plays an important role in the intergenerational transmission of maternal behavior in rodents (Champagne and Curley 2009; see chapter 10).

In mice, genetic lines selected for short and long attack latency (SAL and LAL, also referred to as proactive and reactive coping styles) have been shown to exhibit systematic differences in maternal care (Mendl and Paul 1991). SAL mothers provided more maternal care and their pups spent more time nursing but grew more slowly. Slow offspring growth in the SAL line was hypothesized to be the result of inadequate nutrition, which in turn might induce higher levels of sibling competition and predispose the pups to develop into proactive animals. Benus and Rondigs (1997) performed a reciprocal cross-fostering study using these two lines but did not find an effect of cross-fostering on aggression and other line-typical behavioral traits, except for behavioral flexibility measured in a Y-maze with changing cues. As discussed before, selection might have made these mice insensitive to gene-environment interactions (Stamps and Groothuis 2010a, b). In a later study, Benus and Henkelman (1998) manipulated the sex ratio of the litters of the LAL line, and this manipulation had age-dependent effects on aggression and behavior toward a novel object. Interestingly, the association between aggression and behavioral flexibility present in the original selection lines was disrupted by the sex ratio manipulation. This suggests that the linkage between personality traits is not invariably fixed, as has also been demonstrated in birds.

The relation between naturally occurring individual differences in maternal behavior and the development of offspring personality has been investigated in many studies of macaques and other cercopithecine monkeys. In these primates, it has been shown that mothers differ from one another in their styles of parenting, and that variation in parenting styles mainly occurs along the two orthogonal dimensions of maternal protectiveness and rejection (Tanaka 1989; Schino et al. 1995; Maestripieri 1998). Maternal protectiveness includes behaviors such as approaching and making contact with the infant, and cradling, grooming, or restraining the infant. Maternal rejection includes behaviors such as breaking contact with and walking away from the infant, and preventing the infant from making or maintaining contact by physically rejecting the infant. Although maternal protectiveness and rejection change as a function of infant age and the mother's own age and experience, individual differences in parenting styles are generally consistent over time and across infants (Fairbanks 1996).

The majority of studies investigating the influence of parenting style on offspring personality have focused on measures of reactivity to the environment (i.e., anxiety or fearfulness) and exploration, while a few others have also considered social dimensions, such as aggressiveness or sociability. Early studies of rhesus macaques showed that exposure to high levels of maternal rejection in the first few months of life was associated with reduced infant's exploration at the end of the first year (e.g., Simpson 1985). Subsequent studies, however, show that infants reared by highly rejecting (or less responsive) mothers explore the environment more and are more sociable (e.g., play more with their peers) than infants reared by less rejecting mothers (Simpson et al. 1989; Simpson and Datta 1990; Bardi and Huffman 2005). In contrast, infants reared by more protective mothers appear to be relatively fearful and cautious in response to novelty or risky situations (Fairbanks and McGuire 1988).

While the above described parental influences on offspring personality were observed in infancy, other studies have shown that such influences are long-lasting. For example, vervet monkey juveniles who were exposed to greater maternal protectiveness in infancy had a higher latency to enter a new enclosure and to approach novel food containers (Fairbanks and McGuire 1993), whereas adolescent males reared by highly rejecting mothers were more willing to approach and challenge a strange adult male (Fairbanks 1996). In Japanese macaques, individuals that were rejected more by their mothers early in life were less likely to respond with submissive signals or with avoidance to an approach from another individual and exhibited lower rates of scratching in the 5-minute period following the receipt of aggression (Schino et al. 2001). Finally, rhesus macaque infants that were rejected more by their mothers in the first 6 months of life engaged more in solitary play and exhibited greater avoidance of other individuals in the second year (Maestripieri et al. 2006a).

It is possible that, in some cases, what have been interpreted as influences of maternal behavior on offspring personality development are actually similarities in behavior or personality between mothers and offspring due to shared genes. A relation between variable maternal behavior and offspring personality traits, however, has also been observed in studies in which maternal behavior was experimentally manipulated (Vochteloo et al. 1993), as well as in studies in which infants were cross-fostered at birth between mothers with different parenting styles (Maestripieri et al. 2006a, b). Therefore, there is some evidence that mothers can influence the personality development of their offspring above and beyond their genetic contributions to them.

Similar to what has been shown in rodents, the mechanisms through which maternal behavior can affect offspring personality in primates may include parenting-induced long-term alterations in neuroendocrine and neurochemical substrates of offspring emotional and behavioral responses. Although relatively few data are available on the relationship between variable maternal behavior and HPA axis hormones in the offspring, these data suggest that high rates of maternal protectiveness may enhance offspring HPA responsiveness (e.g., Bardi and Huffman 2005; Bardi et al. 2005), whereas high rates of maternal rejection (sometimes associated with infant abuse) can result in a short-term increase in HPA responsiveness followed by a long-term decrease (McCormack et al. 2009; Sanchez et al. 2010).

More information is available about the relation between variable maternal behavior and brain neurotransmitter systems involved in emotion regulation, particularly the noradrenergic and the serotonergic system. The brain noradrenergic system is involved in the regulation of arousal and fearful or aggressive responses to novel or threatening stimuli, whereas the serotonergic system plays an important role in impulse control and impulsive aggression (Higley 2003). Maestripieri et al. (2006a) reported that offspring reared by mothers with high levels of maternal rejection exhibited lower cerebrospinal fluid (CSF) levels of serotonin metabolite (5-HIAA), norepinephrine metabolite (MHPG), and dopamine metabolite (HVA) in the first 3 years of life than offspring reared by mothers with low levels of rejection. This difference was observed in both cross-fostered and non–cross-fostered infants. Furthermore, CSF MHPG levels in the second year of life were negatively correlated with solitary play and avoidance of other individuals, while CSF 5-HIAA levels were negatively correlated with scratching rates, suggesting that individuals with low CSF 5-HIAA had higher anxiety (Maestripieri et al. 2006b). In contrast, variation in maternal protectiveness early in life did not predict later variation in CSF monoamine metabolite levels or offspring behavior (Maestripieri et al. 2006a). Other studies of rhesus and vervet monkeys have shown that low levels of CSF 5-HIAA and MHPG are associated, at least in males, with high impulsivity, risk-taking behavior, and propensity to engage in severe forms of aggression (see Higley 2003). Therefore, long-term alterations in the offspring monoaminergic neurotransmitter systems induced by varying levels of maternal rejection can alter offspring's responsiveness to the environment later in life.

Long-term changes in the offspring serotonergic system and concomitant changes in anxiety and impulsivity resulting from early exposure to variable maternal rejection may also be responsible for the intergenerational transmission of maternal rejection in primates. Maestripieri et al. (2007) found significant similarities in maternal rejection rates between mothers and daughters for both nonfostered and cross-fostered rhesus females, suggesting that the daughters' behavior was affected by exposure to their mothers' rejection in their first 6 months of life. Both nonfostered and cross-fostered rhesus females reared by mothers with high rates of maternal rejection had significantly lower CSF concentrations of the serotonin metabolite 5-HIAA in their first 3 years of life than females reared by mothers with lower (below the median) rates of maternal rejection, and that low CSF 5-HIAA was associated with high rejection rates when the daughters produced and reared their first offspring (Maestripieri et al. 2006a; 2007). Related data analyses involving some of the same rhesus females also showed that abusive parenting (i.e., pathological maternal behavior that includes infant dragging, hitting, and other harmful patterns of behavior) is transmitted across generations, from mothers to daughters, through experiential mechanisms (Maestripieri 2005), and that the females that were abused by their mothers in infancy and became abusive mothers themselves had lower CSF 5-HIAA than the abused females that did not become abusive mothers (Maestripieri et al. 2006a). Individual differences in maternal rejection rates may represent adaptations to particular maternal characteristics (e.g., dominance rank, body condition, or age) or demographic and ecological circumstances (e.g., availability of food or social support from relatives). In cercopithecine monkeys such as rhesus macaques, mothers and daughters have very similar dominance ranks and share their environment as well. Therefore, the intergenerational transmission of maternal rejection rates through the shaping of offspring personality may represent an example of nongenomic transmission of behavioral adaptations from mothers to daughters. Abusive parenting, however, is probably maladaptive. Thus, effects of maternal behavior on particular aspects of the offspring personality can also result in the intergenerational transmission of maladaptive behavior.

OVIPAROUS VERTEBRATES

Two, not necessarily independent, maternal effects can profoundly affect the phenotype of the offspring in oviparous vertebrates with parental care: habitat choice and food provisioning (e.g., Price 1998). Habitat choice may affect exposure to predators and intensity of competition with conspecifics. The latter may depend on the breeding density of that species in the habitat of choice. The breeding density may affect interactions among young animals and thereby their hormone production, with long-term consequences for behavior. For example, higher breeding density stimulates testosterone production in gull chicks, which has both short- and long-term effects on aggression (Ros et al. 2002). It is likely that this early hormone exposure organizes other behaviors too, but this remains to be tested.

Some recent studies have investigated personality development in stickleback fishes in relation to predator pressure in the habitat (e.g., Bell and Sih 2007; see Stamps and Groothuis 2010a, b, for review). Previous studies had shown that sticklebacks from habitats that differ in predator exposure show differences in the organization of their personality (Dingemanse et al. 2007). The often-found link between aggressiveness toward conspecifics, general activity, and exploration was found only in habitats with piscivorous predators. To test this relationship experimentally, Bell and Sih (2007) tested boldness (approaching food in the presence of a stuffed predator) and aggression to a conspecific in sticklebacks before and after exposure to a predator. A correlation between boldness and aggression was found only after predation, and it was due to both selective predation (removing bold animals from the population) and behavioral plasticity in the survivors (decreasing their aggression) (see also Dingemanse et al. 2009).

The effect of food availability on the development of personality has been experimentally investigated in birds. Two genetic lines of great tits selected for fast and slow exploring (bold and shy respectively) were used for these studies (for a review see Groothuis and Carere 2005). Tits from the fast line approach a novel object more closely, explore a novel room faster, and are more aggressive than birds from the slow line. In the first experiment, some of the siblings in nests of both selection lines were food deprived, and others served as controls. Following the food deprivation, all the chicks in the nests of bold parents became even more aggressive than their parents, while all the chicks in the nests of shy parents became much faster in exploration but not in aggression, indicating that the lines have different developmental reaction norms (Stamps and Groothuis 2010a, b). This also suggested that food deprivation increased food competition among all siblings (affecting the controls, too) and indeed, begging frequency was higher in these nests than in unmanipulated nests. In another experiment, all chicks in the same nest were either food deprived or used as controls in order to prevent manipulated chicks from influencing their siblings (Carere et al. 2005a). Although only birds from the slow line could be used, the result confirmed the earlier conclusions since now only food-deprived birds became faster in exploration. Aggression scores were again not influenced. This finding highlights the notion that linkages between traits are not fixed, as was the case for the sticklebacks, and that different components of personality are differentially sensitive to maternal effects.

Food provisioning by parents can affect the HPA axis in developing birds, too. In chicks of kittiwakes, a gull species, poor food provisioning enhances corticosterone production (Kitaysky et al. 2002). Experimental long-term elevation of this hormone in young kittiwakes increased begging and aggression and had long-term detrimental effects on cognitive performance both early in life (on learning to open food dishes and associating cues with food location) and at the age of 10 months (spatial learning test) (Kitaysky et al. 2003). In zebra finch chicks, corticosterone elevation reduced neophobia in males and competitive behavior (Spencer and Verhulst 2007). Clearly, parental food provisioning can affect several behavioral traits in the offspring via affecting the HPA axis. Corticosterone production is also affected by the position of the chick in the hatching order (e.g., Love et al. 2003), which is related to the position of the egg in the laying order, another example of a maternal effect.

Conclusions

The evolutionary significance of maternal effects has been increasingly appreciated over the last decade (Mousseau and Fox 1998; Groothuis et al. 2005; Maestripieri and Mateo 2009; Schwabl and Groothuis 2010). Maternal effects can enhance the adaptation of offspring phenotype to prevailing environmental circumstances, profoundly shape speed and direction of evolution, and affect heritability estimates for many phenotypic traits. Maternal effects are still poorly investigated in the context of animal personality development, but our review suggests that such effects are likely to be pervasive in both oviparous and placental vertebrates.

Only a few studies reviewed in this chapter directly addressed maternal effects on behavioral traits that formally fit the criteria of personality. These studies involved, for example, the use of selection lines for different temperaments in sticklebacks or coping styles in birds and rodents. Most studies reviewed here did not specifically address maternal effects on personality. Nevertheless, the behavioral or physiological traits investigated in these studies are relevant to personality. This is because, in some cases, maternal effects resulted in long-term differences between individuals in several different traits that normally cluster together under the umbrella of personality, while in other cases maternal effects involved single traits that are known to contribute to personality in the same or a closely-related species.

In our review of the literature, we have shown that maternal effects on personality development are often mediated by mechanisms involving maternal hormones or maternal behavior. Hormones may be especially powerful pathways for maternal effects, having pleiotrophic and organizing effects on behavior. Glucocorticoid hormones may affect personality by altering an individual's emotional reactivity to the environment, while reproductive hormones may affect aspects of personalities that are sexually differentiated (but also more general behavioral traits such as boldness and risk taking). Environmental influences on personality can also be mediated by epigenetic mechanisms involving DNA silencing and concomitant changes in gene expression. In many cases, the mechanisms underlying maternal effects were investigated by directly manipulating the mother's stress, the mother's behavior or hormones, or the offspring's hormones. Experimental manipulations of the offspring's environment, for example, with regard to food availability, habitat characteristics, or predation exposure, can also result in maternal effects on offspring personality via changes in maternal behavior or hormones.

Taken together, the studies reviewed in this chapter suggest that the study of animal personality should move away from genetic deterministic approaches and take environmental effects as well as gene-environment interactions into serious consideration. Our review also suggests that developmental plasticity plays a more important role in personality than previously thought (see also Bergmüller and Taborsky 2010; Dingemanse et al. 2010); this suggests that personalities should be considered effective vehicles for adaptation rather than constraints on evolution.

Our review revealed some problems in current research on personality development and highlighted some promising avenues for future research. To begin with, there is a clear difference between studies on mammalian species, which focus mainly on the effects of early stress and stress hormones and are often conducted within a health and biomedical perspective, and studies of oviparous vertebrates, which focus mainly on androgen hormones and are conducted within an ecological and adaptive framework. This difference in goals and emphasis can potentially hamper communication between researchers working in different disciplines and impair efforts to uncover general principles that have broad cross-species validity. Related to this problem, laboratory studies of mammals may not always employ ecologically relevant manipulations, and therefore run the risk of focusing on phenomena and mechanisms that may not be of evolutionary relevance for the organism (e.g., artifacts of housing conditions). However, behavioral ecologists sometimes assume adaptation too easily, without the necessary experimental evidence. The hypotheses that different personalities represent adaptations to different environmental situations, and that parents are able to adjust the personality of their offspring to the environment, are reasonable and interesting but still need unequivocal experimental confirmation.

Another problem in current research on personality development is that the underlying mechanisms of many maternal effects are still unknown. For example, are the pleiotropic effects of the hormones due to independent actions of the hormones, or are they the result of interactions between different hormones? Answering this question has important implications for understanding the plasticity of personality (see also Groothuis and Schwabl 2008). Other issues that need to be further addressed concern the relationship between personality, hormones, and sexual differentiation, and whether there is a sensitive period for maternal effects. Finally, it has been suggested that an important difference between personality types depends on the ability to rely on cues from the internal vs. the external environment: proactive personality types may be driven more by internal routines whereas reactive types appear to rely more on external cues. Future research on the development of personality in different vertebrate taxa could help clarify this issue as well as elucidate many other conceptual and empirical aspects of animal personalities.

References

Andrews, M. W., and Rosenblum, L. A. 1991. Attachment in monkey infants raised in variable- and low-demand environments. *Child Development*, 62, 686–693.

Bard, K. A., and Gardner, K. H. 1996. Influences on development in infant chimpanzees: enculturation, temperament, and cognition. In: *Reaching into Thought: The Minds of the Great Apes* (Russon, A. E., Bard, K. A., and Parker, S. T., eds.), pp. 235–256. New York: Cambridge University Press.

Bardi, M., and Huffman, M. A. 2005. Maternal behavior and maternal stress are associated with infant behavioral development. *Developmental Psychobiology*, 48, 1–9.

Bardi, M., Bode, A. E., and Ramirez, S. M. 2005. Maternal care and development of stress responses in baboons. *American Journal of Primatology*, 66, 263–278.

Bell, A. M., and Sih, A. 2007. Exposure to predation generates personality in threespined sticklebacks (*Gasterosteus aculeatus*). *Ecology Letters*, 10, 828–834.

Bell, A. M., and Stamps, J. A. 2004. Development of behavioural differences between individuals and populations of sticklebacks, *Gasterosteus aculeatus*. *Animal Behaviour*, 68, 1339–1348.

Benus, R. F. 1999. Differential effect of handling on adult aggression in male mice bidirectionally selected for attack latency. Aggressive Behavior, 25, 365–368.

Benus, R. F., and Henkelman C. 1998. Litter composition influences the development of aggression and behavioural strategy in male *Mus domesticus*. *Behaviour*, 135, 1229–1249.

Benus, R. F., Den Daas, S. J., Koolhaas, J. M., and Van Oortmerssen, G. A. 1990. Routine formation and flexibility in social and nonsocial behavior of aggressive and nonaggressive male mice. *Behaviour*, 112, 176–193.

Benus, R. F., Koolhaas, J. M., and Van Oortmerssen, G. A. 1987. Individual differences in behavioural reaction to a changing environment in mice and rats. *Behaviour*, 100, 105–122.

Benus, R. F., Koolhaas, J. M., and Van Oortmerssen, G. A. 1988. Aggression and adaptation to the light-dark cycle: role of intrinsic and extrinsic control. *Physiology and Behavior*, 43, 131–137.

Benus, R. F., and Rondigs M. 1997. The influence of the postnatal maternal environment in accounting for differences in aggression and behavioural strategies in *Mus domesticus*. *Behaviour*, 134, 623–641.

Bergmüller, R., and Taborsky, M. 2010. Animal personality due to social niche specialisation. *Trends in Ecology and Evolution*, 25, 504–511.

Berman, C. M. 1989. Trapping activities and mother-infant relationships on Cayo Santiago: a cautionary tale. *Puerto Rican Health Sciences Journal*, 8, 73–78.

Bertin, A., and Richard-Yris, M. A. 2004. Mother's fear of human affects the emotional reactivity of young in domestic Japanese quail. *Applied Animal Behaviour Science*, 89, 215–231.

Bertin, A., Richard-Yris, M. A., Houdelier, C., Richard, S., Lumineau, S., Kotrschal, K., and Möstl, E. 2009. Divergent selection for inherent fearfulness leads to divergent yolk steroid levels in quail. *Behaviour*, 146, 757–770.

Bertin, A., Richard-Yris, M. A., Houdelier, C., Lumineau, S., Möstl, E., Kuchar, A., Hirschenhauser, K., and Kotrschal, K. 2008. Habituation to humans affects yolk steroid levels and offspring phenotype in quail. *Hormones and Behavior*, 54, 396–402.

Bolig, R., Price, C. S., O'Neill, P. L., and Suomi, S. J. 1992. Subjective assessment of reactivity level and personality traits of rhesus monkeys. *International Journal of Primatology*, 13, 287–306.

Buirski, P., Plutchik, R., and Kellerman, H. 1978. Sex differences, dominance, and personality in the chimpanzee. *Animal Behaviour*, 26, 123–129.

Byrne, G., and Suomi, S. J. 1995. Development of activity patterns, social interactions, and exploratory behavior in infant tufted capuchins (*Cebus apella*). *American Journal of Primatology*, 35, 255–270.

Caine, N. G., Earle, H., and Reite, M. 1983. Personality traits of adolescent pig-tailed monkeys (*Macaca nemestrina*): an analysis of social rank and early separation experience. *American Journal of Primatology*, 4, 253–260.

Capitanio, J. P. 1984 Early experience and social processes in rhesus macaques (*Macaca mulatta*), I: dyadic social interaction. *Journal of Comparative Psychology*, 98, 35–44.

Capitanio, J. P., Rasmussen, K. L. R., Snyder, D. S., Laudenslager, M., and Reite, M. 1986. Long-term follow-up of previously separated pigtail macaques: group and individual differences in response to novel situations. *Journal of Child Psychology and Psychiatry*, 27, 531–538.

Capitanio, J. P., and Reite, M. 1984. The roles of early separation experience and prior familiarity in the social relations of pigtail macaques: a descriptive multivariate study. *Primates*, 25, 475–484.

Carere, C., Drent, P. J., Koolhaas, J. M., and Groothuis, T. G. G. 2005a. Epigenetic effects on personality traits: early food provisioning and sibling competition. *Behaviour*, 142, 1329–1355.

Carere, C., Drent, P. J., Privitera, L., Koolhaas, J. M., and Groothuis, T. G. G. 2005b. Personalities in great tits, *Parus major*: stability and consistency. *Animal Behaviour*, 70, 795–805.

Chamove, A. S., Eysenck, H. J., and Harlow, H. F. 1972. Personality in monkeys: factor analyses of rhesus social behavior. *Quarterly Journal of Experimental Psychology*, 24, 496–504.

Champagne, F. A., Francis, D. D., Mar, A., and Meaney, M. J. 2003. Variations in maternal care in the rat as a mediating influence for the effects of environment on development. *Physiology and Behavior*, 79, 359–371.

Champagne, F. A., and Curley, J. P. 2009. The transgenerational influence of maternal care on offspring gene expression and behavior in rodents. In: *Maternal Effects in Mammals* (Maestripieri, D., and Mateo, J. M., eds.), pp. 182–202. Chicago: University of Chicago Press.

Clark, M. M., and Galef, B. G., Jr. 1995. Prenatal influences on reproductive life history strategies. *Trends in Ecology and Evolution*, 10, 151–153.

Clarke, A. S., and Boinski, S. 1995. Temperament in nonhuman primates. *American Journal of Primatology*, 37, 103–125.

Clarke, A. S., and Snipes, M. 1998. Early behavioral development and temperamental traits in mother- vs. peer-reared rhesus monkeys. *Primates*, 39, 433–448.

Cohen-Bendahan, C. C., Van de Beek, C., and Berenbaum S. A. 2005. Prenatal sex hormone effects on child and adult sex-typed behavior: methods and findings *Neuroscience and Biobehavioral Reviews*, 29, 353–384.

Coleman, K., and Wilson, D. S. 1998. Shyness and boldness in pumpkinseed sunfish: individual differences are context-specific. *Animal Behaviour*, 56, 927–936.

Coplan, J. D., Andrews, M. W., Rosenblum, L. A., Owens, M. J., Friedman, S., Gorman, J. M., and Nemeroff, C. B. 1996. Persistent elevations of cerebrospinal fluid concentrations of corticotropin-releasing factor in adult nonhuman primates exposed to early life stressors: implications for the pathophysiology of mood and anxiety disorders. *Proceedings of the National Academy of Sciences U S A*, 93, 1619–1623.

Coplan, J. D., Rosenblum, L. A., and Gorman, J. M. 1995. Primate models of anxiety: longitudinal perspectives. *Psychiatric Clinics of North America*, 18, 727–743.

Coplan, J. D., Trost, R. C., Owens, M. J., Cooper, T. B., Gorman, J. M., Nemeroff, C. B., and Rosenblum, L. A. 1998. Cerebrospinal fluid concentrations of somastotin and biogenic amines in grown primates reared by mothers exposed by manipulated foraging conditions. *Archives of General Psychiatry*, 55, 473–477.

Cote, J., and Clobert, J. 2007. Social personalities influence natal dispersal in a lizard. *Proceedings of the Royal Society of London B*, 274, 383–390.

Crews, D., and Groothuis, T. G. G. 2005. Tinbergen's fourth question, ontogeny: sexual and individual differentiation. *Animal Biology*, 55, 343–370.

Daisley, J., N., Brommundt, V., Mostl, E., and Kotrschal, K. 2005. Enhanced yolk testosterone influences behavioral phenotype independent of sex in Japanese quail chicks *Coturnix japonica*. *Hormones and Behavior*, 47, 185–194.

De Fraipont, M., Clobert, J., John, H., and Alder, S. 2000. Increased prenatal maternal corticosterone promotes philopatry of offspring in common lizards *Lacerta vivipara*. *Journal of Animal Ecology*, 69, 3, 404–413.

Dingemanse, N. J., Both, C., Drent, P. J., Van Oers, K., and Van Noordwijk, A. J. 2002. Repeatability and heritability of exploratory behaviour in great tits from the wild. *Animal Behaviour*, 64, 929–938.

Dingemanse, N. J., Kazem, A. J. N., Réale, D., and Wright, J. 2010. Behavioural reaction norms: animal personality meets individual plasticity. *Trends in Ecology and Evolution*, 25, 81–89.

Dingemanse, N. J., Wright, J., Kazem, A. J. N., Thomas, D. K., Hickling, R., and Dawnay, N. 2007. Behavioural syndromes differ predictably between 12 populations of three-spined stickleback. *Journal of Animal Ecology*, 76, 1128–1138.

Dingemanse, N. J., Van der Plas, F., Wright, J., Réale, D., Schrama, M., Roff, D. A., Van der Zee, E., and Barber, I. 2009. Individual experience and evolutionary history of predation affect expression of heritable variation in fish personality and morphology. *Proceedings of the Royal Society of London B*, 276, 1285–1293.

Drent, P. J., Van Oers, K., and Van Noordwijk, A. J. 2003. Realized heritability of personalities in the great tit (*Parus major*). *Proceedings of the Royal Society of London B*, 270, 45–51.

Eising, C. M., Müller, W., and Groothuis, T. G. G. 2006. Avian mothers create different phenotypes by hormone deposition in their eggs. *Biology Letters*, 2, 20–22.

Fairbanks, L. A. 1996. Individual differences in maternal styles: causes and consequences for mothers and offspring. *Advances in the Study of Behavior*, 25, 579–611.

Fairbanks, L. A., and McGuire, M. T. 1987. Mother-infant relationships in vervet monkeys: response to new adult males. *International Journal of Primatology*, 8, 351–366.

———. 1988. Long-term effects of early mothering behavior on responsiveness to the environment in vervet monkeys. *Developmental Psychobiology*, 21, 711–724.

———. 1993. Maternal protectiveness and response to the unfamiliar in vervet monkeys. *American Journal of Primatology*, 30, 119–129.

Fidler, A. E., Van Oers, K., Drent, P. J., Kuhn, S., Mueller, J. C., and Kempenaers, B. 2007. Drd4 gene polymorphisms are associated with personality variation in a passerine bird. *Proceedings of the Royal Society of London B*, 274, 1685–1691.

Francis, D., Diorio, J., Liu, D., and Meaney, M. J. 1999. Nongenomic transmission across generations of maternal behavior and stress responses in the rat. *Science*, 286, 1155–1158.

Frost, A. J., Winrow-Giffen, A., Ashley, P. J., and Sneddon, L. U. 2007. Plasticity in animal personality traits: does prior experience alter the degree of boldness? *Proceedings of the Royal Society of London B*, 274, 333–339.

Gil, D. 2008. Hormones in avian eggs: physiology, ecology and behavior. *Advances in the Study of Behavior*, 38, 337–398.

Gil, D., and Faure, J. M. 2007. Correlated response in yolk testosterone levels following divergent genetic selection for social behaviour in Japanese quail. *Journal of Experimental Zoology A*, 307A, 91–94.

Goerlich, V. C., Dijkstra, C., Schaafsma, S. M., and Groothuis, T. G. G. 2009. Testosterone has a long-term effect on primary sex ratio of first eggs in pigeons—in search of a mechanism. *General and Comparative Endocrinology*, 163, 184–192.

Gold, K. C., and Maple, T. L. 1994. Personality assessment in the gorilla and its utility as a management tool. *Zoo Biology*, 13, 509–522.

Gosling, S., Lilienfeld, S., and Marino, L. 2003. Personality. In: *Primate Psychology* (Maestripieri, D., ed.), pp. 254–288. Cambridge, MA: Harvard University Press.

Groothuis, T. G. G., and Carere, C. 2005. Avian personalities: characterization and epigenesis. *Neuroscience and Biobehavioral Reviews*, 29, 137–150.

Groothuis, T. G. G., Carere, C., Lipar, J., Drent, P. J., and Schwabl, H. 2008. Selection on personality in a songbird affects maternal hormone levels tuned to its effect on timing of reproduction. *Biology Letters*, 4, 465–467.

Groothuis, T. G. G., Muller, W., Von Engelhardt, N., Carere, C., and Eising, C. 2005. Maternal hormones as a tool to adjust offspring phenotype in avian species. *Neuroscience and Biobehavioral Reviews*, 29, 329–352.

Groothuis, T. G. G., and Schwabl, H. 2008. Hormone-mediated maternal effects in birds: mechanisms matter but what do we know about them? *Philosophical Transactions of the Royal Society of London B*, 363, 1647–1661.

Hauser, M. D., and Fairbanks, L. A. 1998. Mother-offspring conflict in vervet monkeys: variation in response to ecological conditions. *Animal Behaviour*, 36, 802–813.

Hayward, L.S., Richardson, J.B., Grogan, M.N., and Wingfield, J.C. 2006. Sex differences in the organizational effects of corticosterone in the egg yolk of quail. *General and Comparative Endocrinology*, 146, 144–148.

Hayward, L. S., Satterlee, D. G., and Wingfield, J. C. 2005. Japanese quail selected for high plasma corticosterone response deposit high levels of corticosterone in their eggs. *Physiological and Biochemical Zoology*, 78, 6 1026–1031.

Hayward, L. S., and Wingfield, J. C. 2004. Maternal corticosterone is transferred to avian yolk and may alter offspring growth and adult phenotype. *General and Comparative Endocrinology*, 135, 365–371.

Higley, J. D. 2003. Aggression. In: *Primate Psychology* (Maestripieri, D., ed.), pp. 17–40. Cambridge, MA: Harvard University Press.

Higley, J. D., and Suomi, S. J. 1989. Temperamental reactivity in nonhuman primates. In: *Temperament in Childhood* (Kohnstamm, G. A., Bates, J. E., and Rothbart, M. K., eds.), pp. 153–167. New York: Wiley.

Hinde, R. A., Leighton-Shapiro, M. E., and McGinnis, L. 1978. Effects of various types of separation experience on rhesus monkeys five months later. *Journal of Child Psychology and Psychiatry*, 19, 199–211.

Holekamp, K. E., and Dloniak, S. M. 2009. Maternal effects in fissiped carnivores. In: *Maternal Effects in Mammals* (Maestripieri, D., and Mateo, J. M., eds.), pp. 227–255. Chicago: University of Chicago Press.

Janczak, A. M., Heikkilae, M., Valros, A., Torjesen, P., Andersen, I. L., and Bakken, M. 2007. Effects of embryonic corticosterone exposure and post-hatch handling on tonic immobility and willingness to compete in chicks. *Applied Animal Behaviour Science*, 107, 275–286.

Janzen, F. J., Wilson, M. E., Tucker J. K., and Ford, S. P. 1998. Endogenous yolk steroid hormones in turtles with different sex-determining mechanisms. *General and Comparative Endocrinology*, 111, 306–317.

Kapoor, A., Dunn, E., Kostaki, A., Andrews, M. H., and Matthews, S. G. 2006. Fetal programming of hypothalamic-pituitary-adrenal function: prenatal stress and glucocorticoids. *Journal of Physiology*, 572, 31–44.

King, J. E., and Figueredo, A. J. 1997. The Five-Factor Model plus dominance in chimpanzee personality. *Journal of Research in Personality*, 31, 257–271.

Kirkpatrick, M., and Lande, R. 1989. The evolution of maternal characters. *Evolution*, 43, 485–503.

Kitaysky, A. S., Kitaiskaia, E., Piatt, J., and Wingfield, J. C. 2003. Benefits and costs of increased levels of corticosterone in seabird chicks. *Hormones and Behavior*, 43, 140–149.

Kitaysky, A. S., Kitaiskaia, E. V., Wingfield, J. C., and Piatt, J. F. 2002. Dietary restriction causes chronic elevation of corticosterone and enhances stress response in red-legged kittiwake chicks. *Journal of Comparative Physiology B*, 171, 701–709.

Kofman, O. 2002. The role of prenatal stress in the etiology of developmental behavioural disorders. *Neuroscience and Biobehavioral Reviews*, 26, 457–470.

Koolhaas, J. M., Korte, S. M., de Boer, S. F., Van der Vegt, B. J., Van Reenen, C. G., Hopster, H., de Jong, I. C., Ruis, M. A. W., and Blokhuis, H. J. 1999. Coping styles in animals: current status in behavior and stress-physiology. *Neuroscience and Biobehavioral Reviews*, 23, 925–935.

Laudenslager, M., Capitanio, J. P., and Reite, M. 1985. Possible effects of early separation experiences on subsequent immune function in adult macaque monkeys. *American Journal of Psychiatry*, 142, 862–864.

Levine S. 1957. Infantile experience and resistance to physiological stress. *Science*, 126, 405.

Liu, D., Diorio, J., Tannenbaum, B., Caldji, C., Francis, D., Freedman, A., Sharma, S., Pearson, D., Plotsky, P. M., and Meaney, M. J. 1997. Maternal care, hippocampal glucocorticoid receptors, and hypothalamic-pituitary-adrenal responses to stress. *Science*, 277, 1659–1662.

Love, O. P., Bird, D. M., and Shutt, L. J. 2003. Plasma corticosterone in American kestrel siblings: effects of age, hatching order, and hatching asynchrony. *Hormones and Behavior*, 43, 480–488.

Maccari, S., Darnaudery, M., Morley-Fletcher, S., Zuena, A. R., Cinque, C., and Van Reeth, O. 2003. Prenatal stress and long-term consequences: implications of glucorticoid hormones. *Neuroscience and Biobehavioral Reviews*, 27, 119–127.

Macrì, S., Chiarotti, F., and Würbel, H. 2008. Maternal separation and maternal care act independently on the development of HPA responses in male rats. *Behavioural Brain Research*, 191, 227–234.

Macrì, S., Mason, G. J., and Würbel, H. 2004. Dissociation in the effects of neonatal maternal separations on maternal care and the offspring's HPA and fear responses in rats. *European Journal of Neuroscience*, 20, 1017–1024.

Macrì, S., and Würbel, H. 2006. Developmental plasticity of HPA and fear responses in rats: a critical review of the maternal mediation hypothesis. *Hormones and Behavior*, 50, 667–680.

Macrì, S., and Würbel, H. 2007. Effects of variation in postnatal maternal environment on maternal behaviour and fear and stress responses in rats. *Animal Behaviour*, 73, 171–184.

Maestripieri, D. 1998. Social and demographic influences on mothering styles in pigtail macaques. *Ethology*, 104, 379–385.

———. 2000. Measuring temperament in rhesus macaques: consistency and change in emotionality over time. *Behavioural Processes*, 49, 167–171.

———. 2001. Intraspecific variability in parenting styles of rhesus macaques: the role of the social environment. *Ethology*, 107, 237–248.

———. 2005. Early experience affects the intergenerational transmission of infant abuse in rhesus monkeys. *Proceedings of the National Academy of Sciences U S A*, 102, 9726–9729.

———. 2009. Maternal influences on offspring growth, reproduction, and behavior in primates. In: *Maternal Effects in Mammals* (Maestripieri, D., and Mateo, J. M., eds.), pp. 256–291. Chicago: University of Chicago Press.

———. 2010. Neurobiology of social behavior. In: *Primate Neuroethology* (Platt, M., and Ghazanfar, A., eds.), pp. 359–384. Oxford: Oxford University Press.

Maestripieri, D., Higley, J. D., Lindell, S. G., Newman, T. K., McCormack, K. M., and Sanchez, M. M. 2006a. Early maternal rejection affects the development of monoaminergic systems and adult abusive parenting in rhesus macaques. *Behavioral Neuroscience*, 120, 1017–1024.

Maestripieri, D., and Mateo, J. M., eds. 2009. *Maternal Effects in Mammals*. Chicago: University of Chicago Press.

Maestripieri, D., McCormack, K., Lindell, S. G., Higley, J. D., and Sanchez, M. M.

2006b. Influence of parenting style on the offspring's behavior and CSF monoamine metabolite levels in cross-fostered and non–cross-fostered female rhesus macaques. *Behavioural Brain Research*, 175, 90–95.

Maestripieri, D., Lindell, S. G., and Higley, J. D. 2007. Intergenerational transmission of maternal behavior in rhesus monkeys and its underlying mechanisms. *Developmental Psychobiology*, 49, 165–171.

Maestripieri, D., Schino, G., Aureli, F., and Troisi, A. 1992. A modest proposal: displacement activities as an indicator of emotions in primates. *Animal Behaviour*, 44, 967–979.

Maestripieri, D., and Wallen K. 2003. Nonhuman primate models of developmental psychopathology: problems and prospects. In: *Neurodevelopmental Mechanisms in Psychopathology* (Cicchetti, D., and Walker, E., eds.), pp. 187–214. Cambridge: Cambridge University Press.

Martau, P. A., Caine, N. G., and Candland, D. K. 1985. Reliability of the Emotions Profile Index, primate form, with *Papio hamadryas*, *Macaca fuscata*, and two *Saimiri* species. *Primates*, 26, 501–505.

Mateo, J. M. 2009. Maternal influences on development, social relationships, and survival behaviors. In: *Maternal Effects in Mammals* (Maestripieri, D., and Mateo, J. M., eds.), pp. 133–158. Chicago: University of Chicago Press.

McCormack, K., Newman, T. K., Higley, J. D., Maestripieri, D., and Sanchez, M. M. 2009. Serotonin transporter gene variation, infant abuse, and responsiveness to stress in rhesus macaque mothers and infants. *Hormones and Behavior*, 55, 538–547.

McCormick, M. I. 1999. Experimental test of the effect of maternal hormones on larval quality of a coral reef fish. *Oecologia*, 118, 412–422.

McCormick, M. I. 2006. Mothers matter: crowding leads to stressed mothers and smaller offspring in marine fish. *Ecology*, 87, 1104–1109.

McCormick, M. I. 2009. Indirect effects of heterospecific interactions on progeny size through maternal stress. *Oikos*, 118, 744–752.

McGuire, M. T., Raleigh, M. J., and Pollack, D. B. 1994. Personality features in vervet monkeys: the effects of sex, age, social status, and group composition. *American Journal of Primatology*, 33, 1–13.

Meaney, M. J. 2001. Maternal care, gene expression, and the transmission of individual differences in stress reactivity across generations. *Annual Review of Neuroscience*, 24, 1161–1192.

Mendl, M., and Paul, E. S. 1991. Parental care, sibling relationships, and the development of aggressive behavior in 2 lines of wild house mice. *Behaviour*, 116, 11–41.

Meylan, S., Belliure, J., Clobert, J., and de Fraipont, M. 2002. Stress and body condition as prenatal and postnatal determinants of dispersal in the common lizard (*Lacerta vivipara*). *Hormones and Behavior*, 42, 319–326.

Moore, C. M., Hews, D. A., and Knapp, R. 1998. Hormonal control and evolution of alternative male phenotypes: generalizations of models for sexual differentiation. *American Zoologist*, 38, 133–151.

Mousseau, T. A., and Fox, C. W., eds. 1998. *Maternal Effects as Adaptations*. Oxford: Oxford University Press.

Müller, W., Dijkstra, C., and Groothuis, T. G. G. 2009. Maternal yolk androgens stimulate territorial behaviour in black-headed gull chicks. *Biology Letters*, 5, 586–588.

Okuliarova, M., Skrobanek, P., and Zeman, M. E. 2007. Effect of increasing yolk testosterone levels on early behaviour in Japanese quail hatchlings. *Acta Veterinaria Brno*, 76, 325–331.

Paitz, R. T., and Bowden, R. M. 2008. A proposed role of the sulfotransferase/sulfatase pathway in modulating yolk steroid effects. *Integrative and Comparative Biology*, 48, 419–427.

Partecke, J., and Schwabl, H. 2008. Organizational effects of maternal testosterone on reproductive behavior of adult house sparrows. *Developmental Neurobiology*, 68, 1538–1548.

Plotsky, P. M., and Meaney, M. J. 1993. Early, postnatal experience alters hypothalamic corticotropin-releasing factor (CRF) mRNA, median eminence CRF content, and stress-induced release in adult rats. *Molecular Brain Research*, 18, 195–200.

Price, T. 1998. Maternal and paternal effects in birds: effects on offspring fitness. In *Maternal Effects as Adaptations* (Mousseau, T. A., and Fox, C. W., eds.), pp. 202–226. Oxford: Oxford University Press.

Pryce, C. R., Ruedi-Bettschen, D., Dettling, A. C., Weston, A., Russig, H., Ferger, B., and Feldon, J. 2005. Long-term effects of early-life environmental manipulations in rodents and primates: potential animal models in depression research. *Neuroscience and Biobehavioral Reviews*, 29, 649–674.

Reite, M. 1987. Infant abuse and neglect: lessons from the primate laboratory. *Child Abuse and Neglect*, 11, 347–355.

Reite, M., and Short, R. 1980. A biobehavioral developmental profile (BDP) for the pigtailed monkey. *Developmental Psychobiology*, 13, 243–284.

Reite, M., Kaemingk, K., and Boccia, M. L. 1989. Maternal separation in bonnet monkey infants: altered attachment and social support. *Child Development*, 60, 473–480.

Rettenbacher, S., Möstl, E., and Groothuis, T. G. G. 2009. Gestagens and glucocorticoids in chicken eggs. *General and Comparative Endocrinology*, 164, 125–129.

Richard-Yris, M. A., Michel, N., and Bertin, A. 2005. Nongenomic inheritance of emotional reactivity in Japanese quail. *Developmental Psychobiology*, 46, 1–12.

Riedstra, B. J. 2003. Development and social nature of feather pecking. PhD diss., University of Groningen, the Netherlands.

Ros, A. F. H., Dieleman, S. J., and Groothuis, T. G. G. 2002. Social stimuli, testosterone, and aggression in gull chicks: support for the challenge hypothesis. *Hormones and Behavior*, 41, 334–342.

Rosenblum, L. A., and Paully, G. S. 1984. The effects of varying environmental demands on maternal and infant behavior. *Child Development*, 55, 305–314.

Rosenblum, L. A., Coplan, J. D., Friedman, S., Bassoff, T., Gorman, J. M., and Andrews, M. W. 1994. Adverse early experiences affect noradrenergic and serotonergic functioning in adult primates. *Biological Psychiatry*, 35, 221–227.

Saino, N., Romano, M., Ferrari, R. P., Martinella, R., and Moller, A. P. 2005. Stressed mothers lay eggs with high corticosterone levels which produce low-quality offspring. *Journal of Experimental Zoology A*, 303A, 998–1006.

Sanchez, M. M., McCormack, K., Grand, A. P., Fulks, R., Graff, A., and Maestripieri, D. 2010. Effects of early maternal abuse and sex on ACTH and cortisol responses to the CRH challenge during the first 3 years of life in group-living rhesus monkeys. *Development and Psychopathology*, 22, 45–53.

Schino, G., D'Amato, F. R., and Troisi, A. 1995. Mother-infant relationships in Japanese macaques: sources of interindividual variation. *Animal Behaviour*, 49, 151–158.

Schino, G., Speranza, L., and Troisi, A. 2001. Early maternal rejection and later social anxiety in juvenile and adult Japanese macaques. *Developmental Psychobiology*, 38, 186–190.

Schneider, M. L., Moore, C. F., Kraemer, G. W., Roberts, A. D., and DeJesus, O. T. 2002. The impact of prenatal stress, fetal alcohol exposure, or both on development: perspectives from a primate model. *Psychoneuroendocrinology*, 27, 285–298.

Schwabl, H., and Groothuis, T. G .G. 2010. Maternal effects on behavior. In: *Encyclopedia of Animal Behaviour* (Breed, M., and Moore, J., eds.), pp. 71–82. Burlington: Elsevier Academic Press.

Simpson, M. J. A. 1985. Effects of early experience on the behaviour of yearling rhesus monkeys (*Macaca mulatta*) in the presence of a strange object: classification and correlation approaches. *Primates*, 26, 57–72.

Simpson, M. J. A., and Datta, S. B. 1990. Predicting infant enterprise from early relationships in rhesus macaques. *Behaviour*, 116, 42–63.

Simpson, M. J. A., Gore, M. A., Janus, M., and Rayment, F. D. G. 1989. Prior experience of risk and individual differences in enterprise shown by rhesus monkey infants in the second half of their first year. *Primates*, 30, 493–509.

Sluyter, F., Van der Vlugt, J. J., Van Oortmerssen, G. A., Koolhaas, J. M., Van der Hoeve, F., and De Boer, P. 1996. Studies on wild house mice, VII: prenatal maternal environment and aggression. *Behaviour Genetics*, 26, 513–518.

Spencer, K. A., and Verhulst, S. 2007. Delayed behavioral effects of postnatal exposure to corticosterone in the zebra finch (*Taeniopygia guttata*) . *Hormones and Behavior*, 51, 273–280.

Spencer-Booth, Y., and Hinde, R. A. 1969. Tests of behavioural characteristics for rhesus monkeys. *Behaviour*, 33, 180–211.

Stamps, J. A. 2007. Growth-mortality tradeoffs and "personality traits" in animals. *Ecology Letters*, 10, 355–363.

Stamps, J. A., and Groothuis, T. G. G. 2010a. The development of animal personality: relevance, concepts and perspectives. *Biological Reviews*, 85, 301–325.

———. 2010b. Developmental perspectives on personality: implications for ecological and evolutionary studies of individual differences. *Philosophical Transactions of the Royal Society of London B*, 365, 4029–4041.

Stevenson-Hinde, J., Stillwell-Barnes, R., and Zunz, M. 1980a. Subjective assessment of rhesus monkeys over four successive years. *Primates*, 21, 66–82.

———. 1980b. Individual differences in young rhesus monkeys: consistency and change. *Primates*, 21, 498–509.

Suomi, S. J. 1987. Genetic and maternal contributions to individual differences in rhesus monkey biobehavioral development. In: *Perinatal Development: A Psychobiological Perspective* (Krasnegor, N., Blass, E., Hofer, M., and Smotherman, W., eds.), pp. 397–420. New York: Academic Press.

———. 1991. Up-tight and laid-back monkeys: individual differences in the response to social challenger. In: *Plasticity of Development* (Brauth, S., Hall, W., and Dooling, R., eds.), pp. 27–56. Cambridge, MA: MIT Press.

Suomi, S. J., Kraemer, G. W., Baysinger, C. M., and DeLizio, R. D. 1981. Inherited and experiential factors associated with individual differences in anxious behavior displayed by rhesus monkeys. In: *Anxiety: New Research and Changing Concepts* (Klein, D. F., and Rabkin, J., eds.), pp. 179–200. New York: Raven Press.

Tanaka, I. 1989. Variability in the development of mother-infant relationships among free-ranging Japanese macaques. *Primates*, 30, 477–491.

Tschirren, B., Sendecka, J., Groothuis, T. G. G., Gustafsson, L., and Doligez, B. 2009. Heritable variation in maternal yolk hormone transfer in a wild bird population. *American Naturalist*, 174, 557–564.

Vandenbergh, J. G. 2009. Effects of intrauterine position in litter-bearing mammals. In: *Maternal Effects in Mammals* (Maestripieri, D., and Mateo, J. M., eds.), pp. 203–255. Chicago: University of Chicago Press.

Van Hierden, Y. M., Korte, S. M., Ruesink, E. W., Van Reenen, C. G., Engel, B., Korte-Bouws, G. A. H., Koolhaas, J. M., and Blokhuis, H. J. 2002. Adrenocortical reactivity and central serotonin and dopamine turnover in young chicks from a high and low feather-pecking line of laying hens. *Physiology and Behavior*, 75, 653–659.

Van Oers, K., De Jong, G., Drent, P. J., and Van Noordwijk, A. J. 2004. A genetic analysis of avian personality traits: correlated response to artificial selection. *Behaviour Genetics*, 34, 611–619.

Van Oers, K., De Jong, G., Van Noordwijk, A. J., Kempenaers, B., and Drent, P. J. 2005a. Contribution of genetics to the study of animal personalities: a review of case studies. *Behaviour*, 142, 1185–1206.

Van Oers, K., Klunder, M., and Drent, P. J. 2005b. Context dependence of personalities: risk-taking behavior in a social and a nonsocial situation. *Behavioral Ecology*, 16, 716–723.

Vochteloo, J. D., Timmermans, P. J. A., Duijghuisen, J. A. H., and Vossen, J. M. H. 1993. Effects of reducing the mother's radius of action on the development of mother-infant relationships in longtailed macaques. *Animal Behaviour*, 45, 603–612.

Weiss, A., King, J. E., and Hopkins, W. D. 2007. A cross-setting study of chimpanzee (*Pan troglodytes*) personality structure and development: zoological parks and Yerkes National Primate Research Center. *American Journal of Primatology*, 69, 1264–1277.

Welberg, L. A., and Seckl, J. R. 2001. Prenatal stress, glucocorticoids, and the programming of the brain. *Journal of Neuroendocrinology*, 132, 113–128.

12

Neuroendocrine and Autonomic

Correlates of Animal Personalities

DORETTA CARAMASCHI, CLAUDIO CARERE,

ANDREA SGOIFO, AND JAAP M. KOOLHAAS

Introduction

Physiological traits are causally linked to behavioral traits and provide a proximate mechanistic explanation to personality types. Importantly, they can help us understand the evolutionary maintenance and stability of behavioral correlations, as well as the degree of phenotypic plasticity. Two or more behavioral traits may be governed by a common physiological mechanism, for example, as a result of a gene with pleiotropic effects or they may be regulated by independent mechanisms. The correlation between traits with a common underlying mechanism is expected to be stronger than that between traits with independent underlying mechanisms (Sih et al. 2004).

In this chapter we provide an overview of documented links between physiological and behavioral traits commonly considered in animal personality assessments. Most of what we know on this topic comes from research on interindividual variation in responsiveness to stress. We first provide a historical overview of research on stress, different behavioral strategies for coping with stress, and their physiological correlates. We review a number of studies on rodents, birds, nonhuman primates, and humans focusing on the hypothalamic-pituitary-adrenal axis (HPA); the autonomic nervous system, in particular, the sympatho-adrenomedullary pathway (SAM); and the hypothalamic-pituitary-gonadal axis (HPG). We then report available evidence on neurobiological correlates of coping strategies, with special focus on the neurotransmitters serotonin and dopamine, and cortical brain structures such as the hippocampus. Finally, we highlight the evolutionary importance of such "architectural" constraints in canalizing behavioral plasticity.

Physiological responses to stress in different personality types: HPA and SAM

EARLY STUDIES

The relationship between behavioral and physiological responses to stress was elucidated by pioneering work by Cannon. He described how fighting or fleeing, as a result of feeling anger or fear, are both coupled to a physiological activation of the sympathetic branch of the autonomic nervous system and consequent stimulation of the adrenal medulla to release adrenaline and noradrenaline into the blood stream (Cannon 1915). Henry and Stephens (1977) subsequently documented the presence of two distinct physiological response patterns in mice on the basis of their behavior in a social conflict. Mice that tend to win fights are prone to developing cardiovascular problems associated with strong sympathetic activation, and consequent elevation of plasma noradrenaline and adrenaline, whereas mice that become subordinate tend to develop strong activation of the HPA axis, with consequent high levels of blood corticosterone. The latter physiological pattern was attributed to a more general strategy aimed at conservation/withdrawal characterized by behavioral inhibition to avoid energy depletion through a homeostatic balancing mechanism (Engel and Schmale 1972).

RECENT STUDIES ON MICE AND RATS

On the basis of extensive work on variation in mouse and rat behavioral and physiological responses to conflict and stress, which also addressed the functional significance of individual differences in coping, Koolhaas and colleagues (1999) introduced the notion of "coping styles." They argued that there are two main alternative coping strategies, which correspond to basic personality types. The "proactive strategy" consists of aggressiveness, boldness, and inflexibility, and it is characterized by a general fight-or-flight behavioral response indicating a high emotional activation. The "reactive strategy" combines low aggressiveness, risk aversion, and flexibility to changes, and it is dominated by freezing behavior in response to stress. Many studies in a wide range of vertebrate species showed that individuals adopting the proactive strategy manifest an enhanced sympathetic and noradrenergic response to stress, while individuals with the reactive strategy manifest an enhanced parasympathetic activation and a high HPA physiological response (e.g., Hessing et al. 1994; Fokkema et al. 1995; Sgoifo et al. 1996; Korte et al. 1999). Within the two branches of the autonomic nervous system, which together regulate cardiovascular function, a shift toward sympathetic or parasympathetic activation explains a higher or a lower predisposition,

Table 12.1. Summary of the behavioral and physiological traits that define proactive and reactive coping styles in mice and rats.

PROACTIVE		REACTIVE
	Mice	
high	Aggression	low
high	Routine formation	low
low	Flexibility	high
high	Shock-prod burying	high
low	Freezing	high
low	HPA activation	high
	Rats	
high	Aggression	low
high	Shock-prod burying/ Active avoidance	low
low	Freezing	high
high	Sympathetic activation	low
low	Parasympathetic activation	high
low	HPA activation	High

Abbreviation: HPA, hypothalamic-pituitary-adrenal.

respectively, to arrhythmias and long-term cardiovascular problems associated with coping styles (Sgoifo et al. 2005). Table 12.1 summarizes the main characteristics of proactive and reactive coping styles in mice and rats.

In laboratory male mice, the two coping styles have been studied using genetic selection lines for short and long attack latency (SAL and LAL) in response to a male intruder. Compared with LAL, SAL mice are generally more aggressive against male or female conspecifics (Van Oortmerssen and Busser 1989; Benus et al. 1990; Caramaschi et al. 2008) and their attacks are directed toward more vulnerable regions of the opponent's body, therefore causing severe injury (Haller et al. 2006). Defeated SAL mice flee more, while LAL mice freeze more (Benus et al. 1992). In nonsocial situations, SAL mice are unaffected by changes in the environment, and are slower in reversal learning tasks (Benus et al. 1990) and in actively coping with aversive challenges (Sluyter et al. 1996). LAL mice show more anxiety behavior and freezing in several testing paradigms (Hogg et al. 2000; Veenema et al. 2003a,b). The glucocorticoid response to adrenocorticotropic hormone (ACTH), novelty, and forced swimming is significantly higher in LAL than in SAL mice (Veenema et al. 2004; Veenema, Sijtsma, et al. 2005). Furthermore, chronic psychosocial stress induces a long-lasting increase in glucocorticoid production in LAL mice, but not in SAL mice (Veenema et al. 2003a; Veenema, Sijtsma, et al. 2005). These findings indicate that LAL mice have higher stress responsiveness in terms of glucocorticoid production than SAL mice (Veenema et al. 2004).

In laboratory rats, aggressiveness is positively correlated with proactive behavior in responding to a mild aversive stimulus, specifically measured as the ability to bury a shock-prod in the cage after contact and delivery of a mild shock (de Boer and Koolhaas 2003; de Boer et al. 2003). Low aggressiveness is correlated with immobility and other aspects of the passive/reactive response. In physiological terms, aggressive animals display a larger reactivity of the SAM system (both noradrenaline and adrenaline plasma levels are higher, Sgoifo et al. 1996), whereas the low aggressive/passive animals show a higher activation of the parasympathetic system, the other branch of the autonomic nervous system that counteracts the effects of the sympathetic system. Evidence for this comes from individual differences within the Wild-Type-Groningen rats and by comparing Wild-Type-Groningen with Wistar rats (Sgoifo et al. 1998). In laboratory Wistar rats, the amount of measurable aggression may be very low, making studies on aggression difficult to conduct (de Boer et al. 2003). However, Bignami (1965) was able to create two rat lines on the basis of differences in active/passive coping, the Roman-High-Avoidance (RHA) and Roman-Low-Avoidance (RLA). The selection was based on their behavior in a shuttle-box test, a two-compartment box with one side able to deliver a shock and a "safe" side without shock. The RHA rats go quickly to the compartment without shock, whereas the RLA rats remain in the compartment with shock in an immobile state, thus reducing the probability of receiving a shock in different ways. These lines show more general traits of high and low fearfulness, and their HPA reactivity correlates positively with the fearfulness exhibited in the low-avoider rats (Steimer et al. 1997). In other studies, Sprague-Dawley rats have been classified into two distinct behavioral styles based on their responses to novelty: a neophobic style characterized by cautious movements, and a neophilic style characterized by extended locomotion and inspection. Neophobia and neophilia appear to be lifelong stable traits in these rats. When exposed to a novel situation, neophobic rats exhibited a higher elevation in plasma corticosterone, revealing a more responsive HPA axis and a general reactive phenotype (Cavigelli and McClintock 2003).

FARMED MAMMALS

Studies investigating responses to stress in farm animals have produced results similar those obtained in studies of laboratory rodents. Pigs, for example, can be classified as active (A/R) or passive (NA/NR) copers based on aggressiveness and escape attempts in the back test (in which they are put on their backs and restrained supine). These behavioral traits appear early in piglets and persist later in life. When tested in an open field or in the presence of a novel object, A/R pigs make more escape attempts, vocalize less, and approach the novel object earlier compared with the NA/NR pigs, thus showing a fearless phenotype. Similarly to rodents, pigs show an association between the behavioral phenotype and the physiological response to stress, with the active copers having a higher SAM reactivity and the passive copers a higher HPA reactivity to stressors (Hessing et al. 1994). In pigs, dominance and subordination were linked to higher SAM or HPA activation, respectively (Fernandez et al. 1994), suggesting that aggression/dominance and coping strategy might converge through the physiological response to stress.

BIRDS

In birds, a considerable amount of work on the physiological correlates of different coping styles has involved two selection lines of great tits, although such studies have recently stimulated research on several other species (see chapter 3). The lines were selected from wild populations on the basis of their exploratory behavior and called FAST and SLOW to indicate their latency to explore a novel environment and approach a novel object (Groothuis and Carere 2005). The lines were found to differ consistently in other behavioral contexts as well. For example, the FAST line tended to develop behavioral routines, and was more risk-taking and more aggressive than the SLOW line. Similarly to the rodent studies, the physiological response to stress was different in these two passerine lines. Particularly, the SLOW line showed higher HPA activation and higher body temperature in reaction to stress (Carere et al. 2003; Carere and Van Oers 2004).

Coping styles and their physiological correlates have also been studied in farmed chickens. Different chicken lines in the same commercial condition exhibit large differences in feather-pecking behavior. The White Leghorn chicken lines selected for high (HP) and low (LP) feather pecking consistently show this difference already after hatching (Blockhuis and Beutler 1992; Van Hierden et al. 2002a,b). In these lines, high levels of feather pecking are associated with more preening, more tonic immobility in an open field, and more struggling to restraint, thus indicating an aggressive and proactive, but cautious, coping style. In contrast, low levels of feather pecking are associated with more ground pecking, more locomotion and vocalization in an open field, and higher speed in social runway, indicating a nonaggressive and reactive phenotype. Similar to mammals, the proactive and reactive chicken lines show different physiological profiles. The HP line is character-ized by a marked sympathetic, noradrenergic, and tachycardic response to stress, whereas the LP line is characterized by a pronounced corticosterone and parasympathetic response to stress (Korte et al. 1999). Taken together, the studies on great tits and chickens suggest that, similar to rodents, birds can be classified in terms of their behavioral and physiological responses to stress into two groups, of proactive and reactive copers, though the chicken behavioral profiles are less comparable to those of mice and great tits (see Groothuis and Carere 2005).

NONHUMAN PRIMATES

Studies of laboratory rhesus monkeys have identified individuals that show high and low reactivity to social separation stress (Suomi 1997). In general, young monkeys that underwent social separation in infancy developed high reactivity characterized by high sympathoadrenal and hypothalamo-pituitary-adrenocortical activation during a social encounter (Suomi 1997). In another study of free-ranging monkeys, aggressive individuals showed both higher sympathoadrenal and higher HPA reactivity to stress (Higley et al. 1992). Other studies on rhesus monkeys have described an early detect-able and consistent source of variation in more or less anxious and fearful personality that is related to extreme right or left brain frontocortical activ-ity (Kalin and Shelton 2003). In these monkeys the variation in personal-ity is associated with consistent individual differences in activation of the HPA axis in situations of stress. The anxious monkeys show higher cortisol and corticotrophin-releasing hormone (CRH) levels than the less anxious ones. In capuchin monkeys (*Cebus apella*), high cortisol reactivity correlated negatively with personality traits such as Aggressive, Confident, Curious, Effective, and Opportunistic, and correlated positively with Apprehensive, Fearful, Insecure, Submissive, and Tense (Byrne and Suomi 2002). Taken to-gether, such studies provide evidence for a tight link between behavioral and physiological stress reactivity. Differences between two basic personality types explain a substantial fraction of interindividual variation in behavior.

HUMANS

In humans, the physiological response to stress in relation to personality traits has been measured by analyzing cortisol levels for the HPA axis and catecholamine levels, heart rate, or skin conductance for the SAM system. Pioneering work on the physiological basis of human temperamental traits was done by Kagan and coworkers in children (Kagan et al. 1987; 1988). In their longitudinal studies, cohorts of children were identified as being either behaviorally inhibited or uninhibited (i.e., nonreactive or reactive) when exposed to unfamiliar rooms, people, or objects and were then followed up from 2 to 7 years of age. Inhibition and uninhibition were stable traits and were associated with high and low physiological activation, respectively, during stress exposure. Inhibited children showed higher heart rates, lower heart rate variability, higher noradrenaline in their urine, and higher salivary cortisol. Behavioral inhibition, or childhood shyness, is considered a risk factor for adult anxiety and personality disorders (e.g., Nigg 2000). Although inhibition is a stable and clearly measurable temperamental trait, it might relate to more than one component of human personality.

According to Eysenck's study of personality (Eysenck et al. 1979), shyness can be due to either introversion (tendency to avoid unfamiliar challenges, in contrast to approach) or neuroticism (emotional charge of the uncertain situation, leading to fear and anxiety), or both. More recently, as an extension of Eysenck's ideas, personality traits have been measured using questionnaires that identify four or five main independent dimensions of behavioral variation. This Five-Factor Model is widely used and consists of Extraversion, Agreeableness, Conscientiousness, Neuroticism, and Openness (Digman 1990; chapters 4 and 5). In healthy men and women, a blunted cortisol response to the psychosocial stress of speaking in public was found to be negatively correlated with Openness (Oswald et al. 2006); Openness is a generally considered a healthy and positive characteristic that include aesthetic sensitivity, attentiveness to inner feelings, preferences for variety, intellectual curiosity, and independence of judgment. Among women, blunted cortisol responses were correlated with Neuroticism (Anxiety, Anger, Hostility, Depression, and Self-Consciousness), whereas in men blunted responses were associated with low scores of Extraversion (Warmth, Activity, and Positive emotion) (Oswald et al. 2006). Another study reported that people scoring high on Neuroticism responded less to a dexamethasone-CRH test than individuals scoring low on the same factor (McCleery and Goodwin 2001), suggesting chronic hyperactivation of the HPA axis. Blunted cortisol responses are also generally associated with depressive symptoms and high anxiety (Oswald et al. 2006). Studies conducted on aggression in youth have found in some cases a positive and in others a negative correlation between cortisol response to a stressful situation and aggressiveness (Gerra et al. 1997; Suarez et al. 1998; Popma et al. 2006). Furthermore, research on students has reported a positive correlation between Type A personality, which is impulsive, impatient, hostile, and with high need to control others, and high cortisol responses, particularly when the students' academic performance was good (Jones 1986; Spangler 1995). However, it should be noted that high aggressiveness is a characteristic shared by more than one of the Big Five personality dimensions. When aggressiveness is a symptom of anxiety and depression it is probably related to Neuroticism, whereas it can also be an antisocial negative trait of individuals scoring low in the Agreeableness (or scoring high in Antagonism, e.g., being verbally aggressive). Taken together, human studies investigating variability in responses to stress have shown an association between distinct physiological patterns and behavioral traits. Therefore, it seems that incorporating physiological measures in the study of personality would help recognize and identify specific traits that would otherwise be difficult to distinguish.

In summary, there is across-taxa evidence that individuals differ in the way they cope with stress not only behaviorally, but also physiologically. Particularly, individuals with a proactive, risk-prone, aggressive, and bold personality respond to stress with a strong sympathetic activation and increase in noradrenergic stimulation, with potential side effects on the cardiovascular system, among others. Individuals with a passive, risk-averse, nonaggressive, and shy personality respond to stress with a strong hypothalamic-pituitary-adrenocortical stimulation, with consequent increase in circulating glucocorticoids. These links between behavior and physiology in the stress response across individuals and across taxa suggest that there are basic physiological characteristics in vertebrates, which influence and constrain the behavior of individuals in a systematic and detectable manner.

HPG

Along with the SAM and the HPA axis, another physiological pathway is related to personality traits. The hypothalamic-pituitary-gonadal axis, or HPG, operates through the activation of hypothalamus to produce GnRH (gonadotropin-releasing hormone), which in turn stimulates the pituitary gland to release the gonadotropins LH (luteinizing hormone) and FSH (follicle-stimulating hormone). Gonadotropins stimulate the production of androgens and estrogens in both females and males. However, females produce much more estrogens and progesterone from their ovaries, whereas males produce much more testosterone from their testes (for more details on the HPG axis see Nelson 2005). (See figure 12.1.)

MICE AND RATS

There are a number of reports that link testosterone manipulation with aggression in rodents (e.g., Schuurman 1980). The mouse genetic selection lines for coping styles SAL and LAL show a dose-response relationship between

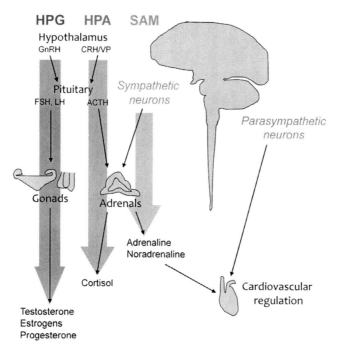

Figure 12.1. Schematic representation of the major neuroendocrine networks associated with behavioral traits. In the middle the hypothalamic-pituitary-adrenal axis (HPA), which consists of release of corticotrophin-releasing (CRH) hormone and vasopressin (VP) from the hypothalamus into the pituitary gland, adrenocorticotropic hormone (ACTH) from the pituitary gland into the blood and release of cortisol (or corticosterone in rodents and birds) from the cortical part of the adrenals into the blood. In light gray the sympatho-adrenomedullary (SAM) system, which consists of release of noradrenaline from the sympathetic branch of the autonomic nervous system and release of adrenaline from the medullary part of the adrenals. The SAM and the parasympathetic branch of the autonomic nervous system are involved in the regulation of cardiovascular physiology. In gray the hypothalamic-pituitary-gonadal (HPG) axis, which consists of release of gonadotropin-releasing hormone (GnRH) from the hypothalamus into the pituitary, folliculus-stimulant hormone (FSH) and luteinizing hormone (LH) from the pituitary gland into the blood, and release of sexual hormones (testosterone, estrogens, and progesterone) from the gonads into the blood.

testosterone and aggression (Van Oortmerssen et al. 1987; Compaan et al. 1992). Low levels of androgens are related to aggressive/proactive coping strategy in adulthood, while naturally high or artificially elevated androgen levels are related to the nonaggressive/reactive coping strategy. Testosterone levels have not been investigated in the rat lines genetically selected for high and low emotionality, that is, the RHA and RLA lines. However, prolactin response to stress was higher in the RLA rats (Steimer et al. 1997), the line of passive copers, and hyperprolactinemia could reduce the FSH and GnRH levels. Therefore, there could be a similar association between active coping and higher HPG functioning.

Classical studies performed in the 1970s and 1980s suggested that testosterone plays a causal role not only in sexual behavior, but also in exploration and emotional behavior in a novel environment (Andrew 1966; Archer 1973; Rogers and Andrew 1989). In these studies specific behavioral effects were observed immediately after birds were injected with testosterone, including increases in sexual motivation, emotional reactivity, and aggression. These findings suggest that some behavioral traits that are usually associated within a personality type may be linked together through testosterone. Unfortunately, the effects of androgens on personality traits have not been studied any further. Studies on maternal effects, however, have shown that testosterone levels in the yolk affect the timing of egg clutching and later personality measures in adulthood, and this mechanism seems to be at the basis of the genetic variation between the selected lines of shy and bold great tits (Groothuis et al. 2008; chapter 11). In particular, the bold offspring develop under lower androgen exposure than the shy offspring. The relationship between androgen levels and boldness during this early developmental phase is opposite to that in adulthood, similar to the findings in the mouse selection lines.

In humans the first link between behavioral traits and HPG functioning was a report on positive correlations between self-reported hostility and aggression and plasma testosterone levels and testosterone production rates (Persky et al. 1971). After that, several studies showed associations between testosterone levels and aggression (Olweus et al. 1988), and between testosterone and sensation-seeking, extraversion, neuroticism, and psychopathy (Westberg et al. 2009). Traits related to dominance are generally associated with higher HPG functioning. High levels of testosterone are also associated with high disinhibition scores and extraversion, sociability, and instrumental traits (Daitzman and Zuckerman 1980; Baucom et al. 1985; Blanco et al. 2001). A genetic variant in the androgen receptor gene suggests an association in the same direction (Westberg et al. 2009). Finally, high scores of spontaneous unprovoked aggression correlate with low levels of sex hormone–binding globulin, the carrier protein for testosterone in the plasma, which limits the amount of testosterone reaching the brain and consequently its functions in modulating the central nervous system (Witte et al. 2009).

Neurobiology of personality types: serotonin

Traditionally, serotonin has been associated negatively with aggressiveness. Initially the association was found in isolated mice that became aggressive after reducing their serotonin turnover, which is an indicator of serotonin neurotransmission and consequent degradation (Garattini 1967; Giacalone et al. 1968; Modigh 1973). Other studies on human and nonhuman primates confirmed a negative correlation between serotonin metabolite levels in the cerebrospinal fluid and impulsive aggression levels (see Higley 2003 for review). The negative correlation between serotonin turnover and aggression became known as the "serotonin-deficiency hypothesis" of aggression. A proper functioning of the serotonergic system involves the integration of several processes: serotonin synthesis, release, receptor activation, re-uptake, and degradation. Research has shown that changes in one or more of these elements generally explain the low serotonin metabolite levels in aggressive/violent individuals (Higley 2003). The relationship between serotonin and personality is more complex, as high and low aggressiveness can contribute to different personality traits.

MICE

Research conducted with the SAL-LAL mouse model for coping styles has suggested an association between proactivity and low serotonergic levels in the brain. The tissue levels of forebrain serotonin (intracellular and extracellular) and its metabolite 5-HIAA are lower in the SAL aggressive line than the LAL counterpart (Olivier et al. 1990; Veenema, Cremers, et al. 2005). In SAL mice, the pre- and the post-synaptic 5-HT_{1A} receptors are more sensitive than in LAL mice (Korte et al. 1996; Van der Vegt et al. 2001; Feldker et al. 2003; Veenema, Cremers, et al. 2005; Caramaschi et al. 2007). Enhanced 5-HT_{1A} autoreceptor functionality may result in inhibited tonic release of serotonin. The consequence is probably an impaired serotonergic neuromodulation in key areas of behavioral control, such as the prefrontal cortex.

In the SAL and LAL selection lines, selection for attack latency, and therefore proneness to attack a conspecific, results in selection for a more general proactive/reactive coping style applicable to other contexts. For example, in the forced swim test, in which a mouse is introduced in a jar of water in which it can swim and climb the walls in an attempt to escape, or alternatively passively float, the SAL mice typically display the former strategy and the LAL mice the latter. In this test, when SAL and LAL mice are injected with a drug mimicking the effects of serotonin on 5-HT_{1A} receptors and thereby activating them, SAL mice reduce their swimming and climbing, while the LAL reduce their immobility levels, consequently reducing the behavioral differences between the two lines (Veenema, Cremers, et al. 2005). Two other mouse selection lines for high and low aggression, the T and the NC lines, do not differ in other behavioral traits and do not show the association between low serotonin levels, enhanced 5-HT_{1A} presynaptic receptor functionality, and aggression. This suggests that low serotonin levels and its strong autoinhibiting mechanism via 5-HT_{1A} presynaptic receptors are more related to the proactive/aggressive coping style than to aggression itself (Caramaschi et al. 2007; 2008). Another line of evidence for the association between serotonin and mouse personality comes from serotonin-transporter knock-out mice ($Slc6a4^{-/-}$). These mice show a plethora of distinctive behavioral features that are primarily attributable to altered intracellular and extracellular serotonin concentrations during development and adulthood (for a review see Murphy and Lesch 2008). The lack of serotonin transporter in these mice is responsible for a stress-prone phenotype, characterized by anxious behavioral traits and exaggerated HPA and SAM response to stressors.

NONHUMAN PRIMATES

In monkeys, as in humans, a polymorphism in the serotonin transporter gene is associated with variable expression of this gene, variation in transporter activity, and consequent variation in intracellular and extracellular serotonin levels. Rhesus monkeys carrying a shorter variant of this gene, and who also experience adverse early life events such as separation from their mothers, develop neurobiological impairments and a hyperaggressive phenotype, together with an exaggerated HPA stress responsiveness (for a review see Suomi 2006).

HUMANS

In the human population, low serotonergic neurotransmission has been loosely associated with impulsivity and sensation seeking (Dolan et al. 2001; Netter et al. 1996). The main association, however, is with traits included in the Neuroticism personality dimension. Traditionally, serotonin has been associated with mood disorders, including depression and anxiety. Polymorphisms in the serotonin transporter gene modulate anxiety-related traits such as harm avoidance and neuroticism (Mandelli et al. 2009; Fox et al. 2009; Gonda et al. 2009). Genotypic differences in the 5-HT_{1A} receptor are also associated with Neuroticism, particularly in proneness to negative emotionality (Lemonde et al. 2003; Strobel et al. 2003). Other studies have found an association between the serotonin transporter polymorphism and Agreeableness (Du et al. 2000). The Agreeableness and Neuroticism dimensions have always been considered independent. However, the fact that they share a robust common genetic basis supports the notion that there is a degree of correlation between them (Jang et al. 2001). Genotypic variation in serotonin-related genes is also involved in the novelty-seeking aspect of personality, particularly through the 5-HT2A receptors (Heck et al. 2009).

In conclusion, serotonin seems a good candidate in the search for neurochemical bases of personality and for its physiological correlates. However, the link with behavioral traits is complicated by the fact that serotonin might exert different roles in different personality dimensions by acting through its different selective receptor subtypes.

Neurobiology of personality types: dopamine, GABA, and noradrenaline

High dopaminergic neurotransmission has been associated with sensation/novelty seeking. In the SAL and LAL mice, the risk-prone SAL show an enhanced behavioral response to dopamine-stimulants compared with the more risk-averse LAL mice (Benus et al. 1991a, b). Similarly, novelty-seeking rats identified as High Responders (HR) show more activation of the dopaminergic system after stimulant drugs such as cocaine and amphetamine (Hooks et al. 1991; 1992). Similarly, in humans individuals more sensitive to dopamine-stimulating drugs are those that score high on sensation seeking (Blanchard et al. 2009). In particular, decreased abundance of autoinhibitory D2 receptors might be a cause for augmented dopaminergic neurotransmission in novelty-seekers (Zald et al. 2008). The basis of the relationship between dopamine physiology and personality traits might lie at least partially in genetic variants. A genetic polymorphism in the gene for D4 receptors seems to underlie the individual variation in sensation seeking (Kreek et al. 2005), whereas angry-impulsive personality traits relate to a genetic polymorphism in the dopamine transporter gene (Joyce et al. 2009).

There is some evidence that GABA and noradrenaline are associated with stress and personality. A gene polymorphism in the GABA receptor is associated with HPA and SAM response to stress and extraversion (Uhart et al. 2004) and to neuroticism (Sen et al. 2004). Genetic variation in the enzyme responsible for the production of noradrenaline is also associated with aggression and neuroticism (Hess et al. 2009).

Hormones, neuroplasticity, and personality types

Circulating hormones such as androgens, estrogens, and corticosteroids are small hydrophobic molecules that easily pass through the endothelial blood-brain barrier and reach the central nervous system. In the brain, the hormones activate specific receptors, which in turn leads to gene expression regulation and therefore neuronal remodeling. This process is called neuroplasticity and it involves structural anatomical changes, such as dendritic growth or branching, and changes in synaptic strength by up- or downregulating the receptor number on the cell surface at the synapses. Different degrees of hormonal activation have important consequences for individual variation in behavioral strategies; hormonal activation, therefore, can be an important cellular/molecular mechanism at the basis of personality. Circulating corticosteroid hormones produced by the adrenals enter the brain and bind to specific receptors, the glucocorticoid receptor (GR) and the mineralocorticoid receptor (MR). Glucocorticoids inhibit vasopressin (Kovács et al. 2000; Watters et al. 1996) and stimulate serotonin neurotransmission (Karten et al. 2001). Earlier in this chapter we mentioned that LAL mice are more reactive to stress in terms of high HPA axis activation than SAL mice. In line with this, the reactive/low-aggressive LAL mice are more responsive to environmental experiences by enhancing their serotonin levels in cortical brain areas, and possibly more generally in other brain areas (Caramaschi et al. 2008). Similarly, vasopressin levels are associated with aggression in rodents (Ferris and Potegal 1988; Veenema and Neumann 2008).

Testosterone and other androgens such as dehydroepiandrosterone (DHEA), androstenediol, androstenedione, and dehydrotestosterone (DHT) can all bind the androgen receptor (AR). Estrogens, which can also be synthesized from androgens, bind the estrogen receptor (ER). Depending on the location of the receptors, hormones can have different effects on different neural networks and consequently behavior. For example, serotonergic agonists at 5-HT1A or 5-HT1B receptors are far more effective in inhibiting offensive aggression in the presence of androgens, while they are little effective in the presence of estrogens (Cologer-Clifford et al. 1997;1999). There is therefore an interaction between sex steroids and serotonin neurotransmission, likely through direct alteration of 5-HT receptors gene expression, as it has been documented for estrogen and 5-HT1A function (Fink et al. 1996; Sumner and Fink 1998). Neurosteroids are also known to affect GABA neurotransmission, which could affect behavioral traits as well (Pinna et al. 2006).

Hippocampus, a crucial neural substrate

One of the main brain areas subject to changes in relation to experience is the hippocampus, in which cholinergic transmission dominates. Its function is mainly related to declarative memory (as opposed to procedural memory, which seems to be stored in other areas related to motor output such as the cerebellum), and in its anatomical structure many morphological changes occur by effect of experience. As early experience shapes the development of personality throughout life (chapters 10 and 11), it is expected that this area may be of crucial importance. The hippocampus projects onto the prefrontal cortex, a key area involved in the regulation of behavior, and receives modulatory input from the serotonin, noradrenaline, and dopamine systems. The 5-HT1A receptor is highly expressed in the hippocampus, making this brain region highly vulnerable to changes in serotonergic neurotransmission. Notably, the hippocampus also receives strong glucocorticoid input especially through the MR receptors.

In the SAL/LAL mouse lines, stress can have a different physiological impact depending on the line and its coping strategy. These differences may be mediated by differences between these lines in the structure and function of the hippocampus. LAL mice have larger hippocampal substructures, such as the infrapyramidal mossy fiber terminal fields (Sluyter et al. 1994). Further, a large-scale gene transcripts analysis conducted on the hippocampus revealed higher expression for many genes involved in neural morphology, such as cytoskeleton genes, genes involved in memory consolidation, such as calmodulin-related genes, and genes for components of the mitogen-activated protein kinase (MAPK) cascade. In contrast, growth-arrest gene 5 (gas5), a gene highly expressed in the growth arrest phase of the cell cycle (Schneider et al. 1988), is highly expressed in the SAL aggressive line (Feldker et al. 2003). Moreover, higher cell proliferation was found in the LAL mice (Veenema et al. 2004).

The higher stress-induced activation in the HPA axis in the LAL mice may be mediated by hippocampal function. After a stressful challenge, the LAL mice upregulated their MR receptors in the hippocampus, and consequently their MR/GR ratio, while the SAL mice did not (Veenema et al. 2003a, b). The MR receptors are in turn responsible for inhibiting tonically the activation of the HPA axis (De Kloet and Reul 1987), therefore providing regulatory feedback and maintaining homeostatic balance. In the LAL mice the hippocampal sensitivity to serotonin was reduced as measured in vitro by electrophysiology, together with a reduced 5-HT1A receptor mRNA expression and an enhanced corticosterone response to novelty stress (Van Riel et al. 2002). Taken together, evidence from the SAL and LAL mice suggests that differences in hormonal responsiveness to stress correlate with differences in neural mechanisms involved in adaptation to stress. At the extremes, two solutions might be equally successful. One strategy, represented by the SAL mice, is aimed at engaging in behaviors that lead to control over the environment in order to reduce potential harm and make the environmental condition more favorable. The other strategy, represented by the LAL mice, is not able to behaviorally master environmental challenges, but has evolved a highly reactive physiological system that is able to maintain the balance between each of its components and buffer potential harm.

Ecological costs and adaptations

In evolutionary terms, the presence of suites of correlated behavioral and physiological traits may have adaptive value (see chapters 6, 7, 8, and 9), which explains why different personalities are maintained in a population. Knowledge of the physiological and neurobiological substrates of personality may enhance our understanding of the ecological costs and benefits of personality variation. For example, behavioral variation in boldness and risk taking is associated with trade-offs in growth rate and mortality (Mangel and Stamps 2001; Stamps 2007). Costs and benefits of the physiological and neurobiological substrates of different personality profiles can also generate tradeoffs. Regarding stress physiology, high activation of sympathetic nervous system and catecholamines might be in the long-term deleterious for the cardiovascular system if uncontrollable stress is repeated. In aggressive individuals, it might be that the benefit of a proactive coping strategy effective in keeping environmental challenges under short-term control and in anticipating adversity offsets the cost on the cardiovascular system. Similarly, the cost of being nonaggressive and submissive can be avoided if an individual is able to physiologically react through the hormone-brain feedback, for example, through the effects of glucocorticoid on brain structure and functioning. Many animal personality studies, such as those on SAL-LAL mice and those on Fast and Slow great tits, suggest that phenotypic behavioral plasticity is a character in its own right; this idea is supported by differences in the gene expression of structures underlying neuronal plasticity in mice adopting different coping strategies (Feldker et al. 2003).

With a highly plastic brain, the reactive/passive individual may cope better in avoiding conflicts by finding new places to get access to food and partners. Suboptimal physiological or behavioral traits can still lead to a successful strategy if more correlates, both behavioral and physiological, are considered altogether.

Physiology sets boundaries to unlimited behavioral plasticity

Physiological processes can contribute to the existence of correlations between behavioral traits if they are the common substrates underlying these traits. For example, since the serotonergic system is involved in the regulation of both aggressive and sexual behavior in male rats (e.g., Veening et al. 2005), the association between high (or low) scores of aggressive and sexual behavior across individuals may be the result of individual differences in serotonergic function. Physiological substrates may act then as evolutionary constraints, reducing variation in behavioral traits that may arise as a result of developmental or selective processes. In some cases, however, selection can act independently on two behavioral traits that share a common physiological substrate by acting on the links between behavioral traits and different components of the substrate, for example, receptor functioning vs. the signaling response. For example, although testosterone affects many different behaviors of male vertebrates, these behaviors may be under different selective pressures because selection can modify different aspects of testosterone physiology in relation to different behaviors (Hau 2007).

Conclusions

Examples from rodents, birds, farmed mammals, nonhuman primates, and humans indicate that personality traits often have identifiable physiological substrates and that these substrates are similar across species. The main peripheral mechanisms underlying animal personalities are those related to stress and sex hormones. In the brain, the neurotransmitters serotonin, dopamine, and GABA are orchestrators and modulators of the different behavioral correlates. At the neuroanatomical level, the hippocampus is a candidate neural substrate for behavioral plasticity, since it is highly responsive to the hormonal fluctuations due to stress. Physiological correlates should be incorporated in the study of animal personalities because they might explain the coexistence of various behavioral types in a population. Moreover, common physiological substrates might represent the mechanism responsible for the correlation of behavioral traits observed in different personalities. A challenge for future research will be the integration of central and peripheral mechanisms in behavioral studies of animal personalities, at all levels of analysis, including genetic, cellular, anatomical, and physiological.

References

Andrew, R. J. 1966. Precocious adult behavior in the young chick. *Animal Behaviour*, 14, 485–500.

Archer, J. A. 1973. Further analysis of responses to a novel environment by testosterone-treated chicks. *Behavioral Biology*, 9, 389–396.

Baucom, D. H., Besch, P. K., and Callahan, S. 1985. Relation between testosterone concentration, sex role identity, and personality among females. *Journal of Personality and Social Psychology*, 48, 1218–1226.

Benus, R. F., Bohus, B., Koolhaas, J. M., and Van Oortmerssen G. A. 1991a. Behavioural differences between artificially selected aggressive and nonaggressive mice: response to apomorphine. *Behavioural Brain Research*, 43, 203–208.

———. 1991b. Heritable variation for aggression as a reflection of individual coping strategies. *Experientia*, 47, 1008–1019.

Benus, R. F., Den Daas, S. J., Koolhaas, J. M., and Van Oortmerssen, G. A. 1990. Routine formation and flexibility in social and nonsocial behaviour of aggressive and nonaggressive male mice. *Behaviour*, 112, 176–193.

Benus, R. F., Koolhaas, J. M., and Van Oortmerssen, G. A. 1992. Individual strategies of aggressive and nonaggressive male mice in encounters with trained aggressive residents. *Animal Behaviour*, 43, 531–540.

Bignami, G. 1965. Selection for high rates and low rates of avoidance conditioning in the rat. *Animal Behaviour*, 13, 221–227.

Blanchard, M. M., Mendelsohn, D., and Stamp, J. A. 2009. The HR/LR model: further evidence as an animal model of sensation seeking. *Neuroscience and Biobehavioral Reviews*, 33, 1145–1154.

Blanco, C., Ibáñez, A., Blanco-Jerez, C. A., Baca-Garcia, E., and Sáiz-Ruiz, J. 2001. Plasma testosterone and pathological gambling. *Psychiatry Research*, 105, 117–121

Blokhuis, H. J., and Beutler, A. 1992. Feather pecking damage and tonic immobility response in two lines of White Leghorn hens. *Journal of Animal Science*, 70, 170.

Byrne, G., and Suomi, S. J. 2002. Cortisol reactivity and its relation to home-cage behavior and personality ratings in tufted capuchin (*Cebus apella*) juveniles from birth to six years of age. *Psychoneuroendocrinology*, 27, 139–154.

Cannon, W. B. 1915. *Bodily Changes in Pain, Hunger, Fear, and Rage: An Account of Recent Researches into the Function of Emotional Excitement*. New York: Appleton and Company.

Caramaschi, D., de Boer, S. F., and Koolhaas, J. M. 2007. Differential role of the 5-HT1A receptor in aggressive and nonaggressive mice: an across-strain comparison. *Physiology and Behavior*, 90, 590–601.

Caramaschi, D., de Boer, S. F., de Vries, H., and Koolhaas, J. M. 2008. Development of violence in mice through repeated victory along with changes in prefrontal cortex neurochemistry. *Behavioural Brain Research*, 189, 263–272.

Carere, C., Groothuis, T. G. G., Möstl, E., Daan, S., and Koolhaas, J. M. 2003. Fecal corticosteroids in a territorial bird selected for different personalities: daily rhythm and the response to social stress. *Hormones and Behavior*, 43, 540–548.

Carere, C., and Van Oers, K. 2004. Shy and bold great tits (*Parus major*): body temperature and breath rate in response to handling stress. *Physiology and Behavior*, 82, 905–912.

Cavigelli, S. A., and McClintock, M. K. 2003. Fear of novelty in infant rats predicts adult corticosterone dynamics and an early death. *Proceedings of the National Academy of Sciences U S A*, 100, 16131–16136.

Cologer-Clifford, A., Simon, N. G., Lu, S. F., and Smoluk, S. A. 1997. Serotonin agonist-induced decreases in intermale aggression are dependent on brain region and receptor subtype. *Pharmacology Biochemistry and Behavior*, 58, 425–430.

Cologer-Clifford, A., Simon, N. G., Richter, M. L., Smoluk, S. A., and Lu, S. 1999. Androgens and estrogens modulate 5-HT1A and 5-HT1B agonist effects on aggression. *Physiology and Behavior*, 65, 823–828.

Compaan, J. C., de Ruiter, A. J., Koolhaas, J. M., Van Oortmerssen, G. A., and Bohus B. 1992. Differential effects of neonatal testosterone treatment on aggression in two selection lines of mice. *Physiology and Behavior*, 51, 7–10.

Daitzman, R., and Zuckerman, M. 1980. Disinhibitory sensation seeking, personality, and gonadal hormones. *Personality and Individual Differences*, 1, 103–110.

de Boer, S. F., and Koolhaas, J. M. 2003. Defensive burying in rodents: ethology, neurobiology, and psychopharmacology. *European Journal of Pharmacology*, 463, 145–161.

de Boer, S. F., Van der Vegt, B., and Koolhaas, J. M. 2003. Individual variation in aggression of feral rodent strains: a standard for the genetics of aggression and violence? *Behaviour Genetics*, 33, 485–501.

de Kloet, E. R., and Reul, J. M. 1987. Feedback action and tonic influence of corticosteroids on brain function: a concept arising from the heterogeneity of brain receptor systems. *Psychoneuroendocrinology*, 12, 83–105.

Digman, J. M. 1990. Personality structure: emergence of the 5-factor model. *Annual Review of Psychology*, 41, 417–440.

Dolan, M. C., Anderson, I. M., and Deakin, J. F. W. 2001. Relationship between 5-HT function and impulsivity and aggression in male offenders with personality disorders, *British Journal of Psychiatry*, 178, 352–359.

Du, L., Bakish, D., and Hrdina, P. D. 2000. Gender differences in association between serotonin transporter gene polymorphism and personality traits. *Psychiatric Genetics*, 10, 159–164.

Engel, G. L., and Schmale, A. H. 1972. Conservation withdrawal: a primary regulatory process for organic homeostasis. In: *Physiology, Emotions, and Psychosomatic Illness* (Porter, R., and Knight, J., eds.), pp. 57–76. New York: Elsevier Science.

Eysenck, H. J., Arnold, W., and Meili, R. 1979. *Encyclopedia of Psychology*. New York: Seabury Press.

Feldker, D. E., Datson, N. A., Veenema, A. H., Meulmeester, E., de Kloet, E. R., and Vreugdenhil, E. 2003. Serial analysis of gene expression predicts structural differences in hippocampus of long attack latency and short attack latency mice. *European Journal of Neuroscience*, 17, 379–387.

Fernandez, X., Meunier-Salaun, M. C., and Mormede, P. 1994. Agonistic behavior, plasma stress hormones, and metabolites in response to dyadic encounters in domestic pigs: interrelationships and effect of dominance status. *Physiology and Behavior*, 56, 841–847.

Ferris, C. F., and Potegal, M. 1988. Vasopressin receptor blockade in the anterior hypothalamus suppresses aggression in hamsters. *Physiology and Behavior*, 44, 235–239.

Fink, G., Sumner, B. E., Rosie, R., Grace, O., and Quinn, J. P. 1996. Estrogen control of central neurotransmission: effect on mood, mental state, and memory. *Cellular and Molecular Neurobiology*, 16, 325–344.

Fokkema, D. S., Koolhaas, J. M., and Van der Gugten, J. 1995. Individual characteristics of behavior, blood pressure, and adrenal hormones in colony rats. *Physiology and Behavior*, 57, 857–862.

Fox, E., Ridgewell, A., and Ashwin, C. 2009. Looking on the bright side: biased attention and the human serotonin transporter gene. *Proceedings of the Royal Society of London B*, 276, 1747–1751.

Garattini, S. 1967. Isolation, aggressiveness, and brain 5-hydroxytryptamine turnover. *Journal of Pharmacy and Pharmacology*, 19, 338–339.

Gerra, G., Zaimovic, A., Avanzini, P., Chittolini, B., Giucastro, G., Caccavari, R., Palladino, M., Maestri, D., Monica, C., Delsignore, R., and Brambilla, F. 1997. Neurotransmitter-neuroendocrine responses to experimentally induced aggression in humans: influence of personality variable. *Psychiatry Research*, 66, 33–43.

Giacalone, E., Tansella, M., Balzelli, L., and Garattini, S. 1968. Brain serotonin metabolism in isolated aggressive mice. *Biochemical Pharmacology*, 17, 1315–1327.

Gonda, X., Fountoulakis, K. N., Juhasz, G., Rihmer, Z., Lazary, J., Laszik, A., Akiska, H. S., and Bagdy, G. 2009. Association of the s allele of the 5-HTTLPR with neuroticism-related traits and temperaments in a psychiatrically healthy population. *European Archives of Psychiatry and Clinical Neuroscience*, 259, 106–113.

Groothuis, T. G. G., and Carere, C. 2005. Avian personalities: characterization and epigenesis. *Neuroscience and Biobehavioral Reviews*, 29, 137–150.

Groothuis, T. G. G., Carere, C., Lipar, J., Drent, P. J., and Schwabl, H. 2008. Selection on personality in a songbird affects maternal hormone levels tuned to its effect on timing of reproduction. *Biology Letters*, 5, 465–467.

Haller, J., Toth, M., Halasz, J., and de Boer, S. F. 2006. Patterns of violent aggression-induced brain c-fos expression in male mice selected for aggressiveness. *Physiology and Behavior*, 88, 173–182.

Hau, M. 2007. Regulation of male traits by testosterone: implications for the evolution of vertebrate life histories. *Bioessays*, 29, 33–44.

Heck, A., Lieb, R., Ellgas, A., Pfister, H., Lucae, S., Roeske, D., Pütz, B., Müller-Myhsok, B., Uhr, M., Holsboer, F., and Ising, M. 2009. Investigation of 17 candidate genes for personality traits confirms effects of the HTR2A gene on novelty seeking. *Genes, Brain, and Behavior*, 8, 464–472.

Henry, J. P., and Stephens, P. M. 1977. *Stress, Health, and the Social Environment: A Sociobiological Approach to Medicine*. New York: Springer Verlag.

Hess, C., Reif, A., Strobel, A., Boreatti-Hümmer, A., Heine, M., Lesch, K. P., and Jacob, C. P. 2009. A functional dopamine-beta-hydroxylase gene promoter polymorphism is associated with impulsive personality styles, but not with affective disorders. *Journal of Neural Transmission*, 116, 121–130.

Hessing, M. J., Hagelsø, A. M., Schouten, W. G., Wiepkema, P. R., and Van Beek, J. A. 1994. Individual behavioral and physiological strategies in pigs. *Physiology and Behavior*, 55, 39–46.

Higley, J. D. 2003. Aggression. In: *Primate Psychology* (Maestripieri, D., ed.), pp. 17–40. Cambridge: Harvard University Press.

Higley, J. D., Mehlman, P. T., Taub, D. M., Higley, S. B., Suomi, S. J., Vickers, J. H., and Linnoila, M. 1992. Cerebrospinal fluid monoamine and adrenal correlates of aggression in free-ranging rhesus monkeys. *Archives of General Psychiatry*, 49, 436–441.

Hogg, S., Hof, M., Würbel, H., Steimer, T., de Ruiter, A., Koolhaas, J. M., and Sluyter, F. 2000. Behavioural profiles of genetically selected aggressive and nonaggressive male wild house mice in two anxiety tests. *Behaviour Genetics*, 30, 439–446.

Hooks, M. S., Jones, G. H., Smith, A. D., Neill, D. B., and Justice, J. B. Jr. 1991. Individual differences in locomotor activity and sensitization. *Pharmacology, Biochemistry, and Behavior*, 38, 467–470.

Hooks, M. S., Colvin, A. C., Juncos, J. L., and Justice, J. B. Jr. 1992. Individual differences in basal and cocaine-stimulated extracellular dopamine in the nucleus accumbens using quantitative microdialysis. *Brain Research*, 587, 306–312.

Jang, K. L., Hu, S., Livesley, W. J., Angleitner, A., Riemann, R., Ando, J., Ono, Y., Vernon, P. A., and Hamer, D. H. 2001. Covariance structure of neuroticism and agreeableness: a twin and molecular genetic analysis of the role of the serotonin transporter gene. *Journal of Personality and Social Psychology*, 81, 295–304.

Jones, K. V., Copolov, D. L., and Outch, K. H. 1986. Type A, test performance, and salivary cortisol. *Journal of Psychosomatic Research*, 30, 699–707.

Joyce, P. R., McHugh, P. C., Light, K. J., Rowe, S., Miller, A. L., and Kennedy, M. A. 2009. Relationships between angry-impulsive personality traits and genetic polymorphisms of the dopamine transporter. *Biological Psychiatry*, 66, 717–721.

Kagan, J., Reznick, J. S., and Snidman, N. 1987. The physiology and psychology of behavioral inhibition in children. *Child Development*, 58, 1459–1473.

———. 1988. Biological bases of childhood shyness. *Science*, 240, 167–171.

Kalin, N. H., and Shelton, S. E. 2003. Nonhuman primate models to study anxiety, emotion regulation, and psychopathology. *Annals of the New York Academy of Science*, 1008, 189–200.

Karten, Y. J., Stienstra, C. M., and Joëls, M. 2001. Corticosteroid effects on serotonin responses in granule cells of the rat dentate gyrus. *Journal of Neuroendocrinology*, 13, 233–288.

Koolhaas, J. M., Korte, S. M., De Boer, S. F., Van Der Vegt, B. J., Van Reenen, C. G., Hopster, H., De Jong, I. C., Ruis, M. A., and Blokhuis, H. J. 1999. Coping styles in animals: current status in behavior and stress-physiology. *Neuroscience and Biobehavioral Reviews*, 23, 925–935.

Korte, S. M., Meijer, O. C., de Kloet, E. R., Buwalda, B., Keijser, J., Sluyter, F., Van Oortmerssen, G., and Bohus, B. 1996. Enhanced 5-HT1A receptor expression in forebrain regions of aggressive house mice. *Brain Research*, 736, 338–343.

Korte, S. M., Ruesink, W., and Blokhuis, H. J. 1999. Heart rate variability during manual restraint in chicks from high- and low-feather-pecking lines of laying hens. *Physiology and Behavior*, 65, 649–652.

Kovács, K. J., Földes, A., and Sawchenko, P. E. 2000. Glucocorticoid negative feedback selectively targets vasopressin transcription in parvocellular neurosecretory neurons. *Journal of Neuroscience*, 20, 3843–3852.

Kreek, M. J., Nielsen, D. A., Butelman, E. R., and LaForge, K. S. 2005. Genetic influences on impulsivity, risk taking, stress responsivity, and vulnerability to drug abuse and addiction. *Nature Neuroscience*, 8, 1450–1457.

Lemonde, S., Turecki, G., Bakish, D., Du, L., Hrdina, P. D., Bown, C. D., Sequeira, A.,

Kushwaha, N., Morris, S. J., Basak, A., Ou, X. M., and Albert, P. R. 2003. Impaired repression at a 5-hydroxytryptamine 1A receptor gene polymorphism associated with major depression and suicide. *Journal of Neuroscience*, 23, 8788–8799.

Mandelli, L., Mazza, M., Martinotti, G., Di Nicola, M., Tavian, D., Colombo, E., Missaglia, S., De Ronchi, D., Colombo, R., and Janiri, L. 2009. Harm avoidance moderates the influence of serotonin transporter gene variants on treatment outcome in bipolar patients. *Journal of Affective Disorders*, 119, 205–209.

Mangel, M., and Stamps, J. 2001. Trade-offs between growth and mortality and the maintenance of individual variation in growth. *Evolutionary Ecology Research*, 3, 583–593.

McCleery, J. M., and Goodwin, G. M. 2001. High and low neuroticism predict different cortisol responses to the combined dexamethasone-CRH test. *Biological Psychiatry*, 49, 410–415.

Modigh, K. 1973. Effects of isolation and fighting in mice on the rate of synthesis of noradrenaline, dopamine and 5-hydroxytryptamine in the brain. *Psychopharmacology*, 33, 1–17.

Murphy, D. L., and Lesch, K. P. 2008. Targeting the murine serotonin transporter: insights into human neurobiology. *Nature Reviews Neuroscience*, 9, 85–96.

Nelson, R. J. 2005. *An Introduction to Behavioral Endocrinology*. 3rd ed. Sunderland, MA: Sinauer.

Netter, P., Hennig, J., and Roed, I. S. 1996. Serotonin and dopamine as mediators of sensation-seeking behavior. *Neuropsychobiology*, 34, 155–165.

Nigg, J. T. 2000. On inhibition/disinhibition in developmental psychopathology: views from cognitive and personality psychology and a working inhibition taxonomy. *Psychological Bulletin*, 126, 220–246.

Olivier, B., Mos, J., Tulp, M., Schipper, J., Den Daas, S. J., and Van Oortmerssen, G. A. 1990. Serotonergic involvement in aggressive behaviour in animals. In: *Violence and Suicidality: Perspectives in Clinical and Psychological Research* (Van Praag, H. M., Plutchik, R., and Apter, A., eds.), pp. 79–84. New York: Brunner/Mazel.

Olweus, D., Mattsson, A., Schalling, D., and Löw, H. 1988. Circulating testosterone levels and aggression in adolescent males: a causal analysis. *Psychosomatic Medicine*, 50, 261–272.

Oswald, L. M., Zandi, P., Nestadt, G., Potash, J. B., Kalaydjian, A. E., and Wand, G. S. 2006. Relationship between cortisol responses to stress and personality. *Neuropsychopharmacology*, 31, 1583–1591.

Persky, H., Smith, K. D., and Basu, G. K. 1971. Relation of psychologic measures of aggression and hostility to testosterone production in man. *Psychosomatic Medicine*, 33, 265–277.

Pinna, G., Costa, E., and Guidotti, A. 2006. Fluoxetine and norfluoxetine stereospecifically and selectively increase brain neurosteroid content at doses that are inactive on 5-HT reuptake. *Psychopharmacology*, 186, 362–372.

Popma, A., Jansen, L. M., Vermeiren, R., Steiner, H., Raine, A., Van Goozen, S. H., Van Engeland, H., and Doreleijers, T. A. 2006. Hypothalamus pituitary adrenal axis and autonomic activity during stress in delinquent male adolescents and controls. *Psychoneuroendocrinology*, 31, 948–957.

Rogers, L. J., and Andrew, R. J. 1989. Frontal and lateral visual field use by chicks after treatment with testosterone. *Animal Behaviour*, 38, 394–405.

Schneider, C., King, R. M., and Philipson, L. 1988. Genes specifically expressed at growth arrest of mammalian cells. *Cell*, 54, 787–793.

Schuurman, T. 1980. Hormonal correlates of agonistic behavior in adult male rats. *Progress in Brain Research*, 53, 415–420.

Sen, S., Villafuerte, S., Nesse, R., Stoltenberg, S. F., Hopcian, J., Gleiberman, L., Weder, A., and Burmeister, M. 2004. Serotonin transporter and GABAA alpha 6 receptor variants are associated with neuroticism. *Biological Psychiatry*, 55, 244–249.

Sgoifo, A., de Boer, S. F., Haller, J., and Koolhaas, J. M. 1996. Individual differences in plasma catecholamine and corticosterone stress responses of wild-type rats: relationship with aggression. *Physiology and Behavior*, 60, 1403–1407.

Sgoifo, A., de Boer, S. F., Buwalda, B., Korte-Bouws, G., Tuma, J., Bohus, B., Zaagsma, J., and Koolhaas, J. M. 1998. Vulnerability to arrhythmias during social stress in rats with different sympathovagal balance. *American Journal of Physiology*, 275, H460–H466.

Sgoifo, A., Costoli, T., Meerlo, P., Buwalda, B., Pico-Alfonso, M. A., de Boer, S. F., Musso, E., and Koolhaas, J. M. 2005. Individual differences in cardiovascular response to social challenge. *Neuroscience and Biobehavioral Reviews*, 29, 59–66.

Sih, A., Bell, A. M., Johnson, J. C., and Ziemba, R. E. 2004. Behavioral syndromes: an integrative overview. *Quarterly Reviews of Biology*, 79, 241–77.

Sluyter, F., Jamot, L., Van Oortmerssen, G. A., and Crusio, W. E. 1994. Hippocampal mossy fiber distributions in mice selected for aggression. *Brain Research*, 646, 145–148.

Sluyter, F., Korte, S. M., Bohus, B., and Van Oortmerssen G. A. 1996. Behavioural stress response of genetically selected aggressive and nonaggressive wild house mice in the shock-probe/defensive burying test. *Pharmacology Biochemistry and Behavior*, 54, 113–116.

Spangler, G. 1995. School performance, Type A behavior, and adrenocortical activity in primary school children. *Anxiety, Stress, and Coping*, 8, 299–310.

Stamps, J. A. 2007. Growth-mortality tradeoffs and "personality traits" in animals. *Ecology Letters*, 10, 355–363.

Steimer T., La Fleur, S., and Schulz, P. E. 1997. Neuroendocrine correlates of emotional reactivity and coping in male rats from the Roman high (RHA/Verh)- and low (RLA/Verh)-avoidance lines. *Behaviour Genetics*, 27, 503–512.

Strobel, A., Gutknecht, L., Rothe, C., Reif, A., Mössner, R., Zeng, Y., Brocke, B., and Lesch, P. K. 2003. Allelic variation in 5-HT1A receptor expression is associated with anxiety- and depression-related personality traits. *Journal of Neural Transmission*, 110, 1445–1453.

Suarez, E. C., Kuhn, C. M., Schanberg, S. M., Williams, R. B., Jr., and Zimmermann, E. A. 1998. Neuroendocrine, cardiovascular, and emotional responses of hostile men: the role of interpersonal challenge. *Psychosomatic Medicine*, 60, 78–88.

Sumner, B. E., and Fink, G. 1998. Testosterone as well as estrogen increases serotonin2A receptor mRNA and binding site densities in the male rat brain. *Molecular Brain Research*, 59, 205–214.

Suomi, S. J. 1997. Early determinants of behaviour: evidence from primate studies. *British Medical Bulletin*, 53, 170–184.

———. 2006. Risk, resilience, and gene x environment interactions in rhesus monkeys. *Annals of the New York Academy of Science*, 1094, 52–62.

Uhart, M., McCaul, M. E., Oswald, L. M., Choi, L., and Wand, G. S. 2004. GABRA6 gene polymorphism and an attenuated stress response. *Molecular Psychiatry*, 9, 998–1006.

Van der Vegt, B., de Boer, S. F., Buwalda, B., de Ruiter, A. J., de Jong, J. G., and Koolhaas, J. M. 2001. Enhanced sensitivity of postsynaptic serotonin-1A receptors in rats and mice with high trait aggression. *Physiology and Behavior*, 74, 205–211.

Van Hierden, Y. M., Korte, S. M., Ruesink, E. W., Van Reenen, C. G., Engel, B., Koolhaas, J. M., and Blokhuis, H. J. 2002a. The development of feather-pecking behaviour and targetting of pecking in chicks from a high- and low-feather-pecking line of hens. *Applied Animal Behaviour Science*, 77, 183–196.

Van Hierden, Y. M., Korte, S. M., Ruesink, E. W., Van Reenen, C. G., Engel, B., Korte-Bouws, G. A., Koolhaas, J. M., and Blokhuis, H. J. 2002b. Adrenocortical reactivity and central serotonin and dopamine turnover in young chicks from a high- and low-feather-pecking line of laying hens. *Physiology and Behavior*, 75, 653–659.

Van Oortmerssen, G. A., Dijk, D. J., and Schuurman, T. 1987. Studies in wild house mice, II: testosterone and aggression. *Hormones and Behavior*, 21, 139–152.

Van Oortmerssen, G. A., and Busser, J. 1989. Studies in wild house mice, 3: disruptive selection on aggression as a possible force in evolution. In: *House Mouse Aggression: A Model for Understanding the Evolution of Social Behaviour* (Brain, P. F., Mainardi, D., and Parmigiani, S., eds.), pp. 87–118. Chur, Switzerland: Harwood Academic Publishers.

Van Riel, E., Meijer, O. C., Veenema, A. H., and Joels, M. 2002. Hippocampal serotonin responses in short and long attack latency mice. *Journal of Neuroendocrinology*, 14, 234–239.

Veenema, A. H., Cremers, T. I., Jongsma, M. E., Steenbergen, P. J., de Boer, S. F., and Koolhaas, J. M. 2005. Differences in the effects of 5-HT(1A) receptor agonists on forced swimming behaviour and brain 5-HT metabolism between low and high aggressive mice. *Psychopharmacology*, 178, 151–160.

Veenema, A. H., Koolhaas, J. M., and de Kloet, E. R. 2004. Basal and stress-induced differences in HPA axis, 5-HT responsiveness, and hippocampal cell proliferation in two mouse lines. *Annals of the New York Academy of Science*, 1018, 255–265.

Veenema, A. H., Meijer, O. C., de Kloet, E. R., and Koolhaas, J. M. 2003a. Genetic selection for coping style predicts stressor susceptibility. *Journal of Neuroendocrinology*, 15, 256–267.

Veenema, A. H., Meijer, O. C., de Kloet, E. R., Koolhaas, J. M., and Bohus, B. G. 2003b. Differences in basal and stress-induced HPA regulation of wild house mice selected for high and low aggression. *Hormones and Behavior*, 43, 197–204.

Veenema, A. H., and Neumann, I. D. 2008. Central vasopressin and oxytocin release: regulation of complex social behaviours. *Progress in Brain Research*, 170, 261–276.

Veenema, A. H., Sijtsma, B., Koolhaas, J. M., and de Kloet, E. R. 2005. The stress response to sensory contact in mice: genotype effect of the stimulus animal. *Psychoneuroendocrinology*, 30, 550–557.

Veening, J. G., Coolen, L. M., de Jong, T. R., Joosten, H. W., de Boer, S. F., Koolhaas, J. M., and Olivier, B. 2005. Do similar neural systems subserve aggressive and sexual behaviour in male rats? Insights from c-fos and pharmacological studies. *European Journal of Pharmacology*, 526, 226–239.

Watters, J. J., Swank, M. W., Wilkinson, C. W., and Dorsa, D. M. 1996. Evidence for glucocorticoid regulation of the rat vasopressin V1a receptor gene. *Peptides*, 17, 67–73.

Westberg, L., Henningsson, S., Landén, M., Annerbrink, K., Melke, J., Nilsson, S., Rosmond, R., Holm, G., Anckarsäter, H., and Eriksson, E. 2009. Influence of androgen receptor repeat polymorphisms on personality traits in men. *Journal of Psychiatry and Neuroscience*, 34, 205–13.

Witte, A. V., Flöel, A., Stein, P., Savli, M., Mien, L. K., Wadsak, W., Spindelegger, C., Moser, U., Fink, M., Hahn, A., Mitterhauser, M., Kletter, K., Kasper, S., and Lanzenberger, R. 2009. Aggression is related to frontal serotonin-1A receptor distribution as revealed by PET in healthy subjects. *Human Brain Mapping*, 30, 2558–2570.

Zald, D. H., Cowan, R. L., Riccardi, P., Baldwin, R. M., Ansari, M. S., Li, R., Shelby, E. S., Smith C. E., McHugo, M., and Kessler, R. M. 2008. Midbrain dopamine receptor availability is inversely associated with novelty-seeking traits in humans. *Journal of Neuroscience*, 28, 14372–14378.

IV

IMPLICATIONS OF PERSONALITY RESEARCH

FOR CONSERVATION BIOLOGY,

ANIMAL WELFARE, AND HUMAN HEALTH

13

Animal Personality and Conservation Biology

The Importance of Behavioral Diversity

BRIAN R. SMITH AND DANIEL T. BLUMSTEIN

Introduction

There is significant concern regarding the widespread loss of biodiversity and the destructive effects of human activity on the natural environment (Carson 1962; Wilson 2002). Over 500 extinctions of animal species have been recorded since 1600 (Frankham et al. 2002), with this rate significantly increasing over the past 150 years (Primack 1998). Furthermore, the International Union for Conservation of Nature (IUCN) has over 9,000 species of animals on their Red List of Threatened Species, representing 22% of the total species evaluated (IUCN 2010). This current extinction rate is 100–1,000 times greater than the estimated natural rate of extinction based on the fossil record (Primack 1998), indicating we are in the midst of the Earth's sixth massive extinction.

Conservation biology is a crisis-oriented discipline designed to minimize human impacts on and sustain biodiversity (Primack 1998). The goals of conservation biology include preserving genetic variability to maintain the evolutionary potential of populations (Moritz 2002), sustaining habitats and ecosystem processes, minimizing the effects of habitat disturbance on populations, and maintaining an adequate number of reproductive individuals (Gilpin and Soulé 1986; Hedrick and Kalinowski 2000).

The study and management of animal personalities can play a key role in conservation biology because the variability of personalities within a population may be related to the genetic diversity of that population. Fluctuating selection pressures maintain personality variation (Réale and Festa-Bianchet 2003; Dingemanse et al. 2004; chapters 7 and 9) and thus preserve higher levels of genetic diversity, which allows populations to adapt to changing environmental conditions and reduces their risk of extinction. Anthropogenic factors are now, however, altering the nature of selection (Stockwell et al. 2003; Carroll 2008; Fenberg and Roy 2008; Hendry et al. 2008), which may lead to a reduction in behavioral variation and exacerbate the current conservation crisis.

In this chapter, we first discuss how personality and, in particular, behavioral diversity are important determinants of genetic diversity and the long-term persistence of populations. We then discuss the anthropogenic factors that can reduce behavioral diversity, both in the wild and in captivity, and follow with a discussion on the role of personality research in identifying potentially invasive species. We conclude with a list of recommendations to managers and evolutionary ecologists for understanding and maintaining behavioral diversity in the face of these novel evolutionary pressures.

Personality and genetic diversity

When natural selection acts on phenotypes, the underlying genotypes are where changes take place and genes are, therefore, considered the fundamental unit of biodiversity (Crozier 1997). Without adequate genetic variation, populations are limited in their ability to respond to environmental stochasticity, thereby increasing extinction risk (Frankham 1995; 1999). Maximizing genetic diversity by increasing heterozygosity and decreasing inbreeding (Frankham et al. 2002) is thus considered a top priority in conservation biology for populations or species to persist indefinitely (Frankel 1974; Allendorf and Leary 1986; Frankham 1999).

Maintaining a sufficient breeding population is an effective strategy to maintain high levels of genetic diversity, and population geneticists use the concept of effective population size (N_e) when calculating the probability of a population persisting and remaining genetically viable over time (Gilpin and Soulé 1986). N_e is the number of individuals breeding and contributing genes to succeeding generations in an idealized population (Frankham et al. 2002; Groom et al. 2006; King et al. 2006). An idealized population is one in which there is no selection and consists of a 1:1 sex ratio with random mating, non-overlapping generations, and roughly equal family size (Groom et al. 2006). Natural populations rarely, if ever, meet all of the assumptions of an idealized population, and N_e is usually an order of magnitude lower than actual census data (Frankham 1995; Frankham et al. 2002; Groom et al. 2006).

Behavioral diversity can impact genetic diversity by being directly linked to this underlying diversity, or by affecting demographic patterns that contribute to N_e (Anthony and Blumstein 2000). Variation in personality, which is defined as consistent differences in behavioral patterns between individuals of the same species or population (e.g., Wilson et al. 1994; Boissy

1995; Gosling 2001; Sih et al. 2004b; see introduction, this volume), can be maintained by fluctuating environmental pressures due to variable resource availability (Dingemanse et al. 2004; 2010; Boon et al. 2007; Wolf et al. 2007; Dingemanse and Wolf 2010; Wolf and Wessing 2010), predation levels (Réale and Festa-Bianchet 2003; Smith and Blumstein 2010), social conditions (Both et al. 2005; Bergmüller and Taborsky 2010; Cote et al. 2010), and micro-habitat niches (Buchholz and Clemmons 1997; Wilson 1998) that balance selection (Futuyma 1998) and maintain alternative phenotypes within and between populations. Changing selection pressures promote high levels of genetic diversity underlying phenotypic traits (Houle 1992; Wilson et al. 1994) and, indeed, low to moderate heritabilities have been found for various personality traits in several species (Benus et al. 1991; Réale et al. 2000; Dingemanse et al. 2002; Drent et al. 2003; Fairbanks et al. 2004b; Bell 2005; Sinn et al. 2006; chapters 6– 9). This suggests a significant amount of genetic variation underlying personality (Frankham et al. 2002; King et al. 2006). Thus, novel directional or stabilizing selection pressures (Futuyma 1998) acting counter to balancing or disruptive selection could lead to a reduction of both behavioral and corresponding genetic diversity.

Furthermore, direct links have been found between personality and fitness correlates in many species (for reviews see Dingemanse and Réale 2005; Smith and Blumstein 2008; chapters 6–9). Variation in behavioral types may be shaped by trade-offs between current and future fitness so that individuals that adopt more risk-taking behaviors may experience more reproductive and foraging opportunities in the short-term, but also suffer increased mortality, compared with risk-adverse individuals (Stamps 2007; Wolf et al. 2007; Smith and Blumstein 2008). In some cases, conditions favoring the development of an optimum strategy for a particular individual may depend on the proportion of individuals adopting alternative strategies, and stability in behavioral traits may therefore be maintained in a frequency-dependent manner (Wilson et al. 1994; Dall et al. 2004; Bergmüller and Taborsky 2010).

Thus, differential reproduction and mortality within populations appear to be based, at least in part, on interindividual differences in personality or behavioral types (e.g., more or less bold or aggressive—Sih et al. 2004b). Reproductive and survival skew can have significant effects on many aspects of ecology and evolution in animal populations, including decreasing N_e (Anthony and Blumstein 2000), and environmental changes that increase skew can result in increased inbreeding and hasten the loss of heterozygosity. In the following sections, we discuss the anthropogenic changes to the environment that may alter selection pressures maintaining behavioral and genetic diversity and therefore further exacerbate the current biodiversity crisis.

Anthropogenic factors influencing behavioral diversity in the wild

Animal populations impacted by human activities often suffer from reduced genetic diversity (Gillespie and Guttman 1989; Caizergues et al. 2003; DiBattista 2008). Mechanisms cited for this phenomenon include reduced population sizes and isolated populations, which increase the effect of genetic drift (Frankham et al. 2002; DiBattista 2008). A reduction in behavioral diversity may also be a factor driving the decrease in genetic diversity. Although selection for and against particular behavioral types may vary in the short-term depending on ecological pressures such as predation (Réale and Festa-Bianchet 2003; Smith and Blumstein 2010), food availability (Dingemanse et al. 2004), and social conditions (Both et al. 2005; Bergmüller and Taborsky 2010), shifting environmental conditions may lead to balancing selection on varying life-history strategies and equalize fitness long-term (Dall et al. 2004; Stamps 2007; Wolf et al. 2007; chapter 9). Selective pressures acting for and against particular behavioral types, however, could be significantly altered by anthropogenic activities. In this section, we discuss how personality variation in populations may shape the level of threat caused by habitat fragmentation, pollution, overharvesting, and general human presence and activity in wilderness areas. Particular behavioral types may be more vulnerable to the effects of human activity, disproportionately reducing individual fitness within certain segments of populations and leading to an overall shift in behavioral patterns that reduces behavioral diversity.

HABITAT FRAGMENTATION

The top threat to biodiversity worldwide is habitat loss and degradation, which affects 88% of threatened amphibians and 86% of threatened birds and mammals (Baillie et al. 2004). The building of roads and the clearing of habitat isolate populations and create dispersal barriers limiting the ability of animals to migrate (Andrén 1994; St. Clair 2003; McDonald and St. Clair 2004). Population fragmentation leads to reduced gene flow and increased inbreeding and thus decreases genetic diversity (Wilcox and Murphy 1985; Caizergues et al. 2003; DiBattista 2008). However, negative fitness consequences due to habitat loss may not be generalized throughout a population but, instead, may be distributed on the basis of variability in behavioral responses to fragmentation, such as willingness to disperse to new habitats.

Dispersal strategies between individuals appears to be nonrandom with respect to phenotype (Bélisle and St. Clair 2001; Garant et al. 2005), and personality may be important in determining the effects of fragmentation on species and populations.

Generalist species, for example, do better than more specialized species in fragmented landscapes because of the ability to exploit disturbed areas (Andrén 1994). Generalists exhibit higher rates of dispersal across degraded habitats and anthropogenic barriers (McDonald and St. Clair 2004; Castellón and Sieving 2006), and studies of birds indicate that generalists are less neophobic and more exploratory than specialized species (Greenberg 1983; 1989; Webster and Lefebvre 2001). These behavioral traits may provide generalists with increased flexibility in adapting to novel environments (Greenberg and Mettke-Hofmann 2001; Sol et al. 2002), and skew selection pressures to favor generalist species in fragmented landscapes.

Variability in personality traits may, therefore, shape how individuals respond to and are affected by habitat fragmentation. For example, exploratory behavior in a novel environment or open field, which is a commonly used measure of animal personality (Réale et al. 2007), has been linked with dispersal patterns. Trinidad killifish (*Rivulus hartii*) that ranked higher in exploration of a test tank dispersed further when released back into streams than individuals ranked lower in exploration (Fraser et al. 2001). In great tits (*Parus major*), fast-exploring parents had offspring that dispersed the furthest, fast-exploring females moved farther in the initial stages of natal dispersal than slower ones, and immigrants into the population were faster explorers than locally born birds (Dingemanse et al. 2003). Exploration in great tits is heritable (Dingemanse et al. 2002; Drent et al. 2003) and correlates with other personality traits, including boldness toward a novel object (Verbeek et al. 1994) and aggression (Verbeek et al. 1996). Individuals with differing genotypes may thus differ in their dispersal strategies, leading to skewed mortality among behavioral types in fragmented habitats.

As global climate change shifts ecosystem boundaries, animals are expected to respond by attempting to migrate with these shifts (Walther et al. 2002). As habitats are fragmented, however, populations may be forced to reside in islands of habitat surrounded by disturbed areas (Andrén 1994). Areas around fragmented habitats often consist of unsuitable open habitat (Saunders et al. 1991), which may increase the risk of predation for individuals traveling through them. Furthermore, cleared areas between optimal habitats may consist of low-quality food patches, which could compromise condition, or may be filled with anthropogenic risks, such as power lines or automobiles, which can directly injure or kill individuals (Clevenger et al.

2003). If bolder and more exploratory individuals are more likely to disperse out of fragmented habitats and into open areas between habitats, this could significantly reduce the fitness of those behavioral types and select for a population of shy and nonexploratory individuals. Alternatively, if dispersers are able to migrate into larger, higher-quality habitat than that found in fragmented islands where nondispersers remain, bolder individuals could be selected for over shyer ones. Either of these scenarios could hasten the loss of genetic diversity by biasing the population toward one behavioral type, and effective population size would be significantly reduced if reproductive success is strongly skewed toward either bold or shy individuals (Smith and Blumstein 2008). Other selective processes, however, such as negative frequency-dependent selection, could counteract this process of diversity reduction and contribute to the maintenance of personality variation (see chapters 7 and 9).

POLLUTION

Toxins and chemical pollutants from agricultural, industrial, and municipal activities have become ubiquitous in the environment as they are often carried by abiotic factors around the world far from emission sources (Simonich and Hites 1995; Wilkening et al. 2000; Ikonomou et al. 2002). Many of these compounds, including organochlorine pesticides, polychlorinated biphenyls (PCBs), and heavy metals (Colborn et al. 1993; Zala and Penn 2004), are lipid soluble and therefore accumulate in the tissues of wildlife and increase in concentration at higher trophic levels (Norstrom et al. 1998; Ross et al. 2000; Kannan et al. 2001a, b). Exposure to pollutants, particularly in the early stages of development (Bergeron et al. 1999), can interfere with normal endocrine functions by causing overproduction or underproduction of hormones and can lead to sexual abnormalities and reproductive impairment (Howell et al. 1980; Spearow et al. 1999; Baatrup and Junge 2001; Hayes et al. 2002), altered growth patterns (Relyea and Mills 2001; Bell 2004), decreased cognitive abilities (Schantz et al. 1995; Rice 2000; Schantz and Widholm 2001), and increased mortality (Relyea and Mills 2001).

Behavior is the end point of many developmental and physiological pathways and has, therefore, proved to be a useful indicator of the presence of chemical contaminants (for reviews see Clotfelter et al. 2004; Zala and Penn 2004). Assays that are now standardized for use in the study of animal personality (Réale et al. 2007), including behavioral responses to conspecific intruders (Palanza et al. 1999; Bell 2001), potential mates (Haegele and Hudson 1977; Bayley et al. 1999), potential predators (Bell 2004), novel objects (Erhard and Rhind 2004), and open fields (Palanza et al. 1999; Erhard and Rhind 2004), have all been used to document changes in aggression, courtship, risky behavior, and general activity levels following exposure to pollutants. These studies indicate that behavioral responses of individuals exposed to contaminants may increase or decrease in frequency and intensity depending on the species and contaminant in question (Clotfelter et al. 2004; Zala and Penn 2004), indicating changes in axes of behavioral variation associated with personality. As anthropogenic contaminants create novel selection pressures in environments, resulting shifts in personality traits not only may impact individual fitness by creating less than optimum phenotypes, but also may lead to demographic and social changes and affect population dynamics that maintain personality variation (Bergmüller and Taborsky 2010; Stamps and Groothuis 2010).

Furthermore, the effects of persistent organic and chemical pollutants are not distributed equally throughout a population, and overall phenotypic variation may decrease owing to novel selection pressures associated with environmental contaminants (Medina et al. 2007). In a wide range of organisms, differences have been found in susceptibility to the effects of pollutants in relation to genotypic differences between strains (Parmigiani et al. 1999; Spearow et al. 1999; 2001) and between individuals within single populations (Chagnon and Guttman 1989; Diamond et al. 1989; Schlueter et al. 1997; Kolok et al. 2004). Moreover, selection for more tolerant individuals has led to reduced genetic diversity in contaminated populations (Gillespie and Guttman 1989; Keklak et al. 1994; Schlueter et al. 1995; Matson et al. 2006). In addition to genetic differences in vulnerability or resistance to the effects of pollutants, nongenetic factors may be important as well (Kolok 2001). Bell (2004), for example, found that threespined stickleback (*Gasterosteus aculeatus*) females exposed to environmentally relevant levels of ethinyl estradiol (an active ingredient of birth control pills and postmenopausal hormone replacement therapy) exhibited increased growth and risky behavior, as well as higher mortality. Subjects originating from a population that evolved under high predation levels were particularly affected (Bell 2004). This result suggests that selection by predators favoring particular phenotypes (Magurran 1990) may have led to individuals that are highly sensitive to predator-induced stress and are therefore more susceptible to the detrimental effects of environmental contaminants (Relyea and Mills 2001).

Studies have shown that environmental stress (e.g., predator-induced stress) increases the negative impacts of pollutants (Relyea and Mills 2001).

It is possible that shyer and more reactive individuals are especially vulnerable to pollutants because they respond more strongly to stress than bolder individuals with increased activity in the hypothalamic-pituitary-adrenal (HPA) axis (Uhde et al. 1984; Kagen et al. 1988; Sapolsky 1990), leading to higher corticosteroid levels (Beuving et al. 1989; Koolhaas et al. 1999; Ruis et al. 2000; Cavigelli and McClintock 2003; Cavigelli 2005). The physiological response of sticklebacks to confinement stress varies depending on an individual's boldness and aggression level (Bell et al. 2010), suggesting that particular behavioral phenotypes may be more sensitive to environmental challenges. Behavioral diversity within populations may, therefore, be directly reduced by novel selection pressures associated with pollution acting against more sensitive phenotypes.

OVERHARVESTING

Throughout the world, humans use wildlife as a natural resource for consumption, sport, and profit (Festa-Bianchet 2003; Hutchings and Fraser 2008). Increasing global demand and improved harvesting techniques have led to the overexploitation of many of these resources, however, and overharvesting is now a major threat to a significant proportion of species threatened with extinction; including 55% of threatened fishes, 33% of threatened mammals, 30% of threatened birds, and 6% of threatened amphibians are threatened with extinction (Baillie et al. 2004). Furthermore, harvesting pressures are often not distributed randomly within populations but, instead, are skewed toward individuals possessing targeted phenotypes (Hutchings and Fraser 2008). Trophy hunting in mammals, for example, typically focuses on the largest individuals or those with the most elaborate ornaments (Milner et al. 2007). Fishing practices also bias catches through management regulations imposing minimum size limits (Conover and Munch 2002) and equipment that tends to selectively harvest larger fish (Bohnsack et al. 1989; Dahm 2000; Law 2000). Anthropogenic selection overriding natural selection on heritable phenotypic traits (Carlson et al. 2007; Edeline et al. 2007; Darimont et al. 2009) may, therefore, lead to the rapid removal of specific alleles or genotypes from a population, reducing variation, and hindering population recovery once harvesting pressures subside (Law 2000; Conover and Munch 2002; Allendorf and Hard 2009).

Studies of exploited species have documented several life-history and demographic changes within populations related to harvesting pressures (Law 2000; Festa-Bianchet 2003; Milner et al. 2007; Fenberg and Roy 2008; Darimont et al. 2009; Wolak et al. 2010). Trophy hunting in mammals has been followed by decreases in body weight and ornament size over time (Jachmann et al. 1995; Coltman et al. 2003; Garel et al. 2007; Proaktor et al. 2007), while periods of intense fishing have been correlated with decreases in body size and age at maturity (Law 2000; Sinclair et al. 2002; Olsen et al. 2004; Hutchings 2005; Edeline et al. 2007). Laboratory experiments with a commercially exploited fish species (*Menidia menidia*) have also shown that selective harvesting for large individuals over as little as four generations leads to lower harvested biomass and mean body weight due to selection for slow-growing genotypes (Conover and Munch 2002). Size-selective harvesting has thus been hypothesized to indirectly select against fast growth because individuals possessing this trait attain harvestable size at a younger age and may have less of an opportunity to reproduce (Law 2000; Olsen et al. 2004; Proaktor et al. 2007).

An alternative explanation, however, is that growth rate may be reduced due to correlational selection of both personality and growth rate. Bolder and more active individuals forage and grow at higher rates (Mangel and Stamps 2001; Sih et al. 2003; Ward et al. 2004; Biro et al. 2006; Stamps 2007), but may also be more vulnerable to harvesting because they take more risks (Dugatkin 1992; Sih et al. 2003; Biro et al. 2004; Stamps 2007). In one of the best examples to date of anthropogenic selection acting directly on animal personality traits, Biro and Post (2008) simulated an intensive gillnet fishery of rainbow trout (*Oncorhynchus mykiss*) and found that fast-growing genotypes were harvested at three times the rate of slow-growing genotypes independent of body size. They attributed these results to fast-growing individuals increasing their encounter rate with fishing equipment by behaving more boldly and actively.

Additional studies are needed to determine whether this same pattern is found in other harvested species. If so, this may be an indication that over-harvesting can unintentionally select for slower-growing and more neophobic populations, including practices that do not target specific sizes, such as fur trapping (Coltman 2008), waterfowl hunting (Fox and Madsen 1997), and predator control (Sacks et al. 1999; Mettler and Shivik 2007).

HUMAN PRESENCE AND ACTIVITY IN WILDERNESS AREAS

As human population expands, more areas formerly considered unoccupied wilderness are being developed and used for economic and recreational activities. Even national parks, which are considered prime habitat for wildlife, were visited by over 281 million people in 2010 in the United States alone. With such a significant human presence, even benign activities can have large impacts on animal populations. Evidence suggests that animals are altering their behavior in these anthropogenically affected environments, with some tolerating, and even taking advantage of, the presence of humans. Pregnant moose (*Alces alces*) within and surrounding Grand Teton National Park have redistributed birth sites closer to paved roads as brown bears (*Ursus arctos*), a potential predator of neonates, are becoming reestablished within the ecosystem (Berger 2007). Brown bears avoid paved roads, so birthing females are able to use roads as protection from potential predation. Personality may be playing a role in such behavioral shifts and conferring fitness advantages to individuals more tolerant of and habituating more rapidly to human activity. It has been shown that personality traits play a significant role in the behavioral and physiological responses of animals to stress (Uhde et al. 1984; Carere et al. 2001; Carere et al. 2003; Fairbanks et al. 2004a; chapter 12), with shy or more inhibited individuals responding more strongly to novel or stressful situations than bold or more proactive individuals (Koolhaas et al. 1999; Ruis et al. 2000; Cavigelli and McClintock 2003; chapter 12 and 15). Elevated stress levels over a prolonged time can compromise body condition and immune system functions and cause individuals to be more susceptible to disease and the negative effects of environmental fluctuation (Capitanio et al. 1998; Cavigelli and McClintock 2003; López et al. 2005; chapter 15). Such a scenario could give individuals with bolder and less inhibited personalities a fitness advantage and could alter selection pressures maintaining personality variation.

Studies are just now beginning to investigate the role of personality in shaping patterns of differential response and tolerance to human presence and activity. In Gault Nature Reserve in Québec, Canada, Martin and Réale (2008) found a higher frequency of explorative and docile eastern chipmunks (*Tamias striatus*) in areas with high human activity, while less explorative and docile individuals maintained home ranges in less frequented areas. The population thus distributed themselves in a nonrandom manner according to personality type and human presence. Such human effects of distribution could potentially lead to patchy, fragmented populations and reduce overall effective population sizes if interbreeding between subpopulations is reduced. To avoid stressful situations, individuals less tolerant and more wary of humans may also be forced into suboptimal areas with lower-quality food patches and a higher quantity of predators (Berger 2007) and suffer reduced fitness. Differential selection on personality types from human pressures may therefore lead to changes in the relative frequency of personality types within a population resulting in reduced behavioral diversity.

Behavioral diversity in captive environments

Two major issues concerning animal populations in captivity are loss of genetic diversity and adaptation to captive environments. Loss of genetic diversity reduces the capacity of populations to cope with environmental change, and adaptation to captivity may select for phenotypes maladaptive in the wild (Frankham 1999; Reed and Frankham 2003; Frankham 2008). Either of these factors can seriously compromise the ultimate goal of many captive-breeding programs, which is to release individuals back into their historic ranges and reestablish populations that have been extirpated or severely reduced (Griffith et al. 1989; Kleiman 1989; Frankham 1999; Seddon et al. 2007). In this section we discuss how selection in captivity for particular personality traits can lead to behavioral types ill-suited for survival in the wild. We also discuss how behavioral diversity can be affected by the reintroduction strategy employed.

SELECTING FOR TAMENESS

The success rate of reintroduction projects has historically been low (Beck et al. 1994; Wolf et al. 1996), and a review found that projects using wild-caught animals were twice as likely to succeed as those using captive-reared animals (Griffith et al. 1989). Research on animal domestication indicates that individuals differ in their ability to cope with and reproduce in captivity (Berry 1969; Price 1984), which can lead to selection of phenotypes more tolerant of humans. Changes in morphology, physiology, behavior, and genetics take place with the selective breeding of tame animals (Price 1970; Belyaev 1979; Belyaev et al. 1985; Marliave et al. 1993; Künzl et al. 2003) and can lead to populations ill-equipped to survive in the wild. Thus, unintentional selection for tame and more risk-taking behavioral types in captivity may be partially to blame for reintroduction failures.

Enclosure characteristics and husbandry procedures, for example, can increase stress levels and select for behavioral types that allow individuals to better cope with these conditions (McDougall et al. 2006; Clubb and Mason 2007; Peng et al. 2007). A study of black rhinoceros (*Diceros bicornis*) in zoos throughout the United States found that reproductive success and mortality correlated respectively with the size of the enclosure in which animals were housed and the amount of viewing area to which the public had access (Carlstead et al. 1999a). Keepers also rated subjects on several different behavioral attributes and found that enclosure design affected levels of dominance, fearfulness, and agitation (Carlstead et al. 1999a), all traits that have been shown to affect reproductive success in captive rhinoceros (Carlstead et al. 1999b). Cheetahs (*Acinonyx jubatus*) in North American breeding facilities also exhibit differential reproductive success depending on their assayed behavioral type, with nonbreeders rated as being more fearful than breeders independent of sex, age, and facility where individuals are housed (Wielebnowski 1999). Thus, selection pressures maintaining personality variation in the wild may be replaced in captivity by selection for more docile and tolerant individuals.

Studies do indicate that captive-bred animals are behaviorally distinct from their wild counterparts for several different personality traits. Individuals bred in captivity over several generations exhibit reduced levels of aggression (Mathews et al. 2005), increased boldness (Yoerg and Shier 1997; McPhee 2003b), and increased exploratory behavior (Price 1970; McPhee 2003a). Captive conditions may thus cause shyer and more neophobic individuals to experience prolonged elevated stress levels (Koolhaas et al. 1999; Carere et al. 2003; Cavigelli and McClintock 2003), which may reduce their fitness relative to bolder individuals. Not only might this lead to a captive population maladapted to fluctuating environmental conditions found in the wild, but also selecting for reduced behavioral variation in heritable traits will reduce genetic diversity.

PERSONALITY AND REINTRODUCTION

High levels of mortality are often seen in the initial stages of reintroduction projects (Seddon et al. 2007; Teixeira et al. 2007). Reintroductions can be costly and time-consuming, particularly those employing soft-release techniques (Kleiman 1989) in which released animals are provided with supplemental food or shelter, and each individual is a valuable commodity. Predation (Short et al. 1992; Miller et al. 1994; Banks et al. 2002) and possibly stress (Teixeira et al. 2007) are major causes of mortality for reintroduced animals, which can significantly delay the reestablishment of populations and remove potentially valuable alleles from the gene pool. Several studies have therefore tested the efficacy of predicting release success of individuals on the basis of pre-release behavioral screening (Van Heezik et al. 1999; Munkwitz et al. 2005) for the purpose of selective release: those behavioral types that respond most appropriately to potentially risky and stressful situations are more likely to survive following reintroduction (Mathews et al. 2005). Supporting this idea is a study with swift foxes (*Vulpes velox*), which found that those individuals that died following reintroduction into the wild were bolder toward novel objects in captivity than those that survived (Bremner-Harrison et al. 2004).

Bolder individuals have higher reproductive success, however, across a number of species (for a review see Smith and Blumstein 2008). Thus, releasing only shy individuals may increase survival in the short-term, but could have negative impacts on reproduction. Furthermore, selectively releasing particular behavioral types would reduce overall behavioral diversity, which may limit the variety of microhabitats available for reintroduced populations to exploit (Watters and Meehan 2007).

A reintroduction study of brown trout (*Salmo trutta*) found that individuals identified as dominant and more aggressive in captivity grew faster in the wild than subordinates, suggesting higher fitness levels (Höjesjö et al. 2002). Smaller fish were more mobile, however, when released back into the wild and exploited faster-flowing habitats closer to the shore than the larger fish. These findings indicate that individuals possessing behavioral types deemed "less suitable" for release by behavioral screening can be successful in heterogeneous natural habitats and allow for the establishment of populations over broader environmental gradients.

An alternative pre-release strategy is to train individuals in the appropriate skills and behavioral responses needed for survival (Biggins et al. 1999; Vargas and Anderson 1999; Griffin et al. 2000). Predator training, for example, involves enhancing antipredator behavior by pairing the sight of a predator with an aversive experience (Chivers and Smith 1994; McLean et al. 2000) or, as is often done with fish, pairing conspecific alarm cues with either predatory visual or chemical cues (Chivers and Smith 1994; Korpi and Wisenden 2001).

Research into the underlying mechanisms of behavioral syndromes suggest that personality traits including boldness and aggression (Huntingford 1976; Riechert and Hedrick 1993; Bell and Stamps 2004) and boldness and exploratory behavior (Van Oers et al. 2004) may be tightly linked owing to underlying physiological (Ketterson and Nolan 1999; chapter 12) or genetic mechanisms (Van Oers et al. 2004; Bell 2005; chapter 12). In populations possessing such rigid syndromes, behaviors may not be able to change independently of each other (Bell and Stamps 2004), and training individuals to be more fearful of predators could unintentionally lead to suboptimal behavior in other contexts (e.g., less aggression when competing with conspecifics for food or potential mates) (Sih et al. 2004a, b).

Tight correlations across population are not always found, however, and varying environmental pressures appear to select for the presence or absence of rigid syndromes (Bell 2005; Dingemanse et al. 2007). Without the necessary evolutionary pressures selecting for genetically linked behaviors

(Bell and Stamps 2004; Bell 2005), populations may evolve more plastic personalities and individuals may not exhibit multicontext behavioral changes following predator training.

Personality and invasion biology: identifying potential threats

Globalization of human activity has promoted both the intentional and accidental spread of species across natural dispersal barriers. Exotic and invasive species can disrupt ecological processes by causing the extinction of native species through competition, predation, and disease (for reviews see Mack et al. 2000; Pimentel et al. 2000). Among the most well-known and notorious examples of invasive animals are the cane toad (*Bufo marinus*) in Australia, the zebra mussel (*Dreissena polymorpha*) in North America, and the brown tree snake (*Boiga irregularis*) in Guam (Lowe et al. 2000). Introduced pests and predators have led to the collapse of agricultural, forestry, and fishery resources in some areas (Mack et al. 2000), and economic losses due to exotic species are in excess of $137 billion per year in the United States alone (Pimentel et al. 2000). A long-standing goal of ecologists has been to identify the biological characteristics that define successful colonists (Lodge 1993; Kolar and Lodge 2001), and personality research may be useful in this regard.

Behavioral attributes are increasingly recognized as useful predictors in identifying the proximate causes of invasion success (Holway and Suarez 1999). Dispersal tendency (Rehage and Sih 2004; Bubb et al. 2006), behavioral flexibility (Sol and Lefebvre 2000; Hughes and Cremer 2007), and competitive ability (Rehage et al. 2005; Duckworth and Badyaev 2007; Maestripieri 2007) have been identified as traits common to successful invaders. Variation in these traits has also been linked to animal personality. High dispersal rates are seen in more exploratory individuals (Fraser et al. 2001; Dingemanse et al. 2003), more aggressive individuals outcompete less aggressive ones (Riechert and Hedrick 1993; Civantos 2000), and increased plasticity has been associated with boldness (Greenberg 1983; 1989). Invasion success depends on the ability to colonize, establish, and spread (Holway and Suarez 1999) and may therefore favor species exhibiting syndromes for boldness, aggression, and exploration (Sih et al. 2004a; b; Maestripieri 2007). For example, populations of the signal crayfish (*Pacifastasus leniusculus*), which is a successful invader in many parts of the world, exhibit an aggression/activity/boldness syndrome that may contribute to their ability to outcompete native species (Pintor et al. 2008).

Therefore, to assess potential threats, natural and life-history traits that are correlated with high levels of boldness, aggression, and exploratory behavior should be identified (Bell and Sih 2007; Johnson and Sih 2007; Polo-Cavia et al. 2008). Such an approach would then allow managers to prioritize control efforts for the multiple exotics often found in single ecosystems. Tame species, for example, may be bigger threats because animals that live in close association with humans are often successful invaders (Newsome and Noble 1986; Sol et al. 2002; Jeschke and Strayer 2006; Maestripieri 2007). Invasive species also tend to grow faster than outcompeted natives (Newsome and Noble 1986; Hill and Lodge 1999; Sakai et al. 2001), which may favor more bold or aggressive behavioral types in the invaders (Stamps 2007).

Within-species personality variation may also account for changes in behavior and demography in invasive populations that affect expansion rate and the probability of reestablishing natives. Once exotics colonize, it's expected that a subset of the bolder and more exploratory individuals within the population would lead the invasion. Selection pressures for more competitive and flexible individuals along the invasion front may lead to the evolution of bolder and more aggressive personality traits, relative to the initial colonizing population, and rapidly increase the rate of expansion (Phillips et al. 2006). Birds that have recently invaded are, indeed, bolder (Martin and Fitzgerald 2005) and more aggressive (Duckworth and Badyaev 2007) than more established populations of the same species. Once the invasion front moves through, however, selection pressures may change and traits adaptive on the invasion front may be selected against in stable environments (Brown et al. 2007). Western bluebirds (*Sialia mexicana*) in newly colonized areas are more aggressive, but overall aggression decreases in only a few generations because aggressive males provide little to no parental care (Duckworth and Badyaev 2007). The intensity of behavioral traits indicative of aggressive invaders may therefore be reduced in more established populations, and it may be more feasible to reestablish native animals in these areas than to slow the invasion front.

Recommendations for managing behavioral diversity

Additional research is needed to further elucidate the effects of human activity on shaping and changing personality traits, and periodic assays of personality will allow managers to monitor for effects in the wild and incorporate personality measurements into captive studbooks to track the breeding success rate of varying behavioral types. What further techniques should be employed, however, to maintain variation, and what else do we need to know to better understand the evolutionary ecology of personality?

We have five recommendations to managers. First, to maintain behavioral diversity, environmental heterogeneity should be preserved when designing reserves (e.g., Smith et al. 1997; 2001). Selection acts in different directions in heterogeneous environments (Frankham et al. 2002), which is the basis for some hypotheses explaining personality variation (Réale and Festa-Bianchet 2003; Dall et al. 2004; Dingemanse et al. 2004; Both et al. 2005; Boon et al. 2007; Wolf et al. 2007; chapters 7 and 9). Thus, by conserving different habitat types and not just those most optimal for a particular species or population, the evolutionary processes shaping behavioral and genetic diversity will be maintained (Schneider et al. 1999; Moritz 2002; Watters et al. 2003; Rocha et al. 2005). Heterogeneous habitats with varying selection pressures may then allow recurrent selection to restore variation and phenotypes previously diminished (Moritz 1999; 2002; Bell and Sih 2007).

Second, for harvested populations, management strategies should be developed that preserve natural variation to maintain long-term sustainable yields (e.g., Conover and Munch 2002). Bolder and more active individuals may be more vulnerable to harvesting because they take more risks (Biro and Post 2008). Minimum-size limits for harvesting would therefore be ineffective in preventing population shifts toward slow-growing genotypes if selection on behavioral vulnerability is the evolutionary mechanism driving this response (Biro and Post 2008). Adaptive management strategies therefore need to be implemented that update patterns of harvesting as stocks evolve to maintain and rebuild behavioral and genetic diversity (Law 2000; Allendorf and Hard 2009). Establishing no-take zones and marine protected areas, for example, may act to maintain variation by allowing a portion of populations to express unconstrained ranges of sizes, growth rates, and behavioral types (Trexler and Travis 2000).

Third, for captive animals that may be later released into the wild, heterogeneous environments and enrichment techniques that facilitate the development and maintenance of behavioral diversity should be used (Watters and Meehan 2007). Although there is a limit in the ability to duplicate the heterogeneity of natural environments, maintaining populations under varying environmental (Korhonen and Niemelä 1996; Biggins et al. 1999; Meehan and Mench 2002; Kelley et al. 2005) and social (Goodey and Liley 1986; Chapman et al. 2008) conditions readily achieved in captivity facilitates variation in boldness and aggression.

Fourth, for reintroduction programs, individuals possessing behavioral types identified as most likely to survive following pre-release screening should be the first to be released. Increasing survivorship of the initial individuals released is key to maximizing the chances of reestablishing a self-sustaining population (Frankham et al. 2002). Shyer individuals may be more likely to evade predation (Dugatkin 1992) and less exploratory individuals are not likely to move far from the release site (Fraser et al. 2001; Dingemanse et al. 2003), which is important when soft-release techniques are being employed.

Fifth, if predator training is conducted before reintroduction, managers should assay multiple behavioral types before and after training to determine how traits across contexts are being shaped. Correlated behaviors across contexts should be able to change independently of each other if they are domain specific (Wilson 1998; Nelson et al. 2008; Sinn et al. 2010). Alternatively, correlated behaviors that are due to underlying physiological (Ketterson and Nolan 1999) or genetic (Van Oers et al. 2004; Bell 2005) mechanisms may be constrained to change together (Bell and Stamps 2004). Populations possessing such rigid behavioral syndromes may exhibit unintended behavioral changes in other ecological contexts following training, which may offset any fitness benefits.

For research biologists, we have two recommendations. First, a formal evolutionary ecology of personality should be developed (see Réale et al. 2010). Differences in behavioral types and behavioral syndromes between related taxa and populations can have major effects on ecology and evolution (Sih et al. 2004b; chapters 6–9), as well as on their susceptibility to anthropogenic threats, and their potential to become invasive species. Predation pressure, for example, is a likely explanation for the presence of a bold-aggression syndrome in some populations of sticklebacks and the lack of one in others (Bell and Stamps 2004; Bell 2005; Bell and Sih 2007; Dingemanse et al. 2007), and species-specific differences in wariness may be based on life history, natural history, and morphological differences (Stankowich and Blumstein 2005; Blumstein 2006). Additional selection pressures shaping personality should, therefore, be identified using techniques developed for measuring selection on quantitative traits and correlated characters (Lande and Arnold 1983; Arnold and Wade 1984a, b; Schluter 1988; Schluter and Nychka 1994). Such approaches may highlight whether particular populations are constrained in their behavioral flexibility. Furthermore, the role of selection for maintaining behavioral variability must be thoroughly studied to understand the potential threats caused by anthropogenic activities. Not only may novel selection pressures arising from such activities overwhelm those mechanisms responsible for shaping and maintaining personality variation, but also, because balancing selection as a mechanism becomes less effective in smaller populations (Wright 1931; Seddon and Baverstock 1999;

Miller and Lambert 2004), endangered species may be particularly at risk of losing behavioral diversity.

Second, a widespread understanding of the fitness consequences of personality is needed. Although few studies exist, evidence does suggest that personality and behavioral types account for differential survival and reproductive success (Smith and Blumstein 2008; 2010; Logue et al. 2009). The strength of these relationships may vary across species and populations, however, owing to potential differences in the mechanisms shaping personality variation (Smith and Blumstein 2008). The effect that personality has on effective population size and, in turn, genetic diversity may be less severe in populations exhibiting a weak link between behavioral type and fitness and more so in populations that show a stronger effect of personality on fitness. Thus, discerning the adaptive mechanisms maintaining personality variation would allow managers and behavioral biologists to identify those populations more at-risk from anthropogenic activities and associated novel selection pressures.

Conclusions

Numerous calls have been made for the incorporation of animal behavior into conservation biology (Clemmons and Buchholz 1997; Caro 1998; Shumway 1999; Gosling and Sutherland 2000), but this desired synthesis remains incomplete (Sutherland 1998; Candland 2005). Recognition that individuals, populations, and species behave in a consistent and variable manner may help bring these disciplines more closely together (e.g., McDougall et al. 2006; Watters and Meehan 2007). Personality is correlated with several life-history traits, and heritability estimates indicate a link to genetic diversity. Consideration of personality traits and of differences in personalities between species may also enhance reintroduction success and help identify potentially invasive species. Evolutionary pressures shaping such personality patterns may be altered, however, by novel anthropogenic pressures and lead to a reduction in behavioral and corresponding genetic diversity. Efficiently managing behavioral diversity should be enhanced by adaptive management (Blumstein 2007; Seddon et al. 2007) and the application of rigorous experimental designs.

References
Allendorf, F. W., and Hard, J. J. 2009. Human-induced evolution caused by unnatural selection through harvest of wild animals. *Proceedings of the National Academy of Sciences USA*, 106, 9987–9994.

Allendorf, F. W., and Leary, R. F. 1986. Heterozygosity and fitness in natural populations of animals. In: *Conservation Biology: The Science of Scarcity and Diversity* (Soulé, M. E., ed.), pp. 57–76. Sunderland, MA: Sinauer Associates.

Andrén, H. 1994. Effects of habitat fragmentation on birds and mammals in landscapes with different proportions of suitable habitat: a review. *Oikos*, 71, 355–366.

Anthony, L. L., and Blumstein, D. T. 2000. Integrating behaviour into wildlife conservation: the multiple ways that behavior can reduce N_e. *Biological Conservation*, 95, 303–315.

Arnold, S. J., and Wade, M. J. 1984a. On the measurement of natural and sexual selection: applications. *Evolution*, 38, 720–734.

———. 1984b. On the measurement of natural and sexual selection: theory. *Evolution*, 38, 709–719.

Baatrup, E., and Junge, M. 2001. Antiandrogenic pesticides disrupt sexual characteristics in the adult male guppy (*Poecilia reticulata*). *Environmental Health Perspectives*, 109, 1063–1070.

Baillie, J. E. M., Hilton-Taylor, C., and Stuart, S. N. 2004. *2004 IUCN Red List of Threatened Species: A Global Species Assessment*. Gland, Switzerland: World Conservation Union.

Banks, P. B., Norrdahl, K., and Korpimäki, E. 2002. Mobility decisions and the predation risks of reintroduction. *Biological Conservation*, 103, 133–138.

Bayley, M., Nielsen, J. R., and Baatrup, E. 1999. Guppy sexual behavior as an effect biomarker of estrogen mimics. *Ecotoxicology and Environmental Safety*, 43, 68–73.

Beck, B. B., Rapaport, L. G., Stanley Price, M. R., and Wilson, A. C. 1994. Reintroduction of captive-born animals. In: *Creative Conservation: Interactive Management of Wild and Captive Animals* (Olney, P. J. S., Mace, G. M., Feistner, A. T. C., eds.), pp. 265–286. London: Chapman and Hall.

Bélisle, M., and St. Clair, C. C. 2001. Cumulative effects of barriers on the movements of forest birds. *Conservation Ecology*, 5, 212–219.

Bell, A. M. 2001. Effects of an endocrine disrupter on courtship and aggressive behaviour of male three-spined stickleback, *Gasterosteus aculeatus*. *Animal Behaviour*, 62, 775–780.

———. 2004. An endocrine disrupter increases growth and risky behavior in threespined stickleback (*Gasterosteus aculeatus*). *Hormones and Behavior*, 45, 108–114.

———. 2005. Behavioural differences between individuals and two populations of stickleback (*Gasterosteus aculeatus*). *Journal of Evolutionary Biology*, 18, 464–473.

Bell, A. M., and Sih, A. 2007. Exposure to predation generates personality in threespined sticklebacks (*Gasterosteus aculeatus*). *Ecology Letters*, 10, 828–834.

Bell, A. M., and Stamps, J. A. 2004. Development of behavioural differences between individuals and populations of sticklebacks, *Gasterosteus aculeatus*. *Animal Behaviour*, 68, 1339–1348.

Bell, A. M., Henderson L., Huntingford F. A. 2010. Behavioral and respiratory responses to stressors in multiple populations of three-spined sticklebacks that differ in predation pressure. *Journal of Comparative Physiology B*, 180, 211–220.

Belyaev, D. K. 1979. Destabilizing selection as a factor in domestication. *Journal of Heredity*, 70, 301–308.

Belyaev, D. K., Plyusnina, I. Z., and Trut, L. N. 1985. Domestication in the silver fox (*Vulpes fulvus* desm): changes in physiological boundaries of the sensitive period of primary socialization. *Applied Animal Behaviour Science*, 13, 359–370.

Benus, R. F., Bohus, B., Koolhaas, J. M., and Van Oortmerssen, G. A. 1991. Heritable variation for aggression as a reflection of individual coping strategies. *Experientia*, 47, 1008–1019.

Berger, J. 2007. Fear, human shields, and the redistribution of prey and predators in protected areas. *Biology Letters*, 3, 620–623.

Bergeron, J. M., Willingham, E., Osborn, C. T., III, Rhen, T., and Crews, D. 1999. Developmental synergism of steroidal estrogens in sex determination. *Environmental Health Perspectives*, 107, 93–97.

Bergmüller, R., and Taborsky, M. 2010. Animal personality due to social niche specialisation. *Trends in Ecology and Evolution*, 25, 504–511.

Berry, R. J. 1969. The genetical implications of domestication in animals. In: *The Domestication and Exploitation of Plants and Animals* (Ucko, P. J., Dimbleby, G. W., eds.) pp. 207–217. Chicago: Aldine.

Beuving, G., Jones, R. B., and Blokhuis, H. J. 1989. Adrenocortical and heterophil/lymphocyte responses to challenge in hens showing short and long tonic immobility reactions. *British Poultry Science*, 30, 175–184.

Biggins, D. E., Vargas, A., Godbey, J. L., and Anderson, S. H. 1999. Influence of pre-release experience on reintroduced black-footed ferrets (*Mustela nigripes*). *Biological Conservation*, 89, 121–129.

Biro, P. A., Abrahams, M. V., Post, J. R., and Parkinson, E. A. 2004. Predators select against high growth rates and risk-taking behaviour in domestic trout populations. *Proceedings of the Royal Society of London B*, 271, 2233–2237.

Biro, P. A., Abrahams, M. V., Post, J. R., and Parkinson, E. A. 2006. Behavioural trade-offs between growth and mortality explain evolution of submaximal growth rates. *Journal of Animal Ecology*, 75, 1165–1171.

Biro, P. A., and Post, J. R. 2008. Rapid depletion of genotypes with fast growth and bold personality traits from harvested fish populations. *Proceedings of the National Academy of Sciences U S A*, 105, 2919–2922.

Blumstein, D. T. 2006. Developing an evolutionary ecology of fear: how life history and natural history traits affect disturbance tolerance in birds. *Animal Behaviour*, 71, 389–399.

———. 2007. Darwinian decision making: putting the adaptive into adaptive management. *Conservation Biology*, 21, 552–553.

Bohnsack, J. A., Sutherland, D. L., Harper, D. E., McClellan, D. B., Hulsbeck, M. W., and Holt, C. M. 1989. The effects of fish trap mesh size on reef fish catch off southeastern Florida. *Marine Fisheries Review*, 51, 36–46.

Boissy, A. 1995. Fear and fearfulness in animals. *Quarterly Review of Biology*, 70, 165–191.

Boon, A. K., Réale, D., and Boutin, S. 2007. The interaction between personality, offspring fitness, and food abundance in North American red squirrels. *Ecology Letters*, 10, 1094–1104.

Both, C., Dingemanse, N. J., Drent, P. J., and Tinbergen, J. M. 2005. Pairs of extreme avian personalities have highest reproductive success. *Journal of Animal Ecology*, 74, 667–674.

Bremner-Harrison, S., Prodohl, P. A., and Elwood, R. W. 2004. Behavioural trait assessment as a release criterion: boldness predicts early death in a reintroduction programme of captive-bred swift fox (*Vulpes velox*). *Animal Conservation*, 7, 313–320.

Brown, G. P., Shilton, C., Phillips, B. L., and Shine, R. 2007. Invasion, stress, and spinal arthritis in cane toads. *Proceedings of the National Academy of Sciences U S A*, 104, 17698–17700.

Bubb, D. H., Thom, T. J., and Lucas, M. C. 2006. Movement, dispersal, and refuge use of co-occurring introduced and native crayfish. *Freshwater Biology*, 51, 1359–1368.

Buchholz, R., and Clemmons, J. R. 1997. Behavioral variation: a valuable but neglected biodiversity. In: *Behavioral Approaches to Conservation in the Wild* (Clemmons, J. R., and Buchholz, R., eds.), pp. 181–208. Cambridge: Cambridge University Press.

Caizergues, A., Rätti, O., Helle, P., Rotelli, L., Ellison, L., and Rasplus, J.-Y. 2003. Population genetic structure of male black grouse (*Tetrao tetrix* L.) in fragmented vs. continuous landscapes. *Molecular Ecology*, 12, 2297–2305.

Candland, D. K. 2005. The animal mind and conservation of species: knowing what animals know. *Current Science*, 89, 1122–1127.

Capitanio, J. P., Mendoza, S. P., Lerche, N. W., and Mason, W. A. 1998. Social stress results in altered glucocorticoid regulation and shorter survival in simian acquired deficiency syndrome. *Proceedings of the National Academy of Sciences U S A*, 95, 4714–4719.

Carere, C., Groothuis, T. G. G., Möstl, E., Daan, S., and Koolhaas, J. M. 2003. Fecal corticosteroids in a territorial bird selected for different personalities: daily rhythm and the response to social stress. *Hormones and Behavior*, 43, 540–548.

Carere, C., Welink, D., Drent, P. J., Koolhaas, J. M., and Groothuis, T. G. G. 2001. Effect of social defeat in a territorial bird (*Parus major*) selected for different coping styles. *Physiology and Behavior*, 73, 427–433.

Carlson, S. M., Edeline, E., Vøllestad, L. A., Haugen, T. O., Winfield, I. J., Fletcher, J. M., James, J. B., and Stenseth, N. C. 2007. Four decades of opposing natural and human-induced artificial selection acting on Windermere pike (*Esox lucius*). *Ecology Letters*, 10, 512–521.

Carlstead, K., Fraser, J., Bennett, C., and Kleiman, D. G. 1999a. Black rhinoceros (*Diceros bicornis*) in US zoos, II: behavior, breeding success, and mortality in relation to housing facilities. *Zoo Biology*, 18, 35–52.

Carlstead, K., Mellen, J., and Kleiman, D. G. 1999b. Black rhinoceros (*Diceros bicornis*) in US zoos, I: individual behavior profiles and their relationship to breeding success. *Zoo Biology*, 18, 17–34.

Caro, T. 1998. *Behavioral Ecology and Conservation Biology*. Oxford: Oxford University Press.

Carroll, S. P. 2008. Facing change: forms and foundations of contemporary adaptation to biotic invasions. *Molecular Ecology*, 17, 361–372.

Carson, R. 1962. *Silent Spring*. Boston: Houghton Mifflin Company.

Castellón, T. D., and Sieving, K. E. 2006. An experimental test of matrix permeability and corridor use by an endemic understory bird. *Conservation Biology*, 20, 135–145.

Cavigelli, S. A. 2005. Animal personality and health. *Behaviour*, 142, 1223–1244.

Cavigelli, S. A., and McClintock, M. K. 2003. Fear of novelty in infant rats predicts adult corticosterone dynamics and an early death. *Proceedings of the National Academy of Sciences U S A*, 100, 16131–16136.

Chagnon, N. L., and Guttman, S. I. 1989. Differential survivorship of allozyme genotypes in mosquitofish populations exposed to copper or cadmium. *Environmental Toxicology and Chemistry*, 8, 319–326.

Chapman, B. B., Morrell, L. J., Benton, T. G., and Krause, J. 2008. Early interactions with adults mediate the development of predator defenses in guppies. *Behavioral Ecology*, 19, 87–93.

Chivers, D. P., and Smith, J. F. 1994. Fathead minnows, *Pimephales promelas*, acquire predator recognition when alarm substance is associated with the sight of unfamiliar fish. *Animal Behaviour*, 48, 597–605.

Civantos, E. 2000. Home range ecology, aggressive behaviour, and survival in juvenile lizards, *Psammodromus algirus*. *Canadian Journal of Zoology*, 78, 1681–1685.

Clemmons, J. R., and Buchholz, R., eds. 1997. *Behavioral Approaches to Conservation in the Wild*. Cambridge: Cambridge University Press.

Clevenger, A. P., Chruszcz, B., and Gunson, K. E. 2003. Spatial patterns and factors influencing small vertebrate fauna road-kill aggregations. *Biological Conservation*, 109, 15–26.

Clotfelter, E. D., Bell, A. M., and Levering, K. R. 2004. The role of animal behaviour in the study of endocrine-disrupting chemicals. *Animal Behaviour*, 68, 665–676.

Clubb, R., and Mason, G. J. 2007. Natural behavioural biology as a risk factor in carnivore welfare: how analysing species differences could help zoos improve enclosures. *Applied Animal Behaviour Science*, 102, 303–328.

Colborn, T., vom Saal, F. S., and Soto, A. M. 1993. Developmental effects of endocrine-disrupting chemicals in wildlife and humans. *Environmental Health Perspectives*, 101, 378–384.

Coltman, D. W. 2008. Molecular ecological approaches to studying the evolutionary impact of selective harvesting in wildlife. *Molecular Ecology*, 17, 221–235.

Coltman, D. W., O'Donoghue, P., Jorgenson, J. T., Hogg, J. T., Strobeck, C., and Festa-Bianchet, M. 2003. Undesirable evolutionary consequences of trophy hunting. *Nature*, 426, 655–658.

Conover, D. O., and Munch, S. B. 2002. Sustaining fisheries yields over evolutionary time scales. *Science*, 297, 94–96.

Cote, J., Clobert, J., Brodin, T., Fogarty, S., and Sih, A. 2010. Personality-dependent dispersal: characterization, ontogeny, and consequences for spatially structured populations. *Philosophical Transactions of the Royal Society of London B*, 365, 4065–4076.

Crozier, R. H. 1997. Preserving the information content of species: genetic diversity, phylogeny, and conservation worth. *Annual Review of Ecology and Systematics*, 28, 243–268.

Dahm, E. 2000. Changes in the length compositions of some fish species as a consequence of alterations in the groundgear of the GOV-trawl. *Fisheries Research*, 49, 39–50.

Dall, S. R. X., Houston, A. I., and McNamara, J. M. 2004. The behavioural ecology of personality: consistent individual differences from an adaptive perspective. *Ecology Letters*, 7, 734–739.

Darimont, C. T., Carlson, S. M., Kinnison, M. T., Paquet, P. C., Reimchen, T. E., and Wilmers, C. C. 2009. Human predators outpace other agents of trait change in the wild. *Proceedings of the National Academy of Sciences U S A*, 106, 952–954.

Diamond, S. A., Newman, M. C., Mulvey, M., Dixon, P. M., and Martinson, D. 1989. Allozyme genotype and time to death of mosquitofish, *Gambusia affinis* (Baird and Girard), during acute exposure to inorganic mercury. *Environmental Toxicology and Chemistry*, 8, 613–622.

DiBattista, J. D. 2008. Patterns of genetic variation in anthropogenically impacted populations. *Conservation Genetics*, 9, 141–156.

Dingemanse, N. J., Both, C., Drent, P. J., and Tinbergen, J. M. 2004. Fitness consequences of avian personalities in a fluctuating environment. *Proceedings of the Royal Society of London B*, 271, 847–852.

Dingemanse, N. J., Both, C., Drent, P. J., Van Oers, K., and Van Noordwijk, A. J. 2002. Repeatability and heritability of exploratory behaviour in great tits from the wild. *Animal Behaviour*, 64, 929–938.

Dingemanse, N. J., Both, C., Van Noordwijk, A. J., Rutten, A. L., and Drent, P. J. 2003. Natal dispersal and personalities in great tits (*Parus major*). *Proceedings of the Royal Society of London B*, 270, 741–747.

Dingemanse, N. J., Kazem, A. J. N., Réale, D., and Wright, J. 2010. Behavioural reaction norms: animal personality meets individual plasticity. *Trends in Ecology and Evolution*, 25, 81–89.

Dingemanse, N. J., and Réale, D. 2005. Natural selection and animal personality. *Behaviour*, 142, 1159–1184.

Dingemanse, N. J., and Wolf, M. 2010. Recent models for adaptive personality differences: a review. *Philosophical Transactions of the Royal Society of London B*, 365, 3947–3958.

Dingemanse, N. J., Wright, J., Kazem, A. J. N., Thomas, D. K., Hickling, R., and Dawnay, N. 2007. Behavioural syndromes differ predictably between 12 populations of three-spined stickleback. *Journal of Animal Ecology*, 76, 1128–1138.

Drent, P. J., Van Oers, K., and Van Noordwijk, A. J. 2003. Realized heritability of personalities in the great tit (*Parus major*). *Proceedings of the Royal Society of London B*, 270, 45–51.

Duckworth, R. A., and Badyaev, A. V. 2007. Coupling of dispersal and aggression facilitates the rapid range expansion of a passerine bird. *Proceedings of the National Academy of Sciences U S A*, 104, 15017–15022.

Dugatkin, L. A. 1992. Tendency to inspect predators predicts mortality risk in the guppy (*Poecilia reticulata*). *Behavioral Ecology*, 3, 124–127.

Edeline, E., Carlson, S. M., Stige, L. C., Winfield, I. J., Fletcher, J. M., James, J. B., Haugen, T. O., Vøllestad, L. A., and Stenseth, N. C. 2007. Trait changes in a harvested population are driven by a dynamic tug-of-war between natural and harvest selection. *Proceedings of the National Academy of Sciences U S A*, 104, 15799–15804.

Erhard, H. W., and Rhind, S. M. 2004. Prenatal and postnatal exposure to environmental pollutants in sewage sludge alters emotional reactivity and exploratory behaviour in sheep. *Science of the Total Environment*, 332, 101–108.

Fairbanks, L. A., Jorgensen, M. J., Huff, A., Blau, K., Hung, Y.-Y., and Mann, J. J. 2004a. Adolescent impulsivity predicts adult dominance attainment in male vervet monkeys. *American Journal of Primatology*, 64, 1–17.

Fairbanks, L. A., Newman, T. K., Bailey, J. N., Jorgensen, M. J., Breidenthal, S. E., Ophoff, R. A., Comuzzie, A. G., Martin, L. J., and Rogers, J. 2004b. Genetic contributions to social impulsivity and aggressiveness in vervet monkeys. *Biological Psychiatry*, 55, 642–647.

Fenberg, P. B., and Roy, K. 2008. Ecological and evolutionary consequences of size-selective harvesting: how much do we know? *Molecular Ecology*, 17, 209–220.

Festa-Bianchet, M. 2003. Exploitative wildlife management as a selective pressure for life-history evolution of large mammals. In: *Animal Behavior and Wildlife Conservation* (Festa-Bianchet, M., and Apollonio, M., eds.), pp. 191–207. Washington, DC: Island Press.

Fox, A. D., and Madsen, J. 1997. Behavioural and distributional effects of hunting disturbance on waterbirds in Europe: implications for refuge design. *Journal of Applied Ecology*, 34, 1–13.

Frankel, O. H. 1974. Genetic conservation: our evolutionary responsibility. *Genetics*, 78, 53–65.

Frankham, R. 1995. Conservation genetics. *Annual Review of Genetics*, 29, 305–327.

———. 1999. Quantitative genetics in conservation biology. *Genetic Research*, 74, 237–244.

———. 2008. Genetic adaptation to captivity in species conservation programs. *Molecular Ecology*, 17, 325–333.

Frankham, R., Ballou, J. D., and Briscoe, D. A. 2002. *Introduction to Conservation Genetics.* Cambridge: Cambridge University Press.

Fraser, D. F., Gilliam, J. F., Daley, M. J., Le, A. N., and Skalski, G. T. 2001. Explaining leptokurtic movement distributions: intrapopulation variation in boldness and exploration. *American Naturalist*, 158, 124–135.

Futuyma, D. J. 1998. *Evolutionary Biology.* 3rd ed. Sunderland, MA: Sinauer Associates.

Garant, D., Kruuk, L. E. B., Wilkin, T. A., McCleery, R. H., and Sheldon, B. C. 2005. Evolution driven by differential dispersal within a wild bird population. *Nature*, 433, 60–65.

Garel, M., Cugnasse, J.-M., Maillard, D., Gaillard, J.-M., Hewison, A. J. M., and Dubray, D. 2007. Selective harvesting and habitat loss produce long-term life-history changes in a mouflon population. *Ecological Applications*, 17, 1607–1618.

Gillespie, R. B., and Guttman, S. I. 1989. Effects of contaminants on the frequencies of allozymes in populations of the central steamroller. *Environmental Toxicology and Chemistry*, 8, 309–317.

Gilpin, M. E., and Soulé, M. E. 1986. Minimum viable populations: processes of species extinction. In: *Conservation Biology: The Science of Scarcity and Diversity* (Soulé, M. E., ed.), pp. 19–34. Sunderland, MA: Sinauer Associates.

Goodey, W., and Liley, N. R. 1986. The influence of early experience on escape behaviour in the guppy (*Poecilia reticulata*). *Canadian Journal of Zoology*, 64, 885–888.

Gosling, L. M., and Sutherland, W. J., eds. 2000. *Behaviour and Conservation.* Cambridge: Cambridge University Press.

Gosling, S. D. 2001. From mice to men: what can we learn about personality from animal research? *Psychological Bulletin*, 127, 45–86.

Greenberg, R. 1983. The role of neophobia in determining the degree of foraging specialization in some migrant warblers. *American Naturalist*, 122, 444–453.

———. 1989. Neophobia, aversion to open space, and ecological plasticity in song and swamp sparrows. *Canadian Journal of Zoology*, 67, 1194–1199.

Greenberg, R., and Mettke-Hofmann, C. 2001. Ecological aspects of neophobia and neophilia in birds. *Current Ornithology*, 16, 119–178.

Griffin, A. S., Blumstein, D. T., and Evans, C. S. 2000. Training captive-bred or translocated animals to avoid predators. *Conservation Biology*, 14, 1317–1326.

Griffith, B., Scott, J. M., Carpenter, J. W., and Reed, C. 1989. Translocation as a species conservation tool: status and strategy. *Science*, 245, 477–480.

Groom, M. J., Meffe, G. K., and Carroll, C. R. 2006. *Principles of Conservation Biology*. 3rd ed. Sunderland, MA: Sinauer Associates.

Haegele, M. A., and Hudson, R. H. 1977. Reduction of courtship behavior induced by DDE in male ringed turtle doves. *Wilson Bulletin*, 89, 593–601.

Hayes, T. B., Collins, A., Lee, M., Mendoza, M., Noriega, N., Stuart, A. A., and Vonk, A. 2002. Hermaphroditic, demasculinized frogs after exposure to the herbicide atrazine at low ecologically relevant doses. *Proceedings of the National Academy of Sciences U S A*, 99, 5476–5480.

Hedrick, P. W., and Kalinowski, S. T. 2000. Inbreeding depression in conservation biology. *Annual Review of Ecology and Systematics*, 31, 139–162.

Hendry, A. P., Farrugia, T. J., Kinnison, M. T. 2008. Human influences on rates of phenotypic change in wild animal populations. *Molecular Ecology*, 17, 20–29.

Hill, A. M., and Lodge, D. M. 1999. Replacement of resident crayfishes by an exotic crayfish: the roles of competition and predation. *Ecological Applications*, 9, 678–690.

Höjesjö, J., Johnsson, J. I., and Bohlin, T. 2002. Can laboratory studies on dominance predict fitness of young brown trout in the wild? *Behavioral Ecology and Sociobiology*, 52, 102–108.

Holway, D. A., and Suarez, A. V. 1999. Animal behavior: an essential component of invasion biology. *Trends in Ecology and Evolution*, 14, 328–330.

Houle, D. 1992. Comparing evolvability and variability of quantitative traits. *Genetics*, 130, 195–204.

Howell, W. M., Black, D. A., and Bortone, S. A. 1980. Abnormal expression of secondary sex characters in a population of mosquitofish, *Gambusia affinis holbrooki*: evidence for environmentally induced masculinization. *Copeia*, 1980, 676–681.

Hughes, D. P., and Cremer, S. 2007. Plasticity in antiparasite behaviours and its suggested role in invasion biology. *Animal Behaviour*, 74, 1593–1599.

Huntingford, F. A. 1976. The relationship between antipredator behaviour and aggression among conspecifics in the three-spined stickleback, *Gasterosteus aculeatus*. *Animal Behaviour*, 24, 245–260.

Hutchings, J. A. 2005. Life-history consequences of overexploitation to population recovery in Northwest Atlantic cod (*Gadus morhua*). *Canadian Journal of Fisheries and Aquatic Sciences*, 62, 824–832.

Hutchings, J. A., and Fraser, D. J. 2008. The nature of fisheries- and farming-induced evolution. *Molecular Ecology*, 17, 294–313.

Ikonomou, M. G., Rayne, S., and Addison, R. F. 2002. Exponential increases of the brominated flame retardants, polybrominated diphenyl ethers, in the Canadian Arctic from 1981 to 2000. *Environmental Science and Technology*, 36, 1886–1892.

IUCN. 2010. IUCN Red List of Threatened Species. http://www.iucnredlist.org.

Jachmann, H., Berry, P. S. M., and Imae, H. 1995. Tusklessness in African elephants: a future trend. *African Journal of Ecology*, 33, 230–235.

Jeschke, J. M., and Strayer, D. L. 2006. Determinants of vertebrate invasion success in Europe and North America. *Global Change Biology*, 12, 1608–1619.

Johnson, J. C., and Sih, A. 2007. Fear, food, sex, and parental care: a syndrome of boldness in the fishing spider, *Dolomedes triton*. *Animal Behaviour*, 74, 1131–1138.

Kagan, J., Reznick, J. S., and Snidman, N. 1988. Biological bases of childhood shyness. *Science*, 240, 167–171.

Kannan, K., Franson, J. C., Bowerman, W. W., Hansen, K. J., Jones, P. D., and Giesy, J. P. 2001a. Perfluorooctane sulfonate in fish-eating water birds including bald eagles and albatrosses. *Environmental Science and Technology*, 35, 3065–3070.

Kannan, K., Koistinen, J., Beckmen, K., Evans, T., Gorzelany, J. F., Hansen, K. J., Jones, P. D., Helle, E., Nyman, M., and Giesy, J. P. 2001b. Accumulation of perfluorooctane sulfonate in marine mammals. *Environmental Science and Technology*, 35, 1593–1598.

Keklak, M. M., Newman, M. C., and Mulvey, M. 1994. Enhanced uranium tolerance of an exposed population of the eastern mosquitofish (*Gambusia holbrooki* Girard 1859). *Archives of Environmental Contamination and Toxicology*, 27, 20–24.

Kelley, J. L., Magurran, A. E., Macías-Garcia, C. 2005. The influence of rearing experience on the behaviour of an endangered Mexican fish, *Skiffa multipunctata*. *Biological Conservation*, 122, 223–230.

Ketterson, E. D., and Nolan, V., Jr. 1999. Adaptation, exaptation, and constraint: a hormonal perspective. *American Naturalist*, 154, S4-S25.

King, R. C., Stansfield, W. D., and Mulligan, P. K. 2006. *A Dictionary of Genetics*. 7th ed. Oxford: Oxford University Press.

Kleiman, D. G. 1989. Reintroduction of captive mammals for conservation. *BioScience*, 39, 152–161.

Kolar, C. S., and Lodge, D. M. 2001. Progress in invasion biology: predicting invaders. *Trends in Ecology and Evolution*, 16, 199–204.

Kolok, A. S. 2001. Sublethal identification of susceptible individuals: using swim performance to identify susceptible fish while keeping them alive. *Ecotoxicology*, 10, 205–209.

Kolok, A. S., Peake, E. B., Tierney, L. L., Roark, S. A., Noble, R. B., See, K., and Guttman, S. I. 2004. Copper tolerance in fathead minnows, I: the role of genetic and nongenetic factors. *Environmental Toxicology and Chemistry*, 23, 200–207.

Koolhaas, J. M., Korte, S. M., de Boer, S. F., Van der Vegt, B. J., Van Reenen, C. G., Hopster, H., de Jong, I. C., Ruis, M. A. W., and Blokhuis, H. J. 1999. Coping styles in animals: current status in behavior and stress-physiology. *Neuroscience and Biobehavioral Reviews*, 23, 925–935.

Korhonen, H., and Niemelä, P. 1996. Temperament and reproductive success in farmbred silver foxes housed with and without platforms. *Journal of Animal Breeding and Genetics*, 113, 209–218.

Korpi, N. L., and Wisenden, B. D. 2001. Learned recognition of novel predator odour by zebra danios, *Danio rerio*, following time-shifted presentation of alarm cue and predator odour. *Environmental Biology of Fishes*, 61, 205–211.

Künzl, C., Kaiser, S., Meier, E., and Sachser, N. 2003. Is a wild mammal kept and reared in captivity still a wild mammal? *Hormones and Behavior*, 43, 187–196.

Lande, R., and Arnold, S. J. 1983. The measurement of selection on correlated characters. *Evolution*, 76, 1210–1226.

Law, R. 2000. Fishing, selection, and phenotypic evolution. *ICES Journal of Marine Science*, 57, 659–668.

Lodge, D. M. 1993. Biological invasions: lessons for ecology. *Trends in Ecology and Evolution*, 8, 133–137.

Logue, D. M., Mishra, S., McCaffrey, D., Ball, D., and Cade, W. H. 2009. A behavioral syndrome linking courtship behavior toward males and females predicts reproductive success from a single mating in the hissing cockroach, *Gromphadorhina portentosa*. *Behavioral Ecology*, 20, 781–788.

López, P., Hawlena, D., Polo, V., Amo, L., and Martín, J. 2005. Sources of individual shy-bold variations in antipredator behaviour of male Iberian rock lizards. *Animal Behaviour*, 69, 1–9.

Lowe, S., Browne, M., Boudjelas, S., and De Poorter, M. 2000. *100 of the World's Worst Invasive Species: A Selection from the Global Invasive Database*. University of Auckland: Invasive Species Specialist Group (ISSG), International Union for Conservation of Nature (IUCN).

Mack, R. N., Simberloff, D., Lonsdale, W. M., Evans, H., Clout, M., and Bazzaz, F. A. 2000. Biotic invasions: causes, epidemiology, global consequences, and control. *Ecological Applications*, 10, 689–710.

Maestripieri, D. 2007. *Macachiavellian Intelligence: How Rhesus Macaques and Humans Have Conquered the World*. Chicago: University of Chicago Press.

Magurran, A. E. 1990. The inheritance and development of minnow antipredator behaviour. *Animal Behaviour*, 39, 834–842.

Mangel, M., and Stamps, J. 2001. Trade-offs between growth and mortality and the maintenance of individual variation in growth. *Evolutionary Ecology Research*, 3, 583–593.

Marliave, J. B., Gergits, W. F., and Aota, S. 1993. F_{10} pandalid shrimp: sex determination; DNA and dopamine as indicators of domestication; and outcrossing for wild pigment pattern. *Zoo Biology*, 12, 435–451.

Martin, J. G. A., and Réale, D. 2008. Animal temperament and human disturbance: implications for the response of wildlife to tourism. *Behavioural Processes*, 77, 66–72.

Martin, L. B., II, and Fitzgerald, L. 2005. A taste for novelty in invading house sparrows, *Passer domesticus*. *Behavioral Ecology*, 16, 702–707.

Mathews, F., Orros, M., McLaren, G., Gelling, M., and Foster, R. 2005. Keeping fit on the ark: assessing the suitability of captive-bred animals for release. *Biological Conservation*, 121, 569–577.

Matson, C. W., Lambert, M. M., McDonald, T. J., Autenrieth, R. L., Donnelly, K. C., Islamzadeh, A., Politov, D. I., and Bickham, J. W. 2006. Evolutionary toxicology: population-level effects of chronic contaminant exposure on the marsh frogs (*Rana ridibunda*) of Azerbaijan. *Environmental Health Perspectives*, 114, 547–552.

McDonald, W. R., and St. Clair, C. C. 2004. The effects of artificial and natural barriers on the movement of small mammals in Banff National Park, Canada. *Oikos*, 105, 397–407.

McDougall, P. T., Réale, D., Sol, D., and Reader, S. M. 2006. Wildlife conservation and animal temperament: causes and consequences of evolutionary change for captive, reintroduced, and wild populations. *Animal Conservation*, 9, 39–48.

McLean, I. G., Schmitt, N. T., Jarman, P. J., Duncan, C., and Wynne, C. D. L. 2000. Learning for life: training marsupials to recognise introduced predators. *Behaviour*, 137, 1361–1376.

McPhee, M. E. 2003a. Effects of captivity on response to a novel environment in the oldfield mouse (*Peromyscus polionotus subgriseus*). *International Journal of Comparative Psychology*, 16, 85–94.

———. 2003b. Generations in captivity increases behavioral variance: considerations for captive breeding and reintroduction programs. *Biological Conservation*, 115, 71–77.

Medina, M. H., Correa, J. A., and Barata, C. 2007. Micro-evolution due to pollution: possible consequences for ecosystem responses to toxic stress. *Chemosphere*, 67, 2105–2114.

Meehan, C. L., and Mench, J. A. 2002. Environmental enrichment affects the fear and exploratory responses to novelty of young Amazon parrots. *Applied Animal Behaviour Science*, 79, 75–88.

Mettler, A. E., and Shivik, J. A. 2007. Dominance and neophobia in coyote (*Canis latrans*) breeding pairs. *Applied Animal Behaviour Science*, 102, 85–94.

Miller, B., Biggins, D., Hanebury, L., and Vargas, A. 1994. Reintroduction of the black footed ferret (*Mustela nigripes*). In: *Creative Conservation: Interactive Management of Wild and Captive Animals* (Olney, P. J. S., Mace, G. M., Feistner, A. T. C., eds.), pp. 455–464. London: Chapman and Hall.

Miller, H. C., and Lambert, D. M. 2004. Genetic drift outweighs balancing selection in shaping post-bottleneck major histocompatibility complex variation in New Zealand robins (*Petroicidae*). *Molecular Ecology*, 13, 3709–3721.

Milner, J. M., Nilsen, E. B., and Andreassen, H. P. 2007. Demographic side effects of selective hunting in ungulates and carnivores. *Conservation Biology*, 21, 36–47.

Moritz, C. 1999. Conservation units and translocations: strategies for conserving evolutionary processes. *Hereditas*, 130, 217–228.

———. 2002. Strategies to protect biological diversity and the evolutionary processes that sustain it. *Systematic Biology*, 51, 238–254.

Munkwitz, N. M., Turner, J. M., Kershner, E. L., Farabaugh, S. M., and Heath, S. R. 2005. Predicting release success of captive-reared loggerhead shrikes (*Lanius ludovicianus*) using pre-release behavior. *Zoo Biology*, 24, 447–458.

Nelson, X. J., Wilson, D. R., and Evans, C. S. 2008. Behavioral syndromes in stable social groups: an artifact of external constraints? *Ethology*, 114, 1154–1165.

Newsome, A. E., and Noble, I. R. 1986. Ecological and physiological characters of invading species. In: *Ecology of Biological Invasions* (Groves, R. H., and Burdon, J. J., eds.), pp. 1–20. Cambridge: Cambridge University Press.

Norstrom, R. J., Belikov, S. E., Born, E. W., Garner, G. W., Malone, B., Olpinski, S., Ramsay, M. A., Schliebe, S., Stirling, I., Stishov, M. S., Taylor, M. K., and Wiig, Ø. 1998. Chlorinated hydrocarbon contaminants in polar bears from eastern Russia, North America, Greenland, and Svalbard: biomonitoring of Arctic pollution. *Archives of Environmental Contamination and Toxicology*, 35, 354–367.

Olsen, E. M., Heino, M., Lilly, G. R., Morgan, M. J., Brattey, J., Ernande, B., and Dieckmann, U. 2004. Maturation trends indicative of rapid evolution preceded the collapse of northern cod. *Nature*, 428, 932–935.

Palanza, P., Morellini, F., Parmigiani, S., and vom Saal, F. S. 1999. Prenatal exposure to endocrine-disrupting chemicals: effects on behavioral development. *Neuroscience and Biobehavioral Reviews*, 23, 1011–1027.

Parmigiani, S., Palanza, P., Rodgers, J., and Ferrari, P. F. 1999. Selection, evolution of behavior, and animal models in behavioral neuroscience. *Neuroscience and Biobehavioral Reviews*, 23, 957–970.

Peng, J., Jiang, Z., Qin, G., Huang, Q., Li, Y., Jiao, Z., Zhang, F., Li, Z., Zhang, J., Lu, Y., Liu, X., and Liu, J. 2007. Impact of activity space on the reproductive behaviour of giant panda (*Ailuropoda melanoleuca*) in captivity. *Applied Animal Behaviour Science*, 104, 151–161.

Phillips, B. L., Brown, G. P., Webb, J. K., and Shine, R. 2006. Invasion and the evolution of speed in toads. *Nature*, 439, 803.

Pimentel, D., Lach, L., Zuniga, R., and Morrison, D. 2000. Environmental and economic costs of nonindigenous species in the United States. *BioScience*, 50, 53–65.

Pintor, L. M., Sih, A., and Bauer, M. L. 2008. Differences in aggression, activity, and boldness between native and introduced populations of an invasive crayfish. *Oikos*, 117, 1629–1636.

Polo-Cavia, N., López, P., and Martín, J. 2008. Interspecific differences in responses to predation risk may confer competitive advantages to invasive freshwater turtle species. *Ethology*, 114, 115–123.

Price, E. O. 1970. Differential reactivity of wild and semi-domestic deermice (*Peromyscus maniculatus*). *Animal Behaviour*, 18, 747–752.

———. 1984. Behavioral aspects of animal domestication. *Quarterly Review of Biology*, 59, 1–32.

Primack, R. B. 1998. *Essentials of Conservation Biology*. Sunderland, MA: Sinauer Associates.

Proaktor, G., Coulson, T., and Milner-Gulland, E. J. 2007. Evolutionary responses to harvesting in ungulates. *Journal of Animal Ecology*, 76, 669–678.

Réale, D., Dingemanse, N. J., Kazem, A. J. N., and Wright, J. 2010. Evolutionary and ecological approaches to the study of personality. *Philosophical Transactions of the Royal Society of London B*, 365, 3937–3946.

Réale, D., and Festa-Bianchet, M. 2003. Predator-induced natural selection on temperament in bighorn ewes. *Animal Behaviour*, 65, 463–470.

Réale, D., Gallant, B. Y., Leblanc, M., and Festa-Bianchet, M. 2000. Consistency of temperament in bighorn ewes and correlates with behaviour and life history. *Animal Behaviour*, 60, 589–597.

Réale, D., Reader, S. M., Sol, D., Mcdougall, P. T., and Dingemanse, N. J. 2007. Integrating animal temperament within ecology and evolution. *Biological Reviews*, 82, 291–318.

Reed, D. H., and Frankham, R. 2003. Correlation between fitness and genetic diversity. *Conservation Biology*, 17, 230–237.

Rehage, J. S., Barnett, B. K., and Sih, A. 2005. Foraging behaviour and invasiveness: do invasive *Gambusia* exhibit higher feeding rates and broader diets than their noninvasive relatives? *Ecology of Freshwater Fish*, 14, 352–360.

Rehage, J. S., and Sih, A. 2004. Dispersal behavior, boldness, and the link to invasiveness: a comparison of four *Gambusia* species. *Biological Invasions*, 6, 379–391.

Relyea, R. A., and Mills, N. 2001. Predator-induced stress makes the pesticide carbaryl more deadly to gray treefrog tadpoles (*Hyla versicolor*). *Proceedings of the National Academy of Sciences U S A*, 98, 2491–2496.

Rice, D. C. 2000. Parallels between attention deficit hyperactivity disorder and behavioral deficits produced by neurotoxic exposure in monkeys. *Environmental Health Perspectives*, 108 (suppl. 3), 405–408.

Riechert, S. E., and Hedrick, A. V. 1993. A test for correlations among fitness-linked behavioural traits in the spider *Agelenopsis aperta* (*Araneae, Agelenidae*). *Animal Behaviour*, 46, 669–675.

Rocha, L. A., Robertson, D. R., Roman, J., and Bowen, B. W. 2005. Ecological speciation in tropical reef fishes. *Proceedings of the Royal Society of London B*, 272, 573–579.

Ross, P. S., Ellis, G. M., Ikonomou, M. G., Barrett-Lennard, L. G., and Addison, R. F. 2000. High PCB concentrations in free-ranging Pacific killer whales, *Orcinus orca*: effects of age, sex, and dietary preference. *Marine Pollution Bulletin*, 40, 504–515.

Ruis, M. A. W., te Brake, J. H. A., Van de Burgwal, J. A., de Jong, I. C., Blokhuis, H. J., and Koolhaas, J. M. 2000. Personalities in female domesticated pigs: behavioural and physiological indications. *Applied Animal Behaviour Science*, 66, 31–47.

Sacks, B. N., Jaeger, M. M., Neale, J. C. C., and McCullough, D. R. 1999. Territoriality and breeding status of coyotes relative to sheep predation. *Journal of Wildlife Management*, 63, 593–605.

Sakai, A. K., Allendorf, F. W., Holt, J. S., Lodge, D. M., Molofsky, J., With, K. A., Baughman, S., Cabin, R. J., Cohen, J. E., Ellstrand, N. C., McCauley, D. E., O'Neil, P., Parker, I. M., Thompson, J. N., and Weller, S. G. 2001. The population biology of invasive species. *Annual Review of Ecology and Systematics*, 32, 305–332.

Sapolsky, R. M. 1990. Adrenocortical function, social rank, and personality among wild baboons. *Biological Psychiatry*, 28, 862–878.

Saunders, D. A., Hobbs, R. J., and Margules, C. R. 1991. Biological consequences of ecosystem fragmentation: a review. *Conservation Biology*, 5, 18–32.

Schantz, S. L., Moshtaghian, J., and Ness, D. K. 1995. Spatial learning deficits in adult rats exposed to ortho-substituted PCB congeners during gestation and lactation. *Fundamental and Applied Toxicology*, 26, 117–126.

Schantz, S. L., and Widholm, J. J. 2001. Cognitive effects of endocrine-disrupting chemicals in animals. *Environmental Health Perspectives*, 109, 1197–1206.

Schlueter, M. A., Guttman, S. I., Oris, J. T., and Bailer, A. J. 1995. Survival of copper-exposed juvenile fathead minnows (*Pimephales promelas*) differs among allozyme genotypes. *Environmental Toxicology and Chemistry*, 14, 1727–1734.

Schlueter, M. A., Guttman, S. I., Oris, J. T., and Bailer, A. J. 1997. Differential survival of fathead minnows, *Pimephales promelas*, as affected by copper exposure, prior population stress, and allozyme genotypes. *Environmental Toxicology and Chemistry*, 16, 939–947.

Schluter, D. 1988. Estimating the form of natural selection on a quantitative trait. *Evolution*, 42, 849–861.

Schluter, D., and Nychka, D. 1994. Exploring fitness surfaces. *American Naturalist*, 143, 597–616.

Schneider, C. J., Smith, T. B., Larison, B., and Moritz, C. 1999. A test of alternative models of diversification in tropical rainforests: ecological gradients vs. rainforest refugia. *Proceedings of the National Academy of Sciences U S A*, 96, 13869–13873.

Seddon, P. J., Armstrong, D. P., and Maloney, R. F. 2007. Developing the science of reintroduction biology. *Conservation Biology*, 21, 303–312.

Seddon, J. M., and Baverstock, P. R. 1999. Variation on islands: major histocompatibility complex (MHC) polymorphism in populations of the Australian bush rat. *Molecular Ecology*, 8, 2071–2079.

Short, J., Bradshaw, S. D., Giles, J., Prince, R. I. T., and Wilson, G. R. 1992. Reintroduction of macropods (*Marsupialia: Macropodoidea*) in Australia: a review. *Biological Conservation*, 62, 189–204.

Shumway, C. A. 1999. A neglected science: applying behavior to aquatic conservation. *Environmental Biology of Fishes*, 55, 183–201.

Sih, A., Bell, A. M., and Johnson, J. C. 2004a. Behavioral syndromes: an ecological and evolutionary overview. *Trends in Ecology and Evolution*, 19, 372–378.

Sih, A., Bell, A. M., Johnson, J. C., and Ziemba, R. E. 2004b. Behavioral syndromes: an integrative overview. *Quarterly Review of Biology*, 79, 241–277.

Sih, A., Kats, L. B., and Maurer, E. F. 2003. Behavioural correlations across situations and the evolution of antipredator behaviour in a sunfish-salamander system. *Animal Behaviour*, 65, 29–44.

Simonich, S. L., and Hites, R. A. 1995. Global distribution of persistent organochlorine compounds. *Science*, 269, 1851–1854.

Sinclair, A. F., Swain, D. P., and Hanson, J. M. 2002. Measuring changes in the direction and magnitude of size-selective mortality in a commercial fish population. *Canadian Journal of Fisheries and Aquatic Sciences*, 59, 361–371.

Sinn, D. L., Apiolaza, L. A., and Moltschaniwskyj, N. A. 2006. Heritability and fitness-related consequences of squid personality traits. *Journal of Evolutionary Biology*, 19, 1437–1447.

Sinn, D. L., Moltschaniwskyj, N. A., Wapstra, E., and Dall, S. R. X. 2010. Are behavioral syndromes invariant? Spatiotemporal variation in shy/bold behavior in squid. *Behavioral Ecology and Sociobiology*, 64, 693–702.

Smith, B. R., and Blumstein, D. T. 2008. Fitness consequences of personality: a meta-analysis. *Behavioral Ecology*, 19, 448–455.

———. 2010. Behavioral types as predictors of survival in Trinidadian guppies (*Poecilia reticulata*). *Behavioral Ecology*, 21, 919–926.

Smith, T. B., Kark, S., Schneider, C. J., Wayne, R. K., and Moritz, C. 2001. Biodiversity hotspots and beyond: the need for preserving environmental transitions. *Trends in Ecology and Evolution*, 16, 431.

Smith, T. B., Wayne, R. K., Girman, D. J., and Bruford, M. W. 1997. A role for ecotones in generating rainforest biodiversity. *Science*, 276, 1855–1857.

Sol, D., and Lefebvre, L. 2000. Behavioural flexibility predicts invasion success in birds introduced to New Zealand. *Oikos*, 90, 599–605.

Sol, D., Timmermans, S., and Lefebvre, L. 2002. Behavioral flexibility and invasion success in birds. *Animal Behaviour*, 63, 495–502.

Spearow, J. L., Doemeny, P., Sera, R., Leffler, R., and Barkley, M. 1999. Genetic variation in susceptibility to endocrine disruption by estrogen in mice. *Science*, 285, 1259–1261.

Spearow, J. L., O'Henley, P., Doemeny, P., Sera, R., Leffler, R., Sofos, T., and Barkley, M. 2001. Genetic variation in physiological sensitivity to estrogen in mice. *Acta Pathologica, Microbiologica, and Immunologica Scandinavica*, 109, 356–364.

St. Clair, C. C. 2003. Comparative permeability of roads, rivers, and meadows to songbirds in Banff National Park. *Conservation Biology*, 17, 1151–1160.

Stamps, J. A. 2007. Growth-mortality tradeoffs and "personality traits" in animals. *Ecology Letters*, 10, 355–363.

Stamps, J., and Groothuis, T. G. G. 2010. The development of animal personality: relevance, concepts and perspectives. *Biological Reviews*, 85, 301–325.

Stankowich, T., and Blumstein, D. T. 2005. Fear in animals: a meta-analysis and review of risk assessment. *Proceedings of the Royal Society of London B*, 272, 2627–2634.

Stockwell, C. A., Hendry, A. P., and Kinnison, M. T. 2003. Contemporary evolution meets conservation biology. *Trends in Ecology and Evolution*, 18, 94–101.

Sutherland, W. J. 1998. The importance of behavioural studies in conservation biology. *Animal Behaviour*, 56, 801–809.

Teixeira, C. P., de Azevedo, C. S., Mendl, M., Cipreste, C. F., and Young, R. J. 2007. Revisiting translocation and reintroduction programmes: the importance of considering stress. *Animal Behaviour*, 73, 1–13.

Trexler, J. C., and Travis, J. 2000. Can marine protected areas restore and conserve stock attributes of reef fishes? *Bulletin of Marine Science*, 66, 853–873.

Uhde, T. W., Boulenger, J.-P., Post, R. M., Siever, L. J., Vittone, B. J., Jimerson, D. C., and Roy-Byrne, P. P. 1984. Fear and anxiety: relationship to noradrenergic function. *Psychopathology*, 17 (suppl. 3), 8–23.

Van Heezik, Y., Seddon, P. J., and Maloney, R. F. 1999. Helping introduced houbara bustards avoid predation: effective antipredator training and the predictive value of pre-release behaviour. *Animal Conservation*, 2, 155–163.

Van Oers, K., de Jong, G., Drent, P. J., and Van Noordwijk, A. J. 2004. A genetic analysis of avian personality traits: correlated response to artificial selection. *Behavior Genetics*, 34, 611–619.

Vargas, A., and Anderson, S. H. 1999. Effects of experience and cage enrichment on predatory skills of black-footed ferrets (*Mustela nigripes*). *Journal of Mammalogy*, 80, 263–269.

Verbeek, M. E. M., Boon, A., and Drent, P. J. 1996. Exploration, aggressive behaviour, and dominance in pair-wise confrontations of juvenile male great tits. *Behaviour*, 133, 945–963.

Verbeek, M. E. M., Drent, P. J., and Wiepkema, P. R. 1994. Consistent individual differences in early exploratory behavior of male great tits. *Animal Behaviour*, 48, 1113–1121.

Walther, G.-R., Post, E., Convey, P., Menzel, A., Parmesan, C., Beebee, T. J. C., Fromentin, J.-M., Hoegh-Guldberg, O., and Bairlein, F. 2002. Ecological responses to recent climate change. *Nature*, 416, 389–395.

Ward, A. J. W., Thomas, P., Hart, P. J. B., and Krause, J. 2004. Correlates of boldness in three-spined sticklebacks (*Gasterosteus aculeatus*). *Behavioral Ecology and Sociobiology*, 55, 561–568.

Watters, J. V., Lema, S. C., and Nevitt, G. A. 2003. Phenotype management: a new approach to habitat restoration. *Biological Conservation*, 112, 435–445.

Watters, J. V., and Meehan, C. L. 2007. Different strokes: can managing behavioral types increase postrelease success? *Applied Animal Behaviour Science*, 102, 364–379.

Webster, S. J., and Lefebvre, L. 2001. Problem solving and neophobia in a columbiform–passeriform assemblage in Barbados. *Animal Behaviour*, 62, 23–32.

Wielebnowski, N. C. 1999. Behavioral differences as predictors of breeding status in captive cheetahs. *Zoo Biology*, 18, 335–349.

Wilcox, B. A., and Murphy, D. D. 1985. Conservation strategy: the effects of fragmentation on extinction. *American Naturalist*, 125, 879–887.

Wilkening, K. E., Barrie, L. A., and Engle, M. 2000. Trans-Pacific air pollution. *Science*, 290, 65–67.

Wilson, D. S. 1998. Adaptive individual differences within single populations. *Philosophical Transactions of the Royal Society of London B*, 353, 199–205.

Wilson, D. S., Clark, A. B., Coleman, K., and Dearstyne, T. 1994. Shyness and boldness in humans and other animals. *Trends in Ecology and Evolution*, 9, 442–446.

Wilson, E. O. 2002. *The Future of Life*. New York: Alfred A. Knopf.

Wolak, M. E., Gilchrist, G. W., Ruzicka, V. A., Nally, D. M., and Chambers, R. M. 2010. A contemporary, sex-limited change in body size of an estuarine turtle in response to commercial fishing. *Conservation Biology*, 24, 1268–1277.

Wolf, C. M., Griffith, B., Reed, C., and Temple, S. A. 1996. Avian and mammalian translocations: update and reanalysis of 1987 survey data. *Conservation Biology*, 10, 1142–1154.

Wolf, M., Van Doorn, G. S., Leimar, O., and Weissing, F. J. 2007. Life-history trade-offs favour the evolution of animal personalities. *Nature*, 447, 581–584.

Wolf, M., and Weissing, F. J. 2010. An explanatory framework for adaptive personality differences. *Philosophical Transactions of the Royal Society of London B*, 365, 3959–3968.

Wright, S. 1931. Evolution in Mendelian populations. *Genetics*, 16, 97–159.

Yoerg, S. I., and Shier, D. M. 1997. Maternal presence and rearing condition affect responses to a live predator in kangaroo rats (*Dipodomys heermanni arenae*). *Journal of Comparative Psychology*, 111, 362–369.

Zala, S. M., and Penn, D. J. 2004. Abnormal behaviours induced by chemical pollution: a review of the evidence and new challenges. *Animal Behaviour*, 68, 649–664.

14

Personality Variation in Cultured Fish

Implications for Production and Welfare

FELICITY HUNTINGFORD, FLAVIA MESQUITA, AND SUNIL KADRI

Introduction

As discussed elsewhere in this volume, the term *personality* is often applied to nonhuman animals in cases where individual variability in response to a standard situation is both repeatable when the same animals are observed on different occasions and generalizable, in the sense that what an animal does in one context predicts how it will behave in another (see introduction, this volume). Chapter 2 describes various aspects of personality in the stickleback (*Gasterosteus aculeatus*), so here we concentrate on other species of fish. One major dimension of behavioral variation is the shy-bold continuum (e.g., Wilson et al. 1994), reflecting the fact that certain individuals consistently take risks in potentially dangerous situations, while in the same circumstances others consistently avoid risk. As background information, some of the ways in which such variation in risk taking has been quantified in the published literature on fishes are summarized in figure 14.1a. Sometimes differences in risk taking are context specific, in that the relative performance of different individuals, though consistent within a given situation, changes in different contexts. For example, pumpkinseed sunfish (*Lepomis gibbosus*) show consistent individual differences in response to a potentially threatening object and to novel food, but behavior in these two contexts is unrelated (Coleman and Wilson 1998). In other cases, how fish behave in one kind of risky situation predicts their responses in another; thus juvenile smallmouth bass (*Micropterus dolomieui*) that readily explore an unfamiliar environment are also relatively bold when confronted by a predator (Smith et al. 2008).

The bold-shy continuum sometimes forms part of a broader behavioral syndrome (Sih et al. 2004) in which an individual's risk-taking phenotype predicts behavior in functionally distinct contexts, during fights, for example. Again for background information, some of the ways in which levels

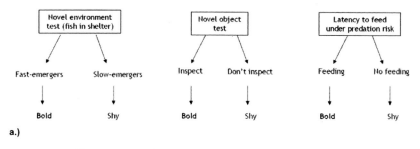

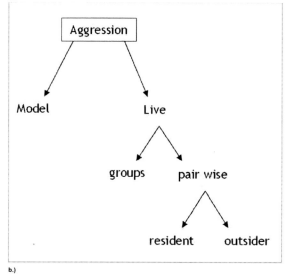

Figure 14.1. Schematic representation of methods used to quantify (a) boldness and (b) aggression in fishes.

of aggression have been quantified in the published literature on fishes are summarized in figure 14.1b. As an example of an association between boldness and aggressiveness, fights between pairs of cichlids (*Nannacara anomala*) classified as bold on the basis of inspection of a model predator escalate to high-intensity actions more quickly than do those involving two timid fish (Brick and Jakobsson 2002). Such cases meet the criteria for using the term *personality* to describe how fish behave.

The zebrafish (*Danio rerio*) is an established model for studying developmental genetics that is becoming increasingly important in the literature on personalities in fish. Striking individual differences along a bold-shy continuum have been demonstrated in terms of inspecting a predator (Dugatkin et al. 2005), approaching a novel object (Wright et al. 2006a), and feeding after a stressful experience (Moretz et al. 2007a). A correlation exists between individual activity levels and tendency to approach a predator (figure 14.2a) and, for some strains, between activity and tendency to feed after disturbance, perhaps because these responses are related through the metabolic costs of activity. In a subset of strains, boldness is positively related to aggressiveness, but in others this relationship is absent or reversed (Moretz et al. 2007a). In other words, in zebrafish as in sticklebacks (chapter 2), a link between boldness and aggression can exist, but this is not an inevitable association.

The main reason for carrying out such studies on zebrafish is that this species has a well-characterized genome, so molecular tools are available for exploring patterns of inheritance for behavioral traits. Crosses between wild and domesticated strains (which differ in boldness measured by response to a novel object; see below; Wright et al. 2003, Robison and Rowland 2005) suggest that fearfulness and boldness are inherited through different genetic mechanisms (Wright et al. 2006a). Exploring such effects further using the genome-wide linkage map for zebrafish, 4 interacting loci with major effects (quantitative trait loci or QTL) on different chromosomes explain over 40% of the variance in latency to explore a novel object (Wright et al. 2006b). Such studies pave the way for a search for candidate genes that may control boldness, aggression, and the link between them (see also chapter 6).

The literature on personality in fishes is mainly based on studies of small, experimentally tractable species such as sticklebacks and zebrafish, but a certain amount is known about behavioral variation in some cultured or potentially culturable species. Table 14.1a lists some cases. By way of example, Eurasian perch (*Perca lucius*) held at high density show marked differences in growth rate (e.g., Melard et al. 1996), and behavioral scientists have explored the possibility that these are the result of differences in personality among the fish concerned. Juvenile perch show clear and consistent differences in response to risk; these responses include spending time in cover as opposed to open water and feeding readily in a novel environment and in the presence of a predator, the latter two measures being correlated (figure 14.2b). Differences in risk taking predict competitive ability, with risk-prone fish gaining the lion's share of presented food (Westerberg et al. 2004). In this shoaling species, social interactions complicate but do not completely obscure individual differences (Magnhagen 2007). These individual differences are consistent over periods of several weeks, but not in the longer term, probably because of intervening changes in nutritional status or life-history events (Staffan et al. 2005).

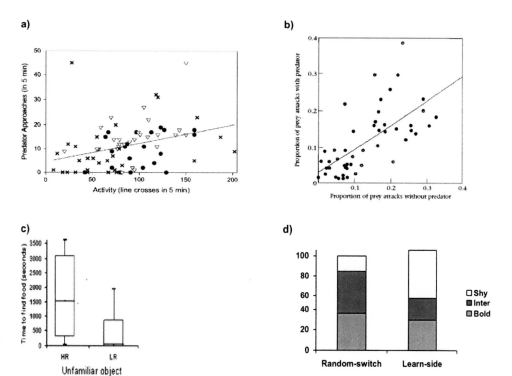

Figure 14.2. Personality in fishes. (a) Correlation between individual activity levels and tendency to approach a predator in individually identified zebrafish (from Moretz et al. 2007a). Symbols represent fish from different strains. (b) Correlation between feeding rates in juvenile perch feeding in the presence and absence of a predator (Westerberg et al. 2004). (c) Effect of novel object on speed of reaching a learned food source in rainbow trout from the high and low responsive strains (Ruiz-Gomez et al. 2011). (d) Learning strategy adopted by common carp classified as bold, intermediate, and shy on the basis of time to emerge from cover and explore a novel environment. LS = fish that learned to use a light cue to find a food source randomly allocated to one of two feeding compartments. RS = fish that chose a compartment at random and switched if no food was found (Mesquita 2010).

Coping styles

Behavioral profiles are often associated with marked differences in physiology, in which case the animals concerned may be described as showing different coping styles (Korte et al. 2005; Koolhaas et al. 2007; Coppens et al. 2010; chapter 12). To reiterate briefly what is covered in chapter 12, so-called proactive individuals typically show an active (fright and flight), adrenaline-based response to various challenges; are aggressive toward competitors;

Table 14.1a. Examples of studies of aspects of personality in cultured fish species.

Species	Observation	Reference
Brown trout (*Salmo trutta*)	Juveniles differ in response to a novel object; bold fish dominate in pairwise fights with size-matched shy fish.	Sundström et al. 2004
	Fish with different poststress catecholamine levels vary in response to a novel environment, with high-catecholamine fish showing more escape attempts.	Brelin et al. 2005, 2008
Grayling (*Thymallus thymallus*)	Juveniles differ in rate of attack to a conspecific and aggressiveness is positively linked to rate of resuming attack following a threat.	Salonen and Peuhkuri 2006
African catfish (*Clarias glariepinus*)	Individuals differ in physiological response to a stressor, fish with high stress responsiveness being less efficient at feed utilization.	Martins et al. 2008
	Fish with low feed efficiency showed active escape in response to chemical alarm cues and the converse. Proactive fish feed sooner in a novel environment and do less well at low but not high stocking densities.	Van de Nieuwegiessen 2009
Atlantic cod (*Gadus morhua*)	Farmed cod vary in probability of escaping through a hole in a net, with some (bold?) individuals repeatedly escaping.	Hansen et al. 2008
	Heritable differences in emergence into and exploration of a novel environment in juvenile fish.	Forbes 2007
	Inherited differences in aggressiveness between wild cod stocks.	Salvanes and Braithwaite 2005
Pufferfish: (*Takifugu rubripes*)	Fish differ in patterns of recovery from an acute stressor, with inactive fish having lower base-level cortisol levels and recovering quicker.	Hosoya et al. 2008
Pentas (*Poecilia parae*)	Fish differ in distance of approach to a model predator and bold fish are more aggressive and explore a T-maze more quickly.	Bourne and Sammons 2008
Guppy (*Poecilia reticulata*)	Strain-related differences in boldness and predator avoidance.	Bleakley et al. 2006

take risks in the face of potential danger, and are relatively inflexible, tending to form behavioral routines. In contrast, so-called reactive individuals show a passive (freeze and hide), cortisol-based response to challenge; avoid risk (including fights), but are responsive to and flexible about environmental change. Different stress coping styles may be associated with differences in energy metabolism, with proactive animals often adopting an energetically expensive strategy and reactive animals being energetically conservative (Korte et al. 2005; Coppens et al. 2010).

Physiological correlates of personality have been relatively poorly studied in fish compared with mammals, but the rainbow trout (*Oncorhynchus mykiss*) has become something of a pioneering model in this context. This follows the successful selective breeding of a domesticated strain for high (HR) and low (LR) cortisol response to a standard stressor (Pottinger and Carrick 1999; 2001). Correlates of stress responsiveness in these strains have been well reviewed (e.g., Øverli et al. 2005; Schjolden and Winberg 2007) and include differences in brain biochemistry (LR fish have higher poststress levels of adrenaline and serotonin than HR fish) and behavior (LR fish are bolder than HR fish and they tend to win fights when paired against an HR opponent). The two strains also differ in flexibility of response to environmental change; for example, after being trained to feed in a particular place, HR fish find food faster than LR fish when it is moved to a new location, but are more disturbed by a novel object placed *en route* to a familiar food supply (Ruiz-Gomez et al. 2011; figure 14.2c). Various studies using unselected strains of rainbow trout have confirmed the existence of a bold-shy syndrome associated with marked physiological differences (e.g., Schjolden et al. 2005). Overall, these results suggest that rainbow trout show inherited coping strategies much as described for mammals and birds and that even in domesticated strains significant variation in stress physiology and risk taking persists.

Different coping styles have also been characterized in the common carp (*Cyprinus carpio*), a strongly schooling freshwater fish that is the oldest cultured fish species (Balon 2004). Individual carp differ strikingly and consistently in boldness (the speed with which they explore a novel, potentially dangerous environment) and in ability to gain access to a spatially restricted food source, with these two traits being positively related. Fish classified as shy on the basis of exploration of a novel environment have lower weight-specific metabolic rates than do bold, risk-taking fish, higher plasma lactate and glucose concentrations, and higher expression of the cortisol receptor gene in both the head kidney and the brain (Huntingford et al. 2010). Shy carp also tend to use a visual landmark to locate a variable food source, whereas bold fish more often follow a simple searching routine (figure 14.2d, Mesquita 2010). Thus a behavioral syndrome based on boldness and aggression exists in common carp, and this is associated with differences in behavioral flexibility, in metabolic rate, and in stress physiology.

Fitness correlates

The previous sections show that aspects of personality have been identified in many species of cultured fish, though details vary from species to species and, in some cases, from strain to strain. Cultured fish therefore bring with them behavioral and physiological profiles that may well influence the effectiveness of intensive production, a topic that we explore in the remainder of this chapter. One important finding in the general literature on animal personalities that is potentially of considerable relevance in this context is the fact that different personalities persist within populations because they grow and survive well in different environmental conditions (see chapters 6–9).

In fish as in other taxa, there are a number of reasons for expecting a relationship between personality and growth, with different theoretical frameworks (which are not mutually exclusive) predicting different relationships. If an association between boldness and aggression is the result of a growth/survival trade off, as suggested by Stamps (Stamps 2007; Biro and Stamps 2008), then individuals following a fast-growth trajectory will take risks to acquire food. Fish that are bold and aggressive are therefore likely to be those that are growing faster and so, at any given time point, to be larger than their shy, slow-growing counterparts. On the other hand, risk-taking (during foraging or while competing for food) might be adjusted to nutritional status (Brown and Braithwaite 2004). This would predict that larger individuals with good nutrient reserves will be less bold and less aggressive, minimizing predation risk in order to protect their high expected future fitness. To give one example, in rockfish large individuals are slower to start feeding after a simulated predatory attack than are small fish, as they have a lower risk of starvation (Lee and Berijikian 2008). Table 14.1b summarizes the results of a number of studies of the relationship between body size and risk taking. Clearly the results are variable, with some (e.g., Brown and Braithwaite 2004) supporting the link to nutritional status and others the growth/survival trade-off (e.g., Azuma et al. 2005).

Whatever the relationship between size and risk-taking phenotype, several studies have shown that fish with different personalities flourish in different environmental conditions. In perch, individual rates of growth at high densities are negatively correlated with boldness as measured in an aquarium; individuals that do well in small groups are less successful at high densities (Staffan et al. 2005). Similarly, laboratory-reared juvenile salmon that compete aggressively and successfully for food and grow fast when held at high densities in tanks do poorly compared with their nonaggres-

Table 14.1b. Selected studies of the relationship between body size and risk taking in fish

Paper	Species	Results	Fish characteristics
Brown and Braithwaite 2004 Brown et al. 2007	*Brachyraphis episcopi*	Smaller fish emerge from shelter sooner. Bold fish heavier than shy fish at any given length and gain mass faster.	Wild fish, male and female, mature
Laakkonen and Hirvonen 2007	*Salvelinus alpinus*	Boldness to predator unrelated to individual growth rate.	Young, hatchery-reared
Azuma et al. 2005	*Oncorhynchus mykiss*	Bold fish grow faster, gaining in length rather than condition.	Hatchery-reared, 4-year-old male and female
Magnhagen and Borcherding 2008	*Perca fluviatilis*	Starved perch bolder when foraging under risk than are well-fed conspecifics.	Wild, juvenile
Ward et al. 2004	*Gasterosteus aculeatu*	Bold fish dominant and with higher rates of growth than shy fish.	Wild, 36–44 mm
Riesch et al. 2009	*Poecilia* spp.	In one of 3 species, smaller individuals are bolder	Wild, male and female

sive companions when transferred to more stream channels where food is spatially and temporally unpredictable (Huntingford and Garcia de Leaniz 1997).

Since the relative performance of fish with different personalities is environment dependent, it is to be expected that distributions of personality types will vary in fish from environments exposed to different selective regimens. For example, more risk-avoiding perch are found in lakes with high levels of cannibalism on small perch and the converse (Magnhagen 2006). The proportion of brown trout (*Salmo trutta*) with high-catecholamine responsiveness (proactive individuals) is higher in populations derived from large, stable rivers than in those from small streams (Brelin et al. 2008).

Domestication, captive rearing, and fish personalities

Research on fishes supports the general conclusion that different personalities flourish in different environments. The conditions in which fish are cultured differ strikingly from those they would experience in their natural environment, with high densities, abundant and predictable food, no predators, and protection against certain diseases (Huntingford 2004), so it is not surprising that the distribution of personality types is often found to vary when wild fish and cultured fish are compared (Huntingford 2004; Brannas and Johnsson 2008). Rearing fish in culture can cause their behavior to deviate from that of wild fish in several ways. Differential mortality by behavioral phenotype may cause differences during a single generation of captive rearing; over successive generations, it can also cause long-term, inherited effects on behavior arising either through domestication (the inadvertent favoring of traits that promote survival and growth in culture conditions) or through deliberate selection for desirable traits such as fast growth. In addition to such inherited effects, the strikingly different environments experienced by wild and captive reared fish during development may generate differences in behavior. All such effects are of considerable practical importance, since they can potentially have both positive and negative effects on the effectiveness of aquaculture.

Inherited effects of long-term captive rearing are demonstrated by comparing behavior in fish of wild and farm origin reared in standard conditions, ideally for several generations to reduce nongenetic maternal effects. For example, zebrafish from domesticated strains show more risk-taking in various situations (figure 14.3a) and also grow faster in the same conditions than do fish of wild origin reared in standard laboratory conditions (Wright et al. 2006a). QTL analysis shows that a wild-type genotype at a small number of interacting loci determines individual levels of boldness (Wright et al. 2006b). Fish that are cultured for food tend to be longer lived than zebrafish, making precisely controlled experiments difficult, but all the same, numerous studies have examined the effects of domestication on the behavior of cultured species.

One consistent finding is that domestication, with or without deliberate selection for fast growth, favors bold fish (Weber and Fausch 2003). For example, hatchery-reared brown trout of wild origin spend more time inactive following a simulated attack than do identically reared trout from a sea ranched strain, with hybrids resembling fish of ranched origin (Petersson and Jarvi 2006). Differences in risk-taking between fish of wild and domesticated origin reared in standard conditions have been reported for other species, including Masu salmon (*Oncorhynchus masou*, Yamamoto and Reinhardt 2003) and sea bass (*Dicentrarchus labrax*, Millot et al. 2009). A few studies report no differences in boldness between domesticated and wild

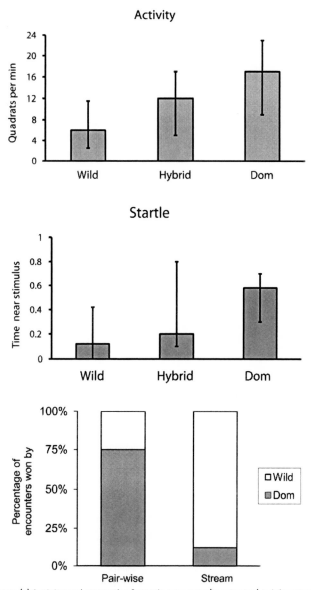

Figure 14.3. (a) Activity and strength of startle response (in minutes) in laboratory-reared zebrafish of wild and domesticated origin (Wright et al. 2003). (b) Percentage of dominance encounters won by hatchery-reared juvenile Atlantic salmon of wild (open bars) and farmed (hatched) origin in pairwise tests in tanks and in stream channels (from Huntingford 2004, after Einum and Fleming 1997; Fleming and Einum 1997).

fish once size is controlled for, but overall, it seems that bold fish flourish in intensive production systems, whereas shy fish fail to thrive and are selected out (Huntingford 2004; Brannas and Johnsson 2008).

Effects of long-term captive rearing on aggression are more complex. Domestication accompanied by selection for rapid growth can result in increased aggressiveness (Doyle and Talbot 1986), possibly linked to higher metabolic rates (as described by Cutts et al. 1999), higher pituitary growth hormone activity (Fleming et al. 2002), and consequently greater appetite. Arctic charr (*Salvelinus alpinus*) selected for fast growth and delayed maturation for four generations are more aggressive during feeding than are fish from the original wild strain (Brannas et al. 2005). Domestication without active selection for rapid growth does not favor aggressiveness in any simple way. Higher levels of aggression in domesticated fish have been found in some studies (e.g., brook trout, *Salvelinus fontinalis*, Moyle 1969; cutthroat trout, *Oncorhynchus clarki*, Mesa 1991; brown trout, Sundström et al. 2004; grayling, *Thymallus thymallus*, Salonen and Peuhkuri 2006; butterfly splitfin, *Ameca splendens*, Kelley et al. 2006). In contrast, for other species decreased aggression in domesticated fish has been reported (medaka, *Oryzias latipes*, Ruzzante and Doyle 1991; tilapia, *Oreochromis mossambicus* × *hornorum*, Robinson and Doyle 1990). Thus the effect of domestication on aggressiveness in cultured fish depends on species and also on the conditions in which behavioral screening takes place; for example, domesticated Atlantic salmon are more aggressive than their wild counterparts in tanks at high densities, but less so in artificial streams (figure 14.3b). The culture conditions in which fish are held are also critical (Ruzzante 1994). If parental fish have been raised with limited food restricted in space and time, domestication is likely to lead to increased aggressiveness; if feeding has been frequent and to excess (for example, to maximize growth), less aggressive fish may well be favored.

The very different environment experienced by wild and cultured fish can also have profound effects on behavior through a variety of nongenetic developmental processes; where such effects are long-lasting they are part of the process that determines the distribution of personalities in cultured fish. For example, brief exposure to simulated predatory attack produces a long-lasting reduction in boldness in cultured Nile tilapia (*Oreochromis niloticus*, Mesquita and Young 2007); zebrafish become more aggressive when reared or held as adults in groups of mixed strains than in single-strain groups (Moretz et al. 2007b) and less aggressive when reared and tested under hypoxic as opposed to normoxic conditions (Marks et al. 2005).

Studies of the lasting effects of captive rearing on the development of fish personalities are difficult for long-lived species, but a number of studies have found differences in behavior between farmed fish of the same genetic origin reared in hatchery and natural conditions. For example, wild brown trout show stronger suppression of heart rate following a simulated predatory attack than do captive reared fish of the same stock, though this effect is short-lived (Sundström et al. 2005). Hatchery-reared juvenile steelhead trout are marginally more aggressive and less likely to settle fights by display than are their wild-reared counterparts (Berejikian et al. 2001), and hatchery-reared juvenile brown trout initiate contests sooner and fight for longer and more fiercely than do size-matched fish captured in the wild (Sundström et al. 2003). Such studies rarely look at long-term effects, so it is not clear whether these can be considered as genuine effects of personality.

Criteria for effective fish culture

GENERAL CRITERIA FOR EFFECTIVE PRODUCTION

Fish are cultured for various reasons, including for food, for the ornamental fish trade, for scientific purposes, for restocking to support recreational fisheries, and for conservation-based reintroductions. Whatever the eventual use to which cultured fish are to be put, during the period that they are being grown to a usable size there is a set of common requirements for sustainable culture, as well as a set of common problems that have to be solved. Briefly, for effective and economically viable fish culture, farmers are concerned to ensure that their fish survive (which means keeping them free from injury and disease), that they grow well (which means achieving fast and reasonably uniform increase in size), and that growth is efficient (which means maximizing the flesh produced for feed delivered, feed costs often making up more than half the cost of producing intensively reared fish). In addition, it is now recognized as important that cultured fish enjoy good welfare; this is a complex topic that we do not have space to review here (see Chandroo et al. 2004; Huntingford et al. 2006), but for the most part in aquaculture there is not a great deal of conflict between welfare and production, since fish held in conditions that promote their wellbeing tend to survive and grow well. A final criterion for sustainable fish culture is that adverse impacts on the environment are kept to an acceptable minimum. Here, too, conflicts with production requirements are less than might be expected; for example, the periodic moving of cage sites, as sometimes required for environmental reasons, can promote the health of cultured fish. In addition, in some places good welfare and environmentally friendly rearing are marketable traits for which farmers can charge a premium. After production is complete, criteria for success are still mainly common; for example, good body condition and general health at the end of the production phase are a target in all cases. Arguably flesh quality is a special requirement for food fish culture, though to the extent that this reflects general health, practices that promote good flesh quality may also promote the effectiveness of culture for nonfood purposes.

Any factor that makes it difficult for farmers to meet these various criteria constitutes a problem for aquaculture and there are many such factors. Some are particular; for example, failure to provide appropriate nutrients and specific departures from good water quality will compromise growth and survival in precise ways. Others are more general; for example, many different kinds of challenge can induce a physiological stress response in cultured fish that, if sustained, can compromise many different body functions, including digestion, assimilation, and immune function. Fish personalities are relevant here both because coping style will mediate the effects of such stressors and because the behavior of farmed fish can itself be a cause of stress. The examples given in the previous section show some of the ways in which, through genetic and nongenetic effects, the environmental conditions that prevail on fish farms can alter the distribution of personality types among the fish concerned. This will determine the material that fish farmers have at their disposal; arguably, by its very nature, intensive culture may generate fish with behavioral profiles that cause them problems.

BEHAVIORAL CRITERIA FOR EFFECTIVE CULTURE

The various different reasons for which fish are cultured have implications for what is required of their behavior. Where fish are cultured for food, the main criterion for effective aquaculture is the efficient and cost-effective production of good quality flesh in a form that is readily marketable. For brood stock, unless artificial fertilization is possible, it is clearly important that fish show effective reproductive behavior. Apart from this, how the fish we farm for food behave is not directly relevant to whether culture is effective. However, their behavior is indirectly very important, because it can influence the growth, health, or welfare of the fish concerned and, potentially, the environmental impact of producing them. For example, exactly how a cultured fish consumes a food pellet may affect how much food is wasted and the efficiency with which it converts nutrients into flesh; whether a fish fights may affect how fast it grows and its physical condition; and whether it is easily frightened may affect its growth, health, and welfare.

In fish cultured for the ornamental trade, the behavioral criteria for effective aquaculture relate to the aesthetic and intellectual motives for keeping fish. Attractive traits, whether morphological or behavioral, are best appreciated if aquarium fish are in good physical condition and also if they behave in an appropriate way. Fish that consistently hide, however healthy, are less rewarding that those that emerge into open water where they can be seen. Arguably, fish that fail to engage in social interactions are also less rewarding than those that interact with conspecifics, since the bright colors that make species of ornamental fish attractive are displayed during and often augmented by social interactions. Over and above such considerations, for many people simply watching the behavior of their fish is a key part of the rewards of maintaining an aquarium. For all these reasons, behavior is of direct importance for the effective culture of ornamental fish. It is also indirectly important, since having ornamental fish in good physical condition is critical and, as discussed above, behavior can promote or detract from fish health. Where fish are reared for scientific purposes, how they behave is directly important if they are used to study neurobiology or behavior and indirectly important wherever fish in good general health and of appropriate maturation status are needed.

In fish cultured for release into the wild, the success of restocking will depend on how well they survive, grow, and in some cases reproduce after being released. Indirect effects of behavior are important here, as these will determine the condition and health status of the fish at release and hence their survival prospects. Direct effects are also extremely important; efficiency when cultured fish feed, compete with conspecifics, and avoid predators is critical for postrelease survival. It is therefore important that cohorts of cultured fish and individuals within them show a natural spectrum of effective species-typical behavior. Where the aim is simply to generate a population of fish for subsequent angling, although the spectrum of behavior in individuals and populations is important, exactly how these develop is not. If the aim is to generate or augment self-sustaining wild populations for conservation, then clearly the inherited behavioral profiles of released fish (individuals and cohorts) are important.

Fish personalities and effective culture

We have argued above that, in general, the behavior of cultured fish is important, indirectly and directly, if the aims of different kinds of aquaculture are to be realized. Here we draw on the material presented in earlier sections to explore the different ways in which one specific aspect of behavior, namely the patterns of variability characterized by the term *fish personality*, is important for fish culture.

Fish personalities may have indirect effects on several different aspects of behavior in culture. First of all, personality may influence how efficient fish are at foraging; for example, being less disturbed by novelty, bold fish may be quicker at switching to new food types than their shy counterparts. In brown trout, boldness in response to a novel object is associated with a higher probability of eating novel prey (Sundström et al. 2004). Similarly, juvenile cod classified as bold on the basis of response to a novel object are more likely than shy conspecifics to activate a trigger to release food in a demand-feeding system (figure 14.4; Ablitt 2009). Personality may also influence how effective cultured fish are in competition for food. Under intensive production conditions with a clumped and predictable food supply, bold, aggressive fish may be better at acquiring food through more effective competition; they may therefore enjoy better growth and welfare in intensive production than do shy fish. Atlantic salmon that fail to grow under intensive production conditions have relatively high rates of serotonin turnover in their brains (Cubitt et al. 2008), comparable with those of reactive trout. However, it is not clear whether these are permanently unaggressive fish or fish that have become subordinate as a result of previous fights. In groups of European seabass fed with a demand feeder, fish that show little or no triggering have higher turnover rates of brain serotonin than do actively triggering fish, but again it is not clear whether nontriggering fish are inherently reactive or subordinate through experience (Di-Poï et al. 2007).

Almost by definition, personality will influence the extent to which fish are stressed in intensive culture, with bold fish being less disturbed by husbandry practices and therefore likely to use energy more efficiently, to grow faster, and to enjoy enhanced immunocompetence, health, and welfare. Rainbow trout classified as proactive on the basis of their behavioral (and physiological) response to low oxygen levels are more likely to die of hypoxia than are their more passive, reactive counterparts (Van Raaij et al. 1996). Under unstressed conditions, high responsive (HR) and low responsive (LR) rainbow trout show the same average growth rates, though groups of LR fish show more marked variation in growth rates (Pottinger 2006). In contrast, following transportation, HR fish are more stressed and show suppressed appetite and lower feed conversion efficiency than do LR fish (Øverli et al. 2007; Ruiz-Gomez et al. 2008). Bold and shy carp show strikingly different patterns of gene expression following a simulated bacterial challenge (MacKenzie et al. 2009; figure 14.4b), and in standard conditions bold fish

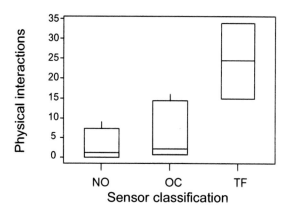

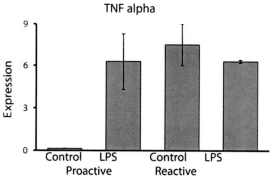

Figure 14.4. (a) The relationship between the number of times a novel object was approached and touched and triggering of a demand feeder in juvenile cod. NO = fish classified as never activating the trigger, OC = fish that occasionally activated the trigger, and TF = the single fish in each group that did most of the triggering (Ablitt 2009). (b) Gene expression (mRNA for the "TNF alpha" cytokine) in the brain of carp, *Cyprinus carpio*, screened for proactive and reactive behavior following a simulated bacterial challenge (MacKenzie et al. 2009).

tend to grow in length whereas shy fish tend to grow in weight, perhaps as a consequence of higher cortisol reactivity. African catfish (*Clarius glariepinus*) that show strong physiological stress responses have poor food conversion efficiencies (Martins et al. 2008; Van de Nieuwegiessen 2009). Overall, for food culture, a bold-unaggressive phenotype might be the best personality, since such fish will be resistant to husbandry stress but will not cause stress and damage to other fish by excessive fighting. The ease with which such fish can be found or produced will depend on the existence and strength of any link between boldness and aggressiveness.

Achieving a natural and effective repertoire of species-typical behavior (in the form of a natural distribution of personality types) is of direct importance for some kinds of fish culture. In the case of ornamental fish, bold fish are probably the most suitable, since they are more likely to enjoy better health and to make themselves more readily available for viewing, both by being active rather than hiding and by engaging in the social interactions in which color patterns are often displayed. On the other hand, if boldness and aggression are linked in a behavioral syndrome, bold fish may also be more aggressive. For species in which colors are displayed during aggressive displays, this may be an advantage, but high levels of aggression may detract from the physical condition, increase stress levels, and impair the health status of ornamental fish. Here again there may be a conflict between requirements for boldness and for low aggression, and resolving this conflict depends on how easy it is to modify these two dimensions of behavioral variation independently.

In the case of fish cultured for release, postrelease survival is often very poor, as cultured fish generally fail to avoid predators and to acquire food effectively compared with wild fish (Olla et al. 1998; Weber and Fausch 2003; Salvanes and Braithwaite 2006). For example, survival in hatchery-reared juvenile pufferfish released into ponds is extremely poor when predators are present (Shimizu et al. 2006), up to 49% of Atlantic salmon fry stocked in natural and artificial streams are eaten by piscivorous fish within the first 2 days of stocking (Henderson and Letcher 2003), and survival from juvenile to adulthood in Coho salmon (*Oncorhynchus kisutch*) ranges from 8% for wild fish to as low as 2% for fish released from hatcheries in private ownership (McNeil 1991). Poor postrelease survival may be due to failure to avoid predators (fish need to be timid enough to survive rather than being bold and aggressive) and/or failure to compete effectively for food in the circumstances prevailing at the release site. As discussed above, different local conditions may favor either the bold/aggressive or the shy/unaggressive phenotype, so success of release programs will be simultaneously environment- and personality-dependent. Where fish are cultured and released for conservation, what matters is that cohorts of released fish produce self-sustaining populations, and they are more likely to do this if they mimic the natural bold-shy and aggressive/nonaggressive spectra of fish populations at the release site.

Using fish personalities to promote effective culture

On the basis of existing knowledge of personalities in fish, including their development and their context-specific implications for fitness, a number of steps can be taken to ensure that fish personalities work for rather than against the aims of effective fish culture. First, it is important to choose the most appropriate fish for rearing, in terms of species, strain, and individual. Second, having chosen which fish to rear, care should be taken that the culture environment does not, through differential growth and selective mortality by personality, favor behavioral phenotypes that are undesirable for the type of culture concerned. A third step is to manipulate personalities and their expression through modifying the rearing environment experienced by individual fish.

CHOOSING APPROPRIATE FISH

As outlined above, arguably when fish are reared for food, fast-growing but bold, unaggressive fish will probably be the most likely to meet the aims of effective culture. Since domestication favors bold fish (see above), part of the solution would seem to be to use domesticated strains where these are available and to select for bold fish where they are not. The existence of a bold-aggression syndrome poses a problem here, since if the association between these aspects of behavior is fixed, bold fish will inevitably also be too aggressive for maximally effective culture. However, the fundamental literature on fish personalities tells us that boldness and aggression are linked only in some strains or populations; in the case of sticklebacks, these are populations that are naturally exposed to piscivores (Bell and Stamps 2004; Dingemanse et al. 2007; and see chapter 2). If natural selection can dissociate boldness and aggression, fish geneticists should be able to do so as well, identifying or developing strains of fish with the ideal personality for culture. That this is indeed possible is indicated by the fact that fish from one of two clonal lines derived from domesticated rainbow trout (D2 in figure 14.5) combined boldness with relatively low levels of aggression (Lucas et al. 2004). Where fish are reared for ornamental purposes, the personality type that would best meet the criteria for effective culture (arguably, bold and reasonably aggressive) might well be found within the natural bold-aggression spectrum for the species concerned. Where fish are reared for release, especially for conservation purposes, effective culture requires a representative range of personalities. In general, a good principle is to use as brood stock fish of wild origin from sites ecologically similar to the proposed release locations, ideally spanning a broad range of personality types.

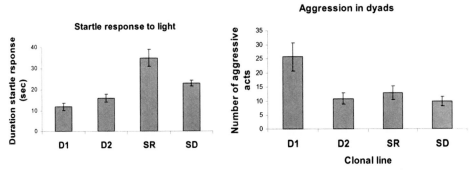

Figure 14.5. (a) Mean (SEM) duration of startle response to a light, and (b) mean (SEM) number of aggressive acts during pairwise fights in rainbow trout from clonal lines derived from 2 domesticated and semidomesticated strains (adapted from Lucas et al. 2004).

It might be possible to identify fish with the optimal personality or range of personalities for culture and, even where domestication is appropriate, to bypass its early stages by screening young fish for favorable phenotypes from the natural range of variants. Screening individual levels of boldness and aggression in thousands of fish is unfeasible, but there are various possible proxies for different personalities that might be used. For example, in rainbow trout a particular haplotype at one major histocompatibility complex (MHC) locus is associated with high levels of aggression, success in food competition, and boldness in the face of danger. If these are required phenotypes, MHC haplotyping could provide a useful molecular marker for suitability for food aquaculture (Azuma et al. 2005). For reasons of cost, this is probably feasible only as part of a selective breeding program, but there are other possibilities. For example, date of first feeding depends on standard metabolic rate (SMR) in several species (Atlantic salmon, for example; Cutts et al. 1999) and where metabolic rate is predictive of personality, date of first feeding and thus SMR could be used to identify fish that would flourish when cultured for food (Vaz-Serrano et al. 2011).

AVOIDING PERSONALITY-DEPENDENT GROWTH
AND SURVIVAL IN CULTURE

Having chosen suitable stock, a second step toward ensuring that personalities work for rather than against the fish farmer is to develop culture environments that do not generate undesirable behavioral phenotypes through selective growth and mortality by personality within a single generation. If, for example, fish that grow poorly or die in culture are reactive fish with strong stress responses, any released populations may be lacking in the very fish that are likely to survive well in natural environments in which predators are found. When fish are cultured for restocking, culture conditions should be sufficiently varied to produce the spectrum of personalities that mirrors that found in the natural environment of the species concerned. This might be achieved by providing several feeding stations with different levels of profitability and perceived risk or by using feeding regimens that allow different kinds of fish to feed at different times. For example, when fed by self-feeding systems, smaller Arctic charr (which might be reactive fish adopting a low risk–low gain strategy) are able to avoid feeding during the day, choosing instead to feed at night and so avoiding competition with larger conspecifics (Brannas 2008).

MANIPULATING PERSONALITIES BY MODIFYING THE REARING ENVIRONMENT

A third step toward ensuring that personalities work for rather than against the fish farmer is to modify the rearing environment experienced by individual fish and hence the way their behavioral profiles develop. Many studies have demonstrated that exposure to a predator makes hatchery-reared fish less bold in subsequent predatory encounters (see Huntingford 2004 for a review), but observed effects are often short-lived (though see Mesquita and Young 2007), so not relevant to the development of fish personalities. In the case of aggression, Atlantic cod, *Gadus morhua*, reared in plain plastic tanks are more likely to flee when attacked than are those reared in more complex tanks with landmarks, possibly because these complex tanks allow the fish to defend territories and to learn how to avoid escalated fights (Salvanes and Braithwaite 2005). Effects of enrichment on aggression seem to be longer-lasting that those on boldness, so here there is some scope for manipulating personalities by modifying rearing conditions. Additionally, it seems clear that boldness and aggression can be coupled or uncoupled by altering the rearing environment, specifically by exposing fish to direct predation as described for sticklebacks in chapter 2. The relevant processes include both selective predation on certain behavioral phenotypes and individual behavioral adjustments to the experience of predatory attacks (Bell and Sih 2007; Brelin et al. 2008).

It is worth noting that the success of such attempts to manipulate personality traits in cultured fish by altering rearing conditions may be compromised by the personalities of the fish concerned where these include differences in flexibility and learning ability. Thus it is possible that proactive fish, whose high levels of boldness and/or aggression might be particularly desirable to reduce, are less amenable to behavioral modification than their nonaggressive, reactive counterparts.

Conclusions

Compared with the literature on other vertebrate taxa, the literature on personalities in fish is less deep; it is, however, broader, in the sense that elements of personality have been described for many species, including a number that are cultured for various purposes. Scrutiny of this literature reveals that, while there are very many gaps in our knowledge, the fitness enjoyed by individuals with different personalities is context specific. Because cultured and wild fish are exposed to very different selection regimens, different personality types are likely to flourish in natural water bodies and on farms, and this is indeed what has been reported for a number of species. The existence of fish personalities, their precise nature, and the fact that they are strongly influenced by farm conditions will determine the extent to which farmers can achieve the optimal behavior (from their perspective) for fish cultured for different purposes. Fish personalities often place constraints on what can be achieved; for example, where a fixed bold-aggression syndrome exists, bold fish that are likely to flourish in culture conditions are exactly those that are likely to cause problems through high levels of aggression. However, there are a number of steps that can be taken to mitigate such effects; for example, bold but nonaggressive fish might be produced by selective breeding or by specific manipulation of the relating environment. In such ways we can try to ensure that fish personalities help rather than hinder effective culture.

References

Ablitt, H. 2009. Studies of appetite variation in juvenile Atlantic cod (*Gadus morhua L.*) using demand feeding systems. PhD diss., University of Glasgow.

Azuma, T., Dijkstra, J.M., Kiryu, I., Sekiguchi, T., Terada, Y., Asahina, K., Fischer, U., and Mitsuru, O. 2005. Growth and behavioral traits in Donaldson rainbow trout (*Oncorhynchus mykiss*) cosegregate with classical major histocompatibility complex (MHC) class I genotype. *Behavior Genetics*, 35, 463–478.

Balon, E. K. 2004. About the oldest domesticates among fishes. *Journal of Fish Biology*, 65, 1–27.

Bell, A. M., and Sih, A. 2007. Predation favors behavioral consistency between boldness and aggressiveness. *Ecology Letters*, 10, 828–834.

Bell, A. M., and Stamps, J. A. 2004. The development of behavioral differences between individuals and populations of threespined stickleback. *Animal Behaviour*, 68, 1339–1348.

Berejikian, B. A., Tezak, E. P., Park, L., LaHood, E., Schroder, S. L., and Beall, E. 2001. Male competition and breeding success in captively reared and wild Coho salmon (*Oncorhynchus kisutch*). *Canadian Journal of Fisheries and Aquatic Sciences*, 58, 804–810.

Biro, P. A., and Stamps, J. A. 2008. Are animal personality traits linked to life history productivity? *Trends in Ecology and Evolution*, 23, 361–368.

Bleakley, B. H., Martell, C. M., and Brodie, E. D. 2006. Variation in antipredator behavior among five strains of inbred guppies, *Poecilia reticulata*. *Behavior Genetics*, 36, 783–791.

Bourne, G. R., and Sammons, A. J. 2008. Boldness, aggression, and exploration: evidence for a behavioural syndrome in male pentamorphic livebearing fish, *Poecilia parae*. *AACL Bioflux*, 1, 39–49.

Brannas, E. 2008. Temporal resource partitioning varies with individual competitive ability: a test with Arctic charr *Salvelinus alpinus* visiting a feeding site from a refuge. *Journal of Fish Biology*, 73, 524–535.

Brannas, E., Chaix, T., Nilsson, J., and Eriksson, L.-O. 2005. Has a 4-generation selection programme affected the social behaviour and growth pattern of Arctic charr (*Salvelinus alpinus*)? *Applied Animal Behaviour Science*, 94, 165–178.

Brannas, E., and Johnsson, J. I. 2008. Behaviour and welfare in farmed fish. In: *Fish Behaviour* (Magnhagen, C., Braithwaite, V., Forsgren, E., and Kapoor, B., eds.), pp. 593–628. Enfield, NH: Science Publishers.

Brelin, D., Petersson, E., and Winberg, S. 2006. Divergent stress coping styles in juvenile brown trout (*Salmo trutta*). *Annals of the New York Academy of Sciences*, 1040, 239–245.

Brelin, D., Petersson, E., Dannewitz, J., Dahl, J., and Winberg, S. 2008. Frequency distribution of coping strategies in four populations of brown trout (*Salmo trutta*). *Hormones and Behavior*, 53, 546–556.

Brick, O., and Jakobsson, S. 2002. Individual variation in risk taking: the effect of a predatory threat on fighting behavior in *Nannacara anomala*. *Behavioral Ecology*, 13, 439–442.

Brown, C., and Braithwaite, V. A. 2004. Size matters: a test of boldness in eight populations of the poeciliid *Brachyraphis episcope*. *Animal Behaviour*, 68, 1325–1329.

Brown, C., Jones, F., and Braithwaite, V. A. 2007. Correlation between boldness and body mass in natural populations of the poeciliid *Brachyrhaphis episcope*. *Journal of Fish Biology*, 71, 1590–1601.

Chandroo, K. P., Duncan, I. J. H., and Moccia, R. D. 2004. Can fish suffer? Perspectives on sentience, pain, fear, and stress. *Applied Animal Behaviour Science*, 86, 225–250.

Coleman, K., and Wilson, D. S. 1998. Shyness and boldness in pumpkinseed sunfish: individual differences are context specific. *Animal Behaviour*, 56, 927–936.

Coppens, C. M., de Boer, S. F., and Koolhaas, J. M. 2010. Coping styles and behavioural flexibility: towards underlying mechanisms. *Philosophical Transactions of the Royal Society of London B*, 365, 4021–4028.

Cubitt, K. F., Winberg, S., Huntingford, F. A., Kadri, S., Crampton, V. O., and Øverli, O. 2008. Social hierarchies, growth, and brain serotonin metabolism in Atlantic salmon (*Salmo salar*) kept under commercial rearing conditions *Physiology and Behavior*, 94, 529–535.

Cutts, C. J., Metcalfe, N. B., and Taylor, A. C. 1999. Competitive asymmetries in territorial juvenile Atlantic salmon, *Salmo salar*. *Oikos*, 86, 479–486.

Dingemanse, N. J., Wright, J., Kazem, A. J. N, Thomas, D. K., Hickling, R., and Dawnay, N. 2007. Behavioural syndromes differ predictably between 12 populations of three-spined stickleback. *Journal of Animal Ecology*, 76, 1128–1138.

Di-Poï, C., Attia, J., Bouchut, C., Dutto, G., Covès, D., and Beauchaud, M. 2007. Behavioral and neurophysiological responses of European sea bass groups reared under food constraint. *Physiology and Behavior*, 90, 559–566.

Doyle, R. W., and Talbot, A. J. 1986. Effective population size and selection in variable aquaculture stocks. *Aquaculture*, 57, 27–35.

Dugatkin, L. A., McCall, M. A., Gregg, R. G., Cavanaugh, A., Christensen, C., and Unseld, M. 2005. Zebrafish (*Danio rerio*) exhibit individual differences in risk-taking behaviour during predator inspection. *Ethology, Ecology, and Evolution*, 17, 77–81.

Einum, S., and Fleming, I. A. 1997. Genetic divergence and interactions in the wild among native, farmed, and hybrid Atlantic salmon. *Journal of Fish Biology*, 50, 634–651.

Fleming, I. A., Agustsson, T., Finstad, D., Johnsson, J. I., and Bjornsson, B. T. 2002. Effects of domestication on growth physiology and endocrinology of Atlantic salmon (*Salmo salar*). *Canadian Journal of Fisheries and Aquatic Sciences*, 59, 1323–1330.

Fleming, I. A., and Einum, S. 1997. Experimental tests of genetic divergence of farmed from wild Atlantic salmon due to domestication. *ICES Journal of Marine Science*, 54, 1051–1063.

Forbes, H. 2007. Individual variability in the behaviour and morphology of larval Atlantic cod. PhD diss., University of Glasgow.

Hansen, L. A., Dale, T., Damsgard, B., Uglem, I., Aas, K., and Bjorn, P.-A. 2008. Escape-related behaviour of Atlantic cod, *Gadus morhua* L., in a simulated farm situation. *Aquaculture Research*, 40, 26–34.

Henderson, J. N., and Letcher, B. H. 2003. Predation on stocked Atlantic salmon (*Salmo salar*) fry. *Canadian Journal of Fisheries and Aquatic Sciences*, 60, 32–42.

Hosoya, S., Kaneko, T., Suzuki, Y., and Hino, A. 2008. Individual variations in behavior and free cortisol responses to acute stress in tiger pufferfish *Takifugu rubripes*. *Fisheries Science*, 74, 755–763.

Huntingford, F. A. 2004. Implications of domestication and rearing conditions for the behaviour of cultivated fishes. *Journal of Fish Biology*, 65A, 122–142.

Huntingford, F. A., and Garcia de Leaniz, C. 1997. Social dominance, prior residence, and the acquisition of profitable feeding sites in juvenile Atlantic salmon, *Salmo salar*. *Journal of Fish Biology*, 51, 1009–1014.

Huntingford, F. A., Adams, C. E., Braithwaite, V. A., Kadri, S., Pottinger, T. G., Sandoe, P., and Turnbull, J. F. 2006. Current issues in fish welfare. *Journal of Fish Biology*, 68, 332–372.

Huntingford, F. A, Andrew, G., MacKenzie, S., Morera, D., Coyle, S. M., Pilarczyk, M., and Kadri, S. 2010. Coping strategies in a strongly schooling fish, the common carp (*Cyprinus carpio* L.). *Journal of Fish Biology*, 76, 1576–1591.

Kelley, J. L., Magurran, A. E., and Garcia, C. M. 2006. Captive breeding promotes aggression in an endangered Mexican fish. *Biological Conservation*, 133, 169–177.

Koolhaas, J. M., de Boer, S. F., Buwalda, B., and Van Reenen, K. 2007. Individual variation in coping with stress: a multidimensional approach of ultimate and proximate mechanisms. *Brain, Behavior, and Evolution*, 70, 218–226.

Korte, S. M., Koolhaas, J. M., Wingfield, J. C., and McEwen, B. S. 2005. The Darwinian concept of stress: benefits of allostasis and costs of allostatic load and the trade-offs in health and disease. *Neuroscience and Biobehavioral Reviews*, 29, 3–38.

Laakkonen, M. V. M., and Hirvonen, H. 2007. Is boldness towards predators related to growth rate in naive captive-reared Arctic char (*Salvelinus alpinus*)? *Canadian Journal of Fisheries and Aquatic Sciences*, 64, 665–671.

Lee, J. S. F., and Bereijikian, B. S. 2008. Stability of behavioral syndromes but plasticity in individual behavior: consequences for rockfish stock enhancement. *Environmental Biology of Fish*, 82, 179–186.

Lucas, M. D., Drew, R. E., Wheeler, P. A., Verrell, P. A., and Thorgaard, G. H. 2004. Behavioral differences among rainbow trout clonal lines. *Behavior Genetics*, 34, 355–365.

MacKenzie, S., Ribas, L., Pilarczyk, M., Capdevila, D. M., Kadri, S., and Huntingford, F. A. 2009. Screening for coping style increases the power of gene expression studies. *PLoS One*, 4, e5314.

Magnhagen, C. 2006. Risk-taking behaviour in foraging young-of-the-year perch varies with population size structure. *Oecologia*, 147, 734–743.

———. 2007. Social influence on the correlation between behaviours in young-of-the-year perch. *Behavioral Ecology and Sociobiology*, 61, 525–531.

Magnhagen, C., and Borcherding, J. 2008. Risk-taking behaviour in foraging perch: does predation pressure influence age-specific boldness? *Animal Behaviour*, 75, 509–517.

Marks, C., West, T. N., Bagatto, B., and Moore, F. B. G. 2005. Developmental environment alters conditional aggression in zebrafish. *Copeia*, 4, 901–908.

Martins, C. I. M., Hillen, B., Schrama, J. W., Verreth, J. A. J., and Johan, A. J. 2008. A brief note on the relationship between residual feed intake and aggression behaviour in juveniles of African catfish *Clarias gariepinus*. *Applied Animal Behaviour Science*, 111, 408–413.

McNeil, W. J. 1991. Expansion of cultured Pacific salmon into marine ecosystems. *Aquaculture*, 98, 173–183.

Melard, C., Kestemont, P., and Grignard, J. C. 1996. Intensive culture of juvenile and adult Eurasian perch (*P. fluviatilis*): effects of major biotic and abiotic factors on growth. *Journal of Applied Ichthyology*, 12, 175–180.

Mesa, M. G. 1991. Variation in feeding, aggression, and position choice between hatchery and wild cutthroat trout in an artificial stream. *Transactions of the American Fisheries Society*, 120, 723–727.

Mesquita, F. de O. 2010. Copying styles and learning in fish: developing behavioral tools for welfare-friendly aquaculture. PhD diss., University of Glasgow.

Mesquita, F. de O., and Young, R. J. 2007. The behavioural response of Nile tilapia (*Oreochromis niloticus*) to antipredator training. *Applied Animal Behaviour Science*, 106, 144–154.

Millot, S., Begout, M.-L., and Chatain, B. 2009. Exploration behaviour and flight response towards a stimulus in three sea bass strains (*Dicentrarchus labrax* L.). *Applied Animal Behaviour Science*, 119, 108–114.

Moretz, J. A., Martins, E. P., and Robison, B. D. 2007a. Behavioral syndromes and the evolution of correlated behavior in zebrafish. *Behavioral Ecology*, 18, 556–562.

———. 2007b. The effects of early and adult social environment on zebrafish (*Danio rerio*) behavior. *Environmental Biology of Fishes*, 80, 91–101.

Moyle, P. B. 1969. Comparative behavior of young brook trout of domestic and wild origin. *Progressive Fish Culturist*, 31, 51–58.

Olla, B. L., Davis, M. W., and Ryer, C. H. 1998. Understanding how the hatchery environment represses or promotes the development of behavioural survival skills. *Bulletin of Marine Science*, 62, 531–550.

Øverli, Ø., Winberg, S., and Pottinger, T. G. 2005. Behavioural and neuroendocrine correlates of stress responsiveness in rainbow trout: a review. *Integrative and Comparative Biology*, 45, 463–474.

Petersson, E., and Järvi, T. 2006. Antipredator response in wild and sea-ranched brown trout and their crosses. *Aquaculture*, 253, 218–228.

Pottinger, T. G. 2006. Context dependent differences in growth of two rainbow trout (*Oncorhynchus mykiss*) lines selected for divergent stress responsiveness. *Aquaculture*, 256, 140–147.

Pottinger, T. G., and Carrick, T. R. 1999. Modification of the plasma cortisol response to stress in rainbow trout by selective breeding. *General and Comparative Endocrinology*, 116, 122–132.

Pottinger, T. G., and Carrick, T. R. 2001. Stress responsiveness affects dominant-subordinate relationships in rainbow trout. *Hormones and Behavior*, 40, 419–427.

Riesch, R., Duwe, V., Herrmann, N., Padur, L., Ramm, A., Scharnweber, K., Schulte, M., Schulz-Mirbach, T., Ziege, M., and Plath, M. 2009. Variation along the shy-bold continuum in extremophile fishes (*Poecilia mexicana, Poecilia sulphuraria*). *Behavioral Ecology and Sociobiology*, 63, 1515–1526.

Robinson, B. W., and Doyle, R. W. 1990. Phenotypic correlations among behavior and growth variables in tilapia: implications for domestication selection. *Aquaculture*, 85, 177–186.

Robison, B. D., and Rowland, W. 2005. A potential model system for studying the genetics of domestication: behavioral variation among wild and domesticated strains of zebrafish (*Danio rerio*). *Canadian Journal of Fisheries and Aquatic Sciences*, 62, 2046–2054.

Ruiz-Gomez, M. de L., Kittilsen, S., Höglund, E., Huntingford, F. A., Sørensen, C., Pottinger, T. G., Bakken, M., Winberg, S., Korzan, W. J., and Ïverli, Ï. 2008. Behavioral plasticity in rainbow trout (*Oncorhynchus mykiss*) with divergent coping styles: when doves become hawks. *Hormones and Behavior*, 54, 534–538.

Ruiz-Gomez, M. de L., Huntingford, F. A, Øverli, Ø., Thörnquist, P. O., and Höglund, E. 2011. Response to environmental change in the rainbow trout selected for divergent stress copying styles. *Physiology and Behavior*, 102, 317–322.

Ruzzante, D. E. 1994. Domestication effects on aggressive and schooling behavior in fish. *Aquaculture*, 120, 1–24.

Ruzzante, D. E., and Doyle, R. W. 1991. Rapid behavioral changes in Medaka (*Oryzias latipes*) caused by selection for competitive and noncompetitive growth. *Evolution*, 45, 1963–1946.

Salonen, A., and Peuhkuri, N. 2006. The effect of captive breeding on aggressive behaviour of European grayling, *Thymallus thymallus*, in different contexts. *Animal Behaviour*, 72, 819–825.

Salvanes, A. G. V., and Braithwaite, V. A. 2005. Exposure to variable spatial information in the early rearing environment generates asymmetries in aggressive and social interactions in cod. *Behavioral Ecology and Sociobiology*, 59, 250–257.

Salvanes, A. G. V., and Braithwaite, V. 2006. The need to understand the behaviour of fish reared for mariculture or restocking. *ICES Journal of Marine Science*, 63, 346–354.

Schjolden, J., and Winberg, S. 2007. Genetically determined variation in stress responsiveness in rainbow trout: behavior and neurobiology. *Brain, Behavior, and Evolution*, 70, 227–238.

Schjolden, J., Stoskhus, A., and Winberg, S. 2005. Does individual variation in stress responses and agonistic behavior reflect divergent stress coping strategies in juvenile rainbow trout? *Physiological and Biochemical Zoology*, 78, 715–723.

Sih, A., Bell, A. M., Johnson, J. C., and Ziemba, R. E. 2004. Behavioral syndromes: an integrative overview. *Quarterly Review of Biology*, 79, 241–277.

Shimizu, D., Sakiyama, K., and Takahashi, Y. I. 2006. Predation of stocked hatchery-reared juveniles of ocellate puffer *Takifugu rubripes* in salt pond mesocosm. *Nippon Suisan Gakkaishi*, 72, 886–893.

Smith, K. L., Miner, J. G., Wiegmann, D. D., and Newman, S. P. 2008. Individual differences in exploratory and antipredator behaviour in juvenile smallmouth bass (*Micropterus dolomieu*). *Behaviour*, 146, 283–294.

Staffan, F., Magnhagen, C., and Alanara, A. 2005. Individual feeding success of juvenile perch is consistent over time in aquaria and under farming conditions. *Journal of Fish Biology*, 66, 798–809.

Stamps, J. A. 2007. Growth-mortality tradeoffs and "personality traits" in animals. *Ecology Letters*, 10, 355–363.

Sundström, L. F., Lohmus, M., and Johnsson, J. I. 2003. Investment in territorial defence depends on rearing environment in brown trout (*Salmo trutta*). *Behavioral Ecology and Sociobiology*, 54, 249–255.

Sundström, L. F., Petersson, E., Höjesjö, J., Johnsson, J. I., and Järvi, T. 2004. Hatchery selection promotes boldness in newly hatched brown trout (*Salmo trutta*): implications for dominance. *Behavioral Ecology*, 15, 192–198.

Sundström, L. F., Petersson, E., Johnsson, J. I., Dannewitz, J., Höjesjö, J., and Järvi, T. 2005. Heart rate responses to predation risk in *Salmo trutta* are affected by the rearing environment. *Journal of Fish Biology*, 67, 1280–1286.

Van de Nieuwegiessen, P. G. 2009. Welfare of African catfish: effects of stocking density. PhD diss., Wageningen University, the Netherlands.

Van Raaij, M. T. M., Pit, D. S. S., Balm, P. H. M., Steffens, A. B., and Van den Thillart, G. E. E. J. M. 1996. Behavioral strategy and the physiological stress response in rainbow trout exposed to severe hypoxia. *Hormones and Behavior*, 30, 85–92.

Vaz-Serrano, J., Ruiz-Gomez, M. L., Gjøen, H. M., Skov, P. V., Huntingford, F. A., Øverli, Ø., and Höglund, E. 2011. Consistent boldness behaviour in early emerging fry of domesticated Atlantic salmon (*Salmo salar*): decoupling of behavioural and physiological traits of the proactive stress coping style. *Physiology and Behaviour*, 103, 359–364.

Ward, A. J. W., Thomas, P., Hart, P. J. B., and Krause, J. 2004. Correlates of boldness in three-spined sticklebacks (*Gasterosteus aculeatus*). *Behavioral Ecology and Sociobiology*, 55, 561–568.

Weber, E. D., and Fausch, K. D. 2003. Interactions between hatchery and wild salmonids in streams: differences in biology and evidence for competition. *Canadian Journal of Fisheries and Aquatic Sciences*, 60, 1018–1036.

Westerberg, M., Staffan, F., and Magnhagen, C. 2004. Influence of predation risk on individual competitive ability and growth in Eurasian perch, *Perca fluviatilis*. *Animal Behaviour*, 67, 273–279.

Wilson, D. S, Clark, A. B., Coleman, K., and Derastyne, T. 1994. Shyness and boldness in humans and other animals. *Trends in Ecology and Evolution*, 9, 442–446.

Wright, D., Butlin, R. K., and Carlborg, O. 2006b. Epistatic regulation of behavioral and morphological traits in the zebrafish (*Danio rerio*). *Behavior Genetics*, 36, 914–922.

Wright, D., Nakamichi, R., Krause, J., and Butlin, R. K. 2006a. QTL analysis of behavioral and morphological differentiation between wild and laboratory zebrafish (*Danio rerio*). *Behavior Genetics*, 36, 271–284.

Wright, D., Rimmer, L. B., Pritchard, V. L., Krause, J., and Butlin, R. K. 2003. Inter- and intrapopulation variation in shoaling and boldness in the zebrafish (*Danio rerio*). *Naturwissenshaften*, 90, 374–378.

Yamamoto, T., and Reinhardt, U. G. 2003. Dominance and predator avoidance in domesticated and wild Masu salmon *Oncorhynchus masou*. *Fisheries Science*, 69, 88–94.

15

Behavioral, Physiological, and

Health Biases in Laboratory Rodents

A Basis for Understanding Mechanistic Links

between Human Personality and Health

SONIA A. CAVIGELLI, KERRY C. MICHAEL, AND CHRISTINA M. RAGAN

Introduction

Animals provide a powerful lens for investigating physiological mechanisms underlying the development of disease. In this chapter, we review the range of behavioral biases (i.e., personality) seen among different laboratory rodent strains to determine whether these biases are systematically associated with physiological and/or disease biases and the extent to which these relationships provide insight into the relationship between human personality, physiology, and disease resistance/susceptibility. Toward this goal, we have conducted a broad review of well-established research rodent strains (mice and rats)—reviewing behavioral, physiological, and health differences among these strains—to identify systematic relationships between personality and health. We have conducted the same review for recently established rodent strains that have been bred specifically for their behavioral traits. A secondary goal of this review is to provide a comprehensive comparison of behavioral, physiological, and health biases across multiple rodent strains to provide a resource for others selecting appropriate strains for their specific research questions. Finally, we compare basic relationships between behavior, physiology, and health in laboratory rodents to what is known on the relationship between human personality and health.

Personality, physiology, and health

In humans, we often study how personality affects health through differential health maintenance behaviors (e.g., diet, drug use, medical visits, and exercise). However, personality may also influence health as a result of differential physiological profiles associated with personality that make individuals more or less susceptible to specific disease processes (see figure 15.1, and Smith and McKenzie 2006). Although personality can be difficult to quantify, there have been many studies linking certain traits to physiology and health. For example, hostility is probably the best-established human behavioral/cognitive trait associated with specific and related physiological and health outcome biases. Early work focused on Type A personality characteristics (hard driving, competitive, and angry) and their association with cardiovascular disease (CVD). Subsequent studies in the 1970s suggested

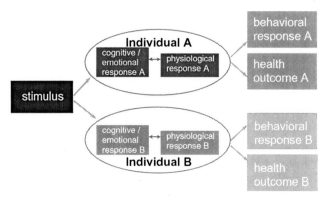

Figure 15.1. A schematic on how physiological processes can underlie systematic relationships between behavioral response biases (temperament/personality) and health outcomes. In the following schematic, two individuals (A and B) have different cognitive or emotional responses to the same stimulus which triggers or is triggered by different physiological responses to this same stimulus in the two individuals (e.g., neuroendocrine, immune, and cardiovascular responses). For example, a fearful individual may have a heightened physiological stress response to an unfamiliar dog whereas a less fearful individual may show no such response to the same dog. These two different physiological responses to the same stimulus will support different behavioral responses to the stimulus— for example, avoidance vs. interaction with the dog. If these two individuals regularly respond to stimuli with these different suites of emotional, cognitive, physiological, and behavioral responses, the assumption is that repeated differential physiological responses can have long-term differential influences on the progression of diseases, such as cardiovascular disease, cancer, and diabetes. This model may be thought of as a combination of the "interactional stress moderation" and the "constitutional predisposition" model of how personality and disease are linked (reviewed in Smith and MacKenzie 2006).

that hostility was the most toxic component of the Type A construct (e.g., Friedman and Rosenman 1971; Williams et al. 1980). For example, recent studies show that hostile/angry individuals exhibit a wide range of coronary risk factors, including platelet activation, elevated night-time blood pressure, elevated triglyceride levels, lowered high-density lipoprotein levels, and elevated pro-inflammatory cytokine and C-reactive protein levels (Niaura et al. 2000; Suarez et al. 2002; Suarez 2003; 2004; Thomas et al. 2004; Bunde and Suls 2006; Linden et al. 2008; Shimbo et al. 2009). These biological biases can clearly influence the progression of cardiovascular disease. For example, blood pressure normally "dips" during the night, and this dip helps protect against target organ damage (Routledge et al. 2007). A lack of this night-time dip in blood pressure is one physiological mechanism that may account for the 2–3 fold increase in CVD morbidity and mortality in hostile individuals (reviewed in Everson-Rose and Lewis 2005). In addition to hostility, other personality traits have been linked to specific physiological biases associated with health (Spangler 1997; Habra et al. 2003; Dawe and Loxton 2004). For example, shy or behaviorally inhibited children had elevated cortisol levels and decreased heart rate variability compared with noninhibited children (Kagan et al. 1987; Kagan and Snidman 1991), and optimistic young adults had greater T-cell counts in circulation and greater natural killer cell cytotoxicity following a stressor compared with nonoptimistic adults (Segerstrom et al. 1998).

Many of the links between human personality and physiological function have focused on stress-related processes (e.g., glucocorticoid production and sympathetic/cardiovascular activity; see chapter 12), and thus we focus on these physiological processes in our review of the data on laboratory rodents. Given the established relationships between behavioral traits and physiological functioning, it is reasonable to hypothesize that consistent, even low-grade, physiological biases associated with personality can have long-lasting influences on the development and progression of diseases, particularly chronic or slow-to-manifest diseases (e.g., cancer, Segerstrom 2003).

In animals, personality-linked behavioral traits have also been associated with specific physiological biases, and to a lesser extent with health outcome biases (Gentsch et al. 1982; Walker et al. 1989; Sandi et al. 1991, Capitanio et al. 1999, 2004; Kavelaars et al. 1999; Koolhaas et al. 1999; Laudenslager et al. 1999; Ruis et al. 2000; Maninger et al. 2003; Cavigelli and McClintock 2003; Carere and Van Oers 2004; Tõnissaar et al. 2004; Coppens et al. 2010; chapter 12). Several of these studies link behavioral traits to differential stress and immune physiological processes and thus provide clear physiological mechanisms by which personality traits can influence health and disease processes. Given the fact that some of the behavioral traits identified in laboratory rodents are similar to traits identified in humans (e.g., aggression and behavioral inhibition), and that physiological processes can be more completely elucidated in animals than in humans, the above array of studies indicate strong potential for application of results with animal models to understanding differential behavior-linked disease susceptibility in humans. Animal research provides an opportunity to examine how personality-linked behavioral traits associate with physiological traits, and how these physiological biases may influence disease progression and health, while holding environment and health-related behaviors relatively constant.

Benefits and limitations of laboratory animal studies

There are several reasons to consider animal research on personality, physiology, and health as a means to better understand human personality and health. Animal studies have some unique attributes that complement human research (Cavigelli 2005). For example, with laboratory animals, systematic and detailed observations of individual behavioral response biases are easily conducted and can be repeated over time and across different environmental contexts to provide systematic, objective, and reliable tests of temperament and trait stability. In addition, animals can be bred according to their behavioral traits (e.g., Driscoll and Battig 1982; Gariépy et al. 2001; Drent et al. 2003; Landgraf 2003; Sluyter et al. 2003; Van Oers and Mueller 2010), thus producing research animals with similar personalities to allow investigations of the interaction between personality, physiology, and environmental conditions on health outcomes. Finally, more invasive and controlled experimental studies can be conducted with laboratory rodents to test the causal relationships and developmental processes linking personality, physiological biases, and environmental conditions to health and disease outcomes. Of course, there are necessary limitations to applying animal research models to understand human personality and health. Determining behavioral analogs between humans and animals is always tricky, and there is no guarantee that these analogs are accurate. In addition, specific rodent strains may be susceptible to certain disease processes as an artifact of the artificial selection involved in strain maintenance. Genetic drift, for example, may arbitrarily link genes associated with a specific disease process and genes associated with a specific behavioral bias, without any underlying physiological mechanism to explain the relation between behavior and health. This same selection process may also make certain laboratory rodents susceptible to certain disease processes that are not regularly seen in humans, thus making them poor models for this kind of work. The potential for rodent studies in this arena grows when behavioral traits in an animal are correlated with physiological biases that are comparable with those seen in humans with similar behavioral traits (e.g., Cavigelli et al. 2007).

Methods used to compare rodent strains

Laboratory researchers select from a variety of well-established rodent strains, many of which were developed well over 50 years ago (referred to as *progenitor strains* here). These progenitor strains were bred based on a variety of characteristics, primarily nonbehavioral traits. However, they are known to exhibit behavioral differences, along with physiological and/or disease biases, and the amount of data on these strains is very large because of the amount of time that they have been available for research and the frequency with which certain strains have been selected for research. However, the relationships among behavioral and physiological/health biases in these strains have not been explicitly reviewed, leaving the possibility that systematic relationships among their behavioral, physiological, and health profiles exist and could inform us about such relationships in other species. Given the great abundance of types and data on rodent strains, we compare several strains to determine whether there are systematic relationships between behavior and physiology and whether these relationships provide physiological mechanisms by which behavioral personality traits associate with resilience or susceptibility to disease. These comparisons were done with a set of the most popular progenitor strains of mice and rats: A/J, BALB/c, C3H/He, C57BL/6, and DBA/2 mice; Fischer 344, Lewis, Brown Norway, Wistar, and Sprague-Dawley rats (see table 15.1).

For each strain we review (1) behavioral responses to a set of common laboratory behavioral tests, (2) information on a subset of physiological processes, and (3) information on resilience and susceptibility to specific disease processes. For behavioral comparisons, we focused on measures from six different established behavioral tests to provide information on activity, anxiety-like behavior, spatial cognition, and aggression. These measures provide a relatively wide array of behavioral categories, and they represent some of the most commonly used laboratory measures and therefore have been tested across a variety of rodent strains, which allows for cross-strain comparisons. In addition to these six tests, we also reviewed responses to two tests that are thought to measure depression/learned helplessness and

Table 15.1. Comparison of behavior, physiology, and health profiles among select mice and rat progenitor strains. (See key at bottom for explanation of symbols.)

Strain	Behavior				Physiology							Immunity	Health Profile
					Glucocorticoids				Metabolism				
	Activ.	Anxiety	Spatial Cog.	Agg.	Basal CORT	Reactive CORT	Heart Rate	Systol. BP	Metab. Rate	Body Weight	Fat		
MICE													
A/J	−	+ / −	−	?	+	−	i	−	−	−	+	+ NK	Short life span Hyperreactive airways Susceptible to diet-induced metabolic syndrome Resistant to type 2 diabetes Lung tumors
BALB/c	− / i	+	+ / −	+	+	+	−	i	+	+	− / i	i / − NK + T + Th2	Intermediate life span Susceptible to induced rheumatoid arthritis Resistant to cerebral malaria Susceptible to herpes simplex virus Lung, kidney, mammary tumors
C3H/He	+	−	− / i	−	−	/i	+	−	− / i	+	i	− NK − B -innate	Intermediate life span Susceptible to induced arthritis/autoimmunity Liver, mammary, leukocyte tumors
C57BL/6	+	−	+	−?	−	− / i	+ / i	i	i	i	− / i	+ NK + Th1	Long life span Susceptible to type 2 diabetes Susceptible to cerebral malaria Resistant to herpes simplex virus Leukocyte, pituitary tumors
DBA/2	+	+	+	+	+	i (long)	i	+ / i		+	+	i NK − T + B	Intermediate life span Seizure-prone Resistant to induced arthritis/autoimmunity Liver, mammary tumors

| Strain | Behavior | | | | Physiology | | | | | | | Immunity | Health Profile |
| | | | Spatial | | Glucocorticoids | | | | --- Metabolism --- | Body | | | |
	Activ.	Anxiety	Cog.	Agg.	Basal CORT	Reactive CORT	Heart Rate	Systol. BP	Metab. Rate	Weight	Fat		
RATS													
Fisher (F344)	−	+ / i	+ / −	−	+	+ (long) + MR	++	+		−	− / +		Intermediate life span Stress ulcers Resistant to autoimmunity/arthritis Cardiomyopathy Mammary, pituitary, testicular tumors
Lewis (LEW)	+ / i	+ / i		−	−	−	−	−		+ / i	+		Long life span Susceptible to Th autoimmunity Susceptible to induced type I diabetes
Brown-Norway (BN)	+ / i	+ / −	+	+ / i	i?	i				−			Long life span Hyperreactive airways Resistant to Th autoimmunity Susceptible to allergies, asthma
Wistar	+	+	−		+		−	−		++			Intermediate life span Resistant to autoimmunity Leukocyte tumors
Sprague-Dawley (SD)	+ / i	+	+		− / i	i	−	−		+	+ / i		Short life span Susceptible to induced arthritis Mammary tumors

Key:

+: greater activity than other strains within same species

−: less activity than other strains within species

i: intermediate activity compared to other strains of same species

NK: natural killer **T**-cells, **T:** T-cells, **B:** B-cells, **Th1/2**: T-helper cells 1/2

MR: mineralocorticoid receptors

coping styles because of the obvious health relevance of these traits. These measures were reviewed in a limited number of strains that had enough comparative data for meaningful interpretation. In the review of physiological profiles for each strain, we focused on neuroendocrine, metabolic, and immune function. Specifically, we compared mean basal and reactive corticosterone (CORT) production, heat rate, systolic blood pressure (BP), metabolic rate, body weight, body fat, and a variety of immune measures (see table 15.1). For an accurate comparison among strains, we selected studies that compared behavioral and/or physiological profiles across more than one progenitor strain—ideally 3 or more (this, however, was not always possible, and thus our comparisons are sometimes based on only two strains.) Because of the vast literature to summarize, we relied on review articles and comparative databases like the Mouse Tumor Biology Database (Krupke et al. 2008) and the Mouse Phenome Database (Grubb et al. 2009). Further, we restricted our description of the many rodent strains to general information on their behavior, physiology, and disease biases. The goal was to identify general personality profiles among strains and to identify global systematic relationships between behavior, physiology, and potentially disease profiles.

In addition to these progenitor strains, we reviewed rodent strains recently identified according to their behavioral responses to specific environmental conditions. There is less information available on these strains, given their recent development, but they are important in understanding how personality traits associate with physiological biases that may predispose individuals to differential health outcomes.

Behavioral tests used to survey rodent progenitor strains

The most common laboratory rodents (mice and rats) are crepuscular and/or nocturnal prey species and thus display behavioral characteristics specific to this ecological niche. For example, rodents avoid brightly lit and open areas (Russell 1979) and display a behavior referred to as *thigmotaxis*—a tendency to remain close to vertical/peripheral objects instead of in open spaces, presumably to reduce the possibility of predation (Grossen and Kelly 1972). These behaviors are adaptive in the natural environment and, in the laboratory, variance in these behaviors among individuals can provide indicators of differential cognitive or emotional processing, and personality or coping styles. Our review of common laboratory behavioral tests includes a short description of the test environment, behavioral measures collected in the environment, and an interpretation of what these behavioral measures are thought to reflect about individual cognitive and emotional processing, personality, and/or coping (see also Crawley and Paylor 1997; Crawley 2003).

ACTIVITY/EXPLORATION: OPEN FIELD (OF)

The open field test presents a mildly noxious situation for a rodent. The open space of the test arena is unprotected, which, in the natural environment would leave a rodent open to overhead predator attacks. On the other hand, for opportunistic foraging species, the space may also hold undiscovered resources. Most researchers test rodents on a brightly lit OF, though some test in low (or red) light. A common behavioral measure of overall activity is distance traveled in this arena. This activity/locomotion measure includes thigmotactic locomotion, and thus is not used to index exploration or boldness. Researchers index fear responses using the relative amount of time an animal spends near the peripheral wall during a testing session (i.e., thigmotaxis) and the amount of defecation in the arena. An animal that is highly active in this environment, but spends all its time near the peripheral walls and/or defecates frequently, is thought to show a high degree of anxiety-related behavior, and this interpretation is supported by the fact that anxiolytic drugs decrease defecation and thigmotaxis (mice: Simon et al. 1994; rats: Hall 1934, Van der Staay et al. 1990).

FEAR/ANXIETY-RELATED BEHAVIOR: ELEVATED PLUS MAZE (EPM) AND LIGHT/DARK BOX (LD)

In addition to thigmotaxis and defecation in the OF, the two most common tests of anxiety-related behavior in laboratory rodents are the elevated plus maze and the light/dark box. Both arenas include open (light) and closed (dark) areas, and the relative frequency and duration of visits to the closed vs. open areas is used to index anxiety-related behavior—with more transitions into or time spent in the closed vs. open areas indicative of a greater fear/anxiety–related response (Pellow et al. 1985; Lister 1987; Costall et al. 1989; Carobrez and Bertoglio 2005). Again, this interpretation is supported by anxiogenic and anxiolytic drugs' influences on these behaviors, although there is still much variance in the validity of these measures (Crawley and Goodwin 1980; Crawley 1981; see Hogg 1996 for full review).

The elevated plus maze draws on rodents' preference for protected areas and avoidance of heights. The maze consists of four narrow runways elevated two to three feet off the floor and oriented 90 degrees from one another in a "plus" formation. Two of the runways have high walls ("closed"), providing a protected space; the other two runways have no walls ("open"), leaving the animal unprotected and elevated above the floor. Usually rodents are tested for 5 minutes on this arena, and frequency and duration of entries into each runway is measured.

The light/dark box consists of two chambers with a small hole between them, allowing a test animal access between chambers. One chamber is brightly lit, while the other is opaque providing a dark environment. Researchers compare frequency and duration of visits in the light and dark chambers.

SPATIAL COGNITION: BARNES MAZE AND MORRIS WATER MAZE

The Barnes maze is an open, brightly lit elevated surface with holes in the floor around the edge of the arena where rodents tend to investigate. One of the holes leads to a box that the rat can crawl into and escape from the open and lit surface. Spatial cognition is judged by how quickly they return to the box on repeated test trials (escape latency, Barnes 1979). Mice do not readily enter holes in the floor, thus, a mouse adaptation to the Barnes maze uses multiple holes in the walls of the maze, one of which lead to the home cage (Koopmans et al. 2003). The Morris water maze involves measuring a rodent's spatial cognition/memory in a challenging situation. The maze consists of a large circular tub filled with water that is mixed with an opaque additive so objects below the surface are not visible. Submerged in the water is a small platform where an animal can stand without swimming. Spatial cognition is measured by how quickly the animal finds the submerged platform (platform latency) after several trials of maze exposure and how much time they spend in the correct quadrant of the arena during a final probe trial (Morris 1984).

AGGRESSION: RESIDENT/INTRUDER TEST

Rats and mice are territorial and will attack unfamiliar intruders in their home territory. The resident-intruder test involves introducing a same-sex intruder into the home cage of the test animal and measuring latency to attack the intruder, as well as the intensity, frequency, and duration of attacks in a given time period (Miczek 1979; Jones and Brain 1987).

DEPRESSION/LEARNED HELPLESSNESS AND COPING STYLES

Given the importance of learned helplessness behavior and proactive vs. reactive coping styles in health trajectories, the following two tests of depression or learned helplessness were reviewed for those strains in which adequate data were available. The forced swim test (FST) involves placing a rodent in water and timing how long it continues to swim. This is thought to be a potent stressor, and most rodents will swim until they are no longer in danger. A short latency to stop swimming (i.e., "giving up") is thought to indicate a high degree of learned helplessness, given that these rats will stop swimming and sink long before their energy stores are exhausted (Porsolt et al. 1977). Learned helplessness is considered an animal model of human major depression. The <u>defensive burying test (DBT)</u> takes advantage of the fact that burying a threatening object is an innate response in rodents. The DBT is a measure of proactive vs. passive coping (Pinel and Treit 1978). This test generally involves a shock-delivering probe, though any noxious stimulus may be used (summarized in de Boer and Koolhaas 2003). Behavioral responses can vary widely, ranging from freezing (passive coping) to defensive burying of the noxious stimulus (active coping). The test is often used to assess the behavioral effects of anxiolytic drugs.

Mice: progenitor strains

Some of the mouse strains reviewed in the following section were developed almost 100 years ago, and their inbreeding was conducted according to specific traits other than behavioral characteristics (e.g., tumor susceptibility; Paigen 2003). Because Jackson Laboratories is a major breeder and distributor of these strains, our review focuses on mice acquired from Jackson labs (indicated with a "J" at the end of the strain name; Staats 1980).

A/J

A/J mice are unique in their very low activity, low spatial abilities, elevated basal glucocorticoid production, low metabolic activity, short life span, and susceptibility to lung and metabolic problems. A/J mice were not very active in the OF and EPM (Trullas and Skolnick 1993; Brown et al. 2004a; Bothe et al. 2005) and had mixed responses to anxiety-related tests. They showed high amounts of anxiety-related behaviors in the OF (thigmotaxis, freezing, and fecal boli), but low levels in the EPM and LD (i.e., proportion time in the closed and dark areas of these test arenas, Wahlsten and Crabbe 2003; Brown et al. 2004a; Trullas and Skolnick 1993). In the Barnes and Morris tests, these mice performed quite poorly—i.e., they took a long time to escape and to reach the hidden platform during acquisition and reversal training, and spent little time in the correct quadrant of the test arena during a final probe trial (Brown et al. 2004b). This strain also showed impaired cued fear memory 24 hours after training, possibly explained by reduced amygdalar long-term potentiation (LTP) involved in learning and memory function (Schimanski and Nguyen 2005). Finally, A/Js spent little time engaged in nonaggressive social interactions when placed in a novel cage with a novel social partner (Bolivar et al. 2007). In line with their very low activity levels, the A/J mice are characterized by relatively low metabolic activity. They have intermediate to low body weights with a relatively high proportion of fat (Kitten et al. 2003; Reed et al. 2007). They have intermediate heart rates with low systolic blood pressure and low daily respiratory exchange ratios (Hampton et al. 2001; Seburn 2001; Kitten et al. 2003; Deschepper and Gallo-Payet 2004; Sugiyama et al. 2007). Compared with the other strains, they have elevated basal corticosterone levels, with low corticosterone responses to uncontrollable foot shock, and intermediate to high adrenal gland volume (Shanks et al. 1990; Trullas and Skolnick 1993; Deschepper et al. 2004). Males have intermediate to high natural killer cell counts in the liver (Rymarchyk et al. 2008). The A/J median life span is the shortest of the 6 strains (median: 20–22 and 21–23 months for males and females respectively, Goodrick 1975; Paigen and Yuan 2007). Symptoms of metabolic syndrome (e.g., weight gain, hyperleptinemia, hypercholesterolemia, impaired glucose tolerance, and insulin resistance) can be induced in these mice when they are exposed to the equivalent of a Western high-fat diet (Gallou-Kabani et al. 2007). However, in these conditions, and despite their slow metabolism, they remain normoglycemic, and therefore appear relatively resistant to diet-induced type 2 diabetes. They are often used to model hyperreactive airways and have a high susceptibility to spontaneous lung tumors (Näf et al. 2002).

BALB/C

The BALB/c mice have high anxiety-like responses, mixed spatial performance, high aggression, elevated basal and reactive glucocorticoid production, relatively high metabolic function, decreased natural killer cell counts, and susceptibility to herpes simplex virus and tumors in multiple locations. They showed low to intermediate activity in the OF and EPM (Trullas and Skolnick 1993; Brown et al. 2004a; Bothe et al. 2005), and relatively high freezing and fecal production in the OF. They also spent a good deal of time in the dark portion of the LD and mixed amounts of time in the closed arms of the EPM (Trullas and Skolnick 1993; Yilmazer-Hanke et al. 2003; Brown et al. 2004a; Brooks et al. 2005). BALB/cJ cognitive abilities differed across spatial tests: they performed relatively well on the Barnes test and poorly in the Morris water maze (Zilles et al. 2000; Brown et al. 2004b; Brooks et al. 2005). This test-specific behavior across the two spatial tests highlights the importance of using multiple testing paradigms to identify generalized behavioral traits. Like the A/J strain, BALB/cJs also had impaired cued fear conditioning and reduced LTP in the amygdala, suggesting learning and memory deficits (Schimanski and Nguyen, 2005). Finally, in the resident-intruder paradigm, BALB/c mice showed an intermediate attack rate with a high degree of overall aggression, indexed by a composite measure of attack frequency, latency, and intensity (Jones and Brain 1987). BALB/cJ mice are relatively large with an intermediate to low proportion of body fat (Kitten et al. 2003) and relatively high metabolic activity. BALB/cByJ mice have very low heart rates, intermediate systolic blood pressure, and a high daily respiratory exchange ratio (Hampton et al. 2001; Seburn 2001; Kitten et al. 2003; Deschepper et al. 2004). They have relatively high basal circulating corticosterone levels and are marked by a very high adrenal response to environmental stress (restraint with BALB/c: Trullas and Skolnick 1993; Harizi et al. 2007; uncontrollable footshock with BALB/cByJ: Shanks et al. 1990) and large adrenal glands (BALB/cJ: Trullas and Skolnick 1993; BALB/cByJ strain, Deschepper et al. 2004). These mice have an intermediate to low proportion of natural killer cells in the liver (Rymarchyk et al. 2008), a high proportion of CD4 and CD8 T cells, and a low proportion of granulocytes in the spleen (Nicholson 2006). They also have relatively high immediate IL-6 responses to restraint stress (Harizi et al. 2007), and they develop a Th2-biased response to *Leishmania major* and suffer from increased mortality from this infection compared with C57BL/6J mice (Scharton-Kersten and Scott 1995). These mice have an intermediate life span (median: 22 and 27 months for males and females respectively; BALB/cJ: Goodrick 1975; BALB/cByJ: Paigen and Yuan 2007). BALB/cJ are prone to chronic pneumonia, arteriosclerosis, and lung, kidney, and mammary tumors (Green 1966; Näf et al. 2002; Szpirer and Szpirer 2007). In addition, they are susceptible to proteoglycan-induced arthritis, a rodent model of rheumatoid arthritis induced by injection of cartilage proteoglycans (Glant et al. 2003).

C3H/HE

These mice are marked by their high activity, low anxiety, and aggression, relatively poor spatial skills, low glucocorticoid production, low metabolic rate, and increased susceptibility to autoimmunity and tumors in multiple locations. In the OF, EPM, and LD they were quite active, with intermediate levels of thigmotaxis and low freezing and fecal production in the OF (Trullas and Skolnick 1993; Wahlsten and Crabbe 2003; Brown et al. 2004a; Bothe et al. 2005) and little time spent in the closed/dark areas of the EPM and LD (Trullas and Skolnick 1993; Yilmazer-Hanke et al. 2003; Brown et al. 2004a; Brooks et al. 2005). Interestingly, this decreased fear response to open spaces and light might fuel their poor to intermediate performance on the Barnes and Morris tests, since both of these tests require animals to be motivated by fear. The C3H/HeJ perform at an intermediate level on the Barnes test and a relatively low level in the Morris water maze (Brown et al. 2004b; Brooks et al. 2005). In the resident-intruder paradigm, CH3/He mice were the least likely of six inbred strains to attack the intruder, and their overall aggression levels were the lowest of all the strains tested (Jones and Brain 1987). Finally, the CH3/HeJ show low to intermediate levels of immobility in the forced swim test (Ducottet and Belzung 2005). C3H/HeJ mice are relatively large with an intermediate proportion of body fat (Reed et al. 2007) and low metabolic activity. Their heart rate is very high, their systolic blood pressure is very low, and their daily respiratory exchange ratio is low to intermediate (Hampton et al. 2001; Seburn 2001; Deschepper et al. 2004; Tsukahara et al. 2004). They have low basal corticosterone levels and low to intermediate corticosterone responses to stress (restraint: Trullas and Skolnick 1993; uncontrollable foot shock: Shanks et al. 1990), and males have relatively small adrenal glands (although females have large glands, Deschepper et al. 2004). These mice have a low proportion of natural killer cells in the liver (Rymarchyk et al. 2008), with a high proportion of monocytes and granulocytes and a low proportion of B cells in the spleen (Nicholson 2006). In addition, their immune response to endotoxin (lipopolysaccharide) is impaired (Rudbach and Reed 1977). C3H/HeJ mice have an intermediate life span (median: 23 and 22 months for males and females respectively, Paigen and Yuan 2007). C3H/HeBFeJ mice are highly sensitive to convulsive effects of kainic acid, but they are relatively protected from pyramidal cell death and synaptic reorganization after kainic acid exposure (McKhann et al. 2003). C3H mice, like BALB/c mice, are susceptible to proteoglycan-induced arthritis (Glant et al. 2003) and alopecia areata, an autoimmune disorder targeting hair follicles in a variety of species including humans (Sundberg et al. 1994). Finally, these mice have a very high susceptibility to spontaneous liver, mammary, and leukocyte tumors (Näf et al. 2002).

C57BL/6

These mice are characterized by high activity, low anxiety, good spatial performance, low basal glucocorticoid production, relatively low body fat, long life span, and resistance to herpes simplex virus and mammary tumors. C57BL/6 were quite active and showed little thigmotaxis, freezing, and fecal production in the OF, spent little time in the dark portion of the LD, and depending on the research lab, spent variable amounts of time in the closed arms of the EPM (Trullas and Skolnick 1993; Wahlsten and Crabbe 2003; Yilmazer-Hanke et al. 2003; Brown et al. 2004a; Brooks et al. 2005; Bothe et al. 2005). In both the Barnes and Morris mazes, they performed well, taking relatively little time to arrive at the test goal (Brown et al. 2004b; Brooks et al. 2005). This profile differs from that of the C3H/HeJ that also had low anxiety-related behaviors (like the C57BL/6J), but slow performance in the cognitive tests (unlike C57BL/6J). Finally, the C57BL/6J had relatively high levels of immobility in the forced swim test (Ducottet and Belzung 2005). Relative to the other strains, the C57BL/6J mice have intermediate body sizes and low to intermediate levels of body fat (Kitten et al. 2003; Reed et al. 2007). Their metabolic activity is intermediate compared with other strains. They maintain intermediate to high heart rates and systolic blood pressure, and their daily respiratory exchange ratio is intermediate (Hampton et al. 2001; Seburn 2001; Kitten et al. 2003; Deschepper et al. 2004; Tsukahara et al. 2004). They have low basal levels of corticosterone and low to intermediate corticosterone responses to uncontrollable foot shock and restraint stress (Shanks et al. 1990; Trullas and Skolnick 1993; Harizi et al. 2007), and very small adrenal glands relative to their body size—particularly the males (Deschepper et al. 2004). They have a high proportion of natural killer cells in the liver (Rymarchyk et al. 2008) and a low proportion of granulocytes in the spleen (Nicholson 2006). In addition, these mice show high levels of immune complement activity and mount a Th1-dominant response to *Leishmania major*, an intracellular parasite which is associated with relatively low mortality rates in C57BL/6 compared with BALB/cJ mice that mount a Th2-dominant response to this antigen (Scharton-Kersten and Scott 1995). Compared with the other six strains, these mice have a long life span (median: 30 and 31 months for males and females respectively, Paigen and Yuan 2007; 28 and 27 months, Goodrick 1975). Like A/J mice, they are susceptible to diet-induced metabolic disorder symptoms, although in the case of the C57BL/6, these mice do develop symptoms of type 2 diabetes (Gallou-Kabani et al. 2007). They are highly susceptible to spontaneous pituitary and leukocyte tumors and are relatively resistant to spontaneous mammary cancer (Näf et al. 2002; Szpirer and Szpirer 2007).

DBA/2

These mice are distinct in their high activity, high anxiety-like behavior, good spatial abilities, and frequent but low-grade aggression. They also have elevated basal glucocorticoid production, elevated metabolic function, and high body fat, and are prone to audiogenic seizures and mammary tumors. DBA/2J mice had high levels of activity on the OF and EPM and intermediate to high anxiety-related behaviors in the OF, EPM, and LD (Trullas and Skolnick 1993; Brown et al. 2004a; Bothe et al. 2005). Specifically, in the OF they had intermediate to high levels of thigmotaxis but low levels of freezing and defecation, and in the EPM and LD they spent a long time on the closed/dark portions of these arenas (Trullas and Skolnick 1993; Wahlsten and Crabbe 2003; Yilmazer-Hanke et al. 2003; Brown et al. 2004a; Brooks et al. 2005). They performed well on the Barnes and Morris tests—with short latencies during acquisition and reversal trials and a high proportion of time spent in the correct quadrant of these tests during the final probe (Brown et al. 2004b; Brooks et al. 2005). These mice also showed deficits in trace fear conditioning, but they had sufficient emotional memory in a fear conditioning model (Holmes et al. 2002). In a resident-intruder test, DBA mice were frequently aggressive; however, they had long attack latencies and mild attack intensity (Jones and Brain 1987). DBA/2J had relatively low levels of immobility in the forced swim test (Ducottet and Belzung 2005). The DBA/2J mice are relatively large with a very high proportion of body fat (Kitten et al. 2003; Reed et al. 2007). Weanling, seizure-prone DBA/2J mice have greater benzodiazepine receptor density than non–seizure-prone C57BL/6J mice (Robertson 1980). These mice have intermediate heart rates, with intermediate to high systolic blood pressure (Hampton et al. 2001; Kitten et al. 2003; Deschepper et al. 2004; Tsukahara et al. 2004). In addition, they have high basal circulating corticosterone levels, with intermediate but long corticosterone responses to environmental stressors (uncontrollable foot shock: Shanks et al. 1990; restraint stress: Harizi et al. 2007), and intermediate to large adrenal weights (Deschepper et al. 2004). Finally, they have intermediate levels of natural killer cells in the liver, a relatively high proportion of B cells, and a low proportion of CD4 and CD8 T cells in the spleen (Nicholson 2006; Rymarchyk et al. 2008). The DBA/2J have an intermediate life span (median: 23 months for both males and females; Goodrick 1975; Paigen and Yuan 2007), and their major documented health issue is a susceptibility to audiogenic and kainic-acid induced seizures and mortality (Schreiber 1986; Neumann and Collins 1991; Ferraro et al. 1995; Royle et al. 1999), which may be related to high benzodiazepine receptor density (Robertson 1980). Unlike BALB/c and C3H/He mice, DBA/2 mice are resistant to proteoglycan-induced arthritis (Glant et al. 2003; Bárdos et al. 2005), and they are susceptible to developing liver and mammary tumors (Näf et al. 2002).

SUMMARY OF MOUSE PROGENITOR STRAINS

The behavioral profiles of different mouse progenitor strains are quite heterogeneous. For example, activity in the OF does not always relate to anxiety-related behavior, and anxiety-like behavior does not always relate to spatial cognition in a systematic way (see also Podhorna and Brown 2002).

Specifically, high active strains show both high and low levels of anxiety-like behavior and spatial performance. For example, DBA/2 mice were highly active in the OF and showed high anxiety-like behaviors and strong spatial cognitive abilities, while C57BL/6 and C3H/He were also highly active, but showed low anxiety-related behaviors and either high (C57BL/6) or intermediate to low spatial cognitive abilities (C3H/He). However, it should be noted that strains that were active in the OF had better performance in the spatial cognition tasks (e.g., C3H/He, C57BL/6, and DBA/2). This may be a result of the testing paradigms used in these spatial tasks, where the measure of performance (e.g., latency to find a target) is dependent on activity levels—with more active animals necessarily having a greater chance of finding a target than less active animals, regardless of spatial cognition. However, we cannot discount the notion that highly active animals may also have enhanced spatial skills and that these two behavioral responses are causally linked. Based on the mouse strains and behavioral tests reviewed, there is no strong evidence that certain behavioral traits were co-selected in mouse progenitors. However, this may be a result of limited strains or limited behavioral sampling. The lack of a systematic relationship in behavioral responses across tests emphasizes the importance of conducting multiple testing paradigms to best characterize complex behavioral profiles of an individual or a strain.

When considering physiological and health correlates of behavior, we do find some consistent associations, and importantly, some of these associations have also been established in humans. For example, highly active strains tended to have rapid heart rates (e.g., C3H/HeJ, C57BL/6, and DBA/2). Strains with high attack frequencies in the resident-intruder paradigm also had elevated systolic blood pressure (BALB/c and DBA/2); this is comparable to the behavior-physiology correlate seen in highly hostile humans. Comparable to human anxiety, strains with elevated anxiety-like responses also have elevated basal glucocorticoid production and enlarged adrenal glands (BALB/c, DBA/2, A/J). In addition, these strains with elevated anxiety and basal corticosterone production also had shorter life spans than the low anxiety, low basal corticosterone strains (C3H/He and C57BL/6). These relationships between anxiety, basal glucocorticoid production, and life span have not been extensively explored in humans, but may prove a promising area for future investigation. Strains with a relatively low proportion of natural killer T cells in the liver had increased risk of autoimmune processes (e.g., BALB/c and C3H/He vs. A/J and C57BL/6). Interestingly, however, there were no common behavioral traits seen within the "low" or "high" natural killer cell strains. Finally, strains prone to lung tumors tended to have low activity levels (A/J, BALB/c), and those prone to mammary tumors had relatively large body sizes (BALB/c, C3H/He, DBA/2). These results relate to studies by Freedman and colleagues (1990), in which they show a relationship to fat intake and incidence of mammary tumors in mice, although they did not find a strong relationship between tumor incidence and body weight. Further analysis of behavioral and physiological traits associated with the development and growth of specific tumors should be examined in both mice and humans.

Rat progenitor strains
FISCHER (F344)

Fischer (F344) rats stand out for their low activity, high anxiety-like behavior, active coping response to novelty, and low levels of aggression. Their physiological profile involves high cardiovascular and HPA activity, with a susceptibility to ulcers, tumors in multiple locations, and resistance to experimentally induced autoimmunity. Compared with other rat strains in the OF, F344s were not very active and had low levels of thigmotaxis and defecation (Harrington 1972; Van der Staay and Blokland 1996; Berton et al. 1997; Ramos et al. 1997). On the EPM and LD, F344s spent intermediate to high amounts of time in the closed/dark areas of the arenas (Van der Staay and Blokland 1996; Berton et al. 1997; Ramos et al. 1997). In conditioned defensive burying tests, F344s buried a shock prod more often than Wistar and Long-Evans rats (Treit et al. 1981), and during 15 minutes of loud noise exposure they engaged in more frequent and active defensive behaviors, like self-grooming, sniffing, rearing and exploring, compared with Lewis rats (Michaud et al. 2003), suggesting they have elevated fear responses and/or use more active coping strategies when faced with potentially harmful situations. F344s have long latencies to find the platform in the Morris water maze, but during the final probe trial, they spend a great deal of time in the target quadrant of the arena suggesting they understand the task but may have long latencies due to their low activity levels (Van der Staay and Blokland 1996). In the resident-intruder test, less than 20% of F344s attacked the intruder, and the number of fights they initiated was very low despite frequent attacks by the intruder Wistar rat (Berton et al. 1997). When exposed to chronic social stress they lose a large amount of body weight, suggesting a high degree of sensitivity to social stressors (Berton et al. 1997). Finally, F344 show very high struggling and low immobility on the forced swim test (Armario et al. 1995; Lahmame et al. 1997), again suggesting a proactive response to challenges. However, they experience poor adaptation to novelty, which suggests that they may provide a good model for human depression (Nakagawara et al. 1997). F344 rats are relatively small but develop a large abdominal fat mass, with high circulating glucose, insulin, and leptin levels, which can lead to insulin and leptin resistance (Masoro 1980; Van den Brandt et al. 2000; Marissal-Arvy et al. 2007). They have a high resting heart rate and systolic blood pressure (Roberts and Goldberg 1976; Baudrie et al. 2001) and are well recognized for having high HPA activity and reactivity with extended glucocorticoid recovery times. F344 rats have normal sleep-phase nadir levels of corticosterone, but their basal levels during waking hours and their levels in response to stressors were very high (Glowa et al. 1992; Armario et al. 1995; Sarrieau and Mormède 1998). They have high corticosteroid-binding globulin levels in the plasma, spleen, and particularly the thymus (Dhabhar et al. 1993). In addition to high basal and reactive corticosterone levels, F344s do not habituate to repeated stress but rather show a sensitized adrenal response with increased corticosterone secretion with repeated stressors (Dhabhar et al. 1997; Shepard and Myers 2008). They have normal HPA negative feedback, with normal corticosterone and ACTH suppression in response to dexamethasone (Gómez et al. 1998). However, F344 rats show a shift toward more mineralocorticoid vs. glucocorticoid receptor activity in the brain, especially in the hippocampus, during a stress response (Dhabhar et al. 1995; Sarrieau and Mormède 1998; Marissal-Arvy et al. 2000). F344 rats have intermediate life spans (median: 22–28 and 27–29 months for males and females, respectively: Masoro 1980; Ghirardi et al. 1995; Lipman et al. 1996; Altun et al. 2007) and are prone to physical characteristics and disorders associated with glucocorticoid hypersecretion. For example, F344 rats are relatively resistant to experimentally induced autoimmunity and arthritis (Griffin and Whitacre 1991). They also have a high HPA response to immune challenges (Marissal-Arvy et al. 2007), making them less susceptible to autoimmunity. This resilience appears dependent on their glucocorticoid production because a glucocorticoid antagonist abolished this resilience (Sternberg et al. 1989). In accordance with their high cardiovascular activity, they are prone to cardiomyopathy—an inflammation of the heart muscle (Maeda et al. 1985; Levy et al. 2002). Finally, F344s are also prone to stress ulcers (Redei et al. 1994) and to spontaneous mammary, pituitary, and testicular interstitial tumors (Sass et al. 1975).

LEWIS (LEW)

Lewis rats are active with intermediate levels of anxiety and aggression. They have low cardiovascular and HPA activity, a relatively long life span, and susceptibility to experimentally induced autoimmunity. In the OF and in home cages, LEWs had mid to high activity compared with other strains (Flaherty et al. 1979; Sternberg et al. 1992; Berton et al. 1997; Ramos et al. 1997; Hlavacova et al. 2006), and in the OF, they showed intermediate anxiety-like behavior (Flaherty et al. 1979; Sternberg et al. 1992; Berton et al. 1997; Ramos et al. 1997). In the EPM and LD their anxiety-like behavior was relatively high and their acoustic startle and freeze response was larger than F344s; however, they recovered quickly without engaging in self-soothing behaviors like self-grooming (Sternberg et al. 1992; Glowa et al. 1992; Ramos et al. 1997; Berton et al. 1997). During 15 minutes of loud noise they spent a great deal of time freezing and engaged in little active defensive behaviors (e.g., self-grooming, sniffing, rearing and exploring) compared with F344 rats (Michaud et al. 2003). In the forced swim test, LEWs show intermediate levels of immobility and struggling, and in the resident-intruder test, only 20% of LEW rats attacked the intruder and the number of fights was relatively low (Sternberg et al. 1992; Armario et al. 1995; Berton et al. 1997). Finally, in response to a chronic social stressor (a week of daily changes in housing partners) they lose a significant amount of weight, suggesting they are highly sensitive to social stressors (Berton et al. 1997). LEWs have an intermediate to large body size with intermediate circulating glucose levels and high insulin and leptin levels (Van den Brandt et al. 2000). Like F344 rats, they have high fat deposition, but for a different reason; as opposed to the F344s with high corticosterone promotion of central adiposity, LEWs have extremely low HPA activity and reactivity (Marissal-Arvy et al. 2007). Comparable with their low HPA activity/reactivity, the LEW rats also have relatively low resting heart rates and systolic blood pressure (Baudrie et al. 2001). LEWs' circadian corticosterone rhythm was very flat with little variation between waking and sleeping phases (Sarrieau and Mormède 1998), and their ACTH and corticosterone response to different stressors was quite low (acute immobilization: Sternberg et al. 1992; Dhabhar et al. 1993; Pardon et al. 2002; forced swim test: Armario et al. 1995). Their ACTH response to CRH was relatively low (Oitzl et al. 1995; Grota et al. 1997), and their adrenals were hyporesponsive to ACTH (Oitzl et al. 1995). They had low levels of corticosteroid-binding globulin in the plasma, spleen, and thymus, compared with F344 (Dhabhar et al. 1993). In their peripheral tissues, they have low glucocorticoid receptor binding during stress, and like F344 rats, they have a shift toward more mineralocorticoid than glucocorticoid receptors in the brain (Dhabhar et al. 1995; Oitzl et al. 1995). Finally, in response to dexamethasone, their corticosterone and ACTH levels suppress normally (Gómez et al. 1998). LEW rats have a relatively long life span (median: 28 and 33 months for LEW/Han males and females, Baum et al. 1995; 29

months for LEW females, Altun et al. 2007). Because of high amounts of serum insulin compared with other strains, LEWs have been used in diabetes
research. Because they have little corticosterone response to inflammation
and abnormally high levels of circulating corticosterone-binding globulin
(further decreasing what little anti-inflammatory glucocorticoid effects
they have, Dhabhar et al. 1995), they are a common model for autoimmune
disorders. For example, they are commonly used to model rheumatoid arthritis using the adjuvant-induced arthritis method and experimental autoimmune encephalomyelitis (Stefferl et al. 1999). This susceptibility to experimentally induced arthritis is abolished when LEWs are supplemented
with exogenous glucocorticoids (Sternberg et al. 1989). People with certain
autoimmune disorders (i.e., rheumatoid arthritis, Bishop 1988; Cushing's
disease and hypothyroid, Dorn et al. 1995) suffer from an atypical form of
depression; LEW rats may provide an interesting model to understand human autoimmune disorders and their accompanying mood disorders.

BROWN NORWAY (BN)

Brown Norway (BN) rats have intermediate behavioral and physiological
profiles, with a relatively long life span. They stand out for having good spatial ability, being relatively aggressive toward a novel partner, and showing
signs of helplessness. They are also susceptible to experimentally induced
allergies/asthma/atopic disorders and resistant to experimentally induced
autoimmunity. In the OF, BNs were moderately to highly active with high
levels of thigmotaxis and defecation (Van der Staay and Blokland 1996; Berton et al. 1997; Ramos et al. 1997). Their anxiety-related behavior was variable on the EPM and LD, with low to high amounts of time in the closed/
dark areas as compared with other strains (Van der Staay and Blokland 1996;
Berton et al. 1997; Ramos et al. 1997). BNs performed well in the Morris water maze, with short latencies to find the hidden platform and intermediate
to high amounts of time spent in the correct quadrant of the arena during a
final probe trial (Van der Staay and Blokland 1996). In the resident-intruder
test with a Wistar intruder rat, 60% of BN rats initiated aggression, but their
total number of fights was not high, and when placed in a neutral situation
with a novel partner, they were not particularly aggressive (Berton et al.
1997). When exposed to the chronic social stress of daily novel cage mates,
BN rats did not lose much weight, indicating they are not very sensitive to
social stressors (Berton et al. 1997). In the forced swim test, they struggle
very little and spend a great deal of time immobile (Armario et al. 1995).
BN rats are relatively small with low circulating glucose, insulin, and leptin levels (Van den Brandt et al. 2000). They have increased corticotropin-

releasing factor binding in the hippocampus and the prefrontal cortex compared with LEW and two other rat strains (Lahmame et al. 1997). They have intermediate basal corticosterone levels throughout the day, with intermediate levels at both the circadian nadir and acrophase (Sarrieau and Mormède 1998). BN rats also differ from the F344 and LEW rats in that they have more efficient glucocorticoid receptors, as opposed to the mineralocorticoid receptor shift seen in these two strains, and they are slightly resistant to dexamethasone suppression (Gómez et al. 1996; 1998). BN rats have a relatively long life span (median: 29–32 and 31–33 months for males and females, Burek and Hollander 1975; Gilad and Gilad 1995; Lipman et al. 1996). Given their hyper-responsive lung reactions to common allergy tests, BNs are often used to model respiratory and food allergies and asthma. They are relatively resistant to most types of experimental autoimmune encephalomyelitis—a demyelinating brain autoimmune process (Stefferl et al. 1999). However, BN rats have an adaptive immune response that favors the Th2 pathway, which leaves them highly susceptible to myelin-oligodendrocyte-glycoprotein- and mercury-chloride-induced autoimmune diseases (Stefferl et al. 1999), but not susceptible to Th1-mediated arthritic diseases induced by other adjuvants like proteoglycans. Because of this trend toward the allergic/asthmatic/atopic branch of immune dysfunction, the BN rat is a popular model for asthma (Elwood et al. 1991; Huang et al. 1999). BN rats have been characterized as a Th1-autoimmune-resistant control for Th1-autoimmune-prone LEW rats. Divergent HPA axis function is probably involved in this divergent immune function: intermediate HPA-reactive BN rats are susceptible to Th2-mediated-autoimmunity, while HPA-hypoactive LEW rats are prone to Th1-mediated-autoimmunity (Fournié et al. 2001).

WISTAR

Outbred Wistar rats show an anxiety-like phenotype across several behavioral tests, with high activity, low spatial cognition, low social behavior, and mixed metabolic function and a resilience against autoimmunity. On the OF, Wistars had high levels of locomotion and thigmotaxis, and on the EPM and LD they spent a long time in the closed and dark portions compared with F344s (Van der Staay and Blokland 1996). Wistars had relatively long latencies to find the platform in the Morris water maze with little time spent in the target quadrant during a final probe trial (Van der Staay and Blokland 1996). Wistar rats had high levels of D1 receptor binding in the basal ganglia and D2 receptor binding in the caudate-putamen compared with Sprague-Dawley rats, which may underlie differences in locomotion between these strains (Zamudio et al. 2005). Wistars are very large rats with intermediate resting heart rates and low systolic blood pressure and relatively high basal corticosterone levels including relatively large adrenal glands (Roberts and Goldberg 1976, Baudrie et al. 2001; Hlavacova et al. 2006; Malkesman et al. 2006). These rats have been used to breed two lines of rats: High-Anxious and Low-Anxious Behavior rats (HAB/LAB) that differ in metabolic and immune function (reviewed later in this chapter). Wistars have an intermediate life span (median: 22–30 and 21–28 months for males and females, Masoro 1980; Altun et al. 2007), and are relatively resilient to adjuvant-induced arthritis. Some of the inbred Wistar rats develop specific health issues (e.g., aged Hannover Wistar rats develop cataracts, Takemoto et al. 1991; Wistar fatty rats become insulin resistant, Sugiyama et al. 1989), but overall there is little information on their susceptibility/resilience to specific disease states.

SPRAGUE DAWLEY (SD)

Sprague-Dawley rats are outbred and have been used to breed behaviorally specific rat lines that we discuss later in the chapter. SD rats are marked by their relatively high degree of anxiety-like behavior, their good spatial abilities, short life span, and susceptibility to experimentally induced autoimmune arthritis. SD rats had high to intermediate levels of activity and thigmotaxis in the OF and a relatively high degree of anxiety-like behavior in the EPM and LD (much time spent in the closed/dark sections, Flaherty et al. 1979; Rex et al. 2004). SD rats are able to find the platform in the Morris water maze relatively quickly (Andrews et al. 1995). Physiologically, SD rats had low resting heart rates and low systolic blood pressure (Roberts and Goldberg 1976). They also produced intermediate to low baseline and reactivity corticosterone levels compared with F344, LEW, and Wistar rats, and low levels of corticosteroid-binding-globulin in the plasma, spleen, and thymus, comparable to LEW rats (Dhabhar et al. 1993; Malkesman et al. 2006; Hlavacova et al. 2006). SD rats have short to intermediate life spans (median: 20–29 and 22–30 months for males and females, Masoro 1980; Ghirardi et al. 1995; Altun et al. 2007) and are susceptible to adjuvant-induced arthritis (a model of rheumatoid arthritis) and mammary tumors (Cai et al. 2006). Because SD rats are outbred and display a wide array of behavioral and physiological biases among individuals, they are frequently used to study temperament and physiology (e.g., Curé and Rolinat 1992; Dellu et al. 1996; Cavigelli et al. 2006; 2007; Clinton et al. 2008). For example, Dellu and colleagues have identified High- and Low-Responder (HR vs. LR) SD rats (Dellu et al. 1996) that have divergent HPA responses to a circular corridor and the elevated plus maze (Kabbaj et al. 2000; Márquez et al. 2006). HR and LR rats have been used as behavioral models for understanding variance in drug use, prenatal stress, maternal behavior, and adaptation to stress (Ballaz et al. 2007; Clinton et al. 2007; 2008, Davis et al. 2008; Stoffel and Cunningham 2008; see chapters 10 and 12). They have great potential as a model for temperament-based health outcomes (Dellu et al. 1996), and will be discussed in more detail later in the chapter. One note to keep in mind when using SD rats as a physiological model is their heterogeneity across vendors. Rats from Simonese, Harlan Sprague-Dawley, and Charles River Laboratories are significantly different from each other in terms of HPA responses to stress and certain metabolic processes (Pecoraro et al. 2006).

SUMMARY OF RAT AND MOUSE PROGENITOR STRAINS

In previous work, Hinojosa and colleagues (2006) established a positive relationship between rat strain anxiety-like behavior and immobility in the forced swim test. In the current review, this relationship was not evident. This may be a result of the between-study comparison method used in the current review, a method with less sensitivity and accuracy than a within-study comparison of individual behavioral responses to a variety of test conditions. This difference in findings supports further cross-strain comparison of behavioral responses to a variety of testing paradigms, where the behavioral tests are all conducted in the same laboratory. Patterns of behavioral responses to a variety of testing paradigms are poorly understood, but such within-study comparisons would help identify the genetic basis of personalities (e.g., Sih et al. 2004; Van Oers and Mueller 2010).

In relating behavioral biases to physiological biases, we found several relationships in both rats and mice. For example, the two rodent strains with hyperreactive airways and susceptibility to allergic responses were those with mixed anxiety-like behavioral responses (i.e., A/J and BN). The only other trait these strains had in common was a relatively small body weight relative to the other within-species strains. This is an intriguing relationship that may be worthy of investigation in humans. Does anxiety in one domain (e.g., social vs. physical) relate to allergy susceptibility? In addition, as has been previously noted by other researchers, there does seem to be an association between high anxiety-like behavior, elevated stress physiology, and a shortened life span (Gilad and Gilad 1995). However, data to support this notion are still quite scant, and the only evidence that a potentially related human trait is associated with life span comes from the work of Swan and Carmelli (1996), who studied older community-dwelling men and women and found that participants with low self-reports of curiosity died more rapidly than those with higher levels of curiosity. Other human traits have been associated with life span; the current review suggests further research on the relationship between human anxiety and life span as well as possible underlying physiological mechanisms like basal glucocorticoid production.

The strongest relationship between behavior, physiology, and health was the tendency for strains with high anxiety-related behavior and passive or reactive coping styles to have elevated basal and reactive corticosterone production compared with strains with lower anxiety-like behaviors (e.g., BALB/c, DBA/2, F344, Wistar vs. C3H/He, C57BL/6, LEW). (LEW rats present an interesting comparison strain here, in that they showed intermediate to high anxiety-like behavior with a proactive coping style and very low glucocorticoid production, suggesting that coping style, as opposed to anxiety-related behavior alone, may be a better predictor of glucocorticoid production.) The relationship between anxiety-like behavior and HPA axis activity has been identified previously, and the neurological mechanisms underlying this relationship are relatively well documented (e.g., Kalin et al. 1998; 2000; Takahashi 2001). Notably, the relationship between anxiety and glucocorticoid production was seen across multiple strains in the current review, and elevated anxiety and glucocorticoid production was associated with strain resilience to experimentally induced autoimmunity/arthritis. Specifically, strains with elevated anxiety-like behavior and glucocorticoid production (particularly basal production) were more resilient to autoimmune/arthritic processes (e.g., F344, Wistar, DBA/2) compared with strains with low anxiety-like behavior (or a proactive coping style) and low glucocorticoid production (e.g., LEW, C3H/He, C57BL/6). This relationship between anxiety-like behavior and resistance to autoimmunity is most likely mediated by the anti-inflammatory effects of glucocorticoids; administration of exogenous glucocorticoid to autoimmune-prone strains minimized their susceptibility. Furthermore, LEW rats with intermediate to high anxiety-like behavior, but proactive coping style and low glucocorticoid production, are susceptible to Th1-autoimmune processes like arthritis. We should note exceptions to this trend: BALB/c mice show high anxiety-like behavior with elevated glucocorticoid production but—like low-anxiety/low-glucocorticoid C3H/He mice—are susceptible to experimentally induced arthritis. This may be due to the fact that different underlying mechanisms are involved in very similar clinical symptoms of arthritis (Adarichev et al. 2008). Even with these exceptions and differential mechanisms in mind, the relationship between rodent anxiety-like behavior, coping, glucocorticoid production, and susceptibility to arthritic and autoimmune processes is an intriguing area for further study in humans.

A recent review by Zozulya and colleagues (2008) has postulated that individuals (rodents and humans) with a passive coping style tend to have greater HPA activation relative to sympathetic activation and are more likely to suffer infectious diseases and cancer (e.g., like "Type C" humans), whereas active copers have more sympathetic activation than HPA activation and are more prone to develop autoimmune and/or cardiac problems (e.g., like "Type A" humans). On the basis of the above review of progenitor rodent strains, we found support for a portion of Zozulya's hypothesis. Specifically, those strains with low HPA activation and high cardiovascular activity (heart rate and systolic blood pressure; C3H/He) were more susceptible to experimentally induced autoimmunity than strains with lower cardiovascular activity and heightened HPA activity (Wistar). Again, the exception here is BALB/c mice. Whether these physiological profiles map nicely onto coping strategies in these progenitor strains would require further investigation into coping styles. In addition, on the basis of the review of progenitor strains, there may be other behavioral "types"; for example, individuals that experience both elevated HPA activity and sympathetic activity (e.g., DBA/2, F344), or both dampened HPA and sympathetic activity (e.g., C57BL/6, LEW). These strains/individuals require more study to identify specific behavioral and health biases.

Because the above common progenitor strains were not bred according to behavioral responses to standard tests, it may be difficult to categorize the strains according to a specific behavioral type/profile or personality. However, we have shown that specific behavioral traits are related to specific physiological or disease biases. The strongest of these relationships was that high anxiety-like strains tended to produce more glucocorticoids and were resilient to autoimmune processes.

Rodent strains bred for behavioral traits

Because the progenitor strains reviewed above were not bred on the basis of their specific behavioral responses, we follow here with a review of mice and rats bred specifically according to behavioral traits (see also chapter 12). These strains have been developed more recently and thus less is known about their physiology and health compared with the progenitor strains, but these strains are important because of their behavior-specific breeding. In the following section, we review behavioral, physiological, and health differences in several strains and make reference to potential human analogs for each strain. We have included some rat strains that were not specifically bred for behavior (e.g., Flinders, HR/LR rats) because of the interesting information they provide on relationships between behavior, physiology, and health.

SAL and LAL mice were bred from wild progenitors on the basis of the amount of time taken to attack a novel social partner in a familiar setting (Van Oortmerssen and Bakker 1981; see chapter 12). These mice display an array of other behavioral biases as well. For example, the rapid attacking SAL mice, compared with LAL, were more active on the EPM, were less immobile in the OF and forced swim test, froze less when exposed to unavoidable stress, explored novel social and nonsocial objects in a changing environment less, and appeared to have a more rigid coping style (e.g., less immobility, more defensive burying of a novel noxious object, and more active shock avoidance; Benus et al. 1987; 1989; 1991; Sluyter et al. 1996; Koolhaas et al. 1999; Veenema et al. 2003a; Gasparotto et al. 2007; Veenema and Neumann 2007).

Physiological differences between SAL and LAL mice may account for their behavioral differences (chapter 12). SAL mice had less neostriatal dopamine activity than LAL mice, as indexed by more frequent stereotypic behavior following apomorphine administration (Benus et al. 1991). These lines also show differences in the serotonergic and adrenergic systems: SAL mice had more 5-HT$_{1A}$ autoreceptor sensitivity, lower levels of 5-HT in the prefrontal cortex, and greater adrenergic responses to stress compared with LAL mice (Caramaschi et al. 2007). Together, these differences in dopaminergic, serotonergic, and adrenergic activity may account for observed behavioral differences in responses to novelty and aggression between the two lines.

SAL mice stress physiology is less reactive to social and nonsocial challenges compared with LAL mice (chapter 12). For example, SAL had lower plasma corticosterone levels following forced swim, resident-intruder, repeated social defeat, and sensory contact with a dominant mouse compared with LAL mice (Veenema et al. 2003a, b; Veenema and Neumann 2007). They also had less glucocorticoid receptor density in the dentate gyrus, increased hippocampal cell proliferation, less hippocampal glucocorticoid and mineralocorticoid receptor mRNA levels, and less CRH mRNA in the paraventricular nucleus compared with LAL (Veenema and Neumann 2007). The circadian corticosterone rhythm was altered in SAL mice, with relatively high levels during the light (inactive) phase compared with LAL mice (Veenema et al. 2003a). The SAL had higher basal ACTH levels throughout the day compared with LAL, whereas basal corticosterone levels during the dark (active) phase did not differ between lines, suggesting differential adrenal sensitivity to ACTH between lines (Veenema et al. 2003b). In line with the differential stress physiology, SAL had lower serum antioxidant capacity than LAL mice, which is a physiological bias associated with a longer life span (Costantini et al. 2008). The male SAL mice have been proposed as a model of antisocial personality disorder in men because they display a range of characteristics comparable with human antisocial behavior (e.g., elevated alcohol use, elevated anxiety levels, and more offspring; Sluyter et al. 2003; Caramaschi et al. 2008a). Their autonomic arousal appears similar to that seen in violent humans, and is in accordance with the physiological hypoarousal theory of aggression (Haller and Kruk 2006; Caramaschi et al. 2008b). Given their behavioral profiles, the SAL mice may also be considered to display greater hostility or Type A behavior than LAL mice, although there is no evidence to suggest that the SAL and LAL differ in cardiovascular reactivity to stress.

MICE: HIGH/LOW OPEN-FIELD THIGMOTAXIS (HOFT/LOFT)

HOFT and LOFT mice were derived from a Swiss albino outbred stock based on their level of thigmotaxis in a 60-minute OF test (Leppänen et al. 2005). In the OF, HOFT males also defecated more, were more thigmotactic, and reared less than LOFT, while the two strains did not differ in OF locomotion (Leppänen et al. 2006). Increased thigmotaxis in the HOFT strain was also evident in the home cage, suggesting these males experience both state and trait anxiety (Leppänen et al. 2005). HOFT males are significantly lighter than LOFT males, though they eat and excrete similar amounts (Leppänen et al. 2005). Other physiological differences have not been examined in these mice; however, researchers have found a significant difference in their life spans, with HOFT mice living longer than LOFT mice (mean life span: 23 vs. 20 months, Leppänen et al. 2005). These life span differences are comparable to differences in senescence-accelerated prone vs. resistant mice (SAMP/R) in which early-dying mice also had less anxiety-related behavior (on EPM and in response to novel food) than later-dying mice (SAM-R/1, Miyamoto et al. 1992; Markowska et al. 1998). However, these results differ from those collected from "slow"/PAM vs. "fast"/NPAM mice and neophobic/neophilic male Sprague-Dawley rats (described below), in which rapid exploration mice/rats lived longer than lower exploration mice/rats (Viveros et al. 2001; Cavigelli and McClintock 2003). Tests used to measure exploration, fear, and anxiety-like behavior in these different rodents were quite different, and without information on HOFT/LOFT physiology, it is difficult to determine differential physiological mechanisms that may underlie these differential associations between exploration-related behavior and life span.

Unlike the LAL/SAL mice, PAM/NPAM are not bred according to their behavior, but rather selected from a population of Swiss outbred mice based on their repeated behavioral response to a T-shaped maze (Viveros et al. 2001). Mice are classified as "fast" or "slow" depending on the time it takes them to move from one arm of the maze to the intersection of the three arms: "slow" mice (PAM) take more than 10 seconds to arrive at the intersection, whereas "fast" mice (NPAM) take less than 10 seconds (Viveros et al. 2001). In addition to this behavioral difference, young "slow"/PAM (vs. "fast"/NPAM) displayed behavioral responses often associated with aging: less muscle coordination (in the tightrope test), greater emotionality (indexed by central and internal ambulation and grooming in the holeboard test), less activity (in the OF), and more anxiety-like behavior (less time in the open arms of the EPM; Viveros et al. 2001; Guayerbas et al. 2002; Pérez-Álvarez et al. 2005). PAM also showed a trend toward more immobility in the forced swim test than the NPAM (Pérez-Álvarez et al. 2005).

During young adulthood, "slow"/PAM and "fast"/NPAM had similar body weights, but during later middle age, female "slow"/PAM were found to be significantly lighter than "fast"/NPAM (Viveros et al. 2001; Guayerbas et al. 2002). Female PAM, compared with NPAM, had elevated basal corticosterone production and a blunted corticosterone response to a short bout of forced swim and tail immersion (Pérez-Álvarez et al. 2005). The most studied physiological difference between PAM and NPAM is immunological function. Young PAM mice, compared with NPAM, had lower lymphoproliferative responses to Concanavalin A mitogen, which is an immune bias often seen in aged animals (Pawelec et al. 1999; Viveros et al. 2001). Also, female "slow"/PAM have decreased phagocytic and chemotaxic immune cell activity compared with "fast"/NPAM (Guayerbas et al. 2002). Finally, PAM had greater oxidant levels (malondialdehyde) in the brain and lower antioxidant levels (reduced glutathione) in the spleen compared with NPAM (Viveros et al. 2007).

As indicated by their acronym, the "slow"/PAM mice have a significantly shorter life span than the "fast"/NPAM mice (mean life spans for females: 19 vs. 22 months; Guayerbas et al. 2002). Other health outcomes for these two types of mice have not yet been identified. The PAM have been proposed as a model of premature aging, given their behavioral and physiological profiles in young adulthood that suggest a more aged profile than the NPAM. There is no proposed human analog for the PAM, although they may be a good model of humans who report lower curiosity levels and also seem to age more rapidly than more curious individuals (e.g., Swan and Carmelli 1996; see below). PAM/NPAMs also present an interesting comparison to other strains with differential behavior and life spans (e.g., the HOFT/LOFT mice reviewed above; see below).

RATS: ROMAN HIGH- AND LOW-AVOIDANCE (RHA/RLA)

The Roman high-avoidance (RHA) and low-avoidance (RLA) rats are Wistar-derived rats selectively bred according to behavior in a two-way shuttle box active avoidance task, with RLA rats taking longer to learn to avoid shock than RHAs (Bignami 1965; see chapter 12). RLAs also moved less and engaged in more self-focused coping behavior (e.g., freezing, self-grooming) in the OF and made fewer transitions between the light and dark sides of the LD test compared with the RHAs (Steimer and Driscoll 2005). Finally, RLAs spent more time in slow-wave sleep, spent less time in paradoxical sleep, and entered paradoxical sleep more slowly than RHAs (Steimer et al. 1999). Physiologically, RLAs experienced heightened and prolonged corticosterone and ACTH production following novelty or mild stressors, whereas RHAs quickly adapted to low-grade stressors (Gentsch et al. 1982; Castanon et al. 1994) (this difference in ACTH production following OF was evident in older rats, 26 weeks, but not younger rats, 14 weeks; Castanon et al. 1994). Baseline and stress-response prolactin levels were higher in RLAs compared with RHAs (Castanon et al. 1994). These rats also differed in immune function; RLA and RHA responses to lipopolysaccharide were similar (indicating little difference in innate and B-cell function); however, T-cell activity differed greatly, with elevated natural killer cell activity and lymphocyte mitotic responses in RLA vs. RHA rats, particularly during stressful conditions (Sandi et al. 1991). This difference in T-cell activity may result from the increased prolactin activity in RLAs (Cross et al. 1989; Castanon et al. 1992; Dorshkind and Horseman 2000; Redelman et al. 2008).

Because RLA rats have increased and lengthened behavioral and physiological responses to stress, and because they focus on "displacement" behaviors like self-grooming or redirected aggression instead of investigating the source of stress, it has been proposed that they use a passive coping style compared with the more problem-focused coping style of the RHAs (Steimer et al. 1997; Steimer and Driscoll 2005). The RLA tendency to be more stress-reactive and less likely to explore may explain their longer latency to learn how to avoid a moving shock in the shuttle box originally used during selection. The RLA behavioral profile may be comparable to a human emotional coping style that includes a focus on internal responses rather than external problems (Lazarus and Folkman 1984). These characteristics suggest that RLA rats may mimic human behavioral inhibition whereas RHAs mimic the human trait of sensation seeking (Steimer and Driscoll 2005). Categorization of RHA as sensation/novelty-seekers is further supported by the fact that RHA had visual evoked potentials to light flashes comparable with those in high sensation-seeking humans (Siegel and Driscoll 1996). Finally, differences in RLA and RHA sleep patterns indicate that they may provide an interesting system to study the relationship between sleep, temperament, and health, particularly in relation to human anxiety (Steimer et al. 1999).

RATS: HIGH/LOW ANXIETY BEHAVIOR RATS (HAB/LAB)

HAB/LAB rats are a recent Wistar-derived model of anxiety bred according to percent time they spent on the open arms of the EPM (HAB spend less than 5% of time on open arms while LAB spend greater than 50% of time on open; Landgraf 2003). HAB rats are more susceptible to stress effects and tend to passively cope (e.g., freezing) as compared with LABs and Wistar controls (Landgraf et al. 1999, Koolhaas et al. 1999). The HAB rats showed increased anxiety-related behavior in the OF and LD, visited more holes in a modified hole board test, and spent less time actively interacting with a novel partner compared with LAB rats (Henniger et al. 2000; Ohl et al. 2001; Salomé et al. 2002). Behavioral differences between HABs and LABs persisted across multiple stages of development, suggesting this behavioral trait may be quite stable over time (Wigger et al. 2001; Landgraf and Wigger 2002). Such long-term measures of behavioral stability are important if claims about trait stability (i.e., personality) are to be made with rodent models.

Adult HAB rats are slightly larger than LABs (Salomé et al. 2008). HABs are more responsive to anxiolytic drugs than LABs, as confirmed by a 20-fold increase in the percent time spent on the open arms of the EPM after diazepam administration (Landgraf and Wigger 2002). Baseline plasma corticosterone levels tend to be similar in both lines (Liebsch et al. 1998; Landgraf et al. 1999), but HABs have higher circulating corticosterone levels following EPM testing and exposure to endotoxin (Landgraf et al. 1999; Salomé et al. 2008). The HAB rats appear to have diminished negative feedback of the HPA axis, and this alteration appears dependent on elevated vasopressin production in the paraventricular nucleus (Keck et al. 2002). Specifically, when administered dexamethasone, HAB rats did not decrease ACTH production as well as LABs; they also produced more vasopressin in the paraventricular nucleus than LABs. Furthermore, dexamethasone suppression of the HPA axis was increased by a vasopressin receptor antagonist, indicating a causal role of vasopressin in altered HPA negative feedback in HAB rats. Finally, HAB rats showed signs of reduced serotonin transmission in the hippocampus compared with LABs (Keck et al. 2005). The physiological profile of HAB rats provides an attractive analogue to human depression (Keck et al. 2002). This is an intriguing outcome for rodents specifically bred for high anxiety-related behavior. In particular, the role of vasopressin in anxiety and depression presents an interesting line of inquiry with humans. Because HAB/LAB rats have been developed quite recently, there is little other information on differential health outcomes across the lines. However, these rats provide an interesting new model of human anxiety because they were specifically bred according to their anxiety-related behavioral responses—as opposed to other strains that have been proposed as models of human anxiety but that were not specifically bred according to this trait.

RATS: HIGH/LOW RESPONDER RATS (HR/LR)

Unlike previous strains, high and low responder rats are not bred for their specific behavioral trait, but rather they are identified from a population of outbred male SD rats on the basis of their locomotion in a circular corridor during a 2-hour test period (Piazza et al. 1989). Specifically, HR rats are those with greater than median levels of locomotion in this arena and LR rats are those with lower than median locomotion. Comparable with the HAB/LAB rats, researchers have tested the stability of HR/LR behavior in the circular corridor and found that locomotor responses were not highly stable from 2 or 4 months to 16 months of age (Dellu et al. 1994; 1996). Although this behavioral bias was not highly stable within individuals over time, HR and LR rats showed several other behavioral and physiological differences, some of which emerged later in life. For example, HR rats had shorter latencies to enter and spent more time in the light side of the LD, and spent more time in the open arms of the EPM (Kabbaj et al. 2000), indicating that they are less fearful than LR rats. Also, HR rats experienced a significant decay in recognition memory with age while LR rats did not (Dellu et al. 1994). Finally, and importantly, HR rats self-administered amphetamine more than LR rats (Piazza et al. 1989; 1990).

In terms of physiological profiles, HR rats had greater novelty and cocaine-induced extracellular dopamine concentrations in the nucleus accumbens and greater stress-induced elevations of mesolimbic dopamine neurotransmission compared with LR rats (Hooks et al. 1991; Piazza et al. 1991). HR rats also responded to the circular corridor, to the LD box, and to restraint stress with a greater glucocorticoid response than LR rats (Piazza et al. 1990; Dellu et al. 1996; Kabbaj et al. 2000). This differential glucocorticoid response to restraint stress between HR and LR rats was evident in both young and middle-age rats (4 and 16 months old), but was no longer significant in old-age rats (21 months; Dellu et al. 1996). In addition, HR rats had lower CRF mRNA in the central nucleus of the amygdala, higher levels in the paraventricular nucleus, and lower basal GR mRNA in the hippocampus than LR rats (Kabbaj et al. 2000). Given this physiological profile, researchers have concluded that HR rats may be analogous to high sensation-seeking humans. Humans with characteristics associated with sensation-seeking are more prone to drug use and have a greater response to amphetamines, making the HR/LR rats an appropriate model system (Kelly et al. 2006; White et al. 2006). Other than this difference in drug-taking propensity, no significant health differences have been identified for HR vs. LR rats.

RATS: FLINDERS SENSITIVE/RESISTANT LINES (FSL/FRL)

FSL/FRL rats were not bred according to behavioral traits, but rather according to their sensitivity to an organophosphate that acts as a muscarinic agonist, with the FSL rats much more sensitive than the FRL and progenitor (SD) rats (Russell et al. 1982). FSL/FRL rats have proven useful in studying behavior and health, particularly in understanding chemical sensitivity, allergies, asthma, depression, and other autoimmune disorders, and interestingly, they also display differences in behavioral profiles. For example, compared with FRL, the FSL are more behaviorally sensitive to a variety of drugs (see Overstreet and Djuric 2001 for review), which may be analogous to patients that become sensitive to multiple chemicals after organophosphate exposure (Cone and Sult 1992). FSL rats have also been used as a model for major depression because some research indicates that individuals with depression are more sensitive to cholinergic agents (Janowsky et al. 1994). Indeed, the FSL rats have many symptoms consistent with depression, such as reduced appetite and body weight, reduced activity in novel environments, increased withdrawal and inhibitory behaviors when exposed to stressors, anhedonia (demonstrated by low frequency bar pressing for water or food reward), and slow completion of a food-motivated non-matching-to-sample learning procedure compared with control rats (Overstreet and Russell 1982; Overstreet et al. 1986; 1993; 1995; Bushnell et al. 1995). Additionally, stress increased anhedonic responses (Muscat et al. 1990) and antidepressant drugs normalized these responses, while amphetamines had no influence, indicating that symptoms are indeed driven by anhedonia and not by fatigue or low preference for locomotion (Overstreet et al. 1995). FSL sleep patterns are also somewhat similar to those seen in depressed humans—REM sleep is increased in FSL rats, while latency to REM is decreased (Shiromani et al. 1988; Benca et al. 1996). Depressed humans, however, also display reduced slow-wave sleep (Jindal et al. 2002), and this effect is not mirrored in the FSL rats (Benca et al. 1992).

Physiologically, FSL rats share many traits with depressed humans. Dopamine release in the reward pathway is low in FSL rats and even lower when the rats are under stress (Overstreet et al. 1986; 1989). After stress, they lack CRH activation, which results in a hypoactive HPA stress response (Zambello et al. 2008). Consistent with the serotonin hypothesis of depression, FSL rats show low serotonin synthesis in the raphe and limbic system (Hasegawa et al. 2006). Further support for the FSL rat as a model of depression comes from the finding that that muscarinic receptors play an important role in pre- and postsynaptic serotonin receptor function (Haddjeri et al. 2004). In addition, FSL rats' immune function is comparable with that seen in human depression (e.g., reduced Th1 immune function with normal Th2 function). For example, FSL rats had reduced basal natural killer cell activity, reduced primary antibody response, and splenocyte production of IFN-gamma and IL-6 in response to keyhole limpet hemocyanin, indicating an increased susceptibility to infection (Friedman et al. 1996; 2002). FSL rats had more hyperresponsive airways (a typical allergy symptom) than FRL or control rats (Djurić et al. 1998); however, they had similar IgE levels to FRL rats—a surprising finding because allergies in humans are mediated by IgE. Despite similar IgE activity, exposure to a choline agonist or a generic antigen (ovalbumin) did cause greater inflammation and immune cell proliferation in FSL compared with FRL rats (Djurić et al. 1995).

Care should be taken when extrapolating FSL behavior to human depression. Other research has demonstrated normal hedonic (Pucilowski et al. 1993; Matthews et al. 1996) and cognitive function in FSL rats, in contrast to the mental slowing that is evident in depressed people (Bushnell et al. 1995; Overstreet et al. 2005). In addition, sensitivity to serotonin is elevated in FSL rats, as opposed to depressed humans who are subsensitive (Cleare et al. 1995), and FSL rats have a low preference for alcohol and saccharin, which is not consistent with human depression (Overstreet et al. 1992). Studies on depression-associated drug-seeking may be better modeled with Fawn-Hooded rats (not reviewed in this chapter, but see Overstreet et al. 2007). As opposed to models of human depression, FSL rats may be a better model of human chronic fatigue syndrome. For example, FSL rats share hormonal responses to a cholinergic agent similar to that seen in people with chronic fatigue syndrome (i.e., prolactin, growth hormone, Overstreet and

Djuric 2001; Overstreet et al. 2005). Additionally, chronic fatigue patients are also supersensitive to the effects of serotonin, comparable with FSL rats (Cleare et al. 1995). More research on the relation between the FSL profile and chronic fatigue syndrome is warranted.

Relationship between personality and health in animals and humans

HOSTILITY/AGGRESSION

One of the earliest links discovered between personality and health was the relationship between Type A personality and coronary heart disease. Rosenman and Friedman (1974) noted an abnormal pattern of wear on the chairs in the waiting room of their cardiology clinic. Sometime later, the doctors realized that their cardiac patients made a habit of sitting on the edge of their seats, impatient and harried. Hostility was later identified as a critical trait that negatively influenced cardiovascular health (Dembroski et al. 1985; Krantz et al. 2006). More recent research has implicated social dominance and depression in the pathogenesis of negative cardiac outcomes, with depressive tendencies potentially playing a moderating role in the connection between hostility and cardiac health (Houston et al. 1992; Ravaja et al. 2000; Trigo et al. 2005; Stewart et al. 2007). Interestingly, these are all behavioral traits that have been found to vary among rodent strains. In addition, mice strains that were aggressive in the resident-intruder paradigm (BALB/c and DBA/2) also had elevated blood pressure, which is a risk factor for poor cardiac health, compared with less aggressive mice strains (e.g., C3H/He, see also Lockwood and Turney 1981). As an exception, this relationship did not hold in F344 rats that are not aggressive but have high systolic blood pressure and are prone to a cardiovascular condition—cardiomyopathy. It may be that rodents specifically bred for hostile-like behavior, i.e., the SAL/LAL mice, provide the best model for human hostility. These mice have also been characterized as "active copers" (Koolhaas 2008), and it has been argued that active copers (both human and animal) have greater sympathetic activation than HPA activation in response to challenges (Coppens et al. 2010). This physiological bias may be one mechanism that makes them more prone to cardiovascular, allergic, and autoimmune diseases than "passive copers" (Zozulya et al. 2008). The relationship between human hostility and cardiac health suggests further examination of sympathetic activity in SAL/LAL mice to determine whether they provide a potential model of human hostility and cardiac health. Furthermore, signs of elevated blood pressure in high aggressive mouse strains (BALB/c and DBA/2) suggest that the relationship between human hostility and cardiac health may be modeled in laboratory rodents.

Additional information from a few strains that were not reviewed in this chapter indicates some protective aspects of an aggressive behavioral bias. For example, inbred NC900 mice that are highly aggressive in a social interaction test following isolation developed fewer tumors after chemical carcinogen (3-methycholanthrene) exposure compared with less aggressive NC100 mice. The difference was quite dramatic, with 100% of the low-aggressive NC100 mice developing tumors 17 weeks following injection, while only 44% of the NC900 mice developed tumors. In addition, NC100 mice had diminished natural killer cell cytotoxicity in response to tumor cell exposure compared with NC900 mice (Petitto et al. 1993). These results are comparable with early reports showing that spontaneously aggressive female mice developed smaller tumors in response to murine sarcoma virus compared with age-matched nonfighting females (Amkraut and Solomon 1972). In addition to decreased induced-tumor susceptibility, the aggressive NC900 line had a range of heightened cellular immune responses to Concanavalin A (greater natural killer cell activity, T-cell proliferation, and cytokine responses) compared with the less aggressive NC100 line (Petitto et al. 1994). Another protective aspect of aggression has emerged from two other mouse lines bred according to aggressive tendencies—the Turku Aggressive (TA) vs. Turku Nonaggressive (TNA) lines. In this case, the median life span for males of the aggressive TA line was approximately 300 days longer than that of the nonaggressive TNA line (Ewalds-Kvist and Selander 1996). The possible benefit of a longer life span in aggressive individuals is further supported by the fact that SAL mice (with a rapid attack response) have lower serum antioxidant capacity than LAL (slower attack latency) mice, a trait associated with healthy aging (Costantini et al. 2008). These positive benefits of aggressive behavioral tendencies were not evident from the rodent strains reviewed in this chapter, but these potentially "protective" aspects of aggression (i.e., tumor resilience, lengthened life span) are important to note in comparison with the "negative" health outcomes associated with aggression (i.e., cardiovascular disease processes). This combination of resilience to certain disease processes and susceptibility to other processes associated with the same behavioral trait points to the importance of documenting the full and specific health biases associated with each behavioral trait. By identifying both disease resilience and susceptibility of a given behavioral type, we can more accurately infer possible physiological mechanisms underlying health outcomes and potential interventions for specific traits. With animal models of hostility, we can experimentally determine how interventions that minimize one kind of disease susceptibility might influence resilience to another disease process.

BEHAVIORAL INHIBITION/SHYNESS

Behaviorally inhibited or shy children and adults are more likely to suffer from nasal allergies and anxiety disorders (Bell et al. 1990; 1995; Hirshfeld et al. 1992; Turner et al. 1996). In addition, certain people who are particularly sensitive to chemical exposure (e.g., perfumes, pollutants) also tend to have higher levels of anxiety and depression (Bell et al. 1995). This human condition may be analogous to that seen in Flinders sensitive line rats that are hypersensitive to organophosphate and prone to allergic and autoimmune responses as well as anhedonia. Alternatively, there are other rodent lines that may prove analogous to human behavioral inhibition/shyness. Specifically, RLA rats bred for their slow response to a two-way shuttle box active avoidance task and neophobic Sprague-Dawley identified according to their response to novel complex settings may be models of human behavioral inhibition. At this point, there is no evidence that either of these rodent models show elevated signs of allergies or physiological profiles associated with anxiety disorders. However, both the RLA and the neophobic rats produce elevated levels of glucocorticoid hormones—comparable with behaviorally inhibited children (Gentsch et al. 1982; Kagan et al. 1987; Dettling et al. 2000; Cavigelli et al. 2007)—and in the case of neophobic Sprague-Dawley rats they also have a significantly shorter life span than neophilic rats. This difference in life span is analogous to the difference in life span reported for curious vs. noncurious older adults (Swan and Carmelli 1996). These preliminary findings support further investigation of these two types of rodents as models of human behavioral inhibition, although further work is needed to elucidate their physiological and health profiles to better determine their applicability. Alternatively, inbred strains that show elevated anxiety-like behavior along with elevated basal glucocorticoid production and cardiovascular reactivity, and a propensity toward allergies (as seen in human behavioral inhibition), may be good models. For example, from the progenitor strains, A/J, BALB/c, or BN strains may prove valuable with their elevated anxiety-like behavior, elevated glucocorticoid levels, and immune profiles that may bias them toward allergic responses. On the basis of the comparative analysis from this chapter, several rodent lines may prove useful for modeling human behavioral inhibition.

In humans, it has been well documented that sensation-seeking individuals, particularly adolescents, are more prone to use drugs than others (Newcomb and McGee 1991; Wills et al. 1994; Donohew et al. 1999; Martin et al. 2002). A similar relationship has been identified in Sprague-Dawley rats (Cain et al. 2005) and the novelty-seeking RHA rats that have a higher preference for several drugs compared with RLA rats (Giorgi et al. 1997; 2005; Piras et al. 2003; Corda et al. 2005; Fattore et al. 2009). RLA rats appear to be insensitive to the behavioral sensitization seen in RHA rats after drug administration (Giorgi et al. 2007). Another rodent model that may help understand health outcomes in human sensation-seeking is the HR rats identified by their locomotor response to a circular corridor (Piazza et al. 1989; Dellu et al. 1996). Specifically, HR rats that move a lot in the corridor test will self-administer amphetamines whereas the less active LR rats will not (Dellu et al. 1996). The HR rats share several physiological traits with high sensation-seeking humans, making them a potential model of this human personality trait. Further investigation of their health consequences may provide insight into other health consequences associated with human sensation-seeking.

Conclusions

One area that will require further attention in the study of rodent personality variation and health is the continued use of both repeated and varied behavioral testing to characterize personalities. Repeated and varied testing has been conducted with some of the rodent lines to determine stability of behavioral traits (e.g., Dellu et al. 1994; Wigger et al. 2001; Leppänen et al. 2006; Cavigelli et al. 2007), but not in all. This repeated testing is particularly important because behavioral biases, even within inbred rodent strains, can be highly variable across time and across laboratories (Wahlsten et al. 2006). In addition, because of this cross-laboratory variability, comparisons of behavioral and physiological biases among strains should be conducted within laboratories as much as possible.

Some additional methodological points follow: if a rodent line is bred to mimic a human personality trait, then the behavioral tests used to identify these rodents for selective breeding programs should be carefully identified in such a way that the selected trait mimics the human trait as closely as possible. For example, behavioral inhibition should be characterized on the basis of a rodent's response to multiple novel conditions, as is done in the identification of human inhibition, as opposed to responses to just one condition (Cavigelli et al. 2007; 2009). Identifying traits that are primarily defined in humans according to cognitive processes (e.g., hostility, optimism/ pessimism) may be difficult in rodent models. However, rodent behavioral responses to different situations (e.g., willingness/latency/intensity to attack a social partner, or latency to become immobile in a forced swim test, or responses to ambiguous cues, e.g., Harding et al. 2004) can provide insight into such cognitive processes. The most effective method to identify these traits may be to include multiple testing with both established and novel behavioral testing paradigms, and to establish testing procedures that identify a naturally occurring range of behavioral biases as opposed to pathological extremes. In addition to multiple behavioral testing schemes, analysis of physiological biases that are related to behavioral biases will strengthen arguments that a rodent biobehavioral trait represents the analog of a human biobehavioral trait (e.g., Koolhaas et al. 1999; Landgraf 2003; Zozulya et al. 2008; Coppens et al. 2010; see chapter 12). Testing rodent and human behavioral and physiological responses to multiple and varied conditions will be the most accurate method to distinguish complex traits that are associated with health outcomes (e.g., Kagan et al. 2002).

On the basis of the research with rodent strains reviewed above, in order to best study physiological and health correlates of personality in rodents, we will be best served by studying strains developed specifically according to their behavioral profiles (i.e., the latter strains reviewed in this chapter). More specifically, if the goal is to understand the development, physiological substrate, and health implications of specific personality traits, then the most promising method will be to identify personality traits in rodents that most closely mimic human personality traits. This may be especially effective for traits that exhibit the same behavior-physiology relationships in rodents and humans. A different but complementary approach involves more fully characterizing personality profiles in rodent models that have been developed to exhibit susceptibility or resilience to a certain disease process (e.g., Crawley 2008).

In conclusion, further research with the rodent models discussed in this chapter has the potential to enhance our understanding of the relationships between behavioral, physiological, and genetic components of animal personalities. Furthermore, rodent models provide a unique opportunity to conduct experimental studies aimed at understanding the mechanisms through which personality variation affects health. These experimental studies may be difficult to conduct with humans, but may prove quite useful in developing environmental, behavioral, and physiological manipulations that help minimize negative health outcomes in individuals with particular personalities. Finally, the findings of research with rodent models also indicate promising lines of future research with humans such as the further examination of the following associations between personality traits and health: (1) anxiety (or fear) and life span, (2) aggression and life span, (3) activity levels and tumor development, (4) aggression and tumor development, and (5) anxiety and autoimmunity.

References

Adarichev, V. A., Vegvari, A., Szabo, Z., Kis-Toth, K., Mikecz, K., and Glant, T. T. 2008. Congenic strains displaying similar clinical phenotype of arthritis represent different immunologic models of inflammation. *Genes and Immunity*, 9, 591–601.

Altun, M., Bergman, E., Edström, E., Johnson, H., and Ulfhake, B. 2007. Behavioral impairments of the aging rat. *Physiology and Behavior*, 92, 911–923.

Amkraut, A., and Solomon, G. F. 1972. Stress and murine sarcoma virus (Moloney)–induced tumors. *Cancer Research*, 32, 1428–1433.

Andrews, J. S., Jansen, J. H., Linders, S., Princen, A., and Broekkamp, C. L. 1995. Performance of four different rat strains in the autoshaping, two-object discrimination, and swim maze tests of learning and memory. *Physiology and Behavior*, 57, 785–790.

Armario, A., Gavaldà, A., and Martí, J. 1995. Comparison of the behavioural and endocrine response to forced swimming stress in five inbred strains of rats. *Psychoneuroendocrinology*, 20, 879–890.

Ballaz, S. J., Akil, H., and Watson, S. J. 2007. Previous experience affects subsequent anxiety-like responses in rats bred for novelty seeking. *Behavioral Neuroscience*, 121, 1113–1118.

Bárdos, T., Szabó, Z., Czipri, M., Vermes, C., Tunyogi-Csapó, M., Urban, R. M., Mikecz, K., and Glant, T. T. 2005. A longitudinal study on an autoimmune murine model of ankylosing spondylitis. *Annals of the Rheumatic Diseases*, 64, 981–987.

Barnes, C. A. 1979. Memory deficits associated with senescence: a neurophysiological and behavioral study in the rat. *Journal of Comparative and Physiological Psychology*, 93, 74–104.

Baudrie, V., Laude, D., Chaouloff, F., and Elghozi, J.-L. 2001. Genetic influences on cardiovascular responses to an acoustic startle stimulus in rats. *Clinical and Experimental Pharmacology and Physiology*, 28, 1096–1099.

Baum, A., Pohlmeyer, G., Rapp, K. G., and Deerberg, F. 1995. Lewis rats of the inbred strain LEW/Han: life expectancy, spectrum, and incidence of spontaneous neoplasms. *Experimental Toxicology and Pathology*, 47, 11–18.

Bell, I. R., Hardin, E., Baldwin, C. M., and Schwartz, G. E. 1995. Increased limbic system symptomatology and sensitizability of young adults with chemical and noise sensitivities. *Environmental Research*, 70, 84–97.

Bell, I. R., Jasnoski, M. L., Kagan, J., and King, D. S. 1990. Is allergic rhinitis more frequent in young adults with extreme shyness? A preliminary survey. *Psychosomatic Medicine*, 52, 517–525.

Benca, R. M., Obermeyer, W. H., Thisted, R. A., and Gillin, J. C. 1992. Sleep and psychiatric disorders: a meta-analysis. *Archives of General Psychiatry*, 49, 651–668.

Benca, R. M., Overstreet, D. E., Gilliland, M. A., Russell, D., Bergmann, B. M., and

Obermeyer, W. H. 1996. Increased basal REM sleep but no difference in dark induction or light suppression of REM sleep in Flinders rats with cholinergic supersensitivity. *Neuropsychopharmacology*, 15, 45–51.

Benus, R. F., Bohus, B., Koolhaas, J. M., and Van Oortmerssen, G. A. 1989. Behavioural strategies of aggressive and nonaggressive male mice in active shock avoidance. *Behavioural Processes*, 20, 1–12.

———. 1991. Behavioural differences between artificially selected aggressive and nonaggressive mice: response to apomorphine. *Behavioural Brain Research*, 15, 203–208.

Benus, R. F., Koolhaas, J. M., and Van Oortmerssen, G. A. 1987. Individual differences in behavioural reactions to a change environment in mice and rats. *Behaviour*, 100, 105–122.

Berton, O., Ramos, A., Chaouloff, F., and Mormède, P. 1997. Behavioral reactivity to social and nonsocial stimulations: a multivariate analysis of six inbred rat strains. *Behavior Genetics*, 27, 155–166.

Bignami, G. 1965. Selection for high rates and low rates of avoidance conditioning in the rat. *Animal Behaviour*, 13, 221–227.

Bishop, D. S. 1988. Depression and rheumatoid arthritis. *Journal of Rheumatology*, 15, 888-899.

Bolivar, V. J., Walters, S. R., and Phoenix, J. L. 2007. Assessing autism-like behavior in mice: variations in social interactions among inbred strains. *Behavioural Brain Research*, 176, 21–26.

Bothe, G. W. M., Bolivar, V. J., Vedder, M. J., and Geistfeld, J. G. 2005. Behavioral differences among fourteen inbred mouse strains commonly used as disease models. *Comparative Medicine*, 55, 326–334.

Brooks, S. P., Pask, T., Jones, L., and Dunnett, S. B. 2005. Behavioural profiles of inbred mouse strains used as transgenic backgrounds, II: cognitive tests. *Genes, Brain, and Behavior*, 4, 307–317.

Brown, R. E., Gunn, R. K., Schellinck, H. M., Wong, A. A., and O'Leary, T. P. 2004a. Anxiety, exploratory behavior, and motor activity. MPD: 94. Mouse Phenome Database website, Jackson Laboratory, Bar Harbor, Maine. http://www.jax.org/phenome.

Brown, R. E., Schellinck, H. M., Gunn, R. K., Wong, A. A., and O'Leary, T. P. 2004b. Visual ability and spatial, motor, and olfactory learning and memory. MPD: 225. Mouse Phenome Database website, Jackson Laboratory, Bar Harbor, Maine. http://www.jax.org/phenome.

Bunde, J., and Suls, J. 2006. A quantitative analysis of the relationship between the Cook-Medley Hostility Scale and traditional coronary artery disease risk factors. *Health Psychology*, 25, 493–500.

Burek, J. D., and Hollander, C. F. 1975. Studies of spontaneous lesions in aging BN/Bi rats, I: neoplastic and nonneoplastic lesions. In: *Annual Report, Organization for Health Research*, pp. 235–237. Rijswijk, the Netherlands: TNO.

Bushnell, P. J., Levin, E. D., and Overstreet, D. H. 1995. Spatial working and reference memory in rats bred for autonomic sensitivity to cholinergic stimulation: acquisition, accuracy, speed, and effects of cholinergic drugs. *Neurobiology of Learning and Memory*, 63, 116–132.

Cai, X., Wong, Y. F., Zhou, H., Xie, Y., Liu, Z. Q., Jiang, J. H., Bian, Z. X., Xu, H. X., and

Liu, L. 2006. The comparative study of Sprague-Dawley and Lewis rats in adjuvant-induced arthritis. *Naunyn-Schmiedeberg's Archives of Pharmacology*, 373, 140–147.

Cain, M. E., Saucier, D. A., and Bardo, M. T. 2005. Novelty seeking and drug use: contribution of an animal model. *Experimental and Clinical Psychopharmacology*, 13, 367–375.

Capitanio, J. P., Mendoza, S. P., and Baroncelli, S. 1999. The relationship of personality dimensions in adult male rhesus macaques to progression of simian immunodeficiency virus disease. *Brain, Behavior, and Immunity*, 13, 138–154.

Capitanio, J. P., Mendoza, S. P., and Bentson, K. L. 2004. Personality characteristics and basal cortisol concentrations in adult male rhesus macaques (*Macaca mulatta*). *Psychoneuroendocrinology*, 29, 1300–1308.

Caramaschi, D., de Boer, S. F., and Koolhaas, J. M. 2007. Differential role of the 5-HT$_{1A}$ receptor in aggressive and nonaggressive mice: an across-strain comparison. *Physiology and Behavior*, 90, 590–601.

Caramaschi, D., de Boer, S. F., de Vries, H., and Koolhaas, J. M. 2008a. Development of violence in mice through repeated victory along with changes in prefrontal cortex neurochemistry. *Behavioural Brain Research*, 189, 263–272.

Caramaschi, D., de Boer, S. F., and Koolhaas, J. M. 2008b. Is hyperaggressiveness associated with physiological hypoarousal? A comparative study on mouse lines selected for high and low aggressiveness. *Physiology and Behavior*, 95, 591–598.

Carere, C., and Van Oers, K. 2004. Shy and bold great tits (*Parus major*): body temperature and breath rate in response to handling stress. *Physiology and Behavior*, 82, 905–912.

Carobrez, A. P., and Bertoglio, L. J. 2005. Ethological and temporal analyses of anxiety-like behavior: the elevated plus maze model 20 years on. *Neuroscience and Biobehavioral Reviews*, 29, 1193–1205.

Castanon, N., Dulluc, J., Le Moal, M., and Mormède, P. 1992. Prolactin as a link between behavioral and immune differences between the Roman rat lines. *Physiology and Behavior*, 51, 1235–1241.

———. 1994. Maturation of the behavioral and neuroendocrine differences between the Roman rat lines. *Physiology and Behavior*, 55, 775–782.

Cavigelli, S. A. 2005. Animal personality and health. *Behaviour*, 142, 1223–1244.

Cavigelli, S. A., and McClintock, M. K. 2003. Fear of novelty in infant rats predicts adult corticosterone dynamics and an early death. *Proceedings of the National Academy of Sciences U S A*, 100, 16131–16136.

Cavigelli, S. A., Ragan, C. M., Michael, K. C., Kovacsics, C. E., and Brusche, A. P. 2009. Stable behavioral inhibition and glucocorticoid production as predictors of longevity. *Physiology and Behavior*, 98, 205–214.

Cavigelli, S. A., Stine, M. M., Kovacsics, C. E., Jefferson, A., Diep, M. N., and Barrett, C. E. 2007. Behavioral inhibition and glucocorticoid dynamics in a rodent model. *Physiology and Behavior*, 92, 897–905.

Cavigelli, S. A., Yee, J. R., and McClintock, M. K. 2006. Infant temperament predicts life span in female rats that develop spontaneous tumors. *Hormones and Behavior*, 50, 454–462.

Cleare, A. J., Bearn, J., Allain, T., McGregor, A., Wessely, S., Murray, R. M., and O'Keane, V. 1995. Contrasting neuroendocrine responses in depression and chronic fatigue syndrome. *Journal of Affective Disorders*, 34, 283–289.

Clinton, S., Miller, S., Stanley, S. J., Watson, J., and Akil, H. 2008. Prenatal stress does not alter innate novelty-seeking behavioral traits, but differentially affects individual differences in neuroendocrine stress responsivity. *Psychoneuroendocrinology*, 33, 162–177.

Clinton, S. M., Vázquez, D. M., Kabbaj, M., Kabbaj, M.-H., Watson, S. J., and Akil, H. 2007. Individual differences in novelty seeking and emotional reactivity correlate with variation in maternal behavior. *Hormones and Behavior*, 51, 655–664.

Cone, J. E., and Sult, T. A. 1992. Acquired intolerance to solvents following pesticide/solvent exposure in a building: a new group of workers at risk for multiple chemical sensitivities? *Toxicology and Industrial Health*, 8, 29–39.

Coppens, C. M., de Boer, S. F., and Koolhaas, J. M. 2010. Coping styles and behavioural flexibility: towards underlying mechanisms. *Philosophical Transactions of the Royal Society of London B*, 365, 4021–4028.

Corda, M. G., Piras, G., Lecca, D., Fernández-Teruel, A., Driscoll, P., and Giorgi, O. 2005. The psychogenetically selected Roman rat lines differ in the susceptibility to develop amphetamine sensitization. *Behavioural Brain Research*, 157, 147–156.

Costall, B., Jones, B. J., Kelly, M. E., Naylor, R. J., and Tomkins, D. M. 1989. Exploration of mice in a black and white test box: validation as a model of anxiety. *Pharmacology, Biochemistry, and Behavior*, 32, 777–785.

Costantini, D., Carere, C., Caramaschi, D., and Koolhaas, J. M. 2008. Aggressive and nonaggressive personalities differ in oxidative status in selected lines of mice (*Mus musculus*). *Biology Letters*, 4, 119–122.

Crawley, J. N. 1981. Neuropharmacologic specificity of a simple animal model for the behavioral actions of benzodiazepines. *Pharmacology, Biochemistry, and Behavior*, 15, 695–699.

———. 2003. Behavioral phenotyping of rodents. *Comparative Medicine*, 53, 140–146.

———. 2008. Behavioral phenotyping strategies for mutant mice. *Neuron*, 57, 809–818.

Crawley, J. N., and Goodwin, F. K. 1980. Preliminary report of a simple animal behavior model for the anxiolytic effects of benzodiazepines. *Pharmacology, Biochemistry, and Behavior*, 13, 167–170.

Crawley, J. N., and Paylor, R. 1997. A proposed test battery and constellations of specific behavioral paradigms to investigate the behavioral phenotypes of transgenic and knockout mice. *Hormones and Behavior*, 31, 197–211.

Cross, R. J., Campbell, J. L., and Roszman, T. L. 1989. Potentiation of antibody responsiveness after the transplantation of a syngeneic pituitary gland. *Journal of Neuroimmunology*, 25, 29–35.

Curé, M. M., and Rolinat, J. P. 1992. Behavioral heterogeneity in Sprague-Dawley rats. *Physiology and Behavior*, 51, 771–774.

Davis, B. A., Clinton, S. M., Akil, H., and Becker, J. B. 2008. The effects of novelty-seeking phenotypes and sex differences on acquisition of cocaine self-administration in selectively bred high-responder and low-responder rats. *Pharmacology, Biochemistry, and Behavior*, 90, 331–338.

Dawe, S., and Loxton, N. J. 2004. The role of impulsivity in the development of substance use and eating disorders. *Neuroscience and Biobehavioral Reviews*, 28, 343–351.

De Boer, S. F., and Koolhaas, J. M. 2003. Defensive burying in rodents: ethology, neurobiology, and psychopharmacology. *European Journal of Pharmacology*, 463, 145–161.

Dellu, F., Mayo, W., Vallée, M., Le Moal, M., and Simon, H. 1994. Reactivity to novelty during youth as a predictive factor of cognitive impairment in the elderly: a longitudinal study in rats. *Brain Research*, 653, 51–56.

Dellu, F., Mayo, W., Vallée, M., Maccari, S., Piazza, P. V., Le Moal, M., and Simon, H. 1996. Behavioral reactivity to novelty during youth as a predictive factor of stress-induced corticosterone secretion in the elderly—a life-span study in rats. *Psychoneuroendocrinology*, 21, 441–453.

Dembroski, T. M., MacDougall, J. M., Williams, R. B., Haney, T. L., and Blumenthal, J. A. 1985. Components of Type A, hostility, and anger in relationship to angiographic findings. *Psychosomatic Medicine*, 47, 219–233.

Deschepper, C. F., and Gallo-Payet, N. 2004. Systolic blood pressure and morphologic phenotyping of cardiovascular-related organs. MPD: 104. Mouse Phenome Database website, Jackson Laboratory, Bar Harbor, Maine. http://www.jax.org/phenome.

Deschepper, C. F., Olson, J. L., Otis, M., and Gallo-Payet, N. 2004. Characterization of blood pressure and morphological traits in cardiovascular-related organs in 13 different inbred mouse strains. *Journal of Applied Physiology*, 97, 369–376.

Dettling, A. C., Parker, S. W., Lane, S., Sebanc, A., and Gunnar, M. R. 2000. Quality of care and temperament determine changes in cortisol concentrations over the day for young children in childcare. *Psychoneuroendocrinology*, 25, 819–836.

Dhabhar, F. S., McEwen, B. S., and Spencer, R. L. 1993. Stress response, adrenal steroid receptor levels, and corticosteroid-binding globulin levels: a comparison between Sprague-Dawley, Fischer 344, and Lewis rats. *Brain Research*, 616, 89–98.

———. 1997. Adaptation to prolonged or repeated stress: comparison between rat strains showing intrinsic differences in reactivity to acute stress. *Neuroendocrinology*, 65, 360–368.

Dhabhar, F. S., Miller, A. H., McEwen, B. S., and Spencer, R. L. 1995. Differential activation of adrenal steroid receptors in neural and immune tissues of Sprague Dawley, Fischer 344, and Lewis rats. *Journal of Neuroimmunology*, 56, 77–90.

Djurić, V. J., Cox, G., Overstreet, D. H., Smith, L., Dragomir, A., and Steiner, M. 1998. Genetically transmitted cholinergic hyperresponsiveness predisposes to experimental asthma. *Brain, Behavior, and Immunity*, 12, 272–284.

Djurić, V. J., Overstreet, D. H., Bienenstock, J., and Perdue, M. H. 1995. Immediate hypersensitivity in the Flinders rat: further evidence for a possible link between susceptibility to allergies and depression. *Brain, Behavior, and Immunity*, 9, 196–206.

Donohew, R. L., Hoyle, R. H., Clayton, R. R., Skinner, W. F., Colon, S. E., and Rice, R. E. 1999. Sensation seeking and drug use by adolescents and their friends: models for marijuana and alcohol. *Journal of Studies on Alcohol*, 60, 622–631.

Dorn, L. D., Burgess, E. S., Dubbert, B., Simpson, S. E., Friedman, T., Kling, M., Gold, P. W., and Chrousos, G. P. 1995. Psychopathology in patients with endogenous Cushing's syndrome: "atypical" or melancholic features. *Clinical Endocrinology*, 43, 433–442.

Dorshkind, K., and Horseman, N. D. 2000. The roles of prolactin, growth hormone, insulin-like growth factor-I, and thyroid hormones in lymphocyte development and function: insights from genetic models of hormone and hormone receptor deficiency. *Endocrine Reviews*, 21, 292–312.

Drent, P. J., Van Oers, K., and Van Noordwijk, A. J. 2003. Realised heritability of personalities in the great tit (*Parus major*). *Proceedings of the Royal Society of London B*, 270, 45–51.

Driscoll, P., and Battig, K. 1982. Behavioral, emotional, and neurochemical profiles of rats selected for extreme differences in active, two-way avoidance performance. In: *Genetics of the Brain* (Lieblich, I., ed.), pp. 95–123. Amsterdam: Elsevier.

Ducottet, C., and Belzung, C. 2005. Correlations between behaviours in the elevated plus maze and sensitivity to unpredictable subchronic mild stress: evidence from inbred strains of mice. *Behavioural Brain Research*, 156, 153–162.

Elwood, W., Lötvall, J. O., Barnes, P. J., and Chung, K. F. 1991. Characterization of allergen- induced bronchial hyperresponsiveness and airway inflammation in actively sensitized brown-Norway rats. *Journal of Allergy and Clinical Immunology*, 88, 951–960.

Everson-Rose, S. A., and Lewis, T. T. 2005. Psychosocial factors and cardiovascular disease. *Annual Review of Public Health*, 26, 469–500.

Ewalds-Kvist, S. B., and Selander, R.-K. 1996. Life spans in mice from strains selected for high or low aggression. *Aggressive Behavior*, 22, 457–464.

Fattore, L., Piras, G., Corda, M. G., and Giorni, O. 2009. The Roman high- and low-avoidance rat lines differ in the acquisition, maintenance, extinction, and reinstatement of intravenous cocaine self-administration. *Neuropsychopharmacology*, 34, 1091–1101.

Ferraro, T. N., Golden, G. T., Smith, G. G., and Berrettini, W. H. 1995. Differential susceptibility to seizures induced by systemic kainic acid treatment in mature DBA/2J and C57BL/6J mice. *Epilepsia*, 36, 301–307.

Flaherty, C. F., Powell, G., and Hamilton, L. W. 1979. Septal lesion, sex, and incentive shift effects on open field behavior of rats. *Physiology and Behavior*, 22, 903–909.

Fournié, G. J., Cautain, B., Xystrakis, E., Damoiseaux, J., Mas, M., Lagrange, D., Bernard, I., Subra, J. F., Pelletier, L., Druet, P., and Saoudi, A. 2001. Cellular and genetic factors involved in the difference between Brown Norway and Lewis rats to develop respectively type-2 and type-1 immune-mediated diseases. *Immunology Reviews*, 184, 145–160.

Freedman, L. S., Clifford, C., and Messina, M. 1990. Analysis of dietary fat, calories, body weight, and the development of tumors in rats and mice: a review. *Cancer Research*, 50, 5710–5719.

Friedman, E. M., Irwin, M. R., and Overstreet, D. H. 1996. Natural and cellular immune responses in Flinders sensitive and resistant line rats. *Neuropsychopharmacology*, 15, 314–322.

Friedman, E. M., Becker, K. A., Overstreet, D. H., and Lawrence, D. A. 2002. Reduced primary antibody responses in a genetic animal model of depression. *Psychosomatic Medicine*, 64, 267–273.

Friedman, M., and Rosenman, R. H. 1971. Type A behavior pattern: its association with coronary heart disease. *Annals of Clinical Research*, 3, 300–312.

Gallou-Kabani, C., Vigé, A., Gross, M. S., Rabès, J. P., Boileau, C., Larue-Achagiotis, C., Tomé, D., Jais, J. P., and Junien, C. 2007. C57BL/6J and A/J mice fed a high-fat diet delineate components of metabolic syndrome. *Obesity*, 15, 1996–2005.

Gariépy, J.-L., Bauer, D., and Cairns, R. B. 2001. Selective breeding for differential aggression in mice provides evidence for heterochrony in social behaviours. *Animal Behaviour*, 61, 1–15.

Gasparotto, O. C., Carobrez, S. G., and Bohus, B. G. 2007. Effects of LPS on the behavioural stress response of genetically selected aggressive and nonaggressive wild house mice. *Behavioural Brain Research*, 183, 52–59.

Gentsch, C., Lichtsteiner, M., Driscoll, P., and Feer, H. 1982. Differential hormonal and physiological responses to stress in Roman high- and low-avoidance rats. *Physiology and Behavior*, 28, 259–263.

Ghirardi, O., Cozzolino, R., Guaraldi, D., and Giuliani, A. 1995. Within- and between-strain variability in longevity of inbred and outbred rats under the same environmental conditions. *Experimental Gerontology*, 30, 485–494.

Gilad, G. M., and Gilad, V. H. 1995. Strain, stress, neurodegeneration, and longevity. *Mechanisms of Ageing and Development*, 78, 75–83.

Giorgi, O., Corda, M. G., Carboni, G., Frau, V., Valentini, V., and Di Chiara, G. 1997. Effects of cocaine and morphine in rats from two psychogenetically selected lines: a behavioral and brain dialysis study. *Behavior Genetics*, 27, 537–546.

Giorgi, O., Piras, G., and Corda, M. G. 2007. The psychogenetically selected Roman high- and low-avoidance rat lines: a model to study the individual vulnerability to drug addiction. *Neuroscience and Biobehavioral Reviews*, 31, 148–163.

Giorgi, O., Piras, G., Lecca, D., and Corda, M. G. 2005. Behavioural effects of acute and repeated cocaine treatments: a comparative study in sensitisation-prone RHA rats and their sensitisation-resistant RLA counterparts. *Psychopharmacology*, 180, 530–538.

Glant, T. T., Finnegan, A., and Mikecz, K. 2003. Proteoglycan-induced arthritis: immune regulation, cellular mechanisms, and genetics. *Critical Reviews in Immunology*, 23, 199- 250.

Glowa, J. R., Geyer, M. A., Gold, P. W., and Sternberg, E. M. 1992. Differential startle amplitude and corticosterone response in rats. *Neuroendocrinology*, 56, 719–723.

Gómez, F., Lahmame, A., de Kloet, E. R., and Armario, A. 1996. Hypothalamic-pituitary-adrenal response to chronic stress in five inbred rat strains: differential responses are mainly located at the adrenocortical level. *Neuroendocrinology*, 63, 327–337.

Gómez, F., De Kloet, E. R., Armario, A. 1998. Glucocorticoid negative feedback on the HPA axis in five inbred rat strains. *American Journal of Physiology*, 274, R420–R427.

Goodrick, C. L. 1975. Life span and the inheritance of longevity of inbred mice. *Journal of Gerontology*, 30, 257–263.

Green, E. L., ed. 1966. *Biology of the Laboratory Mouse.* (2nd ed.) New York: McGraw-Hill.

Griffin, A. C., and Whitacre, C. C. 1991. Sex and strain differences in the circadian rhythm fluctuation of endocrine and immune function in the rat: implications for rodent models of autoimmune disease. *Journal of Neuroimmunology*, 35, 53–64.

Grossen, N. E., and Kelly, M. J. 1972. Species-specific behavior and acquisition of avoidance behavior in rats. *Journal of Comparative and Physiological Psychology*, 81, 307–310.

Grota, L. J., Bienen, T., and Felten, D. L. 1997. Corticosterone responses of adult Lewis and Fischer rats. *Journal of Neuroimmunology*, 74, 95–101.

Grubb, S. C., Maddatu, T. P., Bult, C. J., and Bogue, M. A. 2009. Mouse phenome database. *Nucleic Acids Research*, 37, D720–D730.

Guayerbas, N., Catalán, M., Victor, V. M., Miquel, J., and De la Fuente, M. 2002. Relation of behaviour and macrophage function to life span in a murine model of premature immunosenescence. *Behavioural Brain Research*, 134, 41–48.

Habra, M. E., Linden, W., Anderson, J. C., and Weinberg, J. 2003. Type D personality is related to cardiovascular and neuroendocrine reactivity to acute stress. *Journal of Psychosomatic Research*, 55, 235–245.

Haddjeri, N., Faure, C., Lucas, G., Mnie-Filali, O., Chouvet, G., Astier, B., Renaud, B., Blier, P., and Debonnel, G. 2004. In-vivo modulation of central 5-hydroxytryptamine (5-HT1A) receptor-mediated responses by the cholinergic system. *International Journal of Neuropsychopharmacology*, 7, 391–399.

Hall, C. S. 1934. Emotional behavior in the rat: defecation and urination as measures of individual differences in emotionality. *Journal of Comparative Physiology*, 18, 385–403.

Haller, J., and Kruk, M. R. 2006. Normal and abnormal aggression: human disorders and novel laboratory models. *Neuroscience and Biobehavioral Reviews*, 30, 292–303.

Hampton, T. G., Paigen, B., and Seburn, K. L. 2001. Cardiac characterization. MPD: 17. Mouse Phenome Database website, Jackson Laboratory, Bar Harbor, Maine. http://www.jax.org/phenome.

Harding, E. J., Paul, E. S., and Mendl, M. 2004. Cognitive bias and affective state. *Nature*, 427, 312.

Harizi, H., Homo-Delarche, F., Amrani, A., Coulaud, J., and Mormède, P. 2007. Marked genetic differences in the regulation of blood glucose under immune and restraint stress in mice reveals a wide range of corticosensitivity. *Journal of Immunology*, 189, 59–68.

Harrington, G. M. 1972. Strain differences in open-field behavior in the rat. *Psychonomic Sciences*, 27, 51–53.

Hasegawa, S., Nishi, K., Watanabe, A., Overstreet, D. H., and Diksic, M. 2006. Brain 5-HT synthesis in the Flinders sensitive line rat model of depression: an autoradiographic study. *Neurochemistry International*, 48, 358–366.

Henniger, M. S. H., Ohl, F., Hölter, S. M., Weißenbacher, P., Toschi, N., Lörscher, P., Wigger, A., Spanagel, R., and Landgraf, R. 2000. Unconditioned anxiety and social behaviour in two rat lines selectively bred for high and low anxiety-related behaviour. *Behavioural Brain Research*, 111, 153–163.

Hinojosa, F. R., Spricigo, L., Jr., Izídio, G. S., Brüske, G. R., Lopes, D. M., and Ramos, A. 2006. Evaluation of two genetic animal models in behavioral tests of anxiety and depression. *Behavioural Brain Research*, 168, 127–136.

Hirshfeld, D. R., Rosenbaum, J. F., Biederman, J., Bolduc, E. A., Faraone, S. V., Snidman, N., Reznick, J. S., and Kagan, J. 1992. Stable behavioral inhibition and its association with anxiety disorder. *Journal of the American Academy of Child Psychiatry*, 31, 103–111.

Hlavacova, N., Bakos, J., and Jezova, D. 2006. Differences in home cage behavior and endocrine parametres in rats of four strains. *Endocrine Regulations*, 40, 113–118.

Hogg, S. 1996. A review of the validity and variability of the elevated plus maze as an animal model of anxiety. *Pharmacology Biochemistry and Behavior*, 54, 21–30.

Holmes, A., Wrenn, C. C., Harris, A. P., Thayer, K. E., and Crawley, J. N. 2002. Behavioral profiles of inbred strains on novel olfactory, spatial, and emotional tests for reference memory in mice. *Genes, Brain, and Behavior*, 1, 55–69.

Hooks, M. S., Jones, G. H., Smith, A. D., Neill, D. B., and Justice, J. B., Jr. 1991. Response to novelty predicts the locomotor and nucleus accumbens dopamine response to cocaine. *Synapse*, 9, 121–128.

Houston, B. K., Chesney, M. A., Black, G. W., Cates, D. S., and Hecker, H. 1992. Behavioral clusters and coronary heart disease risk. *Psychosomatic Medicine*, 54, 447–461.

Huang, T. J., MacAry, P. A., Kemeny, D. M., and Chung, K. F. 1999. Effect of CD8+ T-cell depletion on bronchial hyperresponsiveness and inflammation in sensitized and allergen- exposed brown-Norway rats. *Immunology*, 96, 416–423.

Janowsky, D. S., Overstreet, D. H., and Nurnberger, J. I., Jr. 1994. Is cholinergic sensitivity a genetic marker for the affective disorders? *American Journal of Medical Genetics*, 54, 335–344.

Jindal, R. D., Thase, M. E., Fasiczka, A. L., Friedman, E. S., Buysse, D. J., Frank, E., and Kupfer, D. J. 2002. Electroencephalographic sleep profiles in single-episode and recurrent unipolar forms of major depression, II: comparison during remission. *Biological Psychiatry*, 51, 230–236.

Jones, S. E., and Brain, P. F. 1987. Performance of inbred and outbred laboratory mice in putative tests of aggression. *Behavior Genetics*, 17, 87–96.

Kabbaj, M., Devine, D. P., Savage, V. R., and Akil, H. 2000. Neurobiological correlates of individual differences in novelty-seeking behavior in the rat: differential expression of stress-related molecules. *Journal of Neuroscience*, 20, 6983–6988.

Kagan, J., Reznick, J. S., and Snidman, N. 1987. The physiology and psychology of behavioral inhibition in children. *Child Development*, 58, 1459–1473.

Kagan, J., and Snidman, N. 1991. Infant predictors of inhibited and uninhibited profiles. *Psychological Sciences*, 2, 40–44.

Kagan, J., Snidman, N., McManis, M., Woodward, S., and Hardway, C. 2002. One measure, one meaning: multiple measures, clearer meaning. *Development and Psychopathology*, 14, 463–475.

Kalin, N. H., Larson, C., Shelton, S. E., and Davidson, R. J. 1998. Asymmetric frontal brain activity, cortisol, and behavior associated with fearful temperaments in rhesus monkeys. *Behavioral Neuroscience*, 112, 286–292.

Kalin, N. H., Shelton, S. E., and Davidson, R. J. 2000. Cerebrospinal fluid corticotropin-releasing hormone levels are elevated in monkeys with patterns of brain activity associated with fearful temperament. *Biological Psychiatry*, 47, 579–585.

Kavelaars, A., Heijnen, C. J., Tennekes, R., Bruggink, J. E., and Koolhaas, J. M. 1999. Individual behavioral characteristics of wild-type rats predict susceptibility to experimental autoimmune encephalomyelitis. *Brain, Behavior, and Immunity*, 13, 279–286.

Keck, M. E., Sartori, S. B., Welt, T., Müller, M. B., Ohl, F., Holsboer, F., Landgraf, R., and Singewald, N. 2005. Differences in serotonergic neurotransmission between rats displaying high or low anxiety/depression–like behaviour: effects of chronic paroxetine treatment. *Journal of Neurochemistry*, 92, 1170–1179.

Keck, M. E., Wigger, A., Welt, T., Müller, M. B., Gesing, A., Reul, J. M., Holsboer, F., Landgraf, R., and Neumann, I. D. 2002. Vasopressin mediates the response of the combined dexamethasone/CRH test in hyperanxious rats: implications for pathogenesis of affective disorders. *Neuropsychopharmacology*, 26, 94–105.

Kelly, T. H., Robbins, G., Martin, C. A., Fillmore, M. T., Lane, S. D., Harrington, N. G., and Rush, C. R. 2006. Individual differences in drug abuse vulnerability: d-amphetamine and sensation-seeking status. *Psychopharmacology*, 189, 17–25.

Kitten, A., Griffey, S., Chang, F., Dixon, K., Browne, C., and Clary, D. 2003. Multisystem analysis of mouse physiology. MPD: 151. Mouse Phenome Database website, Jackson Laboratory, Bar Harbor, Maine. http://www.jax.org/phenome.

Koolhaas, J. M. 2008. Coping style and immunity in animals: making sense of individual variation. *Brain, Behavior, and Immunity*, 22, 662–667.

Koolhaas, J. M., Korte, S. M., De Boer, S. F., Van Der Vegt, B. J., Van Reenen, C. G., Hopster, H., De Jong, I. C., Ruis, M. A., and Blokhuis, H. J. 1999. Coping styles in animals: current status in behavior and stress physiology. *Neuroscience and Biobehavioral Reviews*, 23, 925–935.

Koopmans, G., Blokland, A., Van Nieuwenhuijzen, P., and Prickaerts, J. 2003. Assessment of spatial learning abilities of mice in a new circular maze. *Physiology and Behavior*, 79, 683–693.

Krantz, D. S., Olson, M. B., Francis, J. L., Phankao, C., Merz, C. N. B., Sopko, G., Vido, D. A., Shaw, L. J., Sheps, D. S., Pepine, C. J., and Matthews, K. A. 2006. Anger, hostility, and cardiac symptoms in women with suspected coronary artery disease: the Women's Ischemia Syndrome Evaluation (WISE) Study. *Journal of Women's Health*, 15, 1214–1223.

Krupke, D. M., Begley, D. A., Sundberg, J. P., Bult, C. J., and Eppig, J. T. 2008. The Mouse Tumor Biology database. *Nature Reviews Cancer*, 8, 459–465.

Lahmame, A., Grigoriadis, D. E., De Souza, E. B., and Armario, A. 1997. Brain corticotropin- releasing factor immunoreactivity and receptors in five inbred rat strains: relationship to forced swimming behaviour. *Brain Research*, 750, 285–292.

Landgraf, R. 2003. HAB/LAB rats: an animal model of extremes in trait anxiety and depression. *Clinical Neuroscience Research*, 3, 239–244.

Landgraf, R., and Wigger, A. 2002. High vs. low anxiety-related behavior rats: an animal model of extremes in trait anxiety. *Behavior Genetics*, 32, 301–314.

Landgraf, R., Wigger, A., Holsboer, F., and Neumann, I. D. 1999. Hyperreactive hypothalamo- pituitary-adrenocortical axis in rats bred for high anxiety-related behaviour. *Journal of Neuroendocrinology*, 11, 405–417.

Laudenslager, M. L., Rasmussen, K. L., Berman, C. M., Lilly, A. A., Shelton, S. E., Kalin, N. H., and Suomi, S. J. 1999. A preliminary description of responses of free-ranging rhesus monkeys to brief capture experiences: behavior, endocrine, immune, and health relationships. *Brain, Behavior, and Immunity*, 13, 124–137.

Lazarus, R., and Folkman, S. 1984. *Stress, Appraisal, and Coping*. New York: Springer.

Leppänen, P. K., Ewalds-Kvist, S. B., and Selander, R. K. 2005. Mice selectively bred for open- field thigmotaxis: life span and stability of the selection trait. *Journal of General Psychology*, 132, 187–204.

Leppänen, P. K., Ravaja, N., and Ewalds-Kvist, S. B. M. 2006. Twenty-three generations of mice bidirectionally selected for open-field thigmotaxis: selection response and repeated exposure to the open field. *Behavioural Processes*, 72, 23–31.

Levy, J. R., Davenport, B., Clore, J. N., and Stevens, W. 2002. Lipid metabolism and resistin gene expression in insulin-resistant Fischer 344 rats. *American Journal of Physiology*, 282, E626–E633.

Liebsch, G., Linthorst, A. C., Neumann, I. D., Reul, J. M., Holsboer, F., and Landgraf, R. 1998. Behavioral, physiological, and neuroendocrine stress responses and differential sensitivity to diazepam in two Wistar rat lines selectively bred for high and low anxiety-related behavior. *Neuropsychopharmacology*, 19, 381–396.

Linden, W., Klassen, K., and Phillips, M. 2008. Can psychological factors account for a lack of nocturnal blood pressure dipping? *Annals of Behavioral Medicine*, 36, 253–258.

Lipman, R. D., Chrisp, C. E., Hazzard, D. G., and Bronson, R. T. 1996. Pathologic characterization of Brown Norway, Brown Norway x Fischer 344, and Fischer 344 x Brown Norway rats with relation to age. *Journal of Gerontology*, 51A, B54–B59.

Lister, R. G. 1987. The use of a plus maze to measure anxiety in the mouse. *Psychopharmacology*, 92, 180–185.

Lockwood, J. A., and Turney, T. H. 1981. Social dominance and stress-induced hypertension: strain differences in inbred mice. *Physiology and Behavior*, 26, 547–559.

Maeda, H., Gleiser, C. A., Masoro, E. J., Murata, I., McMahan, C. A., and Yu, B. P. 1985. Nutritional influences on aging of Fischer 344 rats, II: pathology. *Journal of Gerontology*, 40, 671–688.

Malkesman, O., Maayan, R., Weizman, A., and Weller, A. 2006. Aggressive behavior and HPA axis hormones after social isolation in adult rats of two different genetic animal models for depression. *Behavioural Brain Research*, 175, 408–414.

Maninger, N., Capitanio, J. P., Mendoza, S. P., and Mason, W. A. 2003. Personality influences tetanus-specific antibody response in adult male rhesus macaques after removal from natal group and housing relocation. *American Journal of Primatology*, 61, 73–83.

Marissal-Arvy, N., Gaumont, A., Langlois, A., Dabertrand, F., Bouchecareilh, M., Tridon, C., and Mormède, P. 2007. Strain differences in hypothalamic pituitary adrenocortical axis function and adipogenic effects of corticosterone in rats. *Journal of Endocrinology*, 195, 473–484.

Marissal-Arvy, N., Ribot, E., Sarrieau, A., and Mormède, P. 2000. Is the mineralocorticoid receptor in Brown Norway rats constitutively active? *Journal of Neuroendocrinology*, 12, 576–588.

Markowska, A. L., Spangler, E. L., and Ingram, D. K. 1998. Behavioral assessment of the senescence-accelerated mouse (SAM P8 and R1). *Physiology and Behavior*, 64, 15–26.

Márquez, C., Nadal, R., and Armario, A. 2006. Influence of reactivity to novelty and anxiety on hypothalamic-pituitary-adrenal and prolactin responses to two different novel environments in adult male rats. *Behavioural Brain Research*, 168, 13–22.

Martin, C. A., Kelly, T. H., Rayens, M. K., Brogli, B. R., Brenzel, A., Smith, W. J., and Omar, H. A. 2002. Sensation seeking, puberty, and nicotine, alcohol, and marijuana use in adolescence. *Journal of the American Academy of Child and Adolescent Psychiatry*, 41, 1495–1502.

Masoro, E. J. 1980. Mortality and growth characteristics of rat strains commonly used in aging research. *Experimental Aging Research*, 6, 219–233.

Matthews, K., Baldo, B. A., Markou, A., Lown, O., Overstreet, D. H., and Koob, G. F. 1996. Rewarding electrical brain stimulation: similar thresholds for Flinders sensitive line hypercholinergic and Flinders resistant line hypocholinergic rats. *Physiology and Behavior*, 59, 1155–1562.

McKhann, G. M., Wenzel, H. J., Robbins, C. A., Sosunov, A. A., and Schwartzkroin, P. A. 2003. Mouse strain differences in kainic acid sensitivity, seizure behavior, mortality, and hippocampal pathology. *Neuroscience*, 122, 551–561.

Michaud, D. S., McLean, J., Keith, S. E., Ferrarotto, C., Hayley, S., Khan, S. A., Anisman, H., and Merali, Z. 2003. Differential impact of audiogenic stressors on Lewis and Fischer rats: behavioral, neurochemical, and endocrine variations. *Neuropsychopharmacology*, 28, 1068–1081.

Miczek, K. A. 1979. A new test for aggression in rats without aversive stimulation: differential effects of *d*-amphetamine and cocaine. *Psychopharmacology*, 60, 253–259.

Miyamoto, M., Kiyota, Y., Nishiyama, M., and Nagaoka, A. 1992. Senescence-accelerated mouse (SAM): age-related reduced anxiety-like behavior in the SAM-P/8 strain. *Physiology and Behavior*, 51, 979–985.

Morris, R. 1984. Developments of a water-maze procedure for studying spatial learning in the rat. *Journal of Neuroscience Methods*, 11, 47–60.

Muscat, R., Sampson, D., and Willner, P. 1990. Dopaminergic mechanism of imipramine action in an animal model of depression. *Biological Psychiatry*, 28, 223–230.

Näf, D., Krupke, D. M., Sundberg, J. P., Eppig, J. T., and Bult, C. J. 2002. The mouse tumor biology database: a public resource for cancer genetics and pathology of the mouse. *Cancer Research*, 62, 1235–1240.

Nakagawara, M., Kubota, M., Atobe, M., and Kariya, T. 1997. Strain difference in behavioral response to a new environment in rats. *Psychiatry and Clinical Neurosciences*, 51, 167-170.

Neumann, P. E., and Collins, R. L. 1991. Genetic dissection of susceptibility to audiogenic seizures in inbred mice. *Proceedings of the National Academy of Sciences USA*, 88, 5408–5412.

Newcomb, M. D., and McGee, L. 1991. Influence of sensation seeking on general deviance and specific problem behaviors from adolescence to young adulthood. *Journal of Personality and Social Psychology*, 61, 614–628.

Niaura, R., Banks, S. M., Ward, K. D., Stoney, C. M., Spiro, A., Aldwin, C. M., Landsberg, L., and Weiss, S. T. 2000. Hostility and the metabolic syndrome in older males: the normative aging study. *Psychosomatic Medicine*, 62, 7–16.

Nicholson, A. 2006. Flow-cytometric analysis of 11 inbred strains. MPD: 231. Mouse Phenome Database website, Jackson Laboratory, Bar Harbor, Maine. http://www.jax .org/phenome.

Ohl, F., Toschi, N., Wigger, A., Henniger, M. S., and Landgraf, R. 2001. Dimensions of emotionality in a rat model of innate anxiety. *Behavioral Neuroscience*, 115, 429–436.

Oitzl, M. S., Van Haarst, A. D., Sutanto, W., and de Kloet, E. R. 1995. Corticosterone, brain mineralocorticoid receptors (MRs), and the activity of the hypothalamic-pituitary-adrenal (HPA) axis: the Lewis rat as an example of increased central MR capacity and a hyporesponsive HPA axis. *Psychoneuroendocrinology*, 20, 655–675.

Overstreet, D. H., and Djuric, V. 2001. A genetic rat model of cholinergic hypersensitivity: implications for chemical intolerance, chronic fatigue, and asthma. *Annals of the New York Academy of Sciences*, 933, 92–102.

Overstreet, D. H., Friedman, E., Mathé, A. A., and Yadid, G. 2005. The Flinders sensitive line rat: a selectively bred putative animal model of depression. *Neuroscience and Biobehavioral Reviews*, 29, 739–759.

Overstreet, D. H., Janowsky, D. S., Gillin, J. C., Shiromani, P. J., and Sutin, E. L. 1986. Stress-induced immobility in rats with cholinergic supersensitivity. *Biological Psychiatry*, 21, 657–664.

Overstreet, D. H., Janowsky, D. S., and Rezvani, A. H. 1989. Alcoholism and depressive disorders: is cholinergic sensitivity a biological marker? *Alcohol and Alcoholism*, 24, 253–255.

Overstreet, D. H., Kampov-Polevoy, A. B., Rezvani, A. H., Murrelle, L., Halikas, J. A., and Janowsky, D. S. 1993. Saccharin intake predicts ethanol intake in genetically heterogeneous rats as well as different rat strains. *Alcoholism: Clinical and Experimental Research*, 17, 366–369.

Overstreet, D. H., Pucilowski, O., Rezvani, A. H., and Janowsky, D. S. 1995.

Administration of antidepressants, diazepam, and psychomotor stimulants further confirms the utility of Flinders sensitive line rats as an animal model of depression. *Psychopharmacology*, 121, 27–37.

Overstreet, D. H., Rezvani, A. H., Djouma, E., Parsian, A., and Lawrence, A. J. 2007. Depressive-like behavior and high alcohol drinking co-occur in the FH/WJD rat but appear to be under independent genetic control. *Neuroscience and Biobehavioral Reviews*, 31, 103–114.

Overstreet, D. H., Rezvani, A. H., and Janowsky, D. S. 1992. Genetic animal models of depression and ethanol preference provide support for cholinergic and serotonergic involvement in depression and alcoholism. *Biological Psychiatry*, 31, 919–936.

Overstreet, D. H., and Russell, R. W. 1982. Selective breeding for diisopropyl fluorophosphate-sensitivity: behavioural effects of cholinergic agonists and antagonists. *Psychopharmacology*, 78, 150–155.

Paigen, K. 2003. One hundred years of mouse genetics: an intellectual history, I: the classical period (1902–1980). *Genetics*, 163, 1–7.

Paigen, B., and Yuan, R. 2007. Aging study: lifespan and survival curves. MPD: 234. Mouse Phenome Database website, Jackson Laboratory, Bar Harbor, Maine. http://www.jax.org/phenome.

Pardon, M. C., Gould, G. G., Garcia, A., Phillips, L., Cook, M. C., Miller, S. A., Mason, P. A., and Morilak, D. A. 2002. Stress reactivity of the brain noradrenergic system in three rat strains differing in their neuroendocrine and behavioral responses to stress: implications for susceptibility to stress-related neuropsychiatric disorders. *Neuroscience*, 115, 229–242.

Pawelec, G., Effros, R. B., Caruso, C., Remarque, E., Barnett, Y., and Solana, R. 1999. T-cell and aging. *Frontiers in Bioscience*, 4, 216–269.

Pecoraro, N., Ginsberg, A. B., Warne, J. P., Gomez, F., la Fleur, S. E., and Dallman, M. F. 2006. Diverse basal and stress-related phenotypes of Sprague Dawley rats from three vendors. *Physiology and Behavior*, 89, 598–610.

Pellow, S., Chopin, P., File, S. E., and Briley, M. 1985. Validation of open:closed arm entries in an elevated plus maze as a measure of anxiety in the rat. *Journal of Neuroscience Methods*, 14, 149–167.

Pérez-Álvarez, L., Baeza, I., Arranz, L., Marco, E. M., Borcel, E., Guaza, C., Viveros, M. P., and De la Fuente, M. 2005. Behavioral, endocrine, and immunological characteristics of a murine model of premature aging. *Developmental and Comparative Immunology*, 29, 965–976.

Petitto, J. M., Lysle, D. T., Gariepy, J.-L., Clubb, P. H., Cairns, R. B., and Lewis, M. H. 1993. Genetic differences in social behavior: relation to natural killer cell function and susceptibility to tumor development. *Neuropsychopharmacology*, 8, 35–43.

Petitto, J. M., Lysle, D. T., Gariepy, J.-L., and Lewis, M. H. 1994. Association of genetic differences in social behavior and cellular immune responsiveness: effects of social experience. *Brain, Behavior, and Immunity*, 8, 111–122.

Piazza, P. V., Deminière, J.-M., Le Moal, M., and Simon, H. 1989. Factors that predict individual vulnerability to amphetamine self-administration. *Science*, 245, 1511–1513.

Piazza, P. V., Deminière, J.-M., Maccari, S., Mormède, P., Le Moal, M., and Simon, H. 1990. Individual reactivity to novelty predicts probability of amphetamine self-administration. *Behavioural Pharmacology*, 1, 339–345.

Piazza, P. V., Rougé-Pont, F., Deminière, J. M., Kharoubi, M., Le Moal, M., and Simon, H.

1991. Dopaminergic activity is reduced in the prefrontal cortex and increased in the nucleus accumbens of rats predisposed to develop amphetamine self-administration. *Brain Research*, 567, 169–174.

Pinel, J. P. J., and Treit, D. 1978. Burying as a defensive responsive in rats. *Journal of Comparative and Physiological Psychology*, 92, 708–712.

Piras, G., Lecca, D., Corda, M. G., and Giorgi, O. 2003. Repeated morphine injections induce behavioural sensitization in Roman high- but not in Roman low-avoidance rats. *Neuroreport*, 14, 2433–2438.

Podhorna, J., and Brown, R. E. 2002. Strain differences in activity and emotionality do not account for differences in learning and memory performance between C57BL/6 and DBA/2 mice. *Genes, Brain, and Behavior*, 1, 96–110.

Porsolt, R. D., Bertin, A., and Jalfre, M. 1977. Behavioral despair in mice: a primary screening test for antidepressants. *Archives of International Pharmacodynamics and Therapeutics*, 229, 327–336.

Pucilowski, O., Overstreet, D. H., Rezvani, A. H., and Janowsky, D. S. 1993. Chronic mild stress-induced anhedonia: greater effect in a genetic rat model of depression. *Physiology and Behavior*, 54, 1215–1220.

Ramos, A., Berton, O., Mormède, P., and Chaouloff, F. 1997. A multiple-test study of anxiety- related behaviours in six inbred rat strains. *Behavioural Brain Research*, 85, 57–69.

Ravaja, N., Kauppinen, T., and Keltikangas-Järvinen, L. 2000. Relationships between hostility and physiological coronary heart disease risk factors in young adults: the moderating influence of depressive tendencies. *Psychological Medicine*, 30, 381–393.

Redei, E. E., Paré, W. P., Aird, F., and Kluczynski, J. 1994. Strain differences in hypothalamic-pituitary-adrenal activity and stress ulcer. *American Journal of Physiology*, 266, R353–R360.

Redelman, D., Welniak, L. A., Taub, D., and Murphy, W. J. 2008. Neuroendocrine hormones such as growth hormone and prolactin are integral members of the immunological cytokine network. *Cellular Immunology*, 252, 111–121.

Reed, D. R., Bachmanov, A. A., and Tordoff, M. G. 2007. Forty mouse strain survey of body composition. *Physiology and Behavior*, 91, 593–600.

Rex, A., Voigt, J. P., Gustedt, C., Beckett, S., and Fink, H. 2004. Anxiolytic-like profile in Wistar but not Sprague-Dawley rats in the social interaction test. *Psychopharmacology*, 177, 23–34.

Roberts, J., and Goldberg, P. B. 1976. Changes in basic cardiovascular activities during the lifetime of the rat. *Experimental Aging Research*, 2, 487–517.

Robertson, H. A. 1980. Audiogenic seizures: increased benzodiazepine receptor binding in a susceptible strain of mice. *European Journal of Pharmacology*, 66, 249–252.

Rosenman, R. H., and Friedman, M. 1974. Neurogenic factors in pathogenesis of coronary heart disease. *Medical Clinics of North America*, 58, 269–279.

Routledge, F. S., McFetridge-Durdle, J. A., and Dean, C. R. 2007. Night-time blood pressure patterns and target organ damage: a review. *Canadian Journal of Cardiology*, 23, 132- 138.

Royle, S. J., Collins, F. C., Rupniak, H. T., Barnes, J. C., and Anderson, R. 1999. Behavioural analysis and susceptibility to CNS injury of four inbred strains of mice. *Brain Research*, 816, 337–349.

Rudbach, J. A., and Reed, N. D. 1977. Immunological responses of mice to lipopolysaccharide: lack of secondary responsiveness by C3H/HeJ mice. *Infection and Immunity*, 16, 513–517.

Ruis, M. A. W., ten Brake, J. H. A., Van de Burgwal, J. A., de Jong, I. C., Blokhuis, H. J., and Koolhaas, J. M. 2000. Personalities in female domesticated pigs: behavioural and physiological indications. *Applied Animal Behavior Science*, 66, 31–47.

Russell, P. A. 1979. Fear-evoking stimuli. In: *Fear in Animals and Man* (Sluckin, W., ed.), pp. 86–124. New York: Van Nostrand Reinhold Company.

Russell, R. W., Overstreet, D. H., Messenger, M., and Helps, S. C. 1982. Selective breeding for sensitivity to DFP: generalization of effects beyond criterion variables. *Pharmacology, Biochemistry, and Behavior*, 17, 885–891.

Rymarchyk, S. L., Lowenstein, H., Mayette, J., Foster, S. R., Damby, D. E., Howe, I. W., Aktan, I., Meyer, R. E., Poynter, M. E., and Boyson, J. E. 2008. Widespread natural variation in murine natural killer T-cell number and function. *Immunology*, 125, 331–343.

Salomé, N., Tasiemski, A., Dutriez, I., Wigger, A., Landgraf, R., and Viltart, O. 2008. Immune challenge induces differential corticosterone and interleukin-6 responsiveness in rats bred for extremes in anxiety-related behavior. *Neuroscience*, 151, 1112–1128.

Salomé, N., Viltart, O., Darnaudéry, M., Salchner, P., Singewald, N., Landgraf, R., Sequeira, H., and Wigger, A. 2002. Reliability of high and low anxiety-related behaviour: influence of laboratory environment and multifactorial analysis. *Behavioural Brain Research*, 136, 227–237.

Sandi, C., Castanon, N., Vitiello, S., Neveu, P. J., and Morméde, P. 1991. Different responsiveness of spleen lymphocytes from two lines of psychogenetically selected rats (Roman high and low avoidance). *Journal of Neuroimmunology*, 31, 27–33.

Sarrieau, A. A., and Morméde, P. 1998. Hypothalamic-pituitary-adrenal axis activity in the inbred Brown Norway and Fischer 344 rat strains. *Life Sciences*, 62, 1417–1425.

Sass, B., Rabstein, L. S., Madison, R., Nims, R. M., Peters, R. L., and Kelloff, G. J. 1975. Incidence of spontaneous neoplasms in F344 rats throughout the natural life-span. *Journal of the National Cancer Institute*, 54, 1449–1456.

Scharton-Kersten, T., and Scott, P. 1995. The role of the innate immune response in Th1 cell development following *Leishmania* major infection. *Journal of Leukocyte Biology*, 57, 515–522.

Schimanski, L. A., and Nguyen, P. V. 2005. Mouse models of impaired fear memory exhibit deficits in amygdalar LTP. *Hippocampus*, 15, 502–517.

Schreiber, R. A. 1986. Behavior genetics of audiogenic seizures in DBA/2J and Rb-1 mice. *Behavior Genetics*, 16, 365–368.

Seburn, K. L. 2001. Metabolic characterization. MPD: 92. Mouse Phenome Database website, Jackson Laboratory, Bar Harbor, Maine. http://www.jax.org/phenome.

Segerstrom, S. C. 2003. Individual differences, immunity, and cancer: lessons from personality psychology. *Brain, Behavior, and Immunity*, 17, S92–S97.

Segerstrom, S. C., Taylor, S. E., Kemeny, M. E., and Fahey, J. 1998. Optimism is associated with mood, coping, and immune change in response to stress. *Journal of Personality and Social Psychology*, 74, 1646–1655.

Shanks, N., Griffiths, J., Zalcman, S., Zacharko, R. M., and Anisman, H. 1990. Mouse strain differences in plasma corticosterone following uncontrollable footshock. *Pharmacology, Biochemistry, and Behavior*, 36, 515–519.

Shepard, J. D., and Myers, D. A. 2008. Strain differences in anxiety-like behavior: association with corticotropin-releasing factor. *Behavioural Brain Research*, 186, 239–245.

Shimbo, D., Chaplin, W., Kuruvilla, S., Wasson, L. T., Abraham, D., and Burg, M. M. 2009. Hostility and platelet reactivity in individuals without a history of cardiovascular disease events. *Psychosomatic Medicine*, 71, 741–747.

Shiromani, P. J., Overstreet, D., Levy, D., Goodrich, C. A, Campbell, S. S., and Gillin, J. C. 1988. Increased REM sleep in rats selectively bred for cholinergic hyperactivity. *Neuropsychopharmacology*, 1, 127–133.

Siegel, J., and Driscoll, P. 1996. Recent developments in an animal model of visual evoked potential augmenting/reducing and sensation-seeking behavior. *Neuropsychobiology*, 34, 130–135.

Sih, A., Bell, A., and Johnson, J. C. 2004. Behavioral syndromes: an ecological and evolutionary overview. *Trends in Ecology and Evolution*, 19, 372–378.

Simon, P., Dupuis, R., and Costentin, J. J. 1994. Thigmotaxis as an index of anxiety in mice: influence of dopaminergic transmissions. *Behavioural Brain Research*, 61, 59–64.

Sluyter, F., Arseneault, L., Moffitt, T. E., Veenema, A. H., de Boer, S., and Koolhaas, J. M. 2003. Toward an animal model for antisocial behavior: parallels between mice and humans. *Behavior Genetics*, 33, 563–574.

Sluyter, F., Korte, S. M., and Bohus, B. 1996. Behavioral stress response of genetically selected aggressive and nonaggressive wild house mice in the shock-probe/defensive burying test. *Pharmacology, Biochemistry, and Behavior*, 54, 113–116.

Smith, T. W., and MacKenzie, J. 2006. Personality and risk of physical illness. *Annual Review of Clinical Psychology*, 2, 435–467.

Spangler, G. 1997. Psychological and physiological responses during an exam and their relation to personality characteristics. *Psychoneuroendocrinology*, 22, 423–441.

Staats, J. 1980. Standardized nomenclature for inbred strains of mice: seventh listing. *Cancer Research*, 40, 2083–2128.

Stefferl, A., Linington, C., Holsboer, F., and Reul, J. M. 1999. Susceptibility and resistance to experimental allergic encephalomyelitis: relationship with hypothalamic-pituitary-adrenocortical axis responsiveness in the rat. *Endocrinology*, 140, 4932–4938.

Steimer, T., and Driscoll, P. 2005. Interindividual vs. line/strain differences in psychogenetically selected Roman high (RHA)– and low (RLA)–avoidance rats: neuroendocrine and behavioural aspects. *Neuroscience and Biobehavioral Reviews*, 29, 99–112.

Steimer, T., la Fleur, S., and Schulz, P. E. 1997. Neuroendocrine correlates of emotional reactivity and coping in male rats from the Roman high (RHA/Verh)– and low (RLA/Verh)–avoidance lines. *Behavior Genetics*, 27, 503–512.

Steimer, T., Python, A., Driscoll, P., and de Saint Hilaire, Z. 1999. Psychogenetically selected (Roman high- and low-avoidance) rats differ in 24-hour sleep organization. *Journal of Biological Rhythms*, 14, 221–226.

Sternberg, E. M., Glowa, J. R., Smith, M. A., Calogero, A. E., Listwak, S. J., Aksentijevich, S., Chrousos, G. P., Wilder, R. L., and Gold, P. W. 1992. Corticotropin-releasing hormone–related behavioral and neuroendocrine responses to stress in Lewis and Fischer rats. *Brain Research*, 570, 54–60.

Sternberg, E. M., Hill, J. M., Chrousos, G. P., Kamilaris, T., Listwak, S. J., Gold, P. W., and Wilder, R. L. 1989. Inflammatory mediator-induced hypothalamic-pituitary-adrenal axis activation is defective in streptococcal cell wall arthritis-susceptible Lewis rats. *Proceedings of the National Academy of Sciences U S A*, 86, 2374–2378.

Stewart, J. C., Janicki, D. L., Muldoon, M. F., Sutton-Tyrrell, K., and Kamarck, T. W. 2007. Negative emotions and 3-year progression of subclinical atherosclerosis. *Archives of General Psychiatry*, 64, 225–233.

Stoffel, E. C., and Cunningham, K. A. 2008. The relationship between the locomotor response to a novel environment and behavioral disinhibition in rats. *Drug and Alcohol Dependence*, 92, 69–78.

Suarez, E. 2003. Plasma interleukin-6 is associated with psychological coronary risk factors: moderation by use of multivitamin supplements. *Brain, Behavior, and Immunity*, 17, 296–303.

———. 2004. C-reactive protein is associated with psychological risk factors of cardiovascular disease in apparently healthy adults. *Psychosomatic Medicine*, 66, 684–691.

Suarez, E., Lewis, J. G., and Kuhn, C. 2002. The relation of aggression, hostility, and anger to lipopolysaccharide-stimulated tumor necrosis factor (TNF)-alpha by blood monocytes from normal men. *Brain, Behavior, and Immunity*, 16, 675–684.

Sugiyama, F., Tsukahara, C., and Paigen, B. 2007. Blood pressure for 25 strains. MPD: 236. Mouse Phenome Database website, Jackson Laboratory, Bar Harbor, Maine. http://www.jax.org/phenome.

Sugiyama, Y., Shimura, Y., and Ikeda H. 1989. Pathogenesis of hyperglycemia in genetically obese-hyperglycemic rats, Wistar fatty: presence of hepatic insulin resistance. *Endocrinologia Japonica*, 36, 65–73.

Sundberg, J. P., Cordy, W. R., and King, L. E. 1994. Alopecia areata in aging C3H/HeJ mice. *Journal of Investigative Dermatology*, 102, 847–856.

Swan, G. E., and Carmelli, D. 1996. Curiosity and mortality in aging adults: a 5-year follow-up of the Western Collaborative Group Study. *Psychology of Aging*, 11, 449–453.

Szpirer, C., and Szpirer, J. 2007. Mammary cancer susceptibility: human genes and rodent models. *Mammalian Genome*, 18, 817–831.

Takahashi, L. K. 2001. Role of CRF_1 and CRF_2 receptors in fear and anxiety. *Neuroscience and Biobehavioral Reviews*, 25, 627–636.

Takemoto, L. J., Gorthy, W. C, Morin, C. L., and Steward, D. E. 1991. Changes in lens membrane major intrinsic polypeptide during cataractogenesis in aged Hannover Wistar rats. *Investigative Ophthalmology and Visual Science*, 32, 556–561.

Thomas, K. S., Nelesen, R. A., and Dimsdale, J. E. 2004. Relationships between hostility, anger expression, and blood pressure dipping in an ethnically diverse sample. *Psychosomatic Medicine*, 66, 298–304.

Tõnissaar, M., Philips, M. A., Eller, M., and Harro, J. 2004. Sociability trait and serotonin metabolism in the rat social interaction test. *Neuroscience Letters*, 367, 309–312.

Treit, D., Pinel, J. P., and Fibiger, H. C. 1981. Conditioned defensive burying: a new paradigm for the study of anxiolytic agents. *Pharmacology, Biochemistry, and Behavior*, 15, 619–626.

Trigo, M., Silva, D., and Rocha, E. 2005. Psychosocial risk factors in coronary heart disease: beyond Type A behavior. *Revista Portuguesa de Cardiologia*, 24, 261–281.

Trullas, R., and Skolnick, P. 1993. Differences in fear-motivated behaviors among inbred mouse strains. *Psychopharmacology*, 111, 323–331.

Tsukahara, C., Sugiyama, F., Paigen, B., Kunita, S., and Yagami, K. 2004. Blood pressure in 15 inbred mouse strains and its lack of relation with obesity and insulin resistance in the progeny of an NZO/HILtJ x C3H/HeJ intercross. *Mammalian Genome*, 15, 943–950.

Turner, S. M., Beidel, D. C., and Wolff, P. L. 1996. Is behavioral inhibition related to the anxiety disorders? *Clinical Psychological Review*, 16, 157–172.

Van den Brandt, J., Kovács, P., and Klöting, I. 2000. Metabolic variability among disease-resistant inbred rat strains and in comparison with wild rats (*Rattus norvegicus*). *Clinical and Experimental Pharmacology and Physiology*, 27, 793–795.

Van der Staay, F. J., and Blokland, A. 1996. Behavioral differences between outbred Wistar, inbred Fischer 344, Brown Norway, and hybrid Fischer 344 x Brown Norway rats. *Physiology and Behavior*, 60, 97–109.

Van der Staay, F. J., Kerbusch, S., and Raaijmakers, W. 1990. Genetic correlations in validating emotionality. *Behavior Genetics*, 20, 51–62.

Van Oers, K., and Mueller, J. C. 2010. Evolutionary genomics of animal personality. *Philosophical Transactions of the Royal Society of London B*, 365, 3991–4000.

Van Oortmerssen, G. A., and Bakker, T. C. M. 1981. Artificial selection for short and long attack latencies in wild *Mus musculus domesticus*. *Behavior Genetics*, 11, 115–126.

Veenema, A. H., Meijer, O. C., de Kloet, E. R., and Koolhaas, J. M. 2003a. Genetic selection for coping style predicts stressor susceptibility. *Journal of Neuroendocrinology*, 15, 256–267.

Veenema, A. H., Meijer, O. C., de Kloet, E. R., Koolhaas, J. M., and Bohus, B. G. 2003b. Differences in basal and stress-induced HPA regulation of wild house mice selected for high and low aggression. *Hormones and Behavior*, 43, 197–204.

Veenema, A. H., and Neumann, I. D. 2007. Neurobiological mechanisms of aggression and stress coping: a comparative study in mouse and rat selection lines. *Brain, Behavior, and Evolution*, 70, 274–285.

Viveros, M. P., Arranz, L., Hernanz, A., Miquel, J., and De la Fuente, M. 2007. A model of premature aging in mice based on altered stress-related behavioral response and immunosenescence. *Neuroimmunomodulation*, 14, 157–162.

Viveros, M. P., Fernández, B., Guayerbas, N., and De la Fuente, M. 2001. Behavioral characterization of a mouse model of premature immunosenescence. *Journal of Neuroimmunology*, 114, 80–88.

Wahlsten, D., Bachmanov, A., Finn, D. A., and Crabbe, J. C. 2006. Stability of inbred mouse strain differences in behavior and brain size between laboratories and across decades. *Proceedings of the National Academy of Sciences U S A*, 103, 16364–16369.

Wahlsten, D., and Crabbe, J. C. 2003. Studies of activity, behavior, and forebrain morphometry in two laboratories. MPD: 108. Mouse Phenome Database website, Jackson Laboratory, Bar Harbor, Maine. http://www.jax.org/phenome.

Walker, C. D., Rivest, R. W., Meaney, M. J., and Aubert, M. L. 1989. Differential activation of the pituitary-adrenocortical axis after stress in the rat—use of 2 genetically selected lines (Roman low-avoidance and high-avoidance rats) as a model. *Journal of Endocrinology*, 123, 477–485.

White, T. L., Lott, D. C., and de Wit, H. 2006. Personality and the subjective effects of acute amphetamine in healthy volunteers. *Neuropsychopharmacology*, 31, 1064–1074.

Wigger, A., Loerscher, P., Weissenbacher, P., Holsboer, F., and Landgraf, R. 2001. Cross-fostering and cross-breeding of HAB and LAB rats: a genetic rat model of anxiety. *Behavior Genetics*, 31, 371–382.

Williams, R. B., Haney, T. L., Lee, K. L., Kong, Y. H., Blumenthal, J. A., and Whalen, R. E. 1980. Type A behavior, hostility, and coronary atherosclerosis. *Psychosomatic Medicine*, 42, 539–549.

Wills, T. A., Vaccaro, D., and McNamara, G. 1994. Novelty seeking, risk taking, and related constructs as predictors of adolescent substance use: an application of Cloninger's theory. *Journal of Substance Abuse*, 6, 1–20.

Yilmazer-Hanke, D. M., Roskoden, T., Zilles, K., and Schwegler, H. 2003. Anxiety-related behavior and densities of glutamate, $GABA_A$, acetylcholine, and serotonin receptors in the amygdala in seven inbred mouse strains. *Behavioural Brain Research*, 145, 145–159.

Zambello, E., Jiménez-Vasquez, P. A., El Khoury, A., Mathé, A. A., and Caberlotto, L. 2008. Acute stress differentially affects corticotropin-releasing hormone mRNA expression in the central amygdala of the "depressed" Flinders sensitive line and the control Flinders resistant line rats. *Progress in Neuro-Psychopharmacology and Biological Psychiatry*, 32, 651–661.

Zamudio, S., Fregoso, T., Miranda, A., De La Cruz, F., and Flores, G. 2005. Strain differences of dopamine receptor levels and dopamine-related behaviors in rats. *Brain Research Bulletin*, 65, 339–347.

Zilles, K., Wu, J., Crusio, W. E., and Schwegler, H. 2000. Water maze and radial maze learning and the density of binding sites of glutamate, GABA, and serotonin receptors in the hippocampus of inbred mouse strains. *Hippocampus*, 10, 213–225.

Zozulya, A. A., Gabaeva, M. V., Sokolov, O. Y., Surkina, I. D., and Kost, N. V. 2008. Personality, coping style, and constitutional neuroimmunology. *Journal of Immunotoxicology*, 5, 221–225.

Index